Jakob Nielsen: Collected Mathematical Papers

Volume 1 (1913–1932)

Edited for the
Danish Mathematical Society
by
Vagn Lundsgaard Hansen

Birkhäuser
Boston · Basel · Stuttgart
1986

Editor

Vagn Lundsgaard Hansen
Mathematical Institute
The Technical University of Denmark
DK-2800 Lyngby, Copenhagen, Denmark

Library of Congress Cataloging in Publication Data
Nielsen, Jakob, 1890–1959.
Collected mathematical papers.
(Contemporary mathematicians)
Bibliography: p.
1. Groups, Theory of—Collected works.
2. Surfaces—Collected works. I. Hansen, Vagn Lundsgaard. II. Dansk matematisk forening. III. Title. IV. Series.
QA171.N5713 1986 512′.22 86-6137
ISBN 3-7643-3140-2 (Switzerland: v. 1)
ISBN 3-7643-3151-8 (Switzerland: v. 2)

CIP-Kurztitelaufnahme der Deutschen Bibliothek
Nielsen, Jakob:
Collected papers / Jakob Nielsen. Ed. for the Danish Mathemat. Soc. by Vagn Lundsgaard Hansen—
Boston ; Basel ; Stuttgart ; Birkhäuser
(Contemporary mathematicians)
ISBN 3-7643-3152-6 (Basel . . .)
ISBN 0-8176-3152-6 (Boston)
NE: Nielsen, Jakob Samalung
Vol. 1 (1986)
ISBN 3-7643-3140-2 (Basel . . .)
ISBN 0-8176-3140-2 (Boston)

Printed in USA

ISBN 0-8176-3152-6 (set)	ISBN 3-7643-3152-6 (set)
ISBN 0-8176-3140-2 (Vol. 1)	ISBN 3-7643-3140-2 (Vol. 1)
ISBN 0-8176-3151-8 (Vol. 2)	ISBN 3-7643-3151-8 (Vol. 2)

With this publication, the Danish Mathematical Society wishes to honour the memory of Jakob Nielsen, who was an outstanding member of the Society from 1921 to his death in 1959. We are happy, indeed proud, to present to the mathematical community witness of the creativity and foresight of Jakob Nielsen.

The Society expresses its deep gratitude to the many scholars who have worked on the project. Special thanks go to Vagn Lundsgaard Hansen who resumed the job as editor.

The publication has been made possible by generous donations from the Carlsberg Foundation and from the Danish Natural Science Research Council.

On behalf of the
Danish Mathematical Society

Flemming Topsøe

Preface

One of the central figures in Danish mathematics in the first half of this century was Jakob Nielsen (1890–1959), who won international recognition through his pioneering work in group theory and the topology of surfaces. For some time the Danish Mathematical Society has had the desire to publish the collected papers of Jakob Nielsen, but for various reasons it has not been possible to realize the project until now.

The work of Nielsen in group theory won early recognition, and an important theorem on free groups carries his name: the Nielsen–Schreier theorem. His work on surface transformations, to which he devoted much time throughout his life, was highly respected by his contemporaries, but it was not in the main stream of the then burgeoning field of algebraic topology. The more striking of his results in topology were well known and often applied, although in many cases alternative proofs were found. For many years the seemingly complicated methods of Jakob Nielsen were therefore studied seriously only by very few mathematicians. Recently, however, there has been a growing interest in his methods, and this has created a renewed strong interest in his original papers. It is an obstacle to many potential readers that the most important of Nielsen's papers are written in a rather old-fashioned, but perfect, German.

For these reasons, in the Spring of 1980, the Danish Mathematical Society decided to investigate the possibility of publishing the collected papers of Jakob Nielsen, with many of the important papers translated into English. The plans were strongly motivated by Hans Jørgen Munkholm. From the outset Werner Fenchel and Heiner Zieschang were important scientific advisors, and during a visit to Denmark in December 1981 Joan Birman also offered valuable advice. We were very lucky to find an exceptionally competent and effective translator in John Stillwell.

We have included almost all of Nielsen's published papers in these two volumes, and also a set of hitherto unpublished notes from lectures in Breslau in 1921. Not included are his textbooks on mechanics and aerodynamics, two papers dealing with problems in mechanics, a few papers of purely didactic or expository character, and two obituaries. A photograph of Jakob Nielsen and a complete list of his publications can be found in the memorial paper by Werner Fenchel, which opens Volume 1. At the end of Volume 2 are various essays describing the work of Jakob Nielsen and later developments. Bruce Chandler and Wilhelm Magnus comment on Nielsen's work in group theory. In two essays Joan Birman and Heiner Zieschang describe his investigations of surface mapping classes. Robert F. Brown comments on Nielsen fixed point theory. Finally, Werner Fenchel describes Nielsen's contributions to the theory of discontinuous groups of isometries of the hyperbolic plane.

The publication of Jakob Nielsen's collected papers has involved the collaboration of many individuals in Denmark and abroad as well as encouragement from the publishers. As editor appointed by the Danish Mathematical Society, I wish to express my sincere thanks to all of those who have contributed to make it possible. Special thanks go to Werner Fenchel for his never-failing enthusiasm and help.

Lyngby, June 1986 — Vagn Lundsgaard Hansen

Note Concerning References: References to Nielsen's own papers are given in brackets by an N followed by the number which the paper carries in the list of Nielsen's publications in Fenchel's memorial paper. The numbers are also listed in the Table of Contents together with relevant information about each paper. In the essays, other references will be listed alphabetically at the end of each essay and referred to by a number in brackets.

Table of Contents

Volume 1

Volume 2

Essays

JAKOB NIELSEN IN MEMORIAM

BY

WERNER FENCHEL

Copenhagen

Memorial address given at a meeting of the Danish Mathematical Society on 7 December 1959

It is but natural that the Danish Mathematical Society should wish to commemorate Professor Jakob Nielsen by a meeting, for his death meant to our society the loss of one of its most active and prominent members. He joined it in 1921, sat on its committee for eight years, and in no less than 28 flawless and inspiring lectures given to it, he communicated to the mathematical circle the results of his own research, and often also that of others.

0 – 603808 *Acta mathematica*. 103. Imprimé le 29 juin 1960

Not without grave misgivings have I undertaken to speak about Jakob Nielsen, the man and the scientist, fearing that I should not be able to do justice to all essential aspects of his rich personality, but I have done so out of admiration for his work, and out of deep gratitude for his faithful friendship to me through 30 years, good and bad.

Jakob Nielsen was born on the 15th of October 1890 at the village of Mjels on the island of Als (then under Germany) as the youngest of four children. His father owned a small farm, where Jakob Nielsen spent his childhood, attending the village school until 1900. A relative of his, a teacher at Rendsburg, who took charge of him after the early death of his mother, must have become aware of his unusual gifts. At any rate, she took him with her to Rendsburg, where he went to the so-called Realgymnasium. Here, Latin was an item of considerable importance in the curriculum, and throughout his life he retained a deep love of Roman poetry. The relations between him and his aunt deteriorated, however, both being rather uncompromising of character, and at the age of 14 he left her home to take up a job as private tutor. For the remainder of his schooldays, and throughout his years of study, he earned his own living by coaching in a variety of subjects—even Norwegian, as he once said with a broad smile.

In December 1907 his secondary school career was cut short, he and some schoolfellows being expelled because they had founded a (quite harmless) pupil's club, which was against the rules. However, this did no great harm, for he continued his studies on his own, even matriculating at the University of Kiel in the spring of 1908, although he did not take his school-leaving certificate until the autumn of 1909 (privately at Flensborg). A little episode from this examination illustrates his early developed critical faculty: On one of the examiners asking him whether he had read the (as we know now, false) accounts of F. A. Cook's expedition to the North Pole, he said that he had, remarking that he could not understand how it was possible for Cook to fix his position so far to the north with a precision as great as that claimed by him.

Jakob Nielsen spent his years of study at Kiel, except for the summer term of 1910, spent at the University of Berlin. His interests and gifts were many-sided. At the start, he attended lectures in physics, chemistry, geology, biology, and literature. Only after some terms did mathematics take a prominent place, but philosophy was also strongly represented. I cannot tell what made him finally choose mathematics as the science to whose service he wanted to devote himself. Of course, we mathematicians cannot but feel that he chose rightly!

Among the teachers of mathematics of the time at Kiel, Georg Landsberg was undoubtedly the most prominent, and we note that in the short autobiography appended to Jakob Nielsen's thesis for the doctorate, he gives particular thanks to Landsberg, both for his teaching and for his instigation to take up the problems which form the subject of the thesis. The list of lectures on mathematics attended by Jakob Nielsen shows that the instruction he received at Kiel scarcely exceeded what must be looked upon, at the time, as belonging to the general education of a mathematician. The only mathematician who had an appreciable influence on Jakob Nielsen's start as a scientist was Max Dehn, 12 years his senior, who was attached to the University of Kiel at the end of the year 1911, being already then a scientist of repute. Through Dehn he came into contact with the most recent advances and methods in topology and group theory, fields to which Dehn had made fundamental contributions. This contact between the two men developed into a life-long friendship.

In 1913 Jakob Nielsen took his doctor's degree with the thesis already mentioned: "Kurvennetze auf Flächen". Although he himself did not set great store by this work, for reasons to be mentioned presently, it is of importance for the understanding of his later research to have a look at it. The three first sections deal with the connection between the total curvature of a surface and the topological properties of families of (possibly degenerate) curves covering the surface simply and completely. The last section, added later, is but loosely connected with those preceding it. It deals with the topological properties of the closed surface of genus 1, the torus. The fundamental group F, the free abelian group with two generators, is introduced, and the homotopy class of a closed curve on the surface is characterized by the pair of exponents in the expression of the corresponding element of F as a product of powers of the generators. The minimal number of double points of a curve and of the points of intersection of two curves, the curves varying within their homotopy classes, are determined in terms of these exponents. Then topological mappings of the torus onto itself are studied. Each such mapping induces an automorphism of F which depends only on the homotopy class of the mapping and which is characterized by the two pairs of exponents belonging to the images of the generators. The homotopy classes of mappings form a group isomorphic with the group of automorphisms of F. For this group a system of generators and defining relations is found, and it is shown that it is closely related with the modular group, well known from the theory of elliptic functions. Finally the minimal number of fixed points of a topological mapping, varying within a homotopy class, is determined in terms of the above-mentioned exponents. After the composition of this second part, where both the approach, the tools and the methods reflect the then new views in topology, it must have been clear to Jakob Nielsen that cogent answers to the interesting questions dealt with in the first part would require considerably more refined and rigorous notions than he had had at his disposal.

In his little publication "A Mathematician's Apology", G. H. Hardy calls mathematics "a young man's game". Understood as a contention that the main work of a mathematician is done in youth it is obviously wrong; it is easy to name many who have done excellent work at a ripe age. But there may be this much to the truth in it that the world of ideas in which the faculties develop is formed early in life. At any rate, this holds good of Jakob Nielsen. The problems dealt with in the second part of the thesis, generalized to arbitrary surfaces, came to be the governing idea of the greater part of his life-work. Although they were fairly simple in the case of the torus, and to some extent ended in things that were familiar from elsewhere, yet he must at an early stage have realized both the enormous difficulties presented by the general approach, and the fact that the necessary tools had to be created first. But with singular tenacity of purpose he did carry out these investigations after many years' work.

But there was to be a long and almost complete interruption: After taking his German M. A. degree in the summer of 1913, he was called up for service in the German navy. Owing to the outbreak of the 1914–18 war the prescribed year of military service became five years. Attached as he was to the coast defence artillery, he was sent first to Belgium and then, in April 1915, to Constantinople as one of the German officers functioning as advisers to the Turkish government on the defence of the Bosporus and the Dardanelles. He stayed there till the end of the war, finding a little time for scientific work. Besides a paper on a ballistic subject grown out of his work there, he wrote two short papers (published in Mathematische Annalen in 1917 and 1918). It is the problem occurring in the thesis for the (abelian) fundamental group of the torus which is studied here for the finitely generated free groups, namely to find a system of generators for the group of automorphisms of a given group. During his return journey from Turkey through Russia and Poland, in November 1918, at the time of the collapse of Germany, Jakob Nielsen had kept a diary, published in the newspaper Politiken on the tenth anniversary of Armistice Day. I shall quote a few lines, "I walked about here as a stranger the first two years, but in the wonderful summer of 1917 and its echo in 1918, I came to love this country under the all-powerful sway of the sun, in the company of people I liked. I know that I shall often languish for the sun. I want to go there again some day." It is probable that his experiences there contributed to his complete open-mindedness towards people of a background and with an attitude totally different from his own, that open-mindedness of his which qualified him eminently for his achievements in international co-operation, to which he devoted so great a part of his powers in later years.

Jakob Nielsen spent the summer term of 1919 at Göttingen. Among the many mathematicians he met in this centre of mathematical studies, he was specially attracted by Erich Hecke, three years his senior. When in that same year Hecke received nomination to the recently established University of Hamburg, Jakob Nielsen accompanied him as his assistant and "Privatdozent". A close and life-long friendship developed between the two men, not so much on account of mathematical interests they had in common, for there are few and loose points of contact between Hecke's work in number theory and function theory and Nielsen's in topology and group theory, but on account of their spiritual affinity, the depth and seriousness of their attitude to science. We have two papers by Jakob Nielsen from that period, both dealing with the fixed point problem for surface mappings. In the first an investigation by L. E. J. Brouwer, concerning surfaces which admit of topological mappings onto themselves without fixed points, is completed. In the second the fixed point problem for the torus is taken up again, this time with the intuitive arguments occurring in the thesis replaced by rigorous proofs. At the same time the problem is also solved for the non-orientiable torus, the Klein bottle.

Already in 1920 Jakob Nielsen received nomination to a professorship at the Institute of Technology at Breslau. Here he had an opportunity to resume contact with Max Dehn, who had for some years been attached to the university. Of interest are mimeographed notes to some lectures he gave at Breslau in 1921. Here he formulated clearly the central problem he had set himself to solve: To determine and to investigate the group of homotopy classes of topological mappings of a given surface. In the general case, there corresponds to a homotopy class of a topological mapping of the surface not one automorphism, but a coset of the group of inner automorphisms in the group of all automorphisms, a "family of automorphisms". The group in question is thus the quotient group of the latter with respect to the group of inner automorphisms. In order to throw light on his co-operation with Dehn I may point out that one link of this investigation, namely the proof of the fact that every automorphism of the fundamental group of a closed surface is induced by some topological mapping, had been communicated to him by Dehn, who never published it, however. It is characteristic of Jakob Nielsen that whenever he needed this theorem, or merely an idea that resembled the proof, he would stress his debt to Dehn. Although the spheres of the two men's studies were closely related, no further interference proper ever came about. Jakob Nielsen went his own way.

The Breslau stay was to be as brief as the one in Hamburg, for at the plebiscite preceding the Reunion of North Schleswig with Denmark in 1920, Jakob Nielsen opted for Denmark, moving to Copenhagen in 1921 and taking over the vacant lecturership in mathematics at the Royal Veterinary and Agricultural College. There now followed a long succession of happy and fruitful years of work. Let me quote what Harald Bohr says about this in the lecture "Et Tilbageblik" he gave on his 60th birthday: "We all have a very living impression of what Jakob Nielsen means for mathematics in Denmark—for our science and teaching. At the return of North Schleswig to Denmark after the first world war, he came to Copenhagen after having been professor in Breslau. Within the topology of surfaces, the field of mathematics to which he has so purposefully devoted his efforts in science, he is a recognized master who has been the example for many followers and whose influence will surely be of lasting importance for the future development of mathematics here at home. Bonnesen, Jakob Nielsen, and I followed each other's work with keen interest during those years, and many Tuesday evenings Bonnesen and I walked out to Hellerup to visit Jakob Nielsen, who was distinguished among us by being in possession of a blackboard, and in a cosy atmosphere we told each other what was on our minds." Let me here add a few words about the summers on Als, which have played such an important part in the mathematical world of Denmark. In 1919 Jakob Nielsen bought a little house near Fynshav on Als, and a few years later Harald Bohr followed his example. Year after year, in the summer vacation, a group of mathematicians, young and old, Danish and foreign, gathered about those two. Apart from the normal holiday activities, the study of mathematics was pursued. Not a few advances and discoveries were submitted in Bohr's little half-timbered house, in the study remarkable for its blackboard—unforgettable experiences which are remembered with gratitude by all who had the privilege of attending.

Let me return to Jakob Nielsen's scientific work. Starting in 1921 there appeared some purely group theoretic papers. In the first one, published in Matematisk Tidsskrift under the title "Om Regning med ikke-kommutative Faktorer og dens Anvendelse i Gruppeteorien", a major result is that every subgroup of a finitely generated free group is itself free. The proof is based on an ingeneous method of reduction of systems of generators. In 1927 the theorem was extended by Otto Schreier to arbitrary free groups, and under the name of the Nielsen-Schreier Theorem it contributes now one of the bases of the theory of infinite groups. The two other papers, from 1924, continue earlier investigations of the group of automorphisms of a given group. In the cases of the finitely generated free groups and the free abelian group with three generators not only a system of generators, but now also a system of defining relations is found. In the same year he once more resumed the subject of his thesis, the topology of the torus, in the little paper "Ringfladen og Planen". Here appears for the first time the important notion of a fixed point class. Let a continuous mapping t of the torus T into itself be given. The euclidean plane E being the universal covering surface of T, there exist infinitely many continuous mappings of E into itself which "lie over" the given mapping t. Two fixed points p_1 and p_2 of t are said to belong to the same class if at least one of these covering mappings has two fixed points which lie over p_1 and p_2. This is the case if and only if there exists a curve on T joining p_1 and p_2 such that the closed curve consisting of it and its image by t is homotopic to zero. Determining the number of classes of fixed points, which depends only on the homotopy class of the mapping t, Jakob Nielsen obtains his former result concerning the minimal number of fixed points, now for not necessarily topological, continuous mappings, in a most satisfactory manner. Along with these investigations Jakob Nielsen took up the study of discontinuous groups of motions in the non-euclidean plane and devoted several papers (1923, 1925, 1927) to this subject. His interest in it arose from the fact that the fundamental groups of the surfaces of genus greater than 1 admit of representations by such groups. Among other results he obtained the following: On every closed surface with constant negative curvature there exist geodesics which are everywhere dense on the surface and approach every direction. This theorem is of importance in the theory of dynamical systems. But Jakob Nielsen's paper, published as it was in Danish, long remained unnoticed, and the theorem was found independently by Marston Morse. This case, like others, seems to show that in his modesty, Jakob Nielsen was apt to underrate the importance of the results he reached as bi-products on his way to the goal he had set himself.

These apparently somewhat desultory investigations turned out to be stones that went to the erection of an impressive building. Hints of this are to be found in some lectures given in Hamburg in 1924 and in Copenhagen in 1925, at the 6th Scandinavian Congress of Mathematicians. But in its final form it appeared in three long memoirs (300 pages in all) from the years 1927, 1929, and 1932 in Acta Mathematica under the common title "Untersuchungen zur Topologie der geschlossenen zweiseitigen Flächen". Here we find again the notions and methods he had previously used or developed: The universal covering surface interpreted as the non-euclidean plane, the latter represented by the conformal model in the interior of the unit circle, the fundamental group as a discontinuous group of motions in the non-euclidean plane, the mappings of the latter onto itself which lie over a given surface mapping, the automorphisms induced by them. As an essential new tool comes here the following theorem: Every mapping of the non-euclidean plane onto itself which lies over some surface mapping can be extended continuously to the points on the unit circle, representing the points at infinity of the non-euclidean plane, and the mapping of the circumference which arises in this way depends only on the homotopy class of the surface mapping. Hereby a two-dimensional topological problem is reduced to a one-dimensional one. A thorough analysis of this mapping leads to a wealth of information about homotopy classes of surface mappings. Further we find the notion of a class of fixed points generalized to the present case. To each class of fixed points an index is attached which gives the "algebraic number" of fixed points in the class. For a given homotopy class of surface mappings, estimates are found for the number of classes of fixed points and for their indices. Finally many profound results concerning the group of automorphisms and the group of homotopy classes of mappings are obtained. What we see here are results that for penetration surpass what had otherwise been achieved in the topology of surfaces. That he could achieve so much was due to his unique geometrical intuition, his power of concentration, and his immense capacity for work, enabling him to work out examples complicated enough to show what general regularities might be expected. The special cases previously dealt with were far from sufficient for this purpose. With these memoirs Jakob Nielsen had definitely broken new ground, and they gave him great international reputation. Many scientists have in some way combined their investigations with his. It should be particularly emphasised that the concept of a class of fixed points has undergone a great and important generalisation. Jakob Nielsen's results are not easily accessible, and even today no substantially simpler way to them has been found. So his achievement still stands as a challenge to younger scientists.

In 1925 Jakob Nielsen became professor of theoretical mechanics at the Technical University after the retirement of C. Juel. During the first few years he continued to base his instruction on Juel's textbook; but gradually there arose weighty reasons for a revision. To be sure, mechanics of particles and rigid bodies is among the best worked-out parts of applied mathematics, and in the introductory teaching with its century-old tradition the contents are well-defined, determined by their application in physics, astronomy, and technology. Nor was it on this point that the new textbook, which Jakob Nielsen published in two volumes in 1933–34, came to differ considerably from others. Rather, the object was an adaptation to the development that had taken place in mathematics, and especially in the teaching of mathematical analysis. The point was to introduce and exploit recent mathematical tools, such as vectors and matrices. As an example may be mentioned his elucidation of the close connection between on one hand the theory of frameworks and the principle of virtual work, and on the other hand the theory of systems of linear equations, which has given this subject a conciseness hardly to be found in the numerous other textbooks the world over. The book is not easily read, and Jakob Nielsen's lectures demanded much of the students. He had an unusual power of expressing himself with great lucidity, but also with great terseness. It was incredible how much he could tell, and in great detail, in an hour's lecture.

The textbook was published in German in 1935. A second edition in Danish came out in 1943 and 1945 (revised owing to changes in the teaching of geometry), and a third in 1950 and 1952.

Jakob Nielsen was faced with a new task when the teaching of aerodynamics was to be established at the Technical University. His lectures were published in mimeographed form in the years 1940–42. His presentation of this subject, theoretically not easily accessible in parts, is remarkable for its clear distinction between the empirical foundation and the mathematical theory. He published the more theoretical part of these lectures in 1952, as volume three of the textbook of theoretical mechanics.

Immediately after finishing the first edition of the textbook, Jakob Nielsen resumed his studies in topology and group theory. It is not possible here to mention the many papers, about 20, among them several comprehensive ones, which he published in the years after 1935, most of them carrying on his investigations on surface mappings. By means of the powerful tools developed in the previous papers, he succeeded in solving a series of related problems. In 1937 he gave a complete classification of the periodic mappings of a surface onto itself, and in 1942 a fourth great memoir, "Abbildungsklassen endlicher Ordnung", was published in Acta Mathematica. It deals with a problem to which he had been led in the third of the above-mentioned Acta-papers, and which he had solved there in some special cases: Does every homotopy class of surface mappings which is of finite order, in the sense that a certain power of it is the class of the identity mapping, contain a periodic mapping, that is, a mapping the same power of which is the identity? The proof that this is the case is extremely difficult and makes up all the 90 pages long paper. One cannot but admire the intellectual vigour with which this investigation is carried out. Finally I shall mention one more large paper: "Surface transformation classes of algebraically finite type" from 1944, in which more general classes of surface mappings are thoroughly investigated.

On several occasions Jakob Nielsen lectured at the Mathematical Institute of the university to a small circle of young mathematicians on subjects that occupied him in connection with his research. Of special interest is a series of lectures on discontinuous groups of motions in the non-euclidean plane, given in the year 1938–39. Such groups, whose theory goes back to H. Poincaré's fundamental work on automorphic functions, are, as noted, an important tool in Jakob Nielsen's investigations in surface topology. In the above-mentioned lectures he took up the theory for a certain class of these groups for its own sake. He thus arrived at interesting new notions and results. In connection with this, Svend Lauritzen has, in his thesis for the doctorate, carried out a corresponding investigation for another class of these groups. However, it proved desirable and expedient, in view of their many and important fields of application, to take up the theory of discontinuous groups of motions in the non-euclidean plane in its full generality and from the bottom. By degrees it became clear that this task, which Jakob Nielsen took up together with me, was considerable more extensive and timeconsuming than anticipated; many problems, great and small, which turned up on the way had to be elucidated before the monograph aimed at could take shape.

Although Jakob Nielsen's heart was in this task, he could only devote to it a moderate part of his great working power, for since the end of the 1939–45 war he was occupied on tasks of a totally different kind. It was almost a matter of course that his name should come up when positions of trust and responsibility were to be filled, for he possessed to an eminent degree the qualities of integrity and open-mindedness, and he could make up his mind quickly on a problem and act resolutely when necessary. He required much of himself in attending to the duties he took upon himself. I shall not go into detail, but merely mention that his participation in international co-operation, especially the work of UNESCO, occupied a considerable part of his working power. Instead of trying to describe his activities as a delegate in 1945 to the conference which created UNESCO, as the leader of the Danish delegation to several sessions of the General Conference and as a member of the Executive Board of UNESCO from 1952 to 1958, I shall quote from the "Ad Honorem Jakob Nielsen" which the chairman of the Executive Board, Sir Ben Bowen Thomas, wrote after Jakob Nielsen's death:

"Those of us who worked with Professor Nielsen as members of the Board will recall his services to UNESCO with gratitude. We shall treasure his memory as an inspiration. He was punctual at our meetings, he was industrious in mastering our papers, he was able in our discussions, he was disinterested and wise in his approach to our problems, he was completely and utterly devoted to the service of humanity as a whole within the spheres of our responsibilities in Education, Science and Culture."

"I repeat those words: Devoted to the service of humanity as a whole. In a sense he was a keeper of the Board's conscience, unhappy when lack of resources in money, ability or techniques hampered our work, sad when factors other than the merits of the projects being examined, demanded and obtained overriding weight or authority, as it seemed to him, when we reached some decisions; but no matter how the issue was decided, he always marched breast forward to the end, leaving with those of us who knew him the abiding memory of his powerful, energetic frame, his bright blue eyes and ready smile, his high standards, his unselfish willingness to accept the tasks which we only too readily entrusted to him. An admirable representative of his people, a true son of Denmark, a distinguished scholar, a great public servant, a very humble, modest man."

In 1951, Jakob Nielsen was nominated Harald Bohr's successor at the University of Copenhagen. Here he lectured with delight and zeal to young mathematicians on subjects close to his heart. But the growing demands made upon him by his UNESCO work, and the frequent and at times long journeys abroad, which interrupted his lectures, caused him to feel after a few years that he could no longer give satisfaction as a university teacher. Add to this that during these years there was not much time for scientific work, especially on the above-mentioned monograph, although he took it up untiringly and zealously whenever his duties permitted. So already in 1955, he resigned his professorship, and after finishing his UNESCO work in 1958, he could devote himself whole-heartedly to research work. There now followed a period of intense and successful work, especially on the monograph. He succeeded in surmounting a difficulty which had long prevented a satisfactory conclusion. But already in January 1959 he was stricken with the disease which carried him off on the 3rd of August. It is sincerely to be regretted that he did not live to see the publication of the monograph.

Jakob Nielsen was a sterling character and warm-hearted man. He showed life-long gratitude to those to whom he thought himself indebted, whether in the private or the scientific sphere. He was his friends' most faithful friend in times good and bad. He had always praise and encouragement for his collaborators, and if criticism was required, he always knew how to find a form that could not hurt even the most sensitive. Still, in the face of intolerance and narrow-mindedness he could find sharp words.

His memory as a man and as a scientist will be cherished with deep gratitude.

The publications of Jakob Nielsen

Memoirs and Articles

1. *Kurvennetze auf Flächen.* Inaugural-Dissertation, Kiel 1913, 58 pp.
2. Die Isomorphismen der allgemeinen, unendlichen Gruppe mit zwei Erzeugenden. *Math. Ann.*, 78 (1917), 385–397.
3. Über die Isomorphismen unendlicher Gruppen ohne Relation. *Math. Ann.*, 79 (1918), 269–272.
4. Die Berechnung von Rücklaufdiagrammen. *Artilleristische Monatshefte* Nr. 144 (1918), 169–173.
5. Der Ausfluss aus einem ursprünglich nicht vollen Rohr. *S.-B. Akad. Wiss. Wien, Math. nat. Kl., Abt.* IIa, 128 (1919), 1253–1259.
6. Über fixpunktfreie topologische Abbildungen geschlossener Flächen. *Math. Ann.*, 81 (1920), 94–96.
7. Über die Minimalzahl der Fixpunkte bei den Abbildungstypen der Ringflächen. *Math. Ann.*, 82 (1920), 83–93.
8. Om Regning med ikke-kommutative Faktorer og dens Anvendelse i Gruppeteorien. *Mat. Tidsskr. B* 1921, 77–94.
9. Diskontinuerte Flytningsgrupper. *Mat. Tidsskr. B* 1923, 53–61.
10. Ringfladen og Planen. *Mat. Tidsskr. B* 1924, 1–22.
11. Die Gruppe der dreidimensionalen Gittertransformationen. *Math.-fys. Medd. Danske Vid. Selsk.* 5, nr. 12 (1924), 29 pp.
12. Die Isomorphismengruppe der freien Gruppen. *Math. Ann.*, 91 (1924), 169–209.
13. Über topologische Abbildungen geschlossener Flächen. *Abh. Math. Sem. Univ. Hamburg* 3 (1924), 246–260.
14. Om geodætiske Linier i lukkede Mangfoldigheder med konstant negativ Krumning. *Mat. Tidsskr. B* 1925, 37–44.
15. Zur Topologie der geschlossenen Flächen. *Den sjette skandinav. Matematikerkongres, København* 1925, 263–276.

0*† – 603808

16. Om Flytningsgrupper i Planen og deres Anvendelse i Fladetopologien. *Mat. Tidsskr. B* 1927, 53–65.
17. Om Flytningsgrupper i den hyperbolske Plan. *Mat. Tidsskr. B* 1927, 65–75.
18. Untersuchungen zur Topologie der geschlossenen zweiseitigen Flächen. *Acta Math.*, 50 (1927), 189–358.
19. Om topologiske Afbildninger af en Jordankurve paa sig selv. *Mat. Tidsskr. B* 1928, 39–46.
20. Untersuchungen zur Topologie der geschlossenen zweiseitigen Flächen, II. *Acta Math.*, 53 (1929), 1–76.
21. Untersuchungen zur Topologie der geschlossenen zweiseitigen Flächen, III. *Acta Math.*, 58 (1932), 87–167.
22. Om regulære Riemannske Flader. *Mat. Tidsskr. B* 1932, 1–18.
23. Om Strukturen af en Vektormodul med endelig Basis. Med Anvendelse paa diofantiske Approksimationer. *Mat. Tidsskr. B* 1932, 29–42.
24. Einige Sätze über topologiske Flächenabbildungen. *Acta Litt. Scient. Szeged*, 7 (1935), 200–205.
25. Topologie des transformations des surfaces. *Enseign. Math.*, 35 (1936), 269–287.
26. Topologischer Beweis eines Satzes von Wiman. *Mat. Tidsskr. B* 1936, 11–24.
27. Ækvivalensproblemet for periodiske Transformationer. *Mat. Tidsskr. B* 1937, 33–41.
28. Om Tensorbegrebet. *Mat. Tidsskr. A* 1937, 1–16.
29. Die Struktur periodischer Transformationen von Flächen. *Math.-fys. Medd. Danske Vid. Selsk.* 15, nr. 1, 77 pp.
30. Einige Methoden und Ergebnisse aus der Topologie der Flächenabbildungen. *C. R. Congr. Intern. Math. Oslo* 1936, I (1937), 128–139.
31. Den symmetriske og den alternerende Gruppe. *Mat. Tidsskr. B* 1940, 7–18.
32. Über Gruppen linearer Transformationen. *Mitt. Math. Gesellsch. Hamburg* 8, *Festschrift, Teil II* (1940), 82–104.
33. En Sætning fra Fladeafbildningernes Topologi. *Norsk Mat. Tidsskr.* 23 (1941), 96–99.
34. Abbildungsklassen endlicher Ordnung. *Acta Math.*, 75 (1942), 23–115.
35. Fixpunktfrie Afbildninger. *Mat. Tidsskr. B* 1942, 25–41.
36. Surface transformation classes of algebraically finite type. *Mat.-fys. Medd. Danske Vid. Selsk.* 21, nr. 2 (1944), 87 pp.
37. With S. Bundgaard. Forenklede Beviser for nogle Sætninger i Fladetopologien. *Mat. Tidsskr. B* 1946, 1–16.
38. With W. Fenchel. On discontinuous groups of isometric transformations of the non-euclidean plane. *Studies and Essays presented to R. Courant*, New York 1948, 117–128.
39. Kommutatorgruppen for det frie produkt af cykliske grupper. *Mat. Tidsskr. B* 1948, 49–56.
40. A study concerning the congruence subgroups of the modular group. *Mat.-fys. Medd. Danske Vid. Selsk.* 25, nr. 18, 32 pp.
41. Johannes Hjelmslev. *Overs. Danske Vid. Selsk.* 1949–50 (1950), 89–103.
42. Harald Bohr. *Festskrift Kobenhavns Univ. Årsfest* 1951, 95–100.
43. With S. Bundgaard. On normal subgroups with finite index in F-groups. *Mat. Tidsskr. B* 1951, 56–58.
44. Nogle grundlæggende begreber vedrørende diskontinuerte grupper af lineære substitutioner i en komplex variabel. *C. R. 11ieme Congr. Math. Scand. Trondheim* 1949, Trondheim 1952, 61–70.
45. Det internationale samarbejde i naturvidenskaben. *Beretn. 2. Nord. Fysik-, Kemi- og Matematiklærer Kongr. Aarhus* 1954, 45–57.
46. A basis for subgroups of free groups. *Math. Scand.* 3 (1955), 31–43.

Textbooks

47. *Forelæsninger over Rationel Mekanik, I. Statik og Kinematik.* København 1933, 222 pp.
48. *Forelæsninger over Rationel Mekanik, II. Dynamik.* København 1934, 399 pp.
49. *Vorlesungen über elementare Mechanik.* Übersetzt und bearbeitet von Werner Fenchel. Berlin 1935, 500 pp.
50. *Indledning til Kontinuerte Mediers Mekanik.* Mimeographed lecture notes. København 1940, 34 pp. Second edition 1945.
51. *Forelæsninger over Aerodynamik, I.* Mimeographed lecture notes. København 1941, 164 pp. Second edition 1945.
52. *Forelæsninger over Aerodynamik, II.* Mimeographed lecture notes. København 1942, 57 pp.
53. *Lærebog i Rationel Mekanik, I. Statik.* Second edition. København 1943, 233 pp. Third edition 1950.
54. *Lærebog i Rationel Mekanik, II. Dynamik.* Second edition. København 1945, 367 pp. Third edition 1952.
55. *Lærebog i Rationel Mekanik, III. Vektoranalyse. Potentialteori. Kontinuerlige Medier. Strømninger. Komplekst Potential.* København 1952, 197 pp.

Kurvennetze auf Flächen.

Inaugural-Dissertation
zur Erlangung der Doktorwürde
der hohen philosophischen Fakultät
der Königl. Christian-Albrechts-Universität zu Kiel

vorgelegt von

Jakob Nielsen
aus Meels (Alsen).

KIEL 1913.
Druck von C. Schaidt (Inhaber Georg Oheim).

Referent: Prof. Dr. *Landsberg*.

Tag der mündlichen Prüfung: 15. Juni 1912.

KIEL, den 9. Dezember 1912.

Zum Druck genehmigt:

Dr. *C. Harries*,
z. Zt. Dekan.

Meinen Eltern
gewidmet

Inhalt der Arbeit:

---

§ 1.

Die Totalkrümmung als Randintegral.

Gauß bildet ein bestimmtes Gebiet $\mathfrak{J}$ einer krummen Fläche durch Parallele zu den Flächennormalen auf eine Kugel vom Radius 1 ab und nennt das dadurch auf derselben bestimmte, — durch algebraische Summation aus positiven und negativen Überdeckungen entstehende — Gebiet C die »Curvatura integra«, oder Totalkrümmung des Gebietes $\mathfrak{J}$. Durch Anwendung dieser Abbildung auf ein unendlich kleines Gebiet $d\mathfrak{J}$ der krummen Fläche gibt er eine geometrische Definition des Krümmungsmaßes derselben an einem bestimmten Punkt:

$$K = \frac{dC}{d\mathfrak{J}}$$

Heißt hierbei diejenige Seite von $d\mathfrak{J}$, nach der die gerichteten Flächennormalen gezogen wurden, der Außenseite der Einheitskugel zugeordnet und versieht man $d\mathfrak{J}$ mit einer Indikatrix, so ist K positiv oder negativ zu rechnen, je nachdem sich die Indikatrix bei der Abbildung erhält oder umkehrt. Die Totalkrümmung eines beliebigen Gebietes $\mathfrak{J}$ ist also das über dasselbe erstreckte Doppelintegral*)

$$C = \iint K d\mathfrak{J} \tag{1}$$

Das betrachtete Gebiet $\mathfrak{J}$ sei ein Elementargebilde**), d. h. ein einfach zusammenhängendes, von einer doppelpunktlosen, geschlossenen Kurve L umgrenztes Flächenstück. Dann kann man

*) Enthält das Gebiet $\mathfrak{J}$ Singularitäten, in denen $K = \infty$ wird, so versagt für dieselben die obige Definition von C. Wie man diesen Begriff durch den Hilfsbegriff der Äquivalentkrümmung auch auf solche Fälle ausdehnen kann, ist von *Boy* gezeigt: Diss. phil., Göttingen 1901, und Mathematische Annalen 57.

**) cfr. *Dyck*, Math. Ann. 32, p. 472.

ein Gebiet der Fläche, innerhalb dessen L verläuft, mit zwei Kurvenscharen so überziehen, daß dieselben ein reguläres Netz bilden, d. h. daß jede Kurve einer Schar die übrigen Individuen derselben nicht trifft, aber mit jedem Individuum der anderen Schar einen und nur einen gemeinsamen Punkt hat. Durch dieses Netz können dann zwei Parameter u und v über das Flächenstück stetig verteilt gedacht werden.

Mit Hilfe derselben sei das L enthaltende Flächengebiet dargestellt in der Form

$$x = f_1(u,v)$$
$$y = f_2(u,v)$$
$$z = f_3(u,v)$$

und aus den Funktionen f in bekannter Weise die Fundamentalgrößen erster Ordnung und deren Determinante gebildet:

$$E, F, G, \Delta = \sqrt{EG - F^2}$$

und die Fläche sei singularitätenfrei gedacht innerhalb des betrachteten Gebietes, so daß wir diese Größen als stetige Funktionen von u und v mit jedenfalls abteilungsweise stetigen ersten partiellen Ableitungen annehmen dürfen. Dasselbe gilt dann von den Funktionen

$$M = \frac{F}{E \cdot \Delta} \frac{\partial E}{\partial v} - \frac{1}{\Delta} \frac{\partial G}{\partial u}$$

$$N = \frac{2}{\Delta} \frac{\partial F}{\partial u} - \frac{1}{\Delta} \frac{\partial E}{\partial v} - \frac{F}{E \cdot \Delta} \frac{\partial E}{\partial u}$$

Dann ist das Krümmungsmaß der Fläche*)

$$K = \frac{1}{2\Delta} \left(\frac{\partial M}{\partial u} + \frac{\partial N}{\partial v} \right)$$

Das Flächenelement ist gegeben durch

$$d\mathfrak{J} = \Delta \, du \, dv$$

sodaß Formel (1) die Gestalt annimmt

$$C = \iint \frac{1}{2} \left(\frac{\partial M}{\partial u} + \frac{\partial N}{\partial v} \right) du \, dv$$

wobei das Doppelintegral über das von L umgrenzte Flächenstück erstreckt wird, dessen Totalkrümmung bestimmt werden soll.

---

*) Zur Ableitung dieser Relation sehe man z. B. *Bianchi,* Vorl. über Differentialgeometrie § 35.

Dieses Flächenintegral erfüllt die Bedingungen der Transformierbarkeit in ein Randintegral nach dem *Green*schen Satze:

$$C = \frac{1}{2}\int_L (M\,dv - N\,du)$$

wobei das Integral über die Kurve L in einem Sinne zu erstrecken ist, der einer Drehung $< \pi$ entspricht, welche die Richtung der wachsenden u in die Richtung der wachsenden v überführt.

Durch Angabe des Verhältnisses $\frac{du}{dv}$ sei in einem Punkte der Fläche eine Richtung ω definiert. Dieselbe bildet mit den Koordinatenachsen Winkel, deren cos sind:

$$\frac{dx}{ds} = \frac{\partial f_1}{\partial u}\frac{du}{ds} + \frac{\partial f_1}{\partial v}\cdot\frac{dv}{ds}$$

und analog. Die durch denselben Punkt gehende Parameterkurve $v =$ const. hat Richtungskosinus, die man aus diesen spezialisiert durch $dv = 0$, $ds = \sqrt{E}\,du$

$$\frac{dx}{ds} = \frac{1}{\sqrt{E}}\cdot\frac{\partial f_1}{\partial u}$$

und analog. Für den Winkel der Richtung ω mit der Parameterlinie $v =$ const. erhält man daher durch Komposition

$$\cos\omega = \frac{E\,du + F\,dv}{\sqrt{E}\,ds}$$

$$\sin\omega = \frac{\Delta\,dv}{\sqrt{E}\,ds}$$

Die Differentialgleichung der geodätischen Linien auf einer in Parameterform dargestellten Fläche ist:

$$\frac{\partial E}{\partial u}du^2 + 2\frac{\partial F}{\partial u}du\,dv + \frac{\partial G}{\partial u}dv^2 = 2\,ds\,\frac{d(E\,du + F\,dv)}{ds}$$

Bezeichnet also ω den Winkel, den eine geodätische Linie mit den Kurven $v =$ constans bildet, so ist die rechte Seite obiger Differentialgleichung:

$$\begin{aligned} &2\,ds\cdot d(\sqrt{E}\cos\omega) \\ &= 2\,ds\cos\omega\cdot\frac{1}{2}\frac{dE}{\sqrt{E}} - 2\,ds\sqrt{E}\sin\omega\,d\omega \\ &= \left(du + \frac{F\,dv}{E}\right)\left(\frac{\partial E}{\partial u}du + \frac{\partial E}{\partial v}dv\right) - 2\,\Delta\,dv\,d\omega \end{aligned}$$

Substituiert man diesen Ausdruck in der rechten Seite obiger Differentialgleichung, so heben sich die Glieder mit $\mathrm{d}u^2$, und man erhält nach Tilgung des Faktors $\mathrm{d}v$:

$$\mathrm{d}\omega = \frac{1}{2\Delta}\left[\frac{F}{E}\left(\frac{\partial E}{\partial u}\mathrm{d}u + \frac{\partial E}{\partial v}\mathrm{d}v\right) + \left(\frac{\partial E}{\partial v} - 2\frac{\partial F}{\mathrm{d}u}\right)\mathrm{d}u - \frac{\partial G}{\partial u}\mathrm{d}v\right]$$

eine Formel, die das Inkrement des Winkels angibt, den die in einem gegebenen Punkt durch das Verhältnis $\frac{\mathrm{d}u}{\mathrm{d}v}$ bestimmte Linie mit der Kurve $v =$ constans bildet.

Nun seien φ resp. ω die Winkel, welche in einem Punkte P die Kurve L resp. die sie in P berührende geodätische Linie G bildet mit der Richtung $v =$ const. Diese Winkel sind dann stetige Funktionen des Ortes, also nach der *Taylor*schen Reihe zu entwickeln:

$$\varphi(u+\Delta u, v+\Delta v) = \varphi(u,v) + \frac{\partial\varphi}{\partial u}\Delta u + \frac{\partial\varphi}{\partial v}\Delta v + \cdots$$

$$\omega(u+\Delta u, v+\Delta v) = \omega(u,v) + \frac{\partial\omega}{\partial u}\Delta u + \frac{\partial\omega}{\partial v}\Delta v + \cdots$$

Für genügend kleine Inkremente ist die Differenz der linken Seiten der »geodätische Kontingenzwinkel« $\mathrm{d}\tau$, unter dem L von der berührenden geodätischen Linie abweicht. Wegen der Berührung im Punkte $P(u, v)$ ist

$$\varphi(u,v) = \omega(u,v)$$

also ist unter Vernachlässigung höherer Glieder

$$\mathrm{d}\tau = \mathrm{d}\varphi - \mathrm{d}\omega$$

$$\mathrm{d}\omega = \mathrm{d}\varphi - \mathrm{d}\tau$$

wobei alle Winkel in dem Drehungssinn positiv gerechnet sind, welcher die Richtung der wachsenden u durch eine Drehung $> \pi$ in die Richtung der wachsenden v überführt. Und da φ und ω durch dasselbe Verhältnis $\frac{\mathrm{d}u}{\mathrm{d}v}$ bestimmt sind, so wird die Differenz $\mathrm{d}\varphi - \mathrm{d}\tau$ durch den oben für $\mathrm{d}\omega$ aufgestellten Differentialausdruck dargestellt. Derselbe ist aber nichts anderes als der Integrand des die Totalkrümmung darstellenden Kurvenintegrals:

$$^1/_2\,(M\,\mathrm{d}v - N\,\mathrm{d}u)$$

sodaß für dasselbe die Form gewonnen wird

$$C = \int_L (\mathrm{d}\varphi - \mathrm{d}\tau) \tag{2}$$

Diese Formel ist zuerst von *Ossian Bonnet* angegeben in seiner umfassenden Arbeit: »Mémoire sur la théorie générale des surfaces«*), wo er sie im Wesentlichen auf geometrischem Wege gewinnt.

φ ist der Winkel, den in jedem Punkt die Linie L mit $v =$ const. bildet. Durchläuft man L, so stimmt die Endlage der beiden Kurventangenten mit der Anfangslage überein. Das Integral über die Änderung des Winkels zwischen ihnen ist also ein Vielfaches von 2π, ändert sich demnach nicht, wenn man L durch eine Nachbarkurve ersetzt, da das bei unserer Voraussetzung über das Parameternetz eine stetige Änderung des Integrals hervorrufen müßte. Es ist also zulässig, den Weg des Integrals

$$\int d\varphi$$

auf ein unendlich kleines Gebiet zusammenzuziehen und für dasselbe die Richtung der wachsenden u konstant anzunehmen, was nicht möglich wäre, wenn z. B. die Kurven $v =$ const. im Inneren von L einen Pol hätten. So nimmt für das von L begrenzte Elementarflächenstück die Formel die vom Parameternetz unabhängige Gestalt an

$$C = 2\pi - \int_L d\tau$$

In dieser Form ist die Formel lediglich eine Erweiterung des *Gauß*schen Theorems über die Totalkrümmung eines aus geodätischen Linien gebildeten Dreiecks. Denn denkt man sich ein einfach zusammenhängendes, aus geodätischen Linien gebildetes Vieleck durch geodätische Linien, die von einem Punkte des Inneren ausgehen, in geodätische Dreiecke zerlegt und wendet auf diese das *Gauß*sche Theorem an, so bestimmt man leicht die Totalkrümmung des von dem Vieleck umschlossenen Elementarflächenstücks als

$$2\pi - \text{Summe der Außenwinkel}$$

und läßt man nun die Seitenzahl des geodätischen Vielecks unbegrenzt wachsen und schmiegt dasselbe dem Verlauf der Kurve

*) Journal de l'École Polytechnique, XXXIIième Cahier, Tome XIX. Schon *Gauß* hatte das über eine Flächenkurve erstreckte $\int d\tau$ untersucht. cfr. Nachlaß (Werke VIII) p. 386 ff.

L an, so ist im Grenzfall der Außenwinkel mit dem geodätischen Kontingenzwinkel derselben identisch, und man erhält unmittelbar die obige Formel.

Endlich sei die Formel in eine vom Begriff der geodätischen Linie unabhängige Gestalt gebracht durch Verwendung einer bekannten Relation, welche den geodätischen Kontingenzwinkel $d\tau$ einer Flächenkurve durch Krümmungsradius ϱ und Bogenelement ds derselben sowie den Winkel ϑ ausdrückt, welchen ihre Schmiegungsebene (Krümmungsradius) mit der Flächennormalen bildet:

$$d\tau = \frac{\sin\vartheta}{\varrho} ds$$

Da $d\tau$ in der Formel (2) eine mit Vorzeichen begabte Größe bedeutet, ds und ϱ aber absolute Größen sind, so ist der Winkel ϑ mit einem Vorzeichen zu versehen, das früheren Festsetzungen entspricht. Oben (p. 3 u. 4) wurde sowohl für die Durchlaufungsrichtung der Kurve L als auch für die positive Zählrichtung des Winkels $d\tau$ eine Drehungsrichtung festgelegt und zwar — worauf es hier allein noch ankommt, nachdem vom Koordinatensysteme abstrahiert worden ist — für beide der gleiche Drehungssinn. Diese Festsetzung läßt sich, nachdem man die Kurve L gerichtet und der berührenden geodätischen Linie die im Berührungspunkt gleiche Richtung erteilt hat, so formulieren:

Der Winkel $d\tau$ ist positiv oder negativ zu rechnen, je nachdem die variierte Kurventangente gegen die variierte Tangente der berührenden geodätischen Linie gleich oder entgegengesetzt orientiert ist, wie die ins Innere des umlaufenen Gebietes gerichtete Kurvennormale gegen die Kurventangente.

Erwägt man nun, daß die geodätische Linie durch einen Normalschnitt, die Kurve L durch ihre Schmiegungsebene aus der Fläche ausgeschnitten wird, so erkennt man, daß die Kurve L von der berührenden geodätischen Linie nach der Seite hin abweicht, nach welcher die Projektion des vom Berührungspunkt nach dem Krümmungszentrum gerichteten Krümmungsradius auf die Tangentialebene von der Kurve abweicht. Also hat $d\tau$ und damit auch ϑ das positive oder negative Zeichen, je nachdem die Projektion des Krümmungsradius ϱ auf die Tangentialebene mit der nach innen oder außen gerichteten Kurvennormalen in der Tangentialebene zusammenfällt.

Damit nimmt die *Bonnet*sche Formel eine Gestalt an, die lauter an der Kurve L zu messende Größen enthält und keine bestimmte Durchlaufungsrichtung vorschreibt:

$$C = 2\pi - \int_L \frac{\sin\vartheta}{\varrho} ds \qquad (3)$$

Besitzt die Kurve L Ecken E, d. h. Punkte, an denen ihre Tangentenrichtung um einen endlichen Winkel springt, so bedarf die Formel (3) an diesen Punkten der Interpretation. Wird das Integral über eine die Ecke abrundende Kurve ADB anstatt über den vorgeschriebenen Weg AEB erstreckt, so ist der begangene Fehler von der Größe

$$\iint K \, d\mathfrak{J}$$

erstreckt über die Fläche $AEBDA$, also beliebig klein, wenn sich die Kurve ADB genügend in die Ecke AEB hineinschmiegt. Im Grenzfall ist aber die Schmiegungsebene der Kurve ADB die Tangentialebene in E, sodaß

$$\sin\vartheta = \pm 1$$

während

$$\int \frac{ds}{\varrho}$$

über ADB erstreckt gegen die Größe des Außenwinkels $\pi - AEB$ konvergiert, falls A und B genügend nahe an E liegen. Der Betrag, den das Integral an der Ecke E liefert, ist also gleich dem Winkel, um den die Tangente springt, positiv oder negativ gerechnet, je nachdem der Eckenwinkel, den das umlaufene Gebiet bei E besitzt, $< \pi$ oder $> \pi$ ist.

Die *Bonnet*sche Formel gestattet in einfacher Weise die Herstellung der bekannten Beziehung, die zwischen der Totalkrümmung einer geschlossenen Fläche und ihrem topologischen Charakter besteht. In dem Enzyklopädieartikel »Analysis situs« von *Dehn-Heegaard**) wird gezeigt, daß auf einer geschlossenen (einseitigen oder zweiseitigen) Fläche der Charakteristik K eine Kurve L so gezogen werden kann, daß dieselbe die Fläche zerschneidet in

1. ein Elementarflächenstück F_1,
2. ein Flächenstück F_2, welches besteht aus einem Elementarflächenstück mit $K+2$ angehefteten Bändern.

---

*) Math. Enzykl., Band III, 1 Heft 1.

Zerschneidet man diese $K+2$ Bänder durch je einen Querschnitt q_i, der die Ränder eines Bandes verbindet, so wird dadurch F_2 in ein Elementarflächenstück F_2' verwandelt. Die Totalkrümmung der geschlossenen Fläche ist also zu bestimmen durch Anwendung der *Bonnet*schen Formel auf F_1 und F_2'. Dann ist ersichtlich, daß jeder Kurvenbogen des Schnittsystems $L+\Sigma q_i$ zweimal durchlaufen wird, weil er zur Begrenzung der beiderseits liegenden Flächenteile gehört, daß bei diesen beiden Durchlaufungen die »nach dem Inneren des umlaufenen Gebietes« gerichteten Kurvennormalen entgegengesetzt gerichtet sind, daß also die beiden Integralbeiträge sich aufheben. Der Wert der Totalkrümmung C wird also außer von dem Summanden

$$2\cdot 2\pi$$

der zweimal angewendeten Formel nur von den Integralbeiträgen der an den Einmündungspunkten der q_i in L entstehenden Ecken gebildet. Diese Eckwinkel sind alle $<\pi$, geben also positive Beiträge und zwar geben die zwei Nebenwinkel eines Einmündungspunktes durch die Summe ihrer Außenwinkel den Beitrag π, jeder Querschnitt ruft also den Beitrag 2π hervor. Also folgt:

$$C = 2\cdot 2\pi-(K+2)\,2\pi$$

$$C = -K\cdot 2\pi \qquad (4)^{*)}$$

Bei einer berandeten Fläche ist natürlich die Totalkrümmung C durch die topologischen Eigenschaften allein nicht bestimmt. Man versehe die vorher betrachtete Fläche mit r Löchern, die teils F_1 und teils F_2' angehören mögen und deren Randkurven mit $R_1 \cdots R_r$ bezeichnet seien. Dann verbinde man R_i durch einen Querschnitt s_i mit einem Punkt des Schnittsystems $L+\sum_i q_i$, wodurch F_1 und F_2' wieder zu Elementarflächenstücken werden. Bei der Anwendung der Formel kommen also zu den früheren noch als weitere Integralbestandteile

1. die durch jeden Querschnitt s_i hervorgerufenen Eckenbeiträge 2π,
2. der durch jeden Rand erzeugte Beitrag $\int\limits_{R_i} \frac{\sin\vartheta\, ds}{\varrho}$

*) Es ist zu beachten, daß die im Enzyklopädieartikel als »Charakteristik« bezeichnete Größe entgegengesetzt gleich der von *Dyck* (Math. Ann. 32) als »Charakteristik« definierten Zahl ist.

die als negative Summanden in den Wert von C eingehen. Es ergibt sich also, wenn man bedenkt, daß auch die Charakteristik der Fläche um r gestiegen ist:

Eine Fläche der Charakteristik K mit r Randkurven hat die Totalkrümmung:

$$C = -\left(K \cdot 2\pi + \sum_i \int_{R_i} \frac{\sin \vartheta \, ds}{\varrho}\right) \tag{4a}$$

§ 2.

Totalkrümmung einer Fläche, die eine stetige Schar kongruenter Kurven trägt.

Trägt eine berandete oder unberandete Fläche eine stetige Schar kongruenter Kurven, so kann man sich die Fläche beschrieben denken, indem das Anfangselement jener Schar sukzessive durch die Lage der übrigen hindurch in die Lage des Endelements übergeht. Im Folgenden wird nach der Totalkrümmung der durch Angabe der Gestalt und Bewegung der Kurve bestimmten Fläche gefragt.

Da man die allgemeinste Bewegung aus Schraubungen zusammensetzen kann, so werde zunächst eine Kurve

$$\left.\begin{aligned} x &= f_1(t) \\ y &= f_2(t) \\ z &= f_3(t) \end{aligned}\right\} \quad (1)$$

betrachtet, die einer Schraubung um die z-Achse unterworfen ist. Dieselbe habe die Ganghöhe $2\pi \cdot A$ und der Schraubungswinkel werde von der positiven x-Achse aus gegen die positive y-Achse gemessen. Dann entsteht die Fläche:

$$\left.\begin{aligned} x &= f_1(t)\cos\varphi - f_2(t)\sin\varphi \\ y &= f_1(t)\sin\varphi + f_2(t)\cos\varphi \\ z &= f_3(t) + A\cdot\varphi \end{aligned}\right\} \quad (2)$$

Die Totalkrümmung C des Flächenstücks, welches das zwischen den Parameterwerten t_1 und t_2 liegende Stück der Kurve bei der Schraubung von φ_1 bis φ_2 erzeugt, kann mittels der *Bonnet*schen Formel bestimmt werden, wenn das von den Kurven

$$\begin{aligned} t &= t_1 \qquad\qquad & t &= t_2 \\ \varphi &= \varphi_1 \qquad\qquad & \varphi &= \varphi_2 \end{aligned}$$

bestimmte Flächenstück ein Elementarflächenstück ist. Die Kurven

$t = t_1$ und $t = t_2$ schneiden sich als Schraubenlinien nicht. Wohl aber können $\varphi = \varphi_1$ und $\varphi = \varphi_2$ einen gemeinsamen Punkt im gleichen Flächenmantel haben; ein solcher müßte während der ganzen Bewegung in Ruhe bleiben, also der z-Achse angehören, während die Ganghöhe Null ist. Dieser Fall der Rotation einer die Achse schneidenden Kurve werde im Folgenden ausgeschlossen, da sich auch von anderer Seite her zeigt, daß er gesonderte Behandlung verlangt.

Eine Schraubenfläche ist in sich verschiebbar; durch eine solche Verschiebung läßt sich die Kurve $\varphi = \varphi_1$ mit $\varphi = \varphi_2$ zur Deckung bringen. Dann liegen die Teile des Elementarflächenstücks, deren Begrenzung sie bilden, auf verschiedenen Seiten der gemeinsamen Lage, sodaß sich in der *Bonnet*schen Formel die Integralbeiträge dieser beiden Kurven heben. Aus dem gleichen Grunde sind die Eckenwinkel an den 4 Punkten, in denen sich die 4 Randkurven treffen, zu je zweien Nebenwinkel, sodaß ihr Gesamtbeitrag 2π ist, was sich mit dem konstanten Summanden 2π der Formel hebt. Es bleibt also

$$C = -\int_{\varphi_1}^{\varphi_2} \left[\frac{\sin\vartheta\, ds}{\varrho}\right]_{t=t_1} - \int_{\varphi_1}^{\varphi_2} \left[\frac{\sin\vartheta\, ds}{\varrho}\right]_{t=t_2} \tag{3}$$

Um den Integranden rechnungsmäßig aufzustellen, bestimme man aus der Flächengleichung (2) die Fundamentalgrößen erster Ordnung

$$E = f_1'^2 + f_2'^2 + f_3'^2$$
$$F = f_1 f_2' - f_2 f_1' + A f_3'$$
$$G = f_1^2 + f_2^2 + A^2$$

wobei ein Strich die Ableitung nach t bedeutet. Der Abstand eines Punktes von der z-Achse heiße

$$\sqrt{f_1^2 + f_2^2} = r$$

Dann ist ds bei konstantem t, d. h. längs einer Schraubenlinie gemessen:

$$ds = \sqrt{G}\, d\varphi = \sqrt{r^2 + A^2}\, d\varphi$$

Ferner findet man durch Aufstellung der zweiten Ableitungen von x, y, z nach φ und t den Krümmungsradius einer Schraubenlinie ($t = \text{const.}$)

$$\varrho = \frac{r^2 + A^2}{r}$$

und den Winkel ϑ, den derselbe mit der Flächennormalen bildet:

$$\sin\vartheta = \frac{(f_1 f_1' + f_2 f_2')\cdot\sqrt{r^2+A^2}}{r\cdot\Delta}$$

wenn $\Delta = \sqrt{EG-F^2}$, sodaß sich für das gesuchte Differential der Ausdruck ergibt:

$$\frac{\sin\vartheta\,\mathrm{d}s}{\varrho} = \frac{f_1 f_1' + f_2 f_2'}{\Delta}\,\mathrm{d}\varphi$$

der sich auch schreiben läßt

$$\frac{\sin\vartheta\,\mathrm{d}s}{\varrho} = \frac{\frac{1}{2}\,\frac{\mathrm{d}(r^2)}{\mathrm{d}t}\,\mathrm{d}\varphi}{\Delta} \tag{4}$$

Um dieses Differential zur Ausführung der Integration in (3) verwenden zu können, muß man es mit einem Vorzeichen versehen, das der Forderung der *Bonnet*schen Formel entspricht. Längs der Schraubenlinie $t = t_2$ ist das Innere des umlaufenen Flächenstücks diejenige Seite, nach welcher der Parameter t abnimmt. Nun ist der Krümmungsradius einer Schraubenlinie nach der Schraubungsachse (hier z-Achse) gerichtet, seine Projektion auf die Tangentialebene des zugehörigen Kurvenpunktes fällt also in diejenige Hälfte der von der Kurventangente halbierten Tangentialebene, welche von der z-Achse getroffen wird. Diese ist das Innere des umlaufenen Gebietes dann und nur dann, wenn nach ihr hin der Parameter t abnimmt, d. h. wenn $\frac{\mathrm{d}r}{\mathrm{d}t} > 0$ ist. Also ist längs der Schraubenlinie $t = t_2$ das Differential in (4) mit positivem Zeichen zu verwenden. Längs $t = t_1$ ist das Innere des umlaufenen Gebietes diejenige Seite, nach welcher t wächst, dort ist also das Differential entsprechend mit negativem Zeichen zu versehen. Da sowohl Δ wie r von φ unabhängig sind, so ergibt die Integration von (3)

$$C = -(\varphi_2 - \varphi_1)\left[\frac{1}{2\Delta}\cdot\frac{\mathrm{d}(r^2)}{\mathrm{d}t}\right]_{t=t_1}^{t=t_2} \tag{5}$$

Man kann natürlich diese Formel direkt durch Ausrechnung des nach beiden Variabelen ausführbaren Doppelintegrals $\int\limits_{t_1}^{t_2}\int\limits_{\varphi_1}^{\varphi_2} K\,\mathrm{d}\mathfrak{J}$

gewinnen. Dabei hat man auch hier den oben genannten Sonderfall auszuschließen, denn ein Schnittpunkt mit der Achse erzeugt bei Rotation einen Flächenpunkt, für welchen man $K = -\infty$ rechnen muß, wenn man den Fall der Rotation aus der Schraubung durch $A = 0$ spezialisiert.

In der Gleichung (5) ist die Klammergröße der Kosinus des Winkels zwischen der Flächennormalen und der z-Achse, welcher längs einer Schraubenlinie konstant ist. So sieht man fast unmittelbar, wie Formel (5) die ursprüngliche *Gauß*sche Definition der Totalkrümmung veranschaulicht als des Abbildgebietes, das beim Übertragen auf die Einheitskugel durch Parallele zu den Flächennormalen entsteht.

Des Folgenden wegen soll (5) in eine vom Koordinatensystem unabhängige Gestalt gesetzt werden: φ sei der Winkel, $2\pi A$ die Ganghöhe der Schraubung, r der Abstand eines Kurvenpunktes von der Achse, ψ der nur von t abhängige Winkel einer Kurventangente mit der Achse, s der Bogen der Kurve. Striche bedeuten Ableitungen nach dem Kurvenparameter t. Dann ist der Zähler

$$r \cdot r'$$

Ferner ist

$$\begin{aligned} E &= \Sigma f_1'^2 = s'^2 \\ G &= r^2 + A^2 \\ F &= A \frac{\mathrm{d}z}{\mathrm{d}t} + f_1 f_2' - f_2 f_1' \\ &= A s' \cdot \cos\psi + \sqrt{(f_1^2 + f_2^2)(f_1'^2 + f_2'^2) - (f_1 f_1' + f_2 f_2')^2} \\ &= A s' \cos\psi + \sqrt{r^2 s'^2 \sin^2\psi - (rr')^2} \end{aligned}$$

sodaß die Formel (5) wird:

$$C = \tag{6}$$

$$-(\varphi_2 - \varphi_1)\left[\frac{rr'}{\sqrt{s'^2(r^2+A^2) - [As'\cos\psi + \sqrt{r^2 s'^2 \sin^2\psi - (rr')^2}\,]^2}}\right]_{t_1}^{t_2}$$

Es bietet nun keine prinzipiellen Schwierigkeiten mehr, die Formel auf den Fall der allgemeinsten Bewegung einer Kurve zu erweitern. Will man sich dabei auf singularitätenfreie Flächen beschränken, so ist zu berücksichtigen, daß die Tangentialebene der entstehenden Fläche an jedem Punkt bestimmt wird durch die Bewegungsrichtung des zugehörigen Kurvenpunktes und die Richtung der Kurventangente in demselben; daß die Tangentialebene also nur unbestimmt wird, wenn diese beiden Richtungen zusammenfallen oder die zweite unbestimmt wird. Das letztere schließt man aus durch die Forderung einer Kurve mit stetiger Tangentenrichtung, das erstere durch die Festsetzung, daß zwei aufeinander fogende Lagen der bewegten Kurve keinen Punkt so gemeinsam haben sollen, daß sie in der Umgebung desselben demselben Flächenmantel angehören. Dadurch ist zugleich der Spezialfall der Rotation mit schneidender Achse ausgeschlossen.

Man kann nun die allgemeinste Bewegung der Kurve z. B. so beschreiben, daß man dieselbe in einem beweglichen Koordinatentrieder (o, xyz) durch das Gleichungstripel (1) darstellt und dann die momentanen Rotationsgeschwindigkeiten p_1, p_2, p_3 dieses Trieders um seine Achsen sowie die Translationsgeschwindigkeiten π_1, π_2, π_3 in Richtung derselben in Funktion eines Parameters u angibt, dessen Variation also den Verlauf der Bewegung bestimmt.*) Dann stellt sich in jedem Moment die Bewegung als Schraubung um die Achse dar, deren Gleichung ist:

$$\frac{\pi_1 + p_2 z - p_3 y}{p_1} = \frac{\pi_2 + p_3 x - p_1 z}{p_2} = \frac{\pi_3 + p_1 y - p_2 x}{p_3}$$

Nennt man $\Sigma p_i^2 = \omega^2$, so sind die Richtungskosinus dieser Achse

$$\lambda_i = \frac{p_i}{\omega}$$

und der Schraubungswinkel

$$\frac{d\varphi}{du} = \omega$$

Ferner ist die Verschiebungsgeschwindigkeit der Punkte dieser Achse

$$\frac{dv}{du} = \frac{\Sigma p_i \pi_i}{\omega}$$

sodaß sich die Schraubungskonstante A ergibt

$$A = \frac{dv}{d\varphi} = \frac{dv}{du} : \frac{d\varphi}{du} = \frac{\Sigma p_i \pi_i}{\omega^2}$$

*) Man vergleiche zu den im Folgenden angegebenen Formeln *Darboux*, Leçons sur la théorie des surfaces I. Livre 1, Chap. 1.

Bedeutet s den Bogen der bewegten Kurve, so folgt aus (1)

$$\left(\frac{ds}{dt}\right)^2 = s'^2 = \Sigma f_i'^2$$

und für die Richtungskosinus ihrer Tangente

$$\mu_i = \frac{f_i'}{s'}$$

also ist

$$\cos\psi = \frac{\Sigma p_i f_i'}{\omega s'}$$

Außer diesen Größen tritt in (6) nur noch die Entfernung r eines Kurvenpunktes von der Schraubungsachse auf. Bringt man die Schraubungsachse mit der zu ihr senkrechten Ebene

$$p_1(x-f_1) + p_2(y-f_2) + p_3(z-f_3) = 0$$

zum Schnitt und nennt man zur Abkürzung die aus der Matrix

$$\begin{matrix} p_1 & p_2 & p_3 \\ \pi_1 & \pi_2 & \pi_3 \end{matrix}$$

gebildeten Determinanten (zyklisch)

$$d_1\ d_2\ d_3$$

so bestimmt man durch eine elementare Rechnung die Projektion des Kurvenpunktes f_1, f_2, f_3 auf die Schraubungsachse durch die Koordinaten

$$\frac{1}{\omega^2}(d_i + p_i \Sigma p_i f_i) \qquad (i = 1, 2, 3)$$

also die gesuchte Entfernung

$$r^2 = \frac{1}{\omega^4} \cdot \Sigma (f_i \cdot \omega^2 - d_i - p_i \Sigma p_i f_i)^2$$

und da

$$\Sigma (f_i \cdot \omega^2 \cdot p_i \Sigma p_i f_i) = \omega^2 (\Sigma p_i f_i)^2$$

und $\Sigma p_i d_i = 0$ ist,

$$r^2 = \frac{1}{\omega^4} \cdot (\omega^4 \Sigma f_i^2 + \Sigma d_i^2 - \omega^2 (\Sigma p_i f_i)^2 - 2\omega^2 \Sigma f_i d_i$$

und schließlich ist

$$rr' = \frac{1}{\omega^2}(\omega^2 \Sigma f_i f_i' - \Sigma p_i f_i \Sigma p_i f_i' - \Sigma f_i' d_i)$$

Führt man diese Werte in die Formel (6) ein, so ist die gestellte Aufgabe allgemein gelöst. Die durch die Klammergröße gebildete

Funktion von u und t heiße $F(u,t)$. Dann ist die allgemeine Form des Ergebnisses:

$$C = -\int_{u_1}^{u_2} \omega\, du \cdot [F(u,t)]_{t=t_1}^{t=t_2} \quad (7)$$

t tritt in der Funktion F nur in den Verbindungen f_i und deren Ableitungen auf. Man liest aus dieser Formel ab:

> Alle Raumkurven, die zwischen zwei nach Lage und Richtung gegebenen Endelementen verlaufen, beschreiben Flächen gleicher Totalkrümmung, wenn das Ganze als starres System einer gemeinsamen Bewegung unterworfen wird,

d. h. die Totalkrümmung ist allein durch den Flächenstreifen bestimmt, den die Endelemente beschreiben.

Ist die bewegte Kurve geschlossen und bedeutet $t_2 - t_1$ die Periode der Funktionen f_i, so verschwindet der Integrand für jeden Wert von u, und es folgt:

> Läßt sich eine berandete oder unberandete Fläche mit einer stetigen Folge geschlossener kongruenter Kurven so überdecken, daß zwei konsekutive Individuen derselben keinen Punkt gemeinsam haben, in dessen Umgebung sie demselben Flächenmantel angehören, so hat die Fläche die Totalkrümmung Null.

Bei diesen Überlegungen ist die Forderung stetiger Tangentenrichtung für die bewegte Kurve unwesentlich. Ist die Kurve stetig, aber an einem ihrer Punkte die Tangentenrichtung unstetig, so bestimmen die beiden Kurventangenten mit der Bewegungsrichtung zwei bestimmte Tangentialebenen der entstehenden Fläche, auf der also dieser Kurvenpunkt eine Kante hervorruft. Rechnet man deren Äquivalentkrümmung*) zur Totalkrümmung hinzu, so bleibt alles Ausgeführte richtig, da diese Bewertung einer Abrundung der Kante gleichkommt.

Eine spezielle Flächengattung möge die Anwendung der hier vorgetragenen Überlegungen verdeutlichen. Es wird die Frage nach der Totalkrümmung der Regelflächen gestellt.

---

*) Siehe die zitierte Arbeit von *Boy*, Math. Ann. 57.

Man wähle das Gleichungssystem (1) der bewegten Kurve in der einfachen Form

$$x = f_1(t) = t$$
$$y = f_2(t) = 0$$
$$z = f_3(t) = 0$$

sodaß

$$f_1'(t) = 1$$
$$f_2'(t) = f_3'(t) = 0$$

Alsdann vereinfachen sich die obigen Formeln wie folgt:

$$s' = 1$$

$$\cos\psi = \frac{p_1}{\omega}, \quad \sin\psi = \frac{\sqrt{p_2^2 + p_3^2}}{\omega}$$

$$r^2 = \frac{1}{\omega^4}(\omega^4 t^2 + \Sigma d_i^2 - \omega^2 t^2 p_1^2 - 2\omega^2 t d_1)$$

$$rr' = \frac{1}{\omega^2}(\omega^2 t - p_1^2 t - d_1)$$

Mit diesen Ausdrücken bilde man den Integranden $F(u,t)$ der Formel (7) [siehe Formel (6)] und zwar nach Division mit t in der Grenze für $t = \infty$:

$$\lim_{t=\infty} F(u,t) =$$

$$\frac{\frac{1}{\omega^2}\cdot(\omega^2 - p_1^2)}{\sqrt{\frac{1}{\omega^4}(\omega^4 - \omega^2 p_1^2) - \frac{1}{\omega^4}(\omega^4 - \omega^2 p_1^2)\frac{p_2^2 + p_3^2}{\omega^2} + \frac{1}{\omega^4}(\omega^2 - p_1^2)^2}}$$

$$= \frac{\omega^2 - p_1^2}{\sqrt{(\omega^2 - p_1^2)p_1^2 + (\omega^2 - p_1^2)^2}}$$

$$= \frac{\sqrt{\omega^2 - p_1^2}}{\omega} = \frac{\sqrt{p_2^2 + p_3^2}}{\omega}$$

Die Quadratwurzel des Nenners ist mit positivem Zeichen zu rechnen, die Größe $\omega^2 - p_1^2$ ist stets positiv; also ist für genügend großes t

$$\operatorname{sign} F(u,t) = \operatorname{sign} rr' = \operatorname{sign} t$$

daher verdoppelt sich in (7) die Klammergröße bei der Differenzbildung für $t_2 = +\infty$ und $t_1 = -\infty$, und es folgt:

$$C = -2\int_{u_1}^{u_2} \sqrt{p_2^2 + p_3^2}\, du$$

Bezeichnen α_i resp. β_i und γ_i die Richtungskosinus der x- resp. y- und z-Achse des bewegten Trieders gegen drei zu einander senkrechte, im Raume feste Richtungen, so sind diese Größen Funktionen von u, und bezeichnet ein Strich ihre Ableitung nach diesem Parameter, so ist*)

$$p_2 = -\Sigma \gamma_i \alpha_i' \qquad (i = 1, 2, 3)$$
$$p_3 = \Sigma \beta_i \alpha_i'$$

Also ist

$$p_2^2 + p_3^2 = \Sigma \alpha_i'^2 (\beta_i^2 + \gamma_i^2) + \Sigma \alpha_i' \alpha_k' (\beta_i \beta_k + \gamma_i \gamma_k)$$

und da

$$\alpha_i^2 + \beta_i^2 + \gamma_i^2 = 1$$

und

$$\alpha_i \alpha_k + \beta_i \beta_k + \gamma_i \gamma_k = 0$$

ist, so folgt

$$p_2^2 + p_3^2 = \Sigma \alpha_i'^2 (1 - \alpha_i^2) - \Sigma \alpha_i \alpha_i' \alpha_k \alpha_k' = \Sigma \alpha_i'^2 - (\Sigma \alpha_i \alpha_i')^2$$

woraus wegen $\Sigma \alpha_i^2 = 1$ sich schließlich ergibt:

$$C = -2 \int_{u_1}^{u_2} \sqrt{\alpha_1'^2 + \alpha_2'^2 + \alpha_3'^2}\, du$$
$$= -2 \int_{u_1}^{u_2} \sqrt{d\alpha_1^2 + d\alpha_2^2 + d\alpha_3^2}$$

Der Integrand ist der Winkel, den zwei durch die Variation du aus einander hervorgehende Lagen der x-Achse mit einander bilden. Bezeichnet also $d\vartheta$ den unendlich kleinen Winkel, den zwei konsekutive Lagen der bewegten Geraden einschließen, so ist die Totalkrümmung der entstehenden Fläche

$$C = -2 \int d\vartheta \qquad (8)$$

Zieht man durch den Mittelpunkt einer Einheitskugel Parallele zu den Erzeugenden einer Regelfläche, so bestimmen diese auf derselben die sogenannte »sphärische Indikatrix«. Das Ergebnis läßt sich also wie folgt formulieren:

> Die Totalkrümmung einer Regelfläche ist gleich der negativ gerechneten doppelten Bogenlänge der sphärischen Indikatrix.

*) Siehe *Darboux*, Leçons I, p. 5.

Unter den nicht entarteten Flächen 2. Ordnung sind zwei Regelflächen enthalten, deren Totalkrümmung man leicht mit Hilfe der Formel (8) bestimmt. Dieselbe stellt sich beim hyperbolischen Paraboloid als von den Konstanten unabhängig heraus*)

$$C = -2\pi$$

während sie bei dem einschaligen Hyperboloid

$$\frac{x^2}{a^2} + \frac{y^2}{b^2} - \frac{z^2}{c^2} = 1$$

von den Konstanten abhängt:

$$C = -2\int_0^{2\pi} \frac{\sqrt{a^2 b^2 + c^2(a^2\cos^2\varphi + b^2\sin^2\varphi)}}{a^2\sin^2\varphi + b^2\cos^2\varphi + c^2}\, d\varphi \qquad (8a)$$

Ist dasselbe ein Rotationshyperboloid ($a = b$), so folgt:

$$C = -4\pi \cdot \frac{a}{\sqrt{a^2+c^2}}$$

Die Formel (8a) ist zunächst nicht in Übereinstimmung mit der oben zwischen Totalkrümmung und Charakteristik gefundenen Beziehung:

$$C = -K \cdot 2\pi$$

Das einschalige Hyperboloid im projektiven Raum ist eine geschlossene Fläche der Charakteristik $K = 0$, muß also der genannten Beziehung zufolge auch die Totalkrümmung Null haben. Der Widerspruch löst sich, wenn man bedenkt, daß die Fläche im Unendlichen eine Singularität besitzt. Führt man z. B. beim Rotationshyperboloid die Flächennormale längs einer Meridianhyperbel durchs Unendliche, so sieht man, daß sie dortselbst einen Sprung um π erleidet. Die Äquivalentkrümmung dieser im Unendlichen liegenden singulären Kante der Fläche ist in dem Betrage von C in Formel (8a) nicht enthalten; denn die Ableitung dieser Formel entspricht dem folgenden Verfahren: Man lasse ein beliebig großes endliches Stück A_1 einer Geraden die charakterisierte Bewegung ausführen und bestimme C_1 für die entstehende Fläche; dann verlängere man beiderseits A_1 um eine beliebig große Strecke A_2 und bestimme den hinzutretenden Beitrag C_2

---

*) Vergl. *Boy*. In Diss. phil. Göttingen 1901.

usw., dann bestimmt Formel (8a) die Totalkrümmung C als die Summe des konvergenten Prozesses

$$C = C_1 + C_2 + \cdots$$

Offenbar konvergiert diese Summe gegen die Totalkrümmung der Fläche mit Ausschluß der Äquivalentkrümmung der im Unendlichen gelegenen Singularität. Man kann also geradezu den Betrag C' dieser letzteren durch die Beziehung bestimmen

$$C + C' = -K \cdot 2\pi = 0$$

sodaß C' gleich dem mit dem Faktor $+2$ versehenen Integral der Formel (8a) sich ergibt. Eine analoge Überlegung gilt beim hyperbolischen Paraboloid.

Ist λ der konstante Winkel, den im Falle des Rotationshyperboloids die rotierende Gerade mit der z-Achse (Rotationsachse) bildet, so ist

$$\sin\lambda = \frac{a}{\sqrt{a^2 + c^2}}$$

und

$$C = -4\pi \sin\lambda \tag{9}$$

Ist ε der kürzeste Abstand der rotierenden Geraden von der Rotationsachse, so bleibt Formel (9) richtig, wenn ε bei konstantem λ abnimmt. Dabei schmiegt sich das Rotationshyperboloid mehr und mehr einem Kreiskegel an, dessen Erzeugende mit der Achse den Winkel λ einschließt. Der Hauptbeitrag zum Werte von C wird dann durch die Umgebung des Kehlkreises geliefert, während das Krümmungsmaß der übrigen Flächenpunkte gegen Null konvergiert. Im Grenzfall für $\varepsilon = 0$ wird

$$-4\pi \sin\lambda$$

die Äquivalentkrümmung der in einen Punkt zusammengezogenen Kante, in die der Kehlkreis übergeht. Bei dieser Auffassung der durch Rotation einer Geraden um eine schneidende Achse entstehenden Kreiskegelfläche bleibt also die Formel (8) richtig, und ihre Gültigkeit ist damit auf den vorher auszuschließenden Fall ausgedehnt, daß die bewegte Gerade aus einer Lage in die folgende durch eine rotatorische Bewegung übergeht, d. h. auf abwickelbare Regelflächen. Dort erscheint also der Wert der Totalkrümmung

$$C = -2\int d\vartheta$$

als Äquivalentkrümmung der Rückkehrkante, während alle regulären Flächenpunkte das Krümmungsmaß Null haben und daher zum Werte von C keinen Beitrag liefern.

Die Formel (8) gilt also ausnahmslos, und wir können schließen: C wird nur dann gleich Null, wenn dauernd

$$d\vartheta = 0$$

ist, wenn also die Gerade sich parallel mit sich selbst bewegt. Also gilt:

Hat eine Regelfläche die Totalkrümmung Null, so ist sie notwendig ein Zylinder.

Die Darstellung der Totalkrümmung einer Regelfläche durch die Formel (8) läßt eine sehr anschauliche Deutung zu: Die Gesamtheit der Flächennormalen längs einer Erzeugenden bilden eine Wendelfläche, und bei der Abbildung der Fläche auf die Einheitskugel durch Parallele zu den Flächennormalen wird das Bild einer Erzeugenden ein Halbkreis eines größten Kreises. Bei den abwickelbaren Regelflächen ist diese Abbildung insofern singulär, als der Punkt der Rückkehrkante sich auf den ganzen Halbkreis und alle übrigen Punkte der Erzeugenden sich auf einen der Endpunkte abbilden. Die Nachbarerzeugende bildet sich auf einen benachbarten Halbkreis ab, dessen Schnittpunkt mit dem ersteren dem Punkt der Striktionslinie entspricht. Die Pole dieser größten Halbkreise sind die durch die Richtungen der Erzeugenden bestimmten Punkte, also der Winkel zwischen den Halbkreisen gleich dem Winkel $d\vartheta$ zwischen den Erzeugenden. Daher ist der Inhalt des Flächenraums, den der Bildhalbkreis beim Übergang zum benachbarten beschreibt

$$-4\pi \cdot \frac{d\vartheta}{2\pi} = -2\,d\vartheta$$

wobei das negative Zeichen zu wählen ist, weil jeder Punkt der Regelfläche negatives Krümmungsmaß hat. Die Summe dieser Flächenräume ergibt nach der *Gauß*schen Definition die Totalkrümmung, sodaß diese Überlegung die Formel (8) bestätigt. Wie man sieht, gestattet auch dieser Weg, die abwickelbaren Regelflächen von vorneherein einzuschließen.

Es bleibt noch übrig, den zu Anfang dieses Paragraphen ausgeschlossenen Fall der Rotation mit schneidender Achse zu diskutieren. Zunächst werde unter der Voraussetzung, daß eine rotierende Kurve die Achse nicht trifft, die rechnungsmäßige Vereinfachung der Formel (5) gesucht. Da die Rotationsfläche sich nach einem Zuwachs von φ um 2π schließt, so sei dieser Zuwachs von vorneherein angenommen und als rotierende Kurve die Meridiankurve der entstehenden Fläche gewählt:

$$x = t \qquad f_1' = 1$$
$$y = 0 \qquad f_2' = 0$$
$$z = f(t) \qquad f_3' = f'$$

$$E = 1 + f'^2$$
$$F = 0$$
$$G = t^2$$

$$\frac{1}{2}\frac{d(r^2)}{dt} = f_1 f_1' + f_2 f_2' = t$$

also wird

$$C = -2\pi\left[\frac{1}{\sqrt{1+f'^2(t)}}\right]_{t=t_1}^{t=t_2}$$

Der Klammerausdruck ist der sin des Winkels ϑ zwischen der durch den Parameter t gerichteten Kurventangente und der positiven z-Achse, und zwar nach derjenigen Seite der Meridianebene positiv gerechnet, auf welcher der Kurvenpunkt liegt. Also ist

$$C = 2\pi(\sin\vartheta_1 - \sin\vartheta_2) \tag{10}$$

ein Ausdruck, der offenbar von der Durchlaufungsrichtung der Kurve unabhängig ist und aus dem man die Bedeutung der Totalkrümmung als des Abbildungsgebietes unmittelbar abliest.

Schneidet nun ein Kurvenzweig APB die Achse in einem Punkte P, so entsteht bei der Rotation um dieselbe in P eine Singularität, die einer Kegelspitze gleichwertig ist. Dieselbe unterliegt einer Willkür in der Auffassung der Fläche hinsichtlich der hier betrachteten Eigenschaften. Man überzeugt sich leicht davon, daß dieser Fall sich nicht notwendig als Spezialfall der Schraubung für verschwindende Ganghöhe gewinnen läßt. Stellt man in der Tat für einen Punkt P einer Schraubenfläche z. B. das Krümmungsmaß oder die Richtung der Flächennormalen auf und untersucht diese Größen während zweier Prozesse:

1. P fällt in die Achse,
2. die Ganghöhe konvergiert gegen Null,

so gelangt man zu verschiedenen Grenzwerten je nach der Reihenfolge dieser Übergänge. Man kann die Willkür, die hier in der Auffassung der Fläche bleibt, auch so vorstellig machen: Auf dem durch Rotation von AP entstehenden Mantel sei in einem Punkte die Flächennormale nach der Seite errichtet, welche dem die Achse nicht enthaltenden Winkelraum von P entspricht. Dann werde dieselbe durch die Singularität P hindurch auf den anderen Flächenmantel geführt. Dabei ist die Wahl der Seite dieses Mantels willkürlich und nicht durch die Stetigkeit bestimmt, weil die Lage der Flächennormalen in P unbestimmt wird. Läßt man die Flächennormale wieder dem die Achse nicht enthaltenden Winkelraum angehören, so macht sie einen Sprung um π beim Durchgang durch den mit unendlichem Krümmungsmaß behafteten Punkt P. Diese Auffassung der Fläche entspricht dem Spezialfall der Schraubung mit verschwindender Ganghöhe; man vergegenwärtige sich den oben beschriebenen Grenzübergang vom einschaligen Hyperboloid zum Kegel und bewege bei demselben die Flächennormale längs einer Erzeugenden. (Allgemein steht bei einer Schraubenfläche die Flächennormale in Punkten der Achse senkrecht zu dieser, unabhängig von der Ganghöhe.) Bei dieser Auffassung zählt also die Äquivalentkrümmung der an der Singularität entstehenden, in einen Punkt zusammengezogenen Kante zu der Totalkrümmung mit; wir brauchen dann keinerlei Ausnahmen zu machen, die obigen Resultate gelten allgemeiner, und es läßt sich von vorneherein vermuten, daß sich dieselben nun auch auf die Erzeugung von Flächen, die mit beliebigen Singularitäten behaftet sind, erstrecken werden. Doch hat diese Auffassung besonders in dem Fall etwas durchaus Ungewöhnliches, wenn die rotierende Kurve die Achse senkrecht trifft. So muß man z. B. bei der Rotation eines Kreises um einen Durchmesser sagen: Es entsteht eine zweimal punktierte Kugel mit der Totalkrümmung Null, weil die Krümmung der Kugelfläche durch die Äquivalentkrümmung der an den Punktierungen entstehenden Kanten aufgehoben wird.

Läßt man hingegen die Richtung der Flächennormalen sich beim Durchgang durch die Rotationsachse stetig fortsetzen, geht man also an der Kegelspitze aus dem die Achse nicht enthaltenden in den die Achse enthaltenden Winkelraum über, und umgekehrt, so liefert die Singularität keinen Beitrag zur Krümmung. Bedeutet $P_{\mp\varepsilon}$ einen in der Entfernung ε vor resp. nach P gelegenen Punkt der rotierenden Kurve und heißt C_I resp. C_{II} die Totalkrümmung des durch Rotation von $AP_{-\varepsilon}$ resp. $P_{+\varepsilon}B$ entstehenden Flächenstücks, so ist

$$C = \lim_{\varepsilon = 0} (C_I + C_{II})$$

Dieser Wert von C unterscheidet sich von dem der früheren Auffassung dadurch, daß ihm der Summand fehlt, der von der Äquivalentkrümmung der Singularität herrührt. Dieser war, wenn ϑ der Winkel zwischen rotierender Kurve und Achse war, gegeben durch die Formel (9)

$$C = -4\pi \sin \vartheta$$

Also ist bei dieser Auffassung, wenn ϑ_a resp. ϑ_e den Winkel bedeutet, den die Kurventangente am Anfang resp. am Ende mit der Achse bildet, und ϑ_i den Winkel derselben am Schnittpunkt P_i bezeichnet, die Totalkrümmung der Rotationsfläche gegeben durch

$$C = 2\pi (\sin \vartheta_a - \sin \vartheta_e + 2 \Sigma \sin \vartheta_i)$$

In diesem Fall gibt also der Kreis bei einer vollen Rotation um seinen Durchmesser

$$C = 8\pi$$

was mit der gewöhnlichen Auffassung einer zweimal überdeckten Kugel übereinstimmt.

§ 3.

Bestimmung der Totalkrümmung einer Fläche aus einer sie überdeckenden Kurvenfolge.

Es sei eine ganz im Endlichen gelegene, geschlossene oder berandete, zweiseitige oder einseitige Fläche gegeben. Dann ist es stets möglich, dieselbe mit einer stetigen Kurvenfolge so zu überdecken, daß die folgenden Bedingungen erfüllt sind:

1. Jedes Individuum der Folge ist eine geschlossene Kurve oder eine Mehrheit solcher.
2. Jedes Individuum ist in seinem ganzen Verlauf seinem Vorgänger unendlich benachbart und hat mit ihm keinen gemeinsamen Punkt.
3. Das Anfangselement ist eine geschlossene Kurve.

Ein Beispiel einer solchen Kurvenfolge bieten die Schnittlinien einer im Endlichen geschlossenen Fläche mit einer Parallelebenenschar; dieselbe ist so zu legen, daß die erste Ebene, welche die Fläche überhaupt trifft, dieselbe in einem Punkte berührt. Dann ist das Anfangselement ein Punkt und das folgende ein ihn umgebendes Oval.

Hat die Fläche r Randkurven und gestattet sie, ohne zu zerfallen, p Rückkehrschnitte I. Art (längs deren sich die Flächenindikatrix nicht umkehrt), und dazu gleichzeitig p' Rückkehrschnitte II. Art (längs deren dieselbe sich umkehrt), so hat sie die Charakteristik*)

$$K = 2p + p' + r - 2 \tag{1}$$

und ihre Totalkrümmung ergab sich nach Formel (4a) des § 1 durch

$$C = -\left(K \cdot 2\pi + \sum_i \int_{R_i} \frac{\sin \vartheta \, ds}{\varrho}\right) \tag{2}$$

*) cfr. Enzyklopädie III,1 Heft 1, p. 195.

Die nunmehr gestellte Frage ist: Wie läßt sich die Totalkrümmung aus der die Fläche überdeckenden Kurvenfolge ablesen?

Ist im Besonderen diese Kurvenfolge von der Art, daß sich jedes Individuum in das folgende auf der Fläche stetig überführen läßt, so läßt sich auch das Anfangselement in das Endelement überführen, und das Flächenstück ist einfach zusammenhängend. Es ist also

$$p = p' = 0$$
$$r = 2$$
$$K = 0$$
$$C = -\int_{R_1} - \int_{R_2}$$

Einen besonderen Fall dieser Art hatten wir im vorigen Paragraphen in dem Flächenstück, das eine Schar kongruenter geschlossener Kurven trägt. Dessen Totalkrümmung war Null, sodaß sich die Integralbeiträge des Anfangs- und Endelements bei demselben aufheben müssen. Fällt R_2 mit R_1 zusammen, so ist $C = 0$, weil dann die gemeinsame Endlage die Grenze der beiderseits liegenden Gebiete bildet; dann ist

$$p = 1$$
$$r = 0$$

und der eine mögliche Rückkehrschnitt I. Art kann z. B. durch ein bestimmtes Individuum der Folge dargestellt werden.

Es wird also im Allgemeinen bei einer vorgegebenen Fläche nicht möglich sein, die Kurvenfolge so zu legen, daß jedes Individuum in das folgende sich stetig deformieren läßt, sondern es wird einzelne Individuen geben, deren Zusammenhang von dem des konsekutiven verschieden ist. Da aber auch in diesem besonderen Fall die beiden Kurven sich in ihrem ganzen Verlauf unendlich benachbart bleiben, so kann eine solche Umschaltung der Kurventeile nur an einem Doppelpunkte geschehen.

Die Kurvenfolge wurde zur Vereinfachung des Ausdrucks mit einer Wertefolge eines Parameters u charakterisiert, sodaß

$$u = u_1$$

ein bestimmtes Individuum kennzeichnet, das im allgemeinen aus mehreren geschlossenen Kurven besteht.*)

*) cfr. zum Folgenden: *Boy,* Math. Ann. 57, p. 173 ff.

$u = u_1$ sei ein Individuum der Folge, das einen Doppelpunkt P besitzt. Man lege durch die Tangentenrichtungen der sich in P durchsetzenden Kurvenbögen die Ebene und bezeichne die in dieser zyklisch um P angeordneten Kurventeile mit A, B, C, D. Auf denselben werden Strecken $A_1 A_2$, $B_1 B_2$, $C_1 C_2$, $D_1 D_2$ bestimmt, so daß A_2, B_2, C_2 und D_2 mit P zusammenfallen; sodaß also die Kurvenbögen $A_1 P C_1$ und $B_1 P D_1$ sich in P durchsetzen.

ε sei eine Größe, die bestimmt ist, beliebig klein zu werden. Dann betrachte man die Individuen

$$u = u_1 - \varepsilon$$

und

$$u = u_1 + \varepsilon$$

Dieselben mögen in allen nicht der unmittelbaren Umgebung von P angehörigen Punkten stetig ineinander deformierbar sein. Ist nun auf beiden betrachteten Nachbarkurven die Verbindung der Kurventeile in der Nähe von P diejenige, welche der Zusammengehörigkeit von C_2 und A_2 resp. also von D_2 und B_2 entspricht, so sind auch hier die drei konsekutiven Kurven ineinander deformierbar. Dieser Fall bietet also nichts Neues; P ist ein regulärer Punkt einer Doppelkurve der Fläche.

Wir untersuchen also die Erscheinung, daß die Kurven in der Nachbarschaft von P nicht ineinander deformierbar sind, daß also $u_1 - \varepsilon$ und $u_1 + \varepsilon$ oder beide eine andere Verbindung der Kurventeile zeigen als u_1. Die Wirkung der Zusammenhangsänderung hängt dann davon ab, wie die äußeren Enden A_1, B_1, C_1, D_1 der vier Strecken mit einander verbunden sind.

Nehmen wir an, daß das Individuum $u = u_1$ außer den Elementen, zu denen die sich in P durchsetzenden Kurvenbögen gehören, keine weiteren Elemente enthält, so können wir sagen:

1. Ist A_1 mit C_1, also B_1 mit D_1 verbunden, so besteht $u = u_1$ aus zwei geschlossenen, einander in P durchsetzenden Kurven.
2. Ist A_1 mit B_1 (oder D_1), also C_1 mit D_1 (oder B_1) verbunden, so besteht $u = u_1$ aus einer sich in P durchsetzenden Kurve.

Die Wirkung der Zusammenhangsänderung in der Kurvenfolge kann dann sein:

1. Trennung eines geschlossenen Kurvenzuges in zwei.
2. Vereinigung zweier geschlossener Kurvenzüge zu einem.
3. Umschaltung der Teile eines geschlossenen Kurvenzuges unter einander.

Dabei kann für jeden dieser Fälle noch gelten:

1. Für die Individuen $u > u_1$ tritt ein Doppelpunkt auf in der Nähe der fraglichen Stelle, den die Individuen $u < u_1$ nicht hatten.
2. Für $u > u_1$ verschwindet ein Doppelpunkt, den die $u < u_1$ hatten.
3. Weder die $u > u_1$ noch die $u < u_1$ haben in der Nähe von P einen Doppelpunkt.

In dem letzteren Fall hat also nur $u = u_1$ einen Doppelpunkt, der aber darum unwesentlich ist, weil ihm k e i n mehrfacher Punkt der Fläche entspricht.

Im Besonderen ergibt sich diese Tabelle:

Verbindung der äußeren Enden $A_1 B_1 C_1 D_1$	Verbindung der Enden $A_2 B_2 C_2 D_2$ für $u < u_1$	für $u > u_1$	Wirkung	
A_1 mit C_1	A_2 mit B_2	A_2 mit C_2	Trennung;	Entstehg. eines Dp.
		A_2 mit D_2	Umschaltung;	—
	A_2 mit C_2	A_2 mit B_2	Vereinigung;	Vernichtg. eines Dp.
		A_2 mit D_2		
	A_2 mit D_2	A_2 mit C_2	Trennung;	Entstehg. eines Dp.
		A_2 mit B_2	Umschaltung;	—
A_1 mit B_1	A_2 mit B_2	A_2 mit C_2	Vereiniguug;	Entstehg. eines Dp.
		A_2 mit D_2	»	—
	A_2 mit C_2	A_2 mit B_2	Trennung;	Vernichtg. eines Dp.
		A_2 mit D_2	Umschaltung;	»
	A_2 mit D_2	A_2 mit C_2	Umschaltung;	Entstehg. eines Dp.
		A_2 mit B_2	Trennung;	—

Die Verbindung der äußeren Enden A_1 und D_1 liefert keine neuen Erscheinungen.

Nunmehr erwächst uns die Aufgabe, diese einzelnen Vorgänge für die Bestimmung der Totalkrümmung zu bewerten.

Das Anfangselement der Folge ist ein geschlossener Kurvenzug. Sieht man zunächst von Umschaltungen innerhalb der Kurventeile desselben ab, so muß die erste Erscheinung notwendig eine Trennung in zwei Kurvenzüge sein. Dadurch wird zunächst die Ränderzahl der entstehenden Fläche und damit die Charakteristik um 1 erhöht. Es tritt also in der Formel für C der Summand -2π hinzu, während das Integral über 3 Ränder zu erstrecken ist. Dieser Vorgang möge bei $u = u_1$ stattgefunden haben. Bei $u = u_2$ trete die zweite Zusammenhangsänderung ein, die nun von der Art sein kann, daß sie die beiden getrennten Kurvenzüge der Individuen

$$u_2 > u > u_1$$

wieder zu einem vereinigt. Dieser zweite Vorgang vermindert also die Ränderzahl um eine Einheit, erhöht aber gleichzeitig die Zahl p der Rückkehrschnitte I. Art um eine Einheit; denn es ist klar, daß von den beiden Kurvenzügen eines Individuums

$$u_1 < u < u_2$$

der eine als Rückkehrschnitt gewählt werden kann, ohne daß die zwischen $u = u_0$ und $u > u_2$ liegende Fläche dadurch zerfällt. Und dieser Schnitt ist I. Art, da er mit seiner geschlossenen Nachbarkurve keinen Punkt gemeinsam hat. Jede Trennung oder Vereinigung fügt also zu C den Summanden -2π, und man erkennt leicht, daß diese Bewertung von der Reihenfolge der Trennungen und Vereinigungen unabhängig ist.

Eine Umschaltung endlich, die nur die einzelnen Teile eines geschlossenen Kurvenzuges umordnet, läßt die Ränderzahl ungeändert, vermehrt aber die Anzahl p' der Rückkehrschnitte II. Art um eine Einheit. Um sich von der Existenz dieses Schnittes zu überzeugen, konstruiere man in der Umgebung des Umschaltungspunktes in folgender Weise ein geeignetes Kurvennetz auf der Fläche.

$u = u_1$ bestehe aus einem sich in P durchsetzenden Kurvenzug. (Es wird an der Überlegung nichts geändert, wenn man daneben mehrere andere geschlossene Kurvenzüge zuläßt, die von der Umschaltung unberührt bleiben.) In dem Ausschnitt aus der Kurvenfolge, der durch $u = u_1 - \gamma$ und $u = u_1 + \gamma$ begrenzt wird, finde eine Zusammenhangsänderung nur an dem Punkt P der Kurve $u = u_1$ statt, und dieselbe bestehe aus einer Umordnung der Kurventeile ohne Zerfall der Kurve. Dann lege man von $u = u_1 - \gamma$ bis $u = u_1 + \gamma$ eine Trajektorie durch die Kurvenschar, die den Punkt P mit $u = u_1$ gemeinsam hat. Ferner bestimme man die Punkte jedes Individuums u_i durch einen Parameter t

$$t = \frac{s}{S(u_i)}$$

wobei $S(u_i)$ die Gesamtlänge der Kurve $u = u_i$ und s ihren Bogen von der oben gelegten Trajektorie an nach einer an der Kurve $u = u_1 - \gamma$ frei gewählten und auf die übrigen Kurven stetig übertragenen Zählrichtung bedeutet. Die oben gelegte Trajektorie hat dann die Gleichungen

$$t = 0 \text{ oder } t = 1$$

Zum Doppelpunkte P der Kurve $u = u_1$ gehört außer dem Werte $t = 0$ (oder $t = 1$, was dasselbe ist) noch ein zweiter Wert

$$t = t_1$$

Derselbe teilt also $u = u_1$ in zwei Teile

$$T_1: \quad 0 < t < t_1$$

und

$$T_2: \quad t_1 < t < 1$$

Die Trajektorie $t = t_1$ teilt dann jede Kurve $u = u_i$ in zwei Teile $T_1(u_i)$ und $T_2(u_i)$. Die Durchlaufungsrichtung der Teile $T_1(u_i)$ überträgt sich stetig auf alle Kurven der Folge nach der Festsetzung, die über die Zählrichtung des Parameters s getroffen wurde. Alle Trajektorien $t =$ constans, für

$$0 < t < t_1$$

erleiden daher keinerlei Unterbrechung. Dagegen wechseln die Teile $T_2(u_i)$ ihre Durchlaufungsrichtung bei $T_2(u_1)$, da bei $u = u_1$ eine Umschaltung der Kurventeile stattfindet. Zu $T_2(u_1)$ gehören also zwei Durchlaufungsrichtungen, die man erhält für

$$T_2(u_1) = \lim_{\eta = 0} T_2(u_1 - \eta)$$

und

$$T_2'(u_1) = \lim_{\eta = 0} T_2(u_1 + \eta)$$

Trajektorien t = constans für

$$t_1 < t < 1$$

erleiden also bei $u = u_1$ eine Stetigkeitsunterbrechung. Ein solcher Wert t bestimmt also auf $T_2(u_1)$ und $T_2'(u_1)$ zwei verschiedene Punkte. Dadurch ist $T_2(u_1)$ (das ja geschlossen ist) auf $T_2'(u_1)$, d. h. entgegengesetzt auf sich selbst abgebildet; diese Abbildung hat außer dem Punkt $P(t = t_1)$ notwendig noch einen zweiten Fixpunkt, nämlich bei

$$t_\beta = \frac{t_1 + 1}{2}$$

also ist von allen durch die Teile $T_2(u_i)$ gehenden Trajektorien t = constans nur

$$t = t_\beta$$

stetig.

Man baue nun aus Stücken von 4 solchen Parameterkurven einen geschlossenen Rückkehrschnitt A auf, wobei

$$0 < t_\alpha < t_1$$

und $$\varepsilon + \eta < \gamma$$ sei.

1. $t = t_\alpha$; u von $u_1 - \varepsilon$ bis $u_1 + \varepsilon$
2. $u = u + \varepsilon$; t von t_α bis t_β
3. $t = t_\beta$; u von $u_1 + \varepsilon$ bis $u_1 - \varepsilon$
4. $u = u - \varepsilon$; t von t_β bis $1 + t_\alpha$

und konstruiere dazu eine Nachbarkurve B für genügend kleines η:

1'. $t = t_\alpha + \eta$; u von $u_1 - (\varepsilon + \eta)$ bis $u_1 + \varepsilon + \eta$

2'. $u = u_1 + \varepsilon + \eta$; t von $t_\alpha + \eta$ bis $t_\beta + \eta$

3'. a) $t = t_\beta + \eta$; u von $u_1 + \varepsilon + \eta$ bis u_1

b) $t = t_\beta - \eta$; u von u_1 bis $u_1 - (\varepsilon + \eta)$

4'. $u = u_1 - (\varepsilon + \eta)$; t von $t_\beta - \eta$ bis $1 + t_\alpha + \eta$

Die Trajektorien $t = t_\beta + \eta$ und $t = t_\beta - \eta$ schließen sich in ihren Endpunkten auf $u = u_1$ stetig an einander an. Fragen wir nun nach Schnittpunkten von A und B, so ist zunächst klar, daß ein in A enthaltener Trajektorienteil mit einem in B enthaltenen gleichnamigen Trajektorienteil sich nicht schneidet, weil weder die $u = u_i$ noch die $t = t_k$ einander treffen. Ferner können auf den Bögen 2'. und 4'. von B keine Schnittpunkte liegen, weil in keinem Punkte von A $u > u_1+\varepsilon$ oder $u < u_1-\varepsilon$ ist. Alle Schnittpunkte müssen also auf $u = u_1-\varepsilon$ und $u = u_1+\varepsilon$ liegen.

Auf $u = u_1-\varepsilon$ liegen Punkte von A, für welche

$$t_\beta \leqq t \leqq 1+t_\alpha$$

d. h. $\quad 0 \leqq t \leqq t_\alpha \quad$ und $\quad t_\beta \leqq t \leqq 1$

also wird dieser Kurvenbogen weder von $1'$ $(t = t_\alpha+\eta)$ noch von $3'.b)$ $(t = t_\beta-\eta)$ getroffen; und für $3'.a)$ ist $u > u_1$.

Auf $u = u_1+\varepsilon$ liegen Punkte von A, für welche

$$t_\alpha \leqq t \leqq t_\beta$$

Dieser Bogen hat also mit $1'$. den Punkt gemeinsam, für welchen

$$\begin{cases} t = t_\alpha+\eta \\ u = u_1+\varepsilon \end{cases}$$

ist, während er mit $3'.a)$ $(t = t_\beta+\eta)$ keinen Schnittpunkt ergibt.

A und B stellen also zwei in einander deformierbare Kurven mit einem gemeinsamen Punkt dar, sind also notwendig beide Kurven II. Art. Der Rückkehrschnitt A mit umkehrbarer Indikatrix ist in einem beliebig kleinen, die Umschaltung enthaltenden Ausschnitt aus der Kurvenfolge möglich, da man ε beliebig klein wählen kann, und bildet mit einem Individuum $u = u_1$ der Folge n i c h t die Begrenzung eines Flächenstücks, sodaß er frühere Ergebnisse nicht stört.

Auch in diesem Fall erhöht sich also die Charakteristik um eine Einheit, und wir können sagen:

> Für jede Zusammenhangsänderung, welche die Individuen der Folge erleiden, nimmt die Totalkrümmung den Summanden -2π auf.

Es bleibt noch die Frage der Doppelpunkte zu berücksichtigen. *Boy* zeigt*), daß das Auftreten einer endenden Doppelkurve die Totalkrümmung um 4π erhöht. Geht man also von der Vorstellung aus, daß das Anfangs- und ebenso das Schlußindividuum keine Doppelpunkte hat, so bedingt jede offene Doppelkurve der Fläche das Auftreten und das Verschwinden eines Doppelpunkts; diese Erscheinungen sind also mit $+2\pi$ zu bewerten.

*) Math. Annalen 57, p. 166.

Endlich bleibt noch das Verschwinden eines Kurvenzuges und damit ein partielles Abbrechen der Folge zu berücksichtigen, welches eintritt, wenn einem bestimmten geschlossenen Kurvenzug im konsekutiven Individuum kein Nachbar entspricht. Das kann eintreten, wenn zwei Kurvenzüge, die in $u = u_1 - \varepsilon$ unendlich benachbart waren, in $u = u_1$ zusammenfallen. Dann verschwinden 2 Ränder, während die Zahl p sich um den einen Rückkehrschnitt I. Art erhöht, welcher zum Beispiel der gemeinsamen Lage entspricht. Die Charakteristik bleibt also ungeändert. Ferner kann ein Element von $u = u_1$ mit einem Rande der Fläche zusammenfallen und demgemäß in $u = u_1 + \varepsilon$ nicht mehr auftreten. Dann bleibt als dauernder Summand in der Formel für C der Wert des *Bonnet*schen Integrals längs dieses Randes. Zieht sich der betrachtete Kurvenzug in einen Punkt zusammen und verschwindet so, so konvergiert der Integralbeitrag über denselben in den konsekutiven Lagen gegen den Wert -2π, das Verschwinden eines Kurvenzuges durch Zusammenziehung in einen Flächenpunkt ist also mit $+2\pi$ zu bewerten.

Durch Vereinigung dieser Bewertungen gelangt man zu einer Methode, welche die Totalkrümmung einer Fläche aus einem sie überdeckenden Kurvensystem und dem Charakter der Ränder zu bestimmen gestattet. Für geschlossene Flächen lautet dieselbe also:

> Ist eine geschlossene Fläche mit einem Kurvensystem überdeckt, dessen Individuen aus geschlossenen Kurvenzügen bestehen und ihren Nachbarindividuen im ganzen Verlauf unendlich benachbart bleiben, ohne doch mit diesen gemeinsame Punkte zu haben, in deren Umgebung sie demselben Flächenmantel angehören,
>
> so ist die Totalkrümmung der entstehenden Fläche gleich der Anzahl von Punkten, in die sich Elemente des Systems zusammenziehen, vermindert um die Zahl solcher Zusammenhangsänderungen, bei denen nicht ein Doppelpunkt entsteht oder verschwindet, multipliziert mit 2π.
>
> Von dem Entstehen eines Doppelpunktes ist dann zu sprechen, wenn von einem bestimmten Element der Folge an die folgenden Elemente je einen Doppelpunkt besitzen und diese in ihrer Gesamtheit eine Doppelkurve der Fläche

bilden, ohne daß die vorhergehenden Elemente einen benachbarten Doppelpunkt aufweisen.

*Möbius**) gewinnt ein solches Kurvensystem auf der Fläche, indem er sie mit einer Schar sich stetig umschließender Flächen vom Geschlecht Null zum Schnitt bringt. Dann entsprechen die oben charakterisierten Erscheinungen den elliptischen resp. hyperbolischen Berührungen zwischen den Flächen der Schar und der vorgegebenen Fläche, deren Charakteristik gefunden werden soll. *Boy***) wählt als solche Flächenschar speziell eine Parallelebenenschar; dann sind die hier charakterisierten Anzahlen mit den Anzahlen der Extreme resp. Sattelpunkte identisch.

Einer besonderen Interpretation bedarf im Fall der nicht geschlossenen Flächen die Möglichkeit, daß Ränder der Fläche im Unendlichen liegen; daß also ein geschlossener Kurvenzug eines bestimmten Individuums der Folge ohne weitere Zusammenhangsänderungen zu erfahren, sich ins Unendliche fortbewegt. Dann figuriert in dem Werte von C der Betrag des über die unendlich ferne Begrenzung erstreckten *Bonnet*schen Integrals, und es fragt sich, welchen Sinn dasselbe in diesem Falle hat.

Man denke sich den einfachsten Fall, daß die Kurvenfolge mit einem Punkt beginnt, dessen Nachbarkurve ein denselben umgebendes einfaches Oval ist, und daß dieses nun unter Erweiterung und stetiger Deformation, aber ohne Zusammenhangsänderungen zu erfahren, sich ins Unendliche fortbewegt. (Eine solche Erscheinung veranschaulicht auf dem elliptischen Paraboloid

$$\frac{x^2}{a^2} + \frac{y^2}{b^2} - \frac{z}{c} = 0$$

die Schar der Kurven $z =$ constans.) Jede Lage $u = u_i$ der Kurve begrenzt ein Elementarflächenstück, dessen Totalkrümmung somit gegeben ist durch

$$C = 2\pi - \int\limits_{u = u_i} \frac{ds \sin \vartheta}{\varrho}$$

Andererseits folgert man unmittelbar aus der *Gauß*schen Definition dieses Begriffs, daß C gleich dem Inhalte des Gebietes ist, das

*) *Möbius*, Werke II: Theorie der elementaren Verwandtschaft, § 21. Auch: Berichte der Königl. Sächs. Gesellsch. d. W., 1863, Bd. 15.

**) Math. Ann. 57, p. 157. Auch: In. Diss. Göttingen 1901.

auf der Einheitskugel von dem Bild der Begrenzungskurve $u = u_i$ umschlossen wird. Der Wert desselben in dem Fall, daß $u = u_i$ eine unendlich ferne Lage einnimmt, bestimmt sich leicht wie folgt: Der Asymptotenkegel berührt die entstehende Fläche längs $u = u_i$, die Flächennormalen längs $u = u_i$ stehen also dort auch auf dem Asymptotenkegel senkrecht. Dann ändert sich an dem Bild, das diese Flächennormalen bestimmen, nichts, wenn man dieselben längs der Kegelerzeugenden bis zur Spitze desselben verschiebt. Das fragliche Gebiet ist also gleich der Apertur des Polarkegels, die mit A bezeichnet sei. Dieselbe ist positiv und $< 2\pi$ zu wählen, da der Asymptotenkegel nicht überstumpf sein kann und die Äquivalentkrümmung A seiner Spitze stets positiv ist. Es ist also

$$\lim\left(-\int\limits_{u=u_i}\frac{ds \sin\vartheta}{\varrho}\right) = A - 2\pi$$

welcher Betrag also für jeden unendlich fernen Rand der vorgegebenen Fläche zu bestimmen ist.

Eine besondere Flächengattung zeige die Anwendung dieses Ergebnisses: In seiner Arbeit*): »Sur les surfaces à courbures opposées et leurs lignes géodésiques« gibt *Hadamard* über den topologischen Charakter von Flächen überall negativer Krümmung Aufschluß. Entsendet eine solche n Flächenmäntel ins Unendliche, so gehört zu jedem derselben eine bestimmte geschlossene geodätische Linie. Liegt dieselbe selbst im Unendlichen (nappe non évasée), so existiert eine Asymptotenrichtung (p. 35 der Abhandlung); in diesem Fall ist also der Polarkegel eine Ebene und das *Bonnet*sche Integral wird Null. Liegt die geschlossene geodätische Linie im Endlichen und schnürt einen ins Unendliche gehenden Flächenteil (nappe évasée) gegen den Rest der Fläche ab, so ist für diesen Flächenmantel

$$A - 2\pi$$

ein von Null verschiedener negativer Wert.

Überdeckt man nun die Fläche mit einer Kurvenfolge, die man in einem beliebigen Punkt des durch die n genannten geodätischen Linien G_i begrenzten Flächenteils beginnen läßt, so sind

*) Journal d. Math. (Liouville), 5ème série, tome 4, 1898.

mindestens $(n-1)$ Zusammenhangsänderungen nötig, um sie in das System der n Kurven G_i überzuführen und dann ohne weitere Änderungen die n Flächenmäntel erzeugen zu lassen. Also ergibt sich für eine Fläche durchaus negativer Krümmung, die n Flächenmäntel ins Unendliche sendet, ein oberer Wert der Totalkrümmung durch

$$C \leq -(n-2)\,2\pi$$

Verteilt man über ein Flächenstück zwei Parameter u und v so, daß in jedem Punkt ein eindeutiger Fortgang nach jedem Parameter bestimmt ist, so bilden die Kurven $u = \text{const.}$ und $v = \text{const.}$ ein reguläres Netz. Zwei Elemente der gleichen Schar treffen sich nicht und jedes der einen jedes der andern nur einmal. Es entsteht die Frage: Wann läßt sich eine im Endlichen geschlossene Fläche mit einem solchen Netz überziehen, resp. welche Singularitäten muß ein solches Netz auf einer vorgegebenen Fläche notwendig aufweisen.

Die Parameterkurven einer Schar, die keine Singularitäten aufweisen soll, mussen auf einer im Endlichen geschlossenen Fläche geschlossen sein. Sie bilden also auf derselben eine Kurvenfolge der oben benutzten Art. Sollen sie die Kurven $u = \text{constans}$ darstellen, so kann die Verteilung eines Parameters v längs ihrer zum Beispiel durch die früher benutzte Methode geschehen:

$$v = \frac{s}{S(u_i)} \qquad (1)$$

wo s der Bogen und $S(u_i)$ die Länge des betreffenden Kurvenzuges ist und v demnach periodisch wird. Dann ist die zweite Schar durch die Trajektorien dieser Folge, die Kurven

$$v = \text{constans}$$

gegeben. Dieselben laufen, wenn die Schar $u = \text{constans}$ sich bei $u = u_i$ in einen Punkt zusammenzieht, in diesen zusammen und bilden dort einen »Pol«. Erleiden aber die Kurven der ersten Schar bei $u = u_i$ eine Zusammenhangsänderung, so setzen sich, wie oben an einem bestimmten Fall gezeigt wurde, die Kurven $v = \text{const.}$ bei $u = u_i$ unstetig fort, d. h. man darf bei $u = u_i$ nicht bei konstantem v nach u differenzieren.

Wann kann man durch Abänderung der Kurvenfolge solche Kurven, längs deren die Regularität des Netzes gestört ist, vermeiden?

$u = \alpha$ sei ein einfach geschlossenes Element der Folge. Bei $u = \beta$ teile dasselbe sich in zwei ebensolche. Von diesen ziehe das eine sich bei $u = \gamma$ in einen Punkt zusammen und verschwinde damit. $u = \delta$ sei eine spätere Lage des zweiten, vor welcher es keine weiteren Zusammenhangsänderungen erfährt. Dann ist das zwischen $u = \alpha$ und $u = \delta$ liegende Flächenstück einfach zusammenhängend und doppelt berandet, kann also zu dem zwischen $z = \lambda_1$ und $z = \lambda_2$ gelegenen Stück des Kreiszylinders

$$x^2+y^2 = r^2$$

in eineindeutige punktweise Beziehung gesetzt werden. Dann kann man die ursprüngliche Kurvenfolge zwischen $u = \alpha$ und $u = \delta$ ersetzen durch die Folge der Bilder der Kurven $z = \lambda$, also durch eine Folge, die keine Zusammenhangsänderung und keine Zusammenziehung auf einen Punkt mehr enthält. In dieser Weise fahre man in der Variierung der Kurvenfolge fort, indem man, so oft es geht, eine Zusammenhangsänderung gegen eine Zusammenziehung auf einen Punkt hebt. Kann man auf diese Weise eine Folge ohne Zusammenhangsänderungen erzielen, so erleidet deren Trajektorienschar keine Unterbrechungen und die Regularität der Parameterverteilung ist höchstens an einzelnen Punkten aufgehoben. Wir erkennen also

> Verteilt man über eine Fläche positiver, von Null verschiedener Charakteristik*) zwei Parameter, so gibt es stets Linien auf der Fläche, längs deren die Stetigkeit der Parameterverteilung unterbrochen ist.

Da in einer Kurvenfolge ohne Zusammenhängsänderungen ein Zusammenziehen in einen Punkt und Verschwinden nur zweimal auftreten kann (im Anfangs- und Endelement), so kann man leicht alle Typen von regulären Kurvennetzen aufstellen:

1. Anzahl der Pole = 2.
   » » Zusammenhangsänderungen = 0.

Ist $u = u_i$ die Kurvenfolge und bestimmt man durch (1)

*) Bei diesen überwiegt nach dem auf p. 33 formulierten Ergebnis die Anzahl der Zusammenhangsänderungen diejenige der Zusammenziehungen in einen Punkt.

einen Parameter für ihre Punkte von einer bestimmten Trajektorie an, so ist die Fläche mit ihrem Netz von Kurven

$$u = \text{constans}$$

und $$v = \text{constans}$$

der Kugel mit ihrem Netz von Meridianen und Breitenkreisen elementarverwandt.

2. Anzahl der Pole = 2.

» » Zusammenhangsänderungen = 1.

Die Zusammenhangsänderung ist notwendig eine Umschaltung ohne Zerfall der Kurve, also auf die oben beschriebene Weise nicht zu zerstören. Das einfachste Kurvennetz auf diesen einseitigen Flächen der Charakteristik -1 bilden zwei Geradenbüschel durch zwei nicht zusammenfallende Punkte der projektiven Ebene. Jedes Kurvensystem hat also einen Pol, und außerdem haben die beiden Scharen ein Element gemeinsam, d. h. es gibt ein Wertepaar

$$u = \lambda$$
$$v = \mu$$

das nicht eindeutig einen Punkt, sondern alle Punkte einer Geraden bestimmt. Ordnet man diesem Element die Parameterwerte

$$u = \infty$$
$$v = \infty$$

zu, so kann man die Punkte desselben durch den Wert des Verhältnisses $\frac{u}{v}$ in den Nachbarpunkten charakterisieren. Durch Abbildung der projektiven Ebene auf eine im Endlichen geschlossene Fläche kann man ein solches Netz leicht veranschaulichen, z. B. an der aufgeschnittenen und in eine Ebene ausgebreiteten *Boy*schen schen Fläche.*)

3. Anzahl der Pole = 2.

» » Zusammenhangsänderungen = 2.

α) Die letzteren seien Umschaltungen ohne Zerfall; dann gestattet die Fläche zwei Rückkehrschnitte II. Art. Dieselben lassen sich durch einen solchen der I. Art ersetzen und die Fläche durch Bewegung und stetige Deformation desselben erzeugen. Versieht man die Anfangslage dieser Kurve mit einer Indikatrix, so

*) cfr. Math. Ann. 57, p. 183.

kehrt sie am Schluß mit umgekehrter Indikatrix in die Ausgangslage zurück. Die Fläche ist »ein sich durchsetzender Schlauch«. Seien

$$u = 0$$
$$\text{bis } u = 1$$

die aufeinanderfolgenden Lagen der Kurve und $v = \frac{s}{S(u_i)}$ ein Parameterwert für jeden ihrer Punkte, gemessen von einer festen Trajektorie $v = 0$, so tragen die identischen Kurven $u = 0$ und $u = 1$ zwei sich entgegenlaufende Parameterfolgen und nur die beiden Trajektorien

$$v = 0$$
$$\text{und } v = {}^1/_2$$

sind geschlossen. Diese beiden Trajektorien zerschneiden die Fläche in ein im allgemeinen mehrfach gedrehtes geschlossenes Band, sodaß die Fläche durch Bewegung eines ungeschlossenen Kurvenstücks entstehend gedacht werden kann, dessen Enden auf zwei gegebenen geschlossenen Kurven stetig vorwärts gleiten.

Es gibt also in diesem Fall eine geschlossene Kurve auf der Fläche, längs deren die Regularität des Parameternetzes unterbrochen ist.

β) Die beiden Zusammenhangsänderungen seien keine Umschaltungen, sondern also eine Trennung und eine Vereinigung. Dann lassen sie sich gegen die Pole heben und die Fläche durch eine bewegte geschlossene Kurve erzeugen, die am Schluß mit gleicher Indikatrix in die Ausgangslage zurückkehrt. Ihre Trajektorien sind also Rückkehrkurven I. Art und alle oben definierten Kurven v = constans sind geschlossen.

Die Regularität des Kurvennetzes erleidet also keine Unterbrechung, was so formuliert werden kann:

Die zweiseitige Fläche vom Geschlecht eins (Ring) ist die einzige geschlossene Fläche, über die man zwei Parameter u und v stetig verteilen kann, ohne daß in einem Punkte der Fläche eine der Fundamentalgrößen

$$E = \left(\frac{\partial x}{\partial u}\right)^2 + \left(\frac{\partial y}{\partial u}\right)^2 + \left(\frac{\partial z}{\partial u}\right)^2$$
$$G = \left(\frac{\partial x}{\partial v}\right)^2 + \left(\frac{\partial y}{\partial v}\right)^2 + \left(\frac{\partial z}{\partial v}\right)_2$$

Null oder unendlich wird.

§ 4.

Eineindeutige Abbildungen des Ringes auf sich selbst.

Kurvennetze, welche eine vorgegebene Fläche trägt, können bei der Untersuchung eineindeutiger stetiger Abbildungen derselben Verwendung finden. Bei einer solchen geht ein Kurvennetz in ein ebensolches über, und die Abbildung kann durch Zuordnung der Netzelemente angegeben werden.

Bei eineindeutigen Abbildungen der Fläche der Charakteristik —2, der allgemeinen Kugel, auf sich selbst ist wegen der Eigenschaft jeder geschlossenen Kurve, auf einen Punkt zusammenziehbar zu sein, das Bildnetz in das ursprüngliche oder in eine Spiegelung desselben stetig deformierbar, je nachdem die Abbildung die Indikatrix ungeändert läßt oder nicht, oder mit einem neueren Ausdruck *) »den Grad +1 oder —1 besitzt«. Als Grad einer Abbildung definiert *Brouwer* die algebraische Summe aus gleichsinnigen (positiven) und ungleichsinnigen (negativen) Überdeckungen der das Bild tragenden Fläche, welche Summe für alle Punkte derselben den gleichen Wert hat.

Unter einem notwendigen Fixpunkt einer Abbildung soll ein solcher Fixpunkt, d. h. mit seinem Bild zusammenfallender Punkt verstanden werden, der bei stetiger Deformation der Abbildung (d. h. des Bildnetzes) sich auf der Fläche stetig verschiebt, ohne doch zum Verschwinden gebracht werden zu können. Die Anzahl der notwendigen Fixpunkte ist also für alle in einander deformierbaren Abbildungen dieselbe.

Im Folgenden sollen nur nicht ineinander deformierbare Abbildungen als verschieden gelten.

---

*) *Brouwer:* Math. Ann. 71, p. 105.

Brouwer beweist*), daß die Kugel bei einer Abbildung vom Grade -1 keinen, bei einer solchen vom Grade $+1$ einen notwendigen Fixpunkt aufweist.

Daß es tatsächlich eineindeutige Abbildungen der Kugel auf sich selbst gibt, die nicht wie die kongruenten Abbildungen zwei, sondern nur einen Fixpunkt haben, davon überzeugt man sich zum Beispiel, indem man die komplexe Zahlenebene einer Parallelverschiebung unterwirft und vor und nach derselben in bekannter Weise auf eine berührende Kugel stereographisch projiziert. Die dadurch bestimmte eineindeutige Zuordnung der Punkte der letzteren hat nur das Projektionszentrum zum Fixpunkt, das dem Werte ∞ der komplexen Variablen entspricht.

Bei den Flächen der Charakteristik -1 kann man wegen der Einseitigkeit von einer Erhaltung resp. Umkehrung der Flächenindikatrix nicht reden, man hat dort überhaupt nur einen Abbildungstypus; je zwei Abbildungen sind in einander deformierbar. Auch hier ist die Frage der Fixpunkte von *Brouwer* gelöst: **)

»Jede eineindeutige stetige Transformation der projektiven Ebene in sich läßt wenigstens einen Punkt invariant.«

Wenn nun im Folgenden dieselbe Frage für die geschlossene zweiseitige Fläche der Charakteristik 0 aufgeworfen wird, so stellen sich zunächst unendlich viele nicht in einander deformierbare Abbildungen ein, dem Umstand entsprechend, daß es auf dem Ring unbeschränkt viele nicht in einander deformierbare geschlossene Kurven gibt.

Es ist, wie oben gezeigt, möglich, den Ring durch Bewegung und vollkommen stetige Deformation einer geschlossenen Kurve zu erzeugen, also zwei Parameter u und v so über denselben zu verteilen, daß sich die Kurven $u =$ constans, $v =$ constans in jedem Gebiet der Fläche wie Parallelkoordinatenlinien der Cartesischen Ebene verhalten. Die Wertefolgen u und v sind pe-

*) *Brouwer:* Über Abbildung von Mannigfaltigkeiten. Math. Ann. 71, p. 114.

Vergl. auch: Verslagen der Zittingen van de wis- en naturkundige Afdeeling der Koninglyke Akademie van Wetenschappen, deel XVII, 1909, Amsterdam.

**) Math. Ann. 69, p. 180.

riodisch, und man gebe z. B. beiden die Periode 1. Dann ist der Punkt mit den Koordinaten

$$u+\varrho \qquad v+\sigma$$

identisch mit dem Punkt

$$u \qquad v$$

wenn ϱ und σ ganze Zahlen sind. Eine Kurve auf der Fläche kann dargestellt werden durch

$$u = f_1(\lambda)$$
$$v = f_2(\lambda)$$

Ist die Kurve geschlossen, so sind f_1 und f_2 periodisch, und wenn Λ die Periode des Arguments ist, so ist

$$u_0 = f_1(0); \qquad u_0+\varrho = f_1(\Lambda)$$
$$v_0 = f_2(0); \qquad v_0+\sigma = f_2(\Lambda)$$

Man durchläuft also längs der geschlossenen Kurve die Wertefolge u ϱmal und v σmal. Bezeichnet man den Typus der Kurven $u =$ constans resp. $v =$ constans, mit U resp. V, so ist die genannte Kurve stetig aus derjenigen deformierbar, die entsteht, wenn man U σmal und dann V ϱmal durchläuft, ist also darstellbar durch $U^\sigma V^\rho$. Die einfachste Form nehmen die Gleichungen für die Gerade an:

$$u = \varrho\cdot\lambda+u_0$$
$$v = \sigma\cdot\lambda+v_0$$

wobei λ die Periode 1 hat. Soll diese Kurve Doppelpunkte besitzen, so muß für einen rationalen Wert von $\lambda < 1$ die Größe $\frac{1}{\lambda}$ in ϱ und σ ganzzahlig enthalten sein. Man sieht also:

> Die Kurve $U^\sigma V^\rho$ hat keine notwendigen Doppelpunkte, wenn ϱ und σ teilerfremd sind.

Man erkennt leicht, was hier nicht ausgeführt werden soll, daß dieser Satz sich auf die Form erweitern läßt: Die Kurve $U^\sigma V^\rho$ hat $e-1$ notwendige Schnittpunkte, wenn e der größte gemeinsame Teiler von ϱ und σ ist.

Wie groß ist die Minimalzahl der Schnittpunkte zweier Kurven

$$\begin{cases} U^\sigma V^\rho & \text{(I)} \\ U^{\sigma'} V^{\rho'} & \text{(II)} \end{cases}$$

der Ringfläche? Man denke die Kurven in bestimmter Gestalt

vorgegeben und bringe, so oft es geht, je zwei Schnittpunkte derselben durch stetige Deformation der Kurven auf der Fläche über einen Berührungspunkt hin zum Verschwinden. Diese Operation ist möglich, so oft sich zwei Schnittpunkte von der Art finden lassen, daß die von ihnen begrenzten Kurvenbogen — oder vielmehr je einer der beiden, in die sie jede Kurve trennen, — mit einander ein Elementarflächenstück der Fläche begrenzen. Versieht man die beiden Kurven I und II mit einem Durchlaufungssinn und tritt an dem einen Endpunkt eines solchen Elementarzweiecks die Kurve II über I von rechts nach links, so tritt sie am anderen Endpunkt von links nach rechts über. Es lassen sich also keine so begrenzten Elementarflächenstücke mehr auffinden, wenn die Kurven auf der Fläche so liegen, daß die eine stets im gleichen Sinn über die andere tritt, d. h. die Anzahl der Schnittpunkte zweier solcher Kurven ist notwendig.

Dieser Forderung genügen die Kurven I und II, wenn ihre Gleichung linear ist:

$$\text{I} \quad \begin{cases} u = \varrho\lambda \\ v = \sigma\lambda \end{cases}$$

$$\text{II} \quad \begin{cases} u = \varrho'\mu \\ v = \sigma'\mu \end{cases}$$

denn dann entspricht der Umgebung jedes Schnittpunktes der Schnitt zweier Geraden mit den Steigungen $\frac{\sigma}{\varrho}$ resp. $\frac{\sigma'}{\varrho'}$ in einem rechtwinkligen u-v-Koordinatensystem der Cartesischen Ebene.

Für einen Schnittpunkt beider Kurven ist:

$$\left.\begin{aligned} \varrho\lambda &= \varrho'\mu + \alpha \\ \sigma\lambda &= \sigma'\mu + \beta \end{aligned}\right\} \alpha \text{ und } \beta \text{ sind ganze Zahlen}$$

woraus

$$\lambda = \frac{\sigma'\alpha - \varrho'\beta}{d}$$

$$\mu = \frac{\sigma\alpha - \varrho\beta}{d}$$

wo $d = \varrho\sigma' - \sigma\varrho'$ ist. Sind σ' und ϱ' teilerfremd, so kann man zu jedem der Werte

$$i = 0, 1, 2 \cdots d-1$$

ein Zahlenpaar α, β finden, sodaß

$$\sigma'\alpha - \varrho'\beta = i$$

Dasselbe bestimmt einen Wert

$$i' = \sigma\alpha - \varrho\beta$$

Vermehrt oder vermindert man α um ϱ' und β um σ', so ändert sich i nicht, während i' um d ab oder zunimmt. Man kann also für jeden Wert von i ein Zahlenpaar α, β so finden, daß sowohl i als i' zwischen o und $d-1$ liegen. Die beiden Kurven haben demnach d Schnittpunkte.

Haben σ' und ϱ' den größten gemeinsamen Teiler e und ist

$$\sigma' = \sigma'' \cdot e$$
$$\varrho' = \varrho'' \cdot e$$

so ist die Kurve II gleich der $e$mal durchlaufenen Kurve

$$u = \varrho'' \cdot \nu \qquad (\nu = 0 \cdots 1)$$
$$v = \sigma'' \cdot \nu$$

Jeder der $\dfrac{d}{e}$ Schnittpunkte, welche dieselbe der eben gegebenen Ableitung zufolge mit I hat, bedeutet einen $e$fachen Schnitt zwischen I und II, sodaß das Resultat allgemein gilt:

Zwei Kurven vom Typus

$$U^{\sigma}\, V^{\varrho}$$
$$U^{\sigma'}\, V^{\varrho'}$$

haben notwendig $d = \begin{vmatrix} \sigma & \varrho \\ \sigma' & \varrho' \end{vmatrix}$ Schnittpunkte. (Natürlich gilt der absolute Wert der Determinante).

Wird die Ringfläche, die das oben eingeführte Koordinatennetz trägt, einer eineindeutigen und stetigen Abbildung auf sich selbst unterworfen, so geht das Koordinatennetz N in ein ebensolches N' über, das dieselben topologischen Eigenschaften hat, aber im allgemeinen nicht in N stetig deformierbar ist. Das Bild U ist allgemein vom Typus

$$U^{m}\, V^{n}$$

ebenso das Bild von V

$$U^{m'}\, V^{n'}$$

und weil die Eigenschaft von U und V, mit einander einen notwendigen Schnittpunkt zu haben, bei der Abbildung erhalten bleibt, so ist nach dem obigen Schnittpunktsatze

$$\begin{vmatrix} m & n \\ m' & n' \end{vmatrix} = \pm 1$$

und der Typus der Abbildung (der uns hier und im Folgenden allein interessiert) ist durch Angabe dieses Zahlenquadrupels vollkommen charakterisiert. Die Angabe eines solchen bestimmt also zu einem gegebenen Netz ein anderes mit Angabe der Parameterverteilung, während die Angabe zweier Kurvennetze ohne Festsetzung einer zugehörigen Parameterverteilung 8 verschiedene Abbildungen ermöglicht. Man sieht das daran, daß z. B. das ursprüngliche Koordinatennetz N auf 8 Arten auf sich selbst abgebildet werden kann. Diese sind charakterisiert durch die Zahlenquadrupel

$$\begin{matrix} 1 & 0 \\ 0 & 1 \end{matrix} \qquad \begin{matrix} -1 & 0 \\ 0 & -1 \end{matrix} \qquad \begin{matrix} 1 & 0 \\ 0 & -1 \end{matrix} \qquad \begin{matrix} -1 & 0 \\ 0 & 1 \end{matrix}$$

$$\begin{matrix} 0 & 1 \\ 1 & 0 \end{matrix} \qquad \begin{matrix} 0 & -1 \\ -1 & 0 \end{matrix} \qquad \begin{matrix} 0 & 1 \\ -1 & 0 \end{matrix} \qquad \begin{matrix} 0 & -1 \\ 1 & 0 \end{matrix}$$

Da diese Operationen Vertauschung der Durchlaufungsrichtungen der u- und v-Kurven und Vertauschung dieser Kurven selbst bedeuten, so ist ihre Gruppe isomorph mit derjenigen der Transformationen eines rechtwinkligen u-v-Koordinatenkreuzes in sich, oder mit der invarianten Untergruppe vom Index 3 der symmetrischen Gruppe von 4 Elementen. Bezeichnet man die Klappung um die Gerade $u = v$, d. h. die Abbildung $\begin{Bmatrix} 0 & 1 \\ 1 & 0 \end{Bmatrix}$ mit a_1 und die Klappung um $u = 0$, d. h. die Abbildung $\begin{Bmatrix} 1 & 0 \\ 0 & -1 \end{Bmatrix}$ mit a_2, so ist die Abbildung *)

$$a_1 a_2 = \begin{Bmatrix} 0 & 1 \\ -1 & 0 \end{Bmatrix}$$

eine Drehung des Achsenkreuzes um $\frac{\pi}{2}$. Also ist

$$(a_1 a_2)^4 = 1$$

und ferner ist

$$a_1{}^2 = 1$$
$$a_2{}^2 = 1$$

sodaß das *Dehn*sche Gruppenbild **) dieser Gruppe bei dieser Wahl der Erzeugenden ein Achteck ist, dessen Seiten abwechselnd mit

*) Erst soll a_2, dann a_1 ausgeführt werden.

**) Siehe Math. Ann. 69: *Dehn*, Über die Topologie des dreidimensionalen Raumes.

a_1 und a_2 bezeichnet sind. Bei Ausführung dieser Abbildungen erleidet ein allgemeines Netz die Substitutionen

$$a_1 \begin{Bmatrix} m & n \\ m' & n' \end{Bmatrix} = \begin{Bmatrix} n & m \\ n' & m' \end{Bmatrix} = \text{Vertauschung der Kolonnen}$$

$$a_2 \begin{Bmatrix} m & n \\ m' & n' \end{Bmatrix} = \begin{Bmatrix} m & -n \\ m' & -n' \end{Bmatrix} = \text{Zeichenwechsel der 2. Kolonne}$$

$$a_1\, a_2\, a_1 \begin{Bmatrix} m & n \\ m' & n' \end{Bmatrix} = \begin{Bmatrix} -m & n \\ -m' & n' \end{Bmatrix} = \text{Zeichenwechsel der 1. Kolonne}$$

Da die Determinante des Zahlenquadrupels

$$\begin{vmatrix} m & n \\ m' & n' \end{vmatrix} = \pm 1$$

ist, so muß die Anzahl der negativen Zeichen in demselben notwendig gerade sein. Daraus geht hervor, daß man durch Anwendung von a_2 oder $a_1 a_2 a_1$ oder beider das vorgegebene Quadrupel so transformieren kann, daß seine Vorzeichen in einer der folgenden Arten kombiniert sind:

$$\alpha) \begin{Bmatrix} + & + \\ + & + \end{Bmatrix}$$

$$\beta) \begin{Bmatrix} + & + \\ - & - \end{Bmatrix}$$

Neben den erzeugenden Operationen a_1 und a_2 sei nun als weitere solche die Abbildung

$$a_3 = \begin{Bmatrix} 1 & 1 \\ 0 & 1 \end{Bmatrix}$$

gewählt. Sie ersetzt eine Kurve U des Netzes durch eine solche UV, läßt aber den Typus der Kurven V ungeändert. Dann ist

$$a_3^{\nu} = \begin{Bmatrix} 1 & \nu \\ 0 & 1 \end{Bmatrix}$$

und

$$a_3^{\nu} \cdot \begin{Bmatrix} m & n \\ m' & n' \end{Bmatrix} = \begin{Bmatrix} m & n + \nu m \\ m' & n' + \nu m' \end{Bmatrix}$$

Nun sollen auf das ursprüngliche Netz (die identische Substitution $\begin{Bmatrix} 1 & 0 \\ 0 & 1 \end{Bmatrix}$) die folgenden Operationen angewendet werden, wobei die Größen ν positive Zahlen bedeuten:

$a_3^{\nu_1}$: $\begin{Bmatrix} 1 & \nu_1 \\ 0 & 1 \end{Bmatrix}$

$a_3^{\nu_2} a_1 a_3^{\nu_1}$: $\begin{Bmatrix} \nu_1 & \nu_1 \nu_2 + 1 \\ 1 & \nu_2 \end{Bmatrix}$

$a_3^{\nu_3} a_1 a_3^{\nu_2} a_1 a_3^{\nu_1}$: $\begin{Bmatrix} \nu_1 \nu_2 + 1 & (\nu_1 \nu_2 + 1) \nu_3 + \nu_1 \\ \nu_2 & \nu_2 \nu_3 + 1 \end{Bmatrix}$

$a_3^{\nu_4} a_1 a_3^{\nu_3} a_1 a_3^{\nu_2} a_1 a_3^{\nu_1}$: $\begin{Bmatrix} (\nu_1 \nu_2 + 1) \nu_3 + \nu_1 & [(\nu_1 \nu_2 + 1) \nu_3 + 1] \nu_4 + \nu_1 \nu_2 + 1 \\ \nu_2 \nu_3 + 1 & (\nu_2 \nu_3 + 1) \nu_4 + \nu_2 \end{Bmatrix}$

usw.

Man sieht, daß man für eine Abbildung $\begin{Bmatrix} m & n \\ m' & n' \end{Bmatrix}$, die darstellbar ist in der Form

$$a_3^{\nu_i} a_1 a_3^{\nu_{(i-1)}} a_1 \cdots a_3^{\nu_1}$$

die zugehörigen Exponenten $\nu_1 \cdots \nu_i$ findet durch Anwendung eines Teileralgorithmus auf die Zahlenpaare der einzelnen Zeilen des Quadrupels, wobei die erste Zeile in diesem Teilerverfahren einen Schritt mehr gestattet als die zweite. Wendet man dieselbe Reihe von Abbildungen anstatt auf die Identität auf die Abbildung a_1 an, so erhält man einen analogen Teileralgorithmus, der in der zweiten Zeile ein Glied mehr enthält. Wendet man sie endlich auf die Abbildungen a_2 resp. $a_1 a_2$ an, so erhält man Systeme $\begin{Bmatrix} m & n \\ m' & n' \end{Bmatrix}$, in denen m' und n' negative Zahlen sind. Dieselben liefern denselben Algorithmus, wobei die erste resp. die zweite Zeile ein Glied mehr enthält.

Die Frage, ob die drei Operationen a_1, a_2, a_3 zur Gewinnung aller möglichen Abbildungen des Ringes auf sich ausreichen, ist also zu bejahen, falls jedes Zahlenquadrupel mit der Determinante ± 1 einen Teileralgorithmus dieser Art ermöglicht. Das ist nun in der Tat der Fall.

Ist eines der Elemente des Quadrupels Null, so sind die Elemente derjenigen Diagonale, in der es nicht enthalten ist, notwendig ± 1, und es ist unmittelbar ersichtlich, daß das Quadrupel durch eine einzige Potenz von a_3 und evtl. einigen Operationen a_1 und a_2 zu erzeugen ist.

Es sei also $\begin{Bmatrix} m & n \\ m' & n' \end{Bmatrix}$ ein Quadrupel aus lauter von Null verschiedenen Elementen. Es sei von vorneherein angenommen, was wir durch Anwendung von a_1 und a_2 stets erreichen können, daß die Vorzeichen in einer der Kombinationen α oder β auftreten und daß zugleich $m > n$ ist, falls nicht $m = n = 1$.

Die Größen m, n, m', n' seien alle positiv (α) und ihre Determinante $+1$. Dann ist

$$m = \sigma_1 n + r_1$$
$$m' = \sigma_1' n' + r_1'$$

wobei σ_1 und $\sigma_1' \neq 0$ und $(n, r_1) = (n', r_1') = 1$. Es folgt:

$$\begin{vmatrix} r_1 & n \\ r_1' & n' \end{vmatrix} + (\sigma_1 - \sigma_1') n n' = 1$$

Da $\left.\begin{matrix} 0 \leq r_1 < n \\ 0 \leq r_1' < n' \end{matrix}\right\}$ ist, so ist die Determinante um mehr als eine Einheit kleiner als nn'; also muß notwendig sein

$$\sigma_1 = \sigma_1'$$

und

$$\begin{vmatrix} r_1 & n \\ r_1' & n' \end{vmatrix} = 1$$

Dasselbe Verfahren setzt sich fort, bis nach ν Schritten

$$\sigma_\nu = \sigma_\nu'$$

und

$$\begin{vmatrix} r_{\nu-1} & r_\nu \\ r'_{\nu-1} & r'_\nu \end{vmatrix} = 1$$

evtl. nach Vertauschung der Kolumnen, je nachdem ν gerade oder ungerade ist. Hier möge nun einer von zwei Fällen eintreten, die nach einer endlichen Anzahl von Schritten notwendig kommen müssen, weil die Größen r beständig kleiner werden.

1. r_ν und r'_ν werden beide gleichzeitig gleich 1

Dann ist

$$r_{\nu-1} = r'_{\nu-1} + 1$$

und wir können schreiben

$$\sigma_{\nu+1} = \sigma'_{\nu+1} = r'_{\nu-1}$$

$$r_{\nu+1} = 1 \qquad\qquad r'_{\nu+1} = 0$$

$$\sigma_{\nu+2} = 1$$

2. Es wird z. B. $r'_\nu = 1$, während $r_\nu > 1$

Dann ist

$$r_{\nu-1} = r_\nu \cdot r'_{\nu-1} + 1$$

$$\sigma_{\nu+1} = \sigma'_{\nu+1} = r'_{\nu-1}$$

$$\sigma_{\nu+2} = r_\nu$$

Ist der Wert der Determinante nicht $+1$, sondern -1, so erleiden diese letzten Schritte eine geringe Modifikation, während der ganze Gedankengang und das Ergebnis dasselbe bleibt. Und es ist leicht ersichtlich, daß eine analoge Überlegung im Fall der Vorzeichenkombination β Platz greift.

Damit ist also gezeigt:

Die Gesamtheit aller eineindeutigen Abbildungen des Ringes auf sich läßt sich durch successive Ausführung dreier derselben erzeugen;

und es hat sich zugleich für die Darstellung einer gegebenen Abbildung durch die drei Erzeugenden eine Normalform

$$\varepsilon_2 a_3^{\nu_i} a_1 a_3^{\nu_{(i-1)}} a_1 \cdots a_3^{\nu_1} \varepsilon_1$$

ergeben, wobei ε_1 und ε_2 geeignete Kombinationen von a_1 und a_2 bedeuten. Der Typus dieser drei erzeugenden Operationen war

$$a_1: \begin{cases} U' = V \\ V' = U \end{cases}$$

$$a_2: \begin{cases} U' = U \\ V' = V^{-1} \end{cases}$$

$$a_3: \begin{cases} U' = UV \\ V' = V \end{cases}$$

Die beiden ersten Abbildungen sind ungleichsinnig (Grad -1), die letzte gleichsinnig (Grad $+1$); die Determinante des Zahlenquadrupels ist bei den ersten beiden -1, bei der letzten $+1$. Daraus folgt:

Der Grad einer beliebigen eineindeutigen Abbildung ist gleich der Determinante des sie repräsentierenden Zahlenquadrupels.

Die Gesamtheit der nicht in einander deformierbaren Typen eineindeutiger Abbildungen des Ringes auf sich bilden eine unendliche diskontinuierliche Gruppe, deren nahe Verwandtschaft mit der Modulgruppe in die Augen fällt. Erweitert man diese durch Spiegelung, so ist sie definiert als die Gesamtheit der linearen Substitutionen einer komplexen Variablen ω

$$\frac{\alpha\omega+\beta}{\gamma\omega+\delta}, \text{ wo } \alpha\delta-\beta\gamma = \pm 1 \quad (\alpha, \beta, \gamma, \delta \text{ sind ganze Zahlen})$$

und ω durch die konjugierte Größe $\tilde{\omega}$ zu ersetzen ist, wenn die Determinante $= -1$ ist. Die Elemente dieser Gruppe sind durch ein Zahlenquadrupel mit der Determinante ± 1 völlig charakterisiert und somit den einzelnen Elementen unserer Abbildungsgruppe zugeordnet. Und zwar ist der Isomorphismus beider Gruppen dadurch bestimmt, daß einer Substitution der Modulgruppe zwei Abbildungen entsprechen, da die ersteren nicht, wohl aber die letzteren sich ändern, wenn man im repräsentierenden Zahlenquadrupel allen vier Zahlen das entgegengesetzte Zeichen gibt.

Wie unsere Abbildungsgruppe läßt sich auch die erweiterte Modulpruppe wie bekannt aus 3 erzeugenden Substitutionen gewinnen; daher ist auch der oben über ein Zahlenquadrupel mit der Determinante ± 1 rein zahlentheoretisch bewiesene Satz von dieser Seite her bereits bekannt. Wählt man als erzeugende Substitutionen:

$$A(\omega) = -\tilde{\omega}$$
$$B(\omega) = -\tilde{\omega} - 1$$
$$C(\omega) = \frac{1}{\tilde{\omega}}$$

so bestehen die Relationen*)

1. $A^2 = B^2 = C^2 = 1$
2. $(AC)^2 = 1$
3. $(BC)^3 = 1$

Als erzeugende Abbildungen unserer Gruppe wähle man nun je eine der den Modulsubstitutionen A, B, C entsprechenden Abbildungen, z. B.

$$A = \begin{Bmatrix} 1 & 0 \\ 0 & -1 \end{Bmatrix} = a_2$$

$$B = \begin{Bmatrix} 1 & 1 \\ 0 & -1 \end{Bmatrix} = a_3 a_2$$

$$C = \begin{Bmatrix} 0 & 1 \\ 1 & 0 \end{Bmatrix} = a_1$$

*) Vergl. *Klein*, Vorl. über die Theorie der elliptischen Modulfunktionen. Ausgearbeitet von *R. Fricke*. Leipzig 1890. Bd. I, p. 232 und p. 454 unten.

Da der identischen Modulsubstitution außer der identischen Abbildung $\begin{Bmatrix} 1 & 0 \\ 0 & 1 \end{Bmatrix}$ auch die Abbildung $\begin{Bmatrix} -1 & 0 \\ 0 & -1 \end{Bmatrix}$ entspricht, so sind die Relationen zu prüfen. Dabei zeigt sich, daß 1. und 2. für unsere Gruppe ungeändert bleiben, während

$$(AC)^2 = \begin{Bmatrix} -1 & 0 \\ 0 & -1 \end{Bmatrix}$$

also erst $(AC)^4 = 1$ ist. Wir haben also:

Die unendliche Gruppe der Typen eineindeutiger Abbildungen des Ringes ist charakterisiert wie folgt:

Erzeugende Operationen: A, B, C

Relationen:
1. $A^2 = 1$
2. $B^2 = 1$
3. $C^2 = 1$
4. $(AC)^4 = 1$
5. $(BC)^3 = 1$

Konstruiert man das *Dehn*sche Gruppenbild dieser Gruppe*), so kann man es zufolge den Relationen 1., 2., 3. der Einfachheit wegen unterlassen, den Strecken, welche die erzeugenden Operationen darstellen, einen Richtungssinn zu geben, sodaß jede Strecke in beiderlei Sinn durchlaufen werden kann. Dann erhält man das Gruppenbild folgendermaßen: Man nenne die Seiten eines Achtecks abwechselnd A und C. An dasselbe füge man vier Sechsecke, deren Seiten abwechselnd mit B und C bezeichnet sind, und die mit dem Achteck je eine Seite C, im übrigen aber mit demselben und mit einander keinen Punkt gemeinsam haben. An jede freie Seite C der Figur füge man dann wieder ein Achteck $(AC)^4 = 1$, welches außer dieser Seite wieder keinen Punkt mit der bisherigen Figur noch auch mit den anderen neu hinzugefügten Achtecken gemeinsam hat, und fahre so unbegrenzt fort. Metrisch regulär läßt sich das Bild natürlich nur in der hyperbolischen Ebene gestalten.

Das Gruppenbild der erweiterten Modulgruppe bekommt man, indem man anstatt der Achtecke Vierecke verwendet, das der nicht

*) Vergl. *Dehn:* Über die Topologie des dreidimensionalen Raumes, Math. Ann. 69.

erweiterten Modulgruppe, indem man unmittelbar Sechsecke $(BC)^3 = 1$ mit den Seiten C aneinanderfügt unter Innehaltung der oben festgesetzten Regel.

Dieses Gruppenbild wird von Nutzen sein bei der Frage nach »gleichberechtigten«, d. h. in einander mit Hilfe einer dritten transformierbaren Abbildungen; solchen werden gewisse invariante Eigenschaften zukommen, z. B. die gleiche Fixpunktezahl. Der dabei einzuschlagende Weg wäre wohl der von *M. Dehn* in seiner Abhandlung: »Transformation der Kurven auf zweiseitigen Flächen« (Math. Ann. 72 [1912]) durchgeführten Methode nachzubilden.*)

War somit die Frage nach den Typen der nicht in einander deformierbaren Abbildungen hier ungleich verwickelter als bei den Flächen der Charakteristik -1 und -2, so ist die Bestimmung der Minimalzahl von Fixpunkten jedes Abbildungstypus mit viel elementareren, nämlich rein topologischen Mitteln möglich, dank dem allein auf dieser Fläche möglichen, nirgends unstetigen regulären Netz.

Wir betrachten also eine Abbildung $\left\{ \begin{matrix} m\,n \\ m'n' \end{matrix} \right\}$, wo diese Größen positive oder negative ganze Zahlen mit der Determinante ± 1 bedeuten. Bezeichnen U'_α und V'_β die Kurven, in welche $U_\alpha (u = u_\alpha)$ resp. $V_\beta (v = v_\beta)$ bei der Abbildung übergehen, so ist also

$$U' = U^m V^n$$
$$V' = U^{m'} V^{n'}$$

das aus U, V gebildete Netz N geht über in das aus U', V' gebildete Netz N'.

Macht man zunächst die Voraussetzung

$$n \neq 0$$
$$m' \neq 0$$

so schneidet jede Kurve U ihr Bild U' und zwar mindestens n mal, ebenso jede Kurve V das ihrige m'mal. Allgemein hat eine Kurve $U_\alpha (u = u_\alpha)$ mit ihrem Bild U'_α

$$n + 2\varrho$$

*) Auf diese Problemstellung hat mich Herr Prof. *Dehn* freundlichst verwiesen. Sie soll an dieser Stelle nur angedeutet werden.

gemeinsame Punkte, wo ϱ eine positive ganze Zahl ist. $U_{\alpha+\varepsilon}$ und $U_{\alpha-\varepsilon}$ seien die U_α unendlich benachbarten U Kurven, $U'_{\alpha+\varepsilon}$ und $U'_{\alpha-\varepsilon}$ ihre U'_α unendlich benachbarten Bilder. Dann ist im allgemeinen einem Schnittpunkt von U_α und U'_α ein solcher von $U_{\alpha+\varepsilon}$ und $U'_{\alpha+\varepsilon}$ und von $U_{\alpha-\varepsilon}$ und $U'_{\alpha-\varepsilon}$ benachbart; im besonderen Fall, wenn nämlich der gemeinsame Punkt von U_α und U'_α ein Berührungspunkt ist, an dem sich die Kurven nicht durchdringen, hat z. B. $U_{\alpha+\varepsilon}$ mit $U'_{\alpha+\varepsilon}$ keinen Schnittpunkt, dann hat aber $U_{\alpha-\varepsilon}$ mit $U'_{\alpha-\varepsilon}$ zwei einander und dem Berührungspunkte unendlich benachbarte Schnittpunkte. Die Schnittpunkte der U mit dem zugehörigen U' fügen sich also zu geschlossenen Kurven zusammen, die »Kreuzungslinien der U-Kurven« heißen mögen.

Es entsteht die Frage nach dem Verlauf dieser Linien. Man bestimmt denselben ohne Schwierigkeit an einer besonderen, an dem Netz N zu definierenden Abbildung:

$$v' = (m'u + mv)$$
$$u' = (n'u + nv)$$

wobei die runde Klammer hier wie im Folgenden andeuten soll, daß der in ihr enthaltene Ausdruck nur bis auf ganzzahlige Summanden in Betracht kommt, daß also hier im Besonderen derselbe so um eine ganze Zahl zu vermehren oder zu vermindern ist, daß er zwischen 0 und 1 fällt. Die Gleichungen ordnen jedem Punkt u, v einen solchen u', v' desselben Netzes zu, und man sieht, daß wenn man längs einer Kurve U (u = const.) den Parameter v von 0 bis 1 wachsen läßt, v' $m$mal und u' $n$mal von 0 bis 1 wächst, daß also U auf eine Kurve $U^m V^n$ abgebildet ist. Analog für V. Damit $u_1 v_1$ und $u_2 v_2$ in denselben Punkt $u'v'$ übergehen, muß sein

$$(m'u_1 + mv_1) = (m'u_2 + mv_2)$$
$$(n'u_1 + nv_1) = (n'u_2 + nv_2)$$

also

$$([mn' - nm']u_1) = ([mn' - nm']u_2)$$

mithin muß $u_1 - u_2$ und $v_1 - v_2$ eine ganze Zahl sein. Die Abbildung ist demnach eineindeutig.

Der Schnitt einer Kurve U_α mit ihrem Bild U'_α ist bestimmt durch

$$u' = u_\alpha$$
$$u_\alpha = (n' u_\alpha + n v)$$

woraus

$$v = \frac{\varrho - (n'-1)\, u_\alpha}{n}$$

wobei ϱ die Werte $0, 1, \cdots n-1$ annehmen kann. Wir haben also in diesem Fall auf jeder Kurve U nur die Mindestzahl von n Schnittpunkten. Man wähle unter diesen einen bestimmten $\varrho = \varrho_i$ und beobachte seine Bewegung, wenn man die Schar der U durchläuft, also für u_α der Reihe nach alle Werte von 0 bis 1 einsetzt. Kehrt man zur Ausgangskurve zurück, so hat v den Summanden $-\frac{n'-1}{n}$ aufgenommen. Der größte gemeinsame Teiler von $(n'-1)$ und n sei

$$(n, n'-1) = e$$

und es sei

$$n = e \cdot f$$

dann schließt sich also die betrachtete Kreuzungslinie der U Kurven nach f maliger Durchlaufung der U Schar. v hat dabei den Summanden

$$-f \cdot \frac{n'-1}{n} = -\frac{n'-1}{e}$$

aufgenommen, so daß die geschlossene Kreuzungslinie den Typus besitzt:

$$U^{\frac{1-n'}{e}} V^{\frac{n}{e}}$$

Solcher Kurven erhält man e, indem man der Reihe nach ausgeht von den Werten

$$\varrho = 0, \quad \varrho = 1, \cdots \varrho = e-1$$

Diese e geschlossenen Kreuzungslinien der U Kurven treffen einander nicht, da sie die Gleichungen haben

$$v = cu + c'$$

mit gemeinsamen $c = \frac{1-n'}{n}$ und verschiedenem c'.

Ist $(m', m - 1) = e'$, so ergeben sich analog e' Kreuzungslinien der V Kurven vom Typus

$$U^{\frac{m'}{e'}} V^{\frac{1-m}{e'}}$$

Jede Abbildung vom Typus $\begin{Bmatrix} m & n \\ m' & n' \end{Bmatrix}$ kann aus der hier betrachteten speziellen durch stetige Deformation abgeleitet werden. Dabei erleiden auch die Kreuzungslinien eine stetige Deformation, sodaß ihr Typus erhalten bleibt. Je zwei Kreuzungslinien verschiedener Scharen haben mindestens

$$\begin{vmatrix} \frac{1-n'}{e} & \frac{n}{e} \\ \frac{m'}{e'} & \frac{1-m}{e'} \end{vmatrix} = \frac{1}{e \cdot e'} (1 + m n' - n m' - m - n')$$

gemeinsame Punkte, und da sich e Kreuzungslinien der U Schar und e' Kreuzungslinien der V Schar ergaben, so haben dieselben insgesamt

$$Z = 1 + m n' - n m' - m - n'$$

Schnittpunkte, wobei natürlich der absolute Wert dieses Ausdrucks zu rechnen ist.

Jeder einzelne Punkt P dieser Z Schnittpunkte ist Fixpunkt für die Abbildung. Denn die durch P gehende U- resp. V Kurve des Netzes geht über in eine ebenfalls durch P gehende U'- resp. V' Kurve des Bildnetzes. Da nun U' und V' nur einen Punkt gemeinsam haben, so ist dies der Punkt P selbst; derselbe ist also Fixpunkt.

Die Ableitung des Wertes für Z geschah unter der Voraussetzung, daß

$$n \neq 0$$
$$m' \neq 0$$

ist. Die gefundene Formel behält aber ihre Gültigkeit auch, falls diese Voraussetzungen nicht erfüllt sind.

Es sei z. B. $n = 0$. Dann handelt es sich um Abbildungen der Typen:

$$\begin{Bmatrix} \pm 1 & 0 \\ m' & \pm 1 \end{Bmatrix}$$

α) $n' = +1$.

Eine spezielle Abbildung dieses Typus ist

$$v' = (m'u \pm v)$$
$$u' = (u + c)$$

und dieselbe hat keinen Fixpunkt, wenn man nicht c als ganze Zahl wählt. Ob $m = +1$ oder $= -1$ ist, in beiden Fällen ergibt die obige Formel

$$Z = 0$$

bleibt also in diesem Fall richtig. Die Kreuzungslinien der UKurven sind in diesem Fall vom Typus

$$U^{\frac{1-n'}{e}} V^{\frac{n}{e}} = U^0 V^0$$

treten also nicht notwendig auf. Demgemäß ergibt sich auch kein notwendiger Schnittpunkt.

β) $n' = -1$.

Das Bildnetz bestimmt wieder die Verteilung zweier Parameter $\mathfrak{u}$ und $\mathfrak{v}$ über die Fläche, deren Werte von 0 bis 1 gehen mögen. $\mathfrak{u} = \alpha$ sei das Bild von $u = \alpha$ usw. Eine Kurve V ($v =$ constans) trifft jede Kurve U ($u =$ constans) und jede Kurve U' ($\mathfrak{u} =$ constans). Ihrer Punktfolge entspricht also eine stetige und stetig im gleichen Sinne wachsende Wertefolge v und eine ebenfalls stetige, aber im allgemeinen nicht stetig im gleichen Sinne wachsende Wertefolge $\mathfrak{v}$. Indem man nun die Punkte einander zuordnet, für welche $v = \mathfrak{v}$ ist, entsteht eine stetige, im allgemeinen nicht eineindeutige Abbildung der VKurve auf sich selbst. Dieselbe ist ungleichsinnig wegen $n' = -1$, hat also mindestens zwei Fixpunkte, denn ein die Kurve durchlaufender Punkt muß mindestens zweimal seinem Bilde begegnen. Es gibt stets zwei getrennte Fixpunkte, weil die Bögen, in die sie die VKurve teilen, auf einander abgebildet werden. In jedem dieser Punkte schneidet eine UKurve ihr Bild, dies gilt für jede Kurve V ($v =$ constans), und die Gesamtheit dieser Fixpunkte*)

*) Diese Punkte sind natürlich nur für die hier konstruierte Beziehung jeder VKurve auf sich selbst, dagegen im allgemeinen nicht für die betrachtete Abbildung der Gesamtfläche Fixpunkte.

bilden, analog wie oben, die Kreuzungslinien der U Kurven. An einer speziellen Abbildung dieses Typus:

$$v' = (m'u \pm v)$$
$$u' = (c - u)$$

sieht man, daß es Kurven U gibt, die mit ihrem Bild keinen Punkt gemeinsam haben, woraus hervorgeht, daß der Typus der Kreuzungslinien keinen Bestandteil V enthält. In diesem besonderen Fall stellen sie sich als die auf sich abgebildeten Kurven dar

$$u = \left(\frac{c}{2}\right)$$

und

$$u = \left(\frac{c+1}{2}\right)$$

sind also beide vom Typus U^1. Das gilt also auch allgemein.

Die Kreuzungslinien der V Kurven haben nach dem Früheren den Typus

$$U^{\frac{m'}{e}} V^{\frac{1-m}{e}}$$

wo

$$e' = (m', 1-m)$$

Nun ist $m = \pm 1$, wir unterscheiden also:

1. $m = +1$
   $e' = m'$

Es gibt also in diesem Fall

2 Kreuzungslinien der U vom Typus U^1
m' » » V » » U^1

dieselben haben keinen notwendigen Schnittpunkt.

Die oben gefundene Formel liefert in diesem Fall $Z = 0$.

2. $m = -1$
   $e' = 2$, wenn m' gerade
   $e' = 1$, wenn m' ungerade ist.

Wir haben also

2 Kreuzungslinien der U vom Typus U^1
2 » » V » » $U^{\frac{m'}{2}} V$
oder 1 » » V » » $U^{m'} V^2$

in beiden Fällen also 4 Schnittpunkte.

Die oben gefundene Formel liefert $Z = 4$. Damit ist die beschränkende Annahme $n \neq 0$, $m' \neq 0$ aufgehoben.

Also gilt der Satz:

Sind auf einer zweiseitigen Fläche vom Geschlecht Null U und V zwei geschlossene, gerichtete Kurven, die nur einen Schnittpunkt haben, und gehen dieselben bei einer eineindeutigen und stetigen Abbildung der Fläche auf sich in zwei Kurven vom Typus $U^m V^n$ resp. $U^{m'} V^{n'}$ über, so läßt die Abbildung mindestens $|(m+n'-2)|$ oder $|(m+n')|$ Punkte invariant, je nachdem die Determinante $mn'-m'n = +1$ oder $= -1$ ist, d. h. je nachdem die Abbildung gleichsinnig oder ungleichsinnig ist.

Lebenslauf.

Als Sohn des Landmanns *Jörgen Nielsen* bin ich, *Jakob Nielsen,* am 15. Oktober 1890 zu Meels auf der Insel Alsen geboren. Ich bin preußischer Staatsangehörigkeit und evangelisch-lutherischer Konfession.

Bis zu meinem zehnten Lebensjahr besuchte ich die Dorfschule in Meels, dann von Ostern 1900 an das Königliche Realgymnasium zu Rendsburg, das ich im Dezember 1907 als Unterprimaner verließ. Ostern 1908 wurde ich an der Universität Kiel als Student der Naturwissenschaften immatrikuliert, bestand dann im Herbst 1909 die Reifeprüfung als Extraneer am Königlichen Realgymnasium zu Flensburg, studierte im Sommersemester 1910 in Berlin und seit dem Herbst 1910 wieder in Kiel. In meinen späteren Semestern habe ich mich besonders der Mathematik zugewandt.

Meine akademischen Lehrer waren die Herren Prof. Prof. Dr. Dr. *Baumgarten, Bernhard, Biltz, v. Brockdorff, Daenell, Dehn, Deussen, Dieterici, Erdmann, Frobenius, Harzer, Henning, Hettner, Hoffmann, Knoblauch, Kobold, Landsberg, Lasson, Lehmann-Filhés, Martius, Neuendorff, Pochhammer, Reinke, Riehl, Rügheimer, Tönnies, Weber, Wolff, Wülfing, Zahn.*

In ganz besonderem Maße gilt mein Dank Herrn Prof. Dr. *Landsberg,* dessen Unterricht ich lange genossen und von dem ich auch die Anregung zu der vorliegenden Arbeit erhalten habe. Während meiner ganzen Studienzeit habe ich bei ihm fördernde Hilfe und persönliches Wohlwollen gefunden. Sein im September dieses Jahres erfolgter Tod hat in seinem Kieler Schülerkreise wie in der wissenschaftlichen Welt überhaupt tiefe Trauer ausgelöst.

Sonderabdruck aus Mathematische Annalen. LXXVIII. Band. Heft 3/4.
Verlag und Druck von B. G. Teubner in Leipzig.

Die Isomorphismen der allgemeinen, unendlichen Gruppe mit zwei Erzeugenden.

Von

J. Nielsen im Felde.

Für diejenigen unendlichen, diskontinuierlichen Gruppen, welche definiert werden durch eine endliche Anzahl von erzeugenden Operationen, zwischen denen keine Relation besteht, sind die ersten beiden der von M. Dehn gekennzeichneten „Fundamentalprobleme"*), nämlich das Identitäts- und das Transformationsproblem trivial. Das dritte, das Isomorphieproblem hat bei ihnen die Bedeutung: Zu entscheiden, ob die Zuordnung der n Erzeugenden zu n vorgegebenen Elementen der Gruppe eine isomorphe Zuordnung ist oder nicht.

Im folgenden wird dieses Problem für die allgemeine, unendliche Gruppe $G(a, b)$ mit zwei Erzeugenden a und b gelöst, einige Eigenschaften der isomorphen Zuordnungen entwickelt und die Anwendung gemacht auf ein topologisches Problem, die Abbildungsgruppe der einfach berandeten Fläche vom Geschlecht 1.

I.

Zwei Elemente α und β der Gruppe $G(a, b)$ heißen zusammengehörige, primitive Elemente, wenn aus ihnen die gesamte Gruppe erzeugt werden kann. Das ist der Fall, wenn die ursprünglichen Erzeugenden a und b durch α und β ausgedrückt werden können. Die Substitution $S = (\alpha, \beta)$, bei welcher a durch α und b durch β ersetzt wird, definiert allgemein einen Isomorphismus von $G(a, b)$. Wir suchen zunächst eine Methode, um in einer endlichen Anzahl von Schritten zu entscheiden, ob eine vorgegebene Substitution S einen Isomorphismus definiert oder nicht. α und β sind allgemein von der Form

*) „Über unendliche diskontinuierliche Gruppen". Math. Ann. 71.

$$(1)\qquad \begin{cases} \alpha = a^{m_1}\cdot b^{n_1}\ldots a^{m_p}b^{n_p} = \prod\limits_{i=1}^{i=p}(a^{m_i}b^{n_i}),\\ \beta = a^{m_1'}\cdot b^{n_1'}\ldots a^{m_q'}b^{n_q'} = \prod\limits_{i=1}^{i=q}(a^{m_i'}b^{n_i'}),\end{cases}$$

wobei alle Exponenten und die Indizes p und q als endlich vorausgesetzt sind, und sind zusammengehörige primitive Elemente, wenn für ebenfalls endliche Exponenten und Indizes zwei Gleichungen bestehen:

$$(2)\qquad \begin{cases} a = \alpha^{\mu_1}\beta^{\nu_1}\ldots \alpha^{\mu_\pi}\beta^{\nu_\pi} = \prod\limits_{i=1}^{i=\pi}(\alpha^{\mu_i}\beta^{\nu_i}),\\ b = \alpha^{\mu_1'}\beta^{\nu_1'}\ldots \alpha^{\mu_\psi'}\beta^{\nu_\psi'} = \prod\limits_{i=1}^{i=\psi}(\alpha^{\mu_i'}\beta^{\nu_i'}).\end{cases}$$

Daraus ergeben sich die notwendigen, aber nicht hinreichenden Bedingungsgleichungen:

$$(3)\quad \Sigma\mu\cdot\Sigma m + \Sigma\nu\cdot\Sigma m' = 1,\qquad \Sigma\mu'\cdot\Sigma m + \Sigma\nu'\cdot\Sigma m' = 0,\quad (5)$$

$$(4)\quad \Sigma\mu\cdot\Sigma n + \Sigma\nu\cdot\Sigma n' = 0,\qquad \Sigma\mu'\cdot\Sigma n + \Sigma\nu'\cdot\Sigma n' = 1.\quad (6)$$

Es muß mindestens ein Quadrupel endlicher Werte $\Sigma\mu$, $\Sigma\nu$, $\Sigma\mu'$, $\Sigma\nu'$ geben, das diesen vier Gleichungen genügt. Daraus folgt:

Wegen (3) ist nicht gleichzeitig Σm und $\Sigma m'$ gleich 0,
„ (6) „ „ „ Σn „ $\Sigma n'$ „ 0,

und daraus weiter:

Wegen (3) und (4) „ „ „ Σm „ Σn „ 0,
„ (5) „ (6) „ „ „ $\Sigma m'$ „ $\Sigma n'$ „ 0.

Die Determinante

$$\Delta = \begin{vmatrix} \Sigma m & \Sigma n\\ \Sigma m' & \Sigma n'\end{vmatrix}$$

kann also nur verschwinden, indem alle Glieder von 0 verschieden und die Kolonnen einander proportional sind. Dann ergibt sich aus (3) und (4) ein Widerspruch, ebenso aus (5) und (6). Also ist $\Delta \neq 0$ und

$$(7)\qquad \begin{cases} \Sigma\mu = \frac{\Sigma n'}{\Delta}, & \Sigma\nu = \frac{-\Sigma n}{\Delta},\\ \Sigma\mu' = \frac{-\Sigma m'}{\Delta}, & \Sigma\nu' = \frac{\Sigma m}{\Delta}.\end{cases}$$

Da die links stehenden Summen ganzzahlig werden sollen, müssen die rechts stehenden Zähler alle den Faktor Δ enthalten. Also ist Δ selber durch Δ^2 teilbar und es folgt: $\Delta = \pm 1$.

Im folgenden ist die Anzahl der Erzeugenden eines Aggregats A mit $N(\mathsf{A})$ bezeichnet. Es ist dabei vorausgesetzt, daß in A wie auch in den vorgegebenen, durch (1) definierten, zusammengehörigen primitiven Elementen α und β kein Bestandteil der Form BB^{-1} vorkommt. (α, β) läßt dann folgende Eigenschaften erkennen:

1. Auch die Elemente $\begin{cases} \bar{\alpha} = \mathsf{A}\alpha\mathsf{A}^{-1}, \\ \bar{\beta} = \mathsf{A}\beta\mathsf{A}^{-1} \end{cases}$

bilden ein primitives Paar. Denn bilden wir für sie die den Auflösungsgleichungen (2) entsprechenden Ausdrücke, so ergibt sich:

$$\prod_{i=1}^{i=\pi} (\bar{\alpha}^{\mu_i} \bar{\beta}^{\nu_i}) = \mathsf{A}\,a\,\mathsf{A}^{-1}, \qquad \prod_{i=1}^{i=\psi} (\bar{\alpha}^{\mu_i'} \bar{\beta}^{\nu_i'}) = \mathsf{A}\,b\,\mathsf{A}^{-1}.$$

Besteht nun das Aggregat A nur aus einer Erzeugenden, z. B. a, so ist der erste Produktausdruck unmittelbar $= a$, und wir erhalten weiter:

$$\left[\prod_{i=1}^{i=\pi} (\bar{\alpha}^{\mu_i} \bar{\beta}^{\nu_i})\right]^{-1} \prod_{i=1}^{i=\psi} (\bar{\alpha}^{\mu_i'} \bar{\beta}^{\nu_i'}) \prod_{i=1}^{i=\pi} (\bar{\alpha}^{\mu_i} \bar{\beta}^{\nu_i}) = b.$$

Da jede Transformation durch eine Reihe von Transformationen mit je einer Erzeugenden ersetzt werden kann, so gilt der Satz allgemein für jede Form des Aggregats A.

2. Hat ein Element des Paares, z. B. α, die Eigenschaft, daß α und α^{-1} mit der gleichen Erzeugenden beginnen, so muß, damit überhaupt in (2) die Zahl der Erzeugenden sich durch Absorption verkleinern kann, β *oder* β^{-1} mit derselben Erzeugenden anfangen. Also ist das Paar von der Form:

$$\begin{cases} \alpha = c\mathsf{A}c^{-1} \\ \beta = c\mathsf{B} \end{cases} \quad \text{oder} \quad \begin{cases} \alpha = c\mathsf{A}c^{-1} \\ \beta = \mathsf{B}c^{-1}. \end{cases}$$

Dann wird durch Transformation mit c die Zahl $N(\alpha)$ erniedrigt und die Zahl $N(\beta)$ nicht erhöht. Ist auf diese Weise durch Transformation erreicht, daß weder α noch β von der Form $c\mathsf{A}c^{-1}$ sind, so soll der so erhaltene Ausdruck von (α, β) eine „reduzierte Form“ unter allen ineinander transformierbaren Formen des Paares heißen. Im folgenden ist (α, β) reduziert angenommen.

3. Ist $N(\alpha) = N(\beta) = 1$, so kann das Paar (α, β) offenbar nur eine der acht Formen haben:

$$(a, b), \quad (a, b^{-1}), \quad (a^{-1}, b), \quad (a^{-1}, b^{-1}),$$
$$(b, a), \quad (b, a^{-1}), \quad (b^{-1}, a), \quad (b^{-1}, a^{-1}).$$

4. Ist, was im folgenden vorausgesetzt bleibt, nicht gleichzeitig $N(\alpha)$ und $N(\beta) = 1$, so muß (2) ausgeführt von der Form sein

$$\begin{cases} a = \mathsf{\Gamma}_1 a \mathsf{\Gamma}_2, \\ b = \mathsf{\Gamma}_3 b \mathsf{\Gamma}_4. \end{cases}$$

Hierbei sind die einzelnen $\mathsf{\Gamma}_i = 1$, und mindestens eins von ihnen umfaßt mehr als 0 Erzeugende. Jedes $\mathsf{\Gamma}_i$, für welches $N(\mathsf{\Gamma}_i) > 0$ ist, muß mehr als ein vollständiges Element α oder β enthalten. Denn es ist nicht ein Element oder ein Teil eines solchen für sich $= 1$. In jedem solchen Aggregat $\mathsf{\Gamma}$ gibt es mindestens ein Element, das von seinen Nachbarelementen (dem vorausgehenden und dem nachfolgenden oder beiden zugleich, aber nicht mehr als diesen beiden) absorbiert wird, ohne daß an anderer Stelle im Aggregat $\mathsf{\Gamma}$ Absorption stattzufinden braucht. Denn bliebe nach dem Auslöschen von Bestandteilen $\mathsf{A}\mathsf{A}^{-1}$ an allen Stellen im Aggregat $\mathsf{\Gamma}$, wo sich zwei Elemente berühren, in jedem Element ein Rest nach, so bestände $\mathsf{\Gamma}$ aus lauter solchen Resten, zwischen denen keine Absorption mehr stattfindet, wäre also nicht gleich 1.

5. Das in einem Aggregat $\mathsf{\Gamma}$ von seinen beiden Nachbarelementen absorbierte Element sei ein α.

Fall I: Dies α stehe zwischen einem α und einem β. Da Bestandteile $\alpha\alpha^{-1}$ in (2) also auch in $\mathsf{\Gamma}$ nicht vorkommen und α reduziert ist, so muß α von dem benachbarten β-Element allein absorbiert werden. Also ist α oder α^{-1} ein Bestandteil am Anfang oder Ende von β.

Fall II: Dies α stehe zwischen zwei β-Elementen. Es habe die Form $\alpha = \mathsf{A}_1 \mathsf{A}_2$, und A_1 werde durch das vorhergehende, A_2 durch das nachfolgende β ausgelöscht. Dann bestehen vier Möglichkeiten: Die Reihenfolge der Elemente ist $\cdots \beta\alpha\beta \cdots$ oder $\cdots \beta^{-1}\alpha\beta \cdots$ oder $\cdots \beta\alpha\beta^{-1} \cdots$ oder $\cdots \beta^{-1}\alpha\beta^{-1} \cdots$. Im zweiten und dritten Fall muß α, da es reduziert ist, von einem der Nachbarelemente allein ausgelöscht werden, fällt also unter Fall I. Im ersten (und analog im vierten) Fall müßte β von der Form sein, einmal

$$\beta = \mathsf{B}_1 \mathsf{A}_1^{-1},$$

gleichzeitig aber

$$\beta = \mathsf{A}_2^{-1} \mathsf{B}_2,$$

so daß die Erzeugenden von B_1 der Reihe nach mit denen von A_2^{-1} übereinstimmen. Transformieren wir nun das Paar (α, β) mit A_1, wodurch an seinem reduzierten Charakter nichts geändert wird, so wird

$$\bar{\alpha} = \mathsf{A}_1^{-1} \alpha \mathsf{A}_1 = \mathsf{A}_2 \mathsf{A}_1,$$

$$\bar{\beta} = \mathsf{A}_1^{-1} \beta \mathsf{A}_1 = \mathsf{A}_1^{-1} \mathsf{B}_1.$$

Ist nun $N(\alpha) > N(\beta)$, so ist $N(\mathsf{A}_2) > N(\mathsf{B}_1)$. Dann ist $\bar{\beta}^{-1}$ Endbestandteil von $\bar{\alpha}$.

Ist $N(\alpha) < N(\beta)$, so ist $N(\mathsf{A}_2) < N(\mathsf{B}_1)$. Dann ist $\bar{\alpha}^{-1}$ Anfangsbestandteil von $\bar{\beta}$.

Ist endlich $N(\alpha) = N(\beta)$, also $N(\mathsf{A}_2) = N(\mathsf{B}_1)$, so müßte sein $\bar{\alpha} = \bar{\beta}^{-1}$. Dieser Fall ist also unmöglich bei der Voraussetzung, daß nicht gleichzeitig $N(\alpha)$ und $N(\beta) = 1$ sind.

Zusammenfassend haben wir also den Satz:

In jedem primitiven Paar (α, β) ist das Element mit kleinerer Zahl von Erzeugenden, mit dem Exponenten $+1$ oder -1 versehen, Anfangs- oder Endbestandteil des Elements mit der größeren Zahl von Erzeugenden, oder dies ist durch Transformation erreichbar. Haben beide Elemente die gleiche Zahl von Erzeugenden, so besteht jedes aus nur einer Erzeugenden.

6. Es sei $N(\alpha) < N(\beta)$. Dann ist also — eventuell nach einer reduzierten Transformation des Paares — $\beta = \alpha\beta'$ oder $\beta'\alpha$ oder $\alpha^{-1}\beta'$ oder $\beta'\alpha^{-1}$. Es sei zur Fixierung des Ausdrucks z. B. die erste Form angenommen. Nach (2) ist

$$a = \prod_{i=1}^{i=\pi} (\alpha^{\mu_i} \beta^{\nu_i}) = \prod_{i=1}^{i=\pi} (\alpha^{\mu_i} [\alpha\beta']^{\nu_i}) = \prod_{i=1}^{i=\bar{\pi}} (\alpha^{\bar{\mu}_i} \beta'^{\bar{\nu}_i})$$

und entsprechend für b. Also ist auch (α, β') ein primitives Paar. β' muß in reduzierter Form erhalten werden, denn wäre $\beta' = d\mathsf{B}d^{-1}$, wo d aus einer Erzeugenden besteht, so müßte $\alpha = d\mathsf{A}$ oder $\mathsf{A}d^{-1}$ sein. Im ersten Fall wäre β nicht reduziert, im zweiten Fall käme $d^{-1}d$ in β vor.

Ist nun noch $N(\beta') > N(\alpha)$, so ist — eventuell nach einer reduzierten Transformation — $\bar{\alpha}^{\pm 1}$ Anfangs- oder Endbestandteil von $\bar{\beta}'$.

Durch Abspalten von Bestandteilen α kommt man so schließlich zu einem Residuum γ von β, für das $N(\gamma) < N(\alpha)$ ist. Nun kann man in gleicher Weise immer nötigenfalls nach reduzierten Transformationen Anfangs- oder Endbestandteile $\gamma^{\pm 1}$ von α abspalten und fährt analog fort, bis nach einer endlichen Anzahl von Schritten ein Residuum von nur einer Erzeugenden als Element eines neuen Paares S_2 sich ergibt. Indem man nunmehr ohne weitere Transformationen dieses, so oft es geht, von dem anderen Element des Paares abspaltet, bringt man auch dieses auf ein Residuum von nur einer Erzeugenden und hat damit ein letztes Grundpaar S_1 in einer der unter 3. aufgeführten Grundformen erhalten.

7. (ξ, η) sei ein reduziertes Paar, für welches $1 < N(\xi) < N(\eta) < 2N(\xi)$ ist und das im besonderen die Eigenschaft hat, daß $\xi^{\pm 1}$ Anfangs- oder Endbestandteil von η ist und nach jeder reduzierten Transformation das transformierte $\bar{\xi}^{\pm 1}$ immer Anfangs- oder Endbestandteil des gleichzeitig transformierten $\bar{\eta}$ bleibt. Diese letztere Eigenschaft eines Paares soll kurz „Vollkommenheit" heißen.

Nur zur Fixierung des Ausdrucks sei wieder die bestimmte Form angenommen: $(\xi = \xi, \eta = \xi\eta')$. Dann erhält man durch Transformation mit ξ:

$$\begin{cases} \bar{\xi} = \xi, \\ \bar{\eta} = \eta'\xi. \end{cases}$$

Denkt man sich diese Transformation durch eine Folge von reduzierten, „rechts vertauschenden" (d. h. Aggregate von vorne nach hinten versetzenden) Einzeltransformationen mit je einer der aufeinanderfolgenden Erzeugenden von ξ ersetzt, so muß es darunter eine bestimmte Transformation T mit der Erzeugenden e geben, so daß *vor* T das Element $\bar{\bar{\xi}}$ Anfangsbestandteil von $\bar{\bar{\eta}}$, und *nach* T das Element $\bar{\bar{\bar{\xi}}}$ Endbestandteil von $\bar{\bar{\bar{\eta}}}$ ist. Also gibt es eine Darstellung:

$$\xi = \xi' e \xi'', \qquad \eta = \xi' e \xi'' \eta',$$
$$\bar{\bar{\xi}} = e \xi'' \xi', \qquad \bar{\bar{\eta}} = e \xi'' \eta' \xi',$$
$$\bar{\bar{\bar{\xi}}} = \xi'' \xi' e, \qquad \bar{\bar{\bar{\eta}}} = \xi'' \eta' \xi' e.$$

Es müssen danach die Erzeugenden von η' der Reihe nach mit denen von ξ', und diejenigen von η'^{-1} der Reihe nach mit denen von ξ''^{-1} übereinstimmen.

Das reduzierte Paar (sign μ = sign ν):

$$\begin{cases} \xi_1 = \xi, \\ \eta_1 = \xi^{\mu} \eta \xi^{\nu}, \end{cases}$$

wobei $|\mu + \nu| > 0$ sein darf, hat offenbar auch die Eigenschaft der Vollkommenheit. Es gilt dann der Hilfssatz: Dasselbe gilt weiter auch von dem reduzierten Paar (sign μ' = sign ν'):

$$\begin{cases} \xi_2 = \eta_1^{\mu'} \xi_1 \eta_1^{\nu'}, \\ \eta_2 = \eta_1. \end{cases}$$

Für $|\mu' + \nu'| > 1$ ist das unmittelbar ersichtlich, ist aber z. B. $\mu' = 1$, $\nu' = 0$, so ist also

$$\xi_2 = \eta_1 \xi_1 = \xi^{\mu} \eta \xi^{\nu+1} = (\xi' e \xi'')^{\mu+1} \eta' (\xi' e \xi'')^{\nu+1},$$
$$\eta_2 = \eta_1 \quad = \xi^{\mu} \eta \xi^{\nu} \quad = (\xi' e \xi'')^{\mu+1} \eta' (\xi' e \xi'')^{\nu}.$$

Transformieren wir rechts vertauschend mit $(\xi' e \xi'')^{\mu+1}$, so wird

$$\bar{\xi}_2 = \eta' (\xi' e \xi'')^{\mu+\nu+2},$$
$$\bar{\eta}_2 = \eta' (\xi' e \xi'')^{\mu+\nu+1}$$

$\bar{\eta}_2$ steht am Anfang von $\bar{\xi}_2$. Das bleibt bei allen reduzierten links vertauschenden Transformationen erhalten. Nach den Eigenschaften von T bleibt es auch bei weiteren $N(\xi')$ rechts vertauschenden Transformationen erhalten, falls $N(\xi') \leqq N(\eta')$. Schon nach $N(\eta') - N(\xi'')$ von diesen ist aber das erzielte $\overline{\overline{\eta}}_2$ Endbestandteil von $\overline{\overline{\xi}}_2$ und bleibt das bei allen weiteren rechts vertauschenden Transformationen. Da nun nach Voraussetzung $N(\xi') + N(\xi'') \geqq N(\eta')$ ist, so ist damit die Vollkommenheit für das Paar (ξ_2, η_2) erwiesen. Das bleibt richtig, falls $N(\xi) = 1$, $N(\eta) = 2$ ist. Dann ist $N(\eta') = 1$ und $N(\xi') = N(\xi'') = 0$. Die Richtigkeit des Hilfssatzes ist in diesem Fall am Paar (ξ_2, η_2) unmittelbar abzulesen.

Das Paar S_2, auf das in 6. das ursprüngliche primitive Paar (α, β) schrittweise zurückgeführt war, hat nun offenbar die Eigenschaft der Vollkommenheit. Also besteht sie nach dem Hilfssatz auch für alle Paare der Entwicklung rückwärts und schließlich für (α, β) selbst.

Der beschränkende Zusatz im Ergebnis von 5. lautend: „oder dies ist durch Transformation erreichbar" wird hierdurch aufgehoben.

8. Wir können nach dem Bisherigen den Aufbau jedes reduzierten, primitiven Paares durch zwei Rekursionsformeln charakterisieren:

$$(8) \qquad S_{2\varrho}: \begin{cases} \alpha_{2\varrho} = \alpha_{2\varrho-1}, \\ \beta_{2\varrho} = \alpha_{2\varrho-1}^{\varepsilon_{2\varrho-1}} \beta_{2\varrho-1} \alpha_{2\varrho-1}^{\varepsilon'_{2\varrho-1}}, \end{cases}$$

$$(9) \qquad S_{2\varrho+1}: \begin{cases} \alpha_{2\varrho+1} = \beta_{2\varrho}^{\varepsilon_{2\varrho}} \alpha_{2\varrho} \beta_{2\varrho}^{\varepsilon'_{2\varrho}}, \\ \beta_{2\varrho+1} = \beta_{2\varrho} \end{cases}$$

mit den Bedingungen: S_1 hat eine der in 3. angegebenen Grundformen; ε_i und ε_i' haben das gleiche Vorzeichen, sonst wäre α_{i+1} bzw. β_{i+1} nicht reduziert; ε_i und ε_{i-1} haben das gleiche Vorzeichen, sonst gäbe es in α_{i+1} bzw. β_{i+1} einen Bestandteil von der Form $\Gamma\Gamma^{-1}$. Also haben alle ε das gleiche Vorzeichen, und das gleiche gilt von den durch $\mathsf{E}_i = \varepsilon_i + \varepsilon_i'$ definierten Größen E. Diese notwendigen Bedingungen sind zugleich hinreichend. Denn bildet $S_{2\varrho}$ ein primitives Paar, so schreiben wir die Doppelgleichung (9):

$$\begin{cases} \alpha_{2\varrho} = \beta_{2\varrho+1}^{-\varepsilon_{2\varrho}} \alpha_{2\varrho+1} \beta_{2\varrho+1}^{-\varepsilon'_{2\varrho}}, \\ \beta_{2\varrho} = \beta_{2\varrho+1} \end{cases}$$

und folgern, daß auch $S_{2\varrho+1}$ ein primitives Paar ist. Analog ist $S_{2\varrho}$ ein primitives Paar, wenn es $S_{2\varrho-1}$ ist. Und für das Paar S_1 ist es vorausgesetzt.

Das zu Anfang gestellte Problem ist damit vollständig gelöst.

9. Bilden wir für die Rekursionsformeln das charakteristische Zahlenquadrupel wie am Anfang von I, so erscheint es ebenfalls in rekursiver Form:

$$\text{für } S_{2\varrho}: \quad \begin{Bmatrix} \Sigma m_{2\varrho-1} & \Sigma n_{2\varrho-1} \\ \Sigma m_{2\varrho-1}\cdot \mathsf{E}_{2\varrho-1} + \Sigma m'_{2\varrho-1} & \Sigma n_{2\varrho-1}\cdot \mathsf{E}_{2\varrho-1} + \Sigma n'_{2\varrho-1} \end{Bmatrix}$$

$$\text{und für } S_{2\varrho+1}: \quad \begin{Bmatrix} \Sigma m'_{2\varrho}\cdot \mathsf{E}_{2\varrho} + \Sigma m_{2\varrho} & \Sigma n'_{2\varrho}\cdot \mathsf{E}_{2\varrho} + \Sigma n_{2\varrho} \\ \Sigma m'_{2\varrho} & \Sigma n'_{2\varrho} \end{Bmatrix}.$$

Angenommen, es sei:
$$|\Sigma m_{2\varrho}| + |\Sigma n_{2\varrho}| = N(\alpha_{2\varrho}),$$
$$|\Sigma m'_{2\varrho}| + |\Sigma n'_{2\varrho}| = N(\beta_{2\varrho}),$$
so müssen in $\Sigma m'_{2\varrho}$ und ebenso in $\Sigma n'_{2\varrho}$ beide Summanden das gleiche Vorzeichen haben. Entwickelt man also
$$\begin{aligned}\Sigma m_{2\varrho+1} &= \Sigma m'_{2\varrho} \cdot \mathsf{E}_{2\varrho} + \Sigma m_{2\varrho} \\ &= \Sigma m_{2\varrho-1} \cdot \mathsf{E}_{2\varrho-1} \cdot \mathsf{E}_{2\varrho} + \Sigma m'_{2\varrho-1} \cdot \mathsf{E}_{2\varrho} + \Sigma m_{2\varrho-1}\end{aligned}$$
und berücksichtigt, daß $\mathsf{E}_{2\varrho-1} \cdot \mathsf{E}_{2\varrho}$ positiv ist, so folgt, daß diese drei Summanden das gleiche Vorzeichen haben. Also kann man entwickeln:
$$\begin{aligned}|\Sigma m_{2\varrho+1}| + |\Sigma n_{2\varrho+1}| &= |\mathsf{E}_{2\varrho}| \cdot (|\Sigma m'_{2\varrho}| + |\Sigma n'_{2\varrho}|) + |\Sigma m_{2\varrho}| + |\Sigma n_{2\varrho}| \\ &= |\mathsf{E}_{2\varrho}| \cdot N(\beta_{2\varrho}) + N(\alpha_{2\varrho}) = N(\alpha_{2\varrho+1}).\end{aligned}$$
Ferner ist
$$|\Sigma m'_{2\varrho+1}| + |\Sigma n'_{2\varrho+1}| = |\Sigma m'_{2\varrho}| + |\Sigma n'_{2\varrho}| = N(\beta_{2\varrho}) = N(\beta_{2\varrho+1}).$$
Also hat auch $S_{2\varrho+1}$ die von $S_{2\varrho}$ vorausgesetzte Eigenschaft. Analog läßt sie sich von $S_{2\varrho}$ erweisen, wenn sie von $S_{2\varrho-1}$ gilt. Nun gilt sie sicher von S_1, also gilt allgemein:
$$\begin{cases}|\Sigma m| + |\Sigma n| = N(\alpha), \\ |\Sigma m'| + |\Sigma n'| = N(\beta),\end{cases}$$
in Worten: *In einem in reduzierter Form gegebenen, primitiven Paar haben alle in einem Element auftretenden Erzeugenden einer Art das gleiche Exponentenvorzeichen.*

10. Unterwirft man $S_{2\varrho+1}$ der Transformation mit $\beta_{2\varrho}^{\varepsilon'_{2\varrho}}$, so wird:
$$\begin{cases}\bar{\alpha}_{2\varrho+1} = \beta_{2\varrho}^{\mathsf{E}_{2\varrho}} \alpha_{2\varrho}, \\ \bar{\beta}_{2\varrho+1} = \beta_{2\varrho}.\end{cases}$$
Transformiert man dies Paar weiter mit $\alpha_{2\varrho-1}^{\varepsilon'_{2\varrho-1}}$, so bleibt $\alpha_{2\varrho}$ ungeändert, und es wird
$$\bar{\bar{\beta}}_{2\varrho} = \alpha_{2\varrho-1}^{\mathsf{E}_{2\varrho-1}} \beta_{2\varrho-1}.$$
Durch Fortsetzung dieses Verfahrens wird nach 2ϱ Schritten für alle i von 1 bis 2ϱ
$$\varepsilon_i = \mathsf{E}_i, \quad \varepsilon_i' = 0.$$
Die so erhaltene, transformierte Form von $S_{2\varrho+1}$ möge Normalform heißen. Sie ist vollkommen charakterisiert durch Angabe von S_1 und der charakteristischen Reihe der E_i. Die Zahlen Σm, Σn, $\Sigma m'$ und $\Sigma n'$ werden durch die Transformationen nicht geändert. Nun führt aber*) jedes Zahlenquadrupel mit der Determinante ± 1 in eindeutiger Weise zu einem Teileralgorithmus, definiert also eindeutig eine charakteristische Reihe E_i

*) Vgl. z. B. meine Schrift „Kurvennetze auf Flächen". In. Diss., Kiel 1913, S. 47—49.

und ein aus Gliedern 0, $+1$ und -1 bestehendes Grundquadrupel. Also folgt der Satz:

Alle primitiven Paare von $G(a, b)$, die in den vier Zahlen Σm, Σn, $\Sigma m'$, $\Sigma n'$ übereinstimmen, sind ineinander transformierbar.

II.

Das unter I. erzielte Kriterium läßt sich durch das folgende einfache ersetzen:

*Zwei Elemente α und β von $G(a, b)$ bilden dann und nur dann ein Paar zusammengehöriger primitiver Elemente, wenn die Gleichung erfüllt ist:**)

$$\alpha\beta\alpha^{-1}\beta^{-1} = \Gamma(aba^{-1}b^{-1})^{\pm 1}\Gamma^{-1}. \tag{11}$$

1. Die Bedingung (11) ist notwendig.

Erfüllt ein Paar (α, β) die Bedingung (11), so erfüllt sie auch jedes aus (α, β) durch Transformation hervorgehende Paar. Für ein reduziertes Paar setzt man α_i und β_i aus (8) oder (9) in (11) ein und findet, daß S_i der Bedingung (11) genügt, wenn ihr S_{i-1} genügt. S_1 genügt aber der Bedingung, folglich auch S_i.

2. Die Bedingung (11) ist hinreichend.

(α, β) genüge der Bedingung, und jedes Element sei in den Erzeugenden so geschrieben, daß kein Bestandteil $\Gamma\Gamma^{-1}$ vorkommt. Hat z. B. α die Form $\alpha = c\alpha' c^{-1}$, wo c aus einer Erzeugenden besteht, so muß sich der als Zyklus geschriebene Ausdruck $\alpha\beta\alpha^{-1}\beta^{-1}$ (vgl. Fig. 1) im Sinne

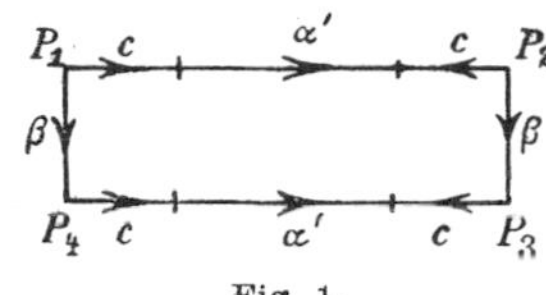

Fig. 1.

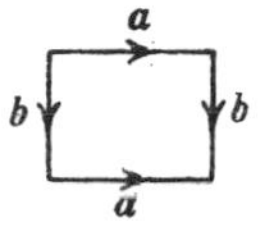

Fig. 2.

$P_1 P_2 P_3 P_4$ gelesen auf $(aba^{-1}b^{-1})^{\pm 1}$ (Fig. 2) reduzieren: Es muß also mindestens an einer Ecke P_i Absorption stattfinden, also β von der Form $c\beta'$ oder $\beta' c^{-1}$ sein. In beiden Fällen kann man durch Transformation mit c $N(\alpha)$ verkleinern, ohne $N(\beta)$ zu vergrößern. (α, β) sei also reduziert vorausgesetzt und da mit (α, β) auch (β, α) die Bedingung (11) erfüllt, können wir, ohne eine Einschränkung zu begehen, $N(\alpha) \leq N(\beta)$ annehmen. Es sei $\alpha = c\alpha' d$. α' besteht aus $0, 1, 2$ usw., c und d je aus 0 oder 1 Erzeugenden. Im Zyklus (Fig. 3) findet Absorption nicht gleichzeitig bei P_1 und P_2 statt, sonst würde $c = d^{-1}$ folgen und α wäre nicht reduziert, ebenso

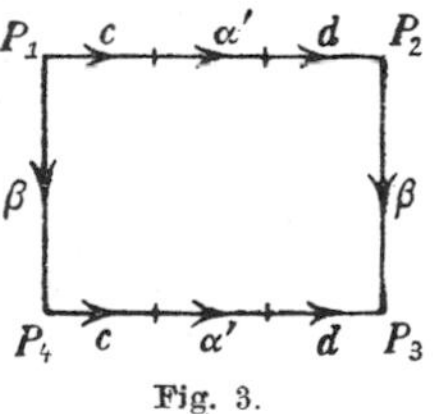

Fig. 3.

*) Dieser Satz ist mir von M. Dehn mitgeteilt worden. Sein Beweis ist von dem folgenden verschieden.

nicht gleichzeitig bei P_1 und P_4, sonst wäre β nicht reduziert. Reduktion findet nur an zwei gegenüberliegenden Ecken statt, z. B. bei P_1 und P_3.

Ist nun $N(\alpha) = N(\beta)$, so kann keine Rechtecksseite der Figur 3 ganz absorbiert werden, denn sonst wäre $\alpha = \beta$ und $\alpha\beta\alpha^{-1}\beta^{-1} = 1$. Findet überhaupt Reduktion statt, so muß nur an P_2 der Streckenzug dc und an P_4 im gleichen Umlaufssinn der Streckenzug $c^{-1}d^{-1}$ nachbleiben. Also ist $\alpha\beta\alpha^{-1}\beta^{-1} = 1$, was ausgeschlossen ist. Es kann also überhaupt keine Reduktion stattfinden, $N(\alpha) = N(\beta) = 1$. Man liest dann aus Fig. 2 ab, daß (α, β) eine der unter I, 3. angegebenen acht Grundformen hat, also ein primitives Paar ist.

Ist aber $N(\alpha) < N(\beta)$, so muß mindestens eine α-Seite ganz absorbiert werden; denn sonst blieben im Zyklus nach der Reduktion sechs oder mehr Erzeugende nach. Also ist $\alpha^{\pm 1}$ Anfangs- oder Endbestandteil von β, z. B. $\beta = \alpha\beta'$. Dann ist $\beta' = \alpha^{-1}\beta$ und

$$\alpha\beta'\alpha^{-1}\beta'^{-1} = \alpha\alpha^{-1}\beta\alpha^{-1}\beta^{-1}\alpha = \alpha^{-1}(\alpha\beta\alpha^{-1}\beta^{-1})\alpha = \Gamma(aba^{-1}b^{-1})^{\pm 1}\Gamma^{-1}.$$

Also erfüllt auch (α, β') die Gleichnng (11); analog für die anderen Spezialfälle. β' wird in reduzierter Form erhalten, denn sonst wäre genau wie unter I, 6. ein Widerspruch zu folgern.

Damit sind die Bedingungen für die gleiche Analyse des Paares (α, β) wie in I mit dem dort als primitiv vorausgesetzten Paar (α, β) gegeben. (α, β) gestattet also eine Auflösung nach den Rekursionsformeln (8) und (9) mit Exponenten ε von gleichem Vorzeichen und führt auf ein primitives Grundpaar S_1, ist also selbst primitiv.

III.

Ist (α, β) ein primitives Paar, so definiert die Substitution S, bei welcher a durch α und b durch β ersetzt wird, einen Isomorphismus von $G(a, b)$. S heiße kurz „isomorphe Substitution“.

Übt man auf eine isomorphe Substitution S_1 eine ebensolche S_2 aus, so entsteht wieder eine isomorphe Substitution S_3. Denn definiert man S_1 und S_2 durch Gleichungssysteme der Form (1), so existiert für jede ein Auflösungssystem (2). Entsteht also S_3, indem man S_2 auf S_1 ausübt, so kann man aus α_3, β_3 mit Hilfe des Auflösungssystems von S_1 die Elemente α_2, β_2 und weiter hieraus mit Hilfe des Auflösungssystems von S_2 die ursprünglichen Erzeugenden a und b ausdrücken.

Schreibt man diese Beziehung

$$S_3 = S_1 \cdot S_2,$$

so läßt sich das assoziative Gesetz:

$$(S_{i1} \cdot S_{i2}) \cdot S_{i3} = S_{i1} \cdot (S_{i2} \cdot S_{i3})$$

leicht nachweisen, indem man die Elemente der durch

$$S_l = (S_{i1} \cdot S_{i2}) \cdot S_{i3},$$
$$S_m = S_{i1} \cdot (S_{i2} \cdot S_{i3})$$

definierten Substitutionen mit Hilfe der Definitionsgleichungen der S_i ausführlich hinschreibt, wodurch dann die Identität von S_l und S_m augenscheinlich wird.

Mit S^{-1} sei eine isomorphe Substitution bezeichnet, welche auf S ausgeübt, diese in die identische Substitution (a, b) überführt. Es wird nach der Existenz und Auffindung von S^{-1} gefragt.

S habe die vier charakteristischen Zahlen Σm, Σn, $\Sigma m'$, $\Sigma n'$ und die aus ihnen gebildete Determinante $\Delta = \pm 1$. Dann sind durch die Gleichungen (7) vier weitere Zahlen definiert, für welche

$$\begin{vmatrix} \Sigma\mu & \Sigma\nu \\ \Sigma\mu' & \Sigma\nu' \end{vmatrix} = \frac{1}{\Delta^2} \begin{vmatrix} \Sigma n' & -\Sigma n \\ -\Sigma m' & \Sigma m \end{vmatrix} = \Delta = \pm 1.$$

Indem man auf dieses Zahlenquadrupel einen Teileralgorithmus anwendet, gelangt man zu einer Reihe von Faktoren $\mathsf{E}_{\sigma-1}$ bis E_1 und einem Anfangsquadrupel $\begin{Bmatrix} \mu_0 & \nu_0 \\ \mu_0' & \nu_0' \end{Bmatrix}$, dessen Glieder in der einen Diagonale 0 und in der andern $+1$ oder -1 sind. Bildet man dann S_1 durch

$$\alpha_1 = a^{\mu_0} b^{\nu_0},$$
$$\beta_1 = a^{\mu_0'} b^{\nu_0'}$$

so kann man weiter mit Hilfe der Rekursionsformeln (8) und (9) S_2 bis S_σ konstruieren. Dabei hat man bei jedem Schritt die Freiheit, E_i in zwei Summanden ε_i und ε_i' von gleichem Vorzeichen zu zerlegen. Übt man dann S_σ auf S aus, so resultiert eine isomorphe Substitution mit den charakteristischen Zahlen

$$\begin{Bmatrix} \Sigma\mu \cdot \Sigma m + \Sigma\mu' \cdot \Sigma n & \Sigma\nu \cdot \Sigma m + \Sigma\nu' \cdot \Sigma n \\ \Sigma\mu \cdot \Sigma m' + \Sigma\mu' \cdot \Sigma n' & \Sigma\nu \cdot \Sigma m' + \Sigma\nu' \cdot \Sigma n' \end{Bmatrix}$$

diese sind aber nach (7) gleich

$$\begin{Bmatrix} 1 & 0 \\ 0 & 1 \end{Bmatrix},$$

die Substitution $S \cdot S_\sigma$ ist also in die Identität transformierbar. Ist

$$S \cdot S_\sigma = (\Gamma a \Gamma^{-1}, \Gamma b \Gamma^{-1}), \quad \text{so ist} \quad S^{-1} = (\Gamma^{-1} \alpha_\sigma \Gamma, \Gamma^{-1} \beta_\sigma \Gamma),$$

also stets nach dieser Methode auffindbar.

Bildet man nun $S^{-1} \cdot S$, so ist das jedenfalls eine isomorphe Substitution, also existiert $(S^{-1} \cdot S)^{-1}$ so, daß

$$(S^{-1} \cdot S) \cdot (S^{-1} \cdot S)^{-1} = 1 \tag{12}$$

Übt man auf S^{-1} die identische Substitution $S \cdot S^{-1}$ aus:

$$S^{-1} \cdot (S \cdot S^{-1}) = S^{-1}$$

und hierauf weiter die Substitution $S \cdot (S^{-1} \cdot S)^{-1}$, so läßt sich das unter Anwendung des assoziativen Gesetzes schreiben:

$$(S^{-1} \cdot S) \cdot (S^{-1} \cdot S) \cdot (S^{-1} \cdot S)^{-1} = (S^{-1} \cdot S) \cdot (S^{-1} \cdot S)^{-1}$$

und mit Hilfe von Gleichung (12) folgt

$$S^{-1} \cdot S = 1.$$

S^{-1} ist also auch definierbar als diejenige Substitution, welche in die Identität übergeht, wenn S auf sie ausgeübt wird. Dies Ergebnis läßt sich formulieren:

Jede isomorphe Substitution ist Auflösungssubstitution zu ihrer Auflösungssubstitution

oder anders gesprochen:

Die Aufgabe, zu einer gegebenen isomorphen Substitution die reziproke zu finden, ist identisch mit der Aufgabe, aus einem gegebenen Paar zusammengehöriger, primitiver Elemente die ursprünglichen Erzeugenden a und b auszudrücken.

IV.

Mit R' sei eine einfach berandete Fläche vom Geschlecht 1 bezeichnet. Sind a und b auf R' zwei geschlossene, doppelpunktlose Kurven mit einem gemeinsamen Punkt, so ist die allgemeine, unendliche Gruppe $G(a, b) = G_{R'}$ die „Fundamentalgruppe von R'"*). R sei die geschlossene Fläche vom Geschlecht 1 (Ring), die entsteht, wenn man die Randkurve von R' durch ein Elementarflächenstück schließt. Die Fundamentalgruppe von R ist die Abelsche Gruppe $G_R = G_A(a, b)$ mit der Relation $a b a^{-1} b^{-1} = 1$. Nun besteht der folgende Hilfssatz:

Zwei geschlossene, doppelpunktlose Kurven, die auf R ineinander transformierbar sind, sind es auch auf R'.**)

Zwei stetige eineindeutige Abbildungen von R bzw. von R' sind ineinander transformierbar, wenn die beiden Bilder von a und gleichzeitig die beiden Bilder von b ineinander transformierbar sind. Jede stetige eineindeutige Abbildung von R' läßt sich zu einer solchen von R erweitern, indem man sie über das die Randkurve schließende Elementarflächenstück fortsetzt. Homotope Abbildungen von R' führen dabei zu homotopen Abbildungen von R. Andererseits können zwei nicht-homotope Abbildungen

*) Vgl. M. Dehn, „Über unendliche diskontinuierliche Gruppen". Math. Ann. 71 (1911).

**) Wegen Raummangels muß dieser Hifssatz hier ohne Beweis angegeben werden. Ich behalte mir vor, den Beweis später in anderem Zusammenhange mitzuteilen.

von R' auch nicht zu homotopen Abbildungen von R erweitert werden, denn zwei einander entsprechende Bildkurven, die auf R' nicht homotop sind, können es nach dem Hilfssatz auch auf R nicht sein. Es findet also eine umkehrbar eindeutige Zuordnung der topologisch unterscheidbaren Abbildungstypen von R und R' statt. Die (unendliche, diskontinuierliche) Abbildungsgruppe von R' ist also wie die von R durch einen zweistufigen Isomorphismus mit der erweiterten Modulgruppe verbunden.*)

Ein Unterschied zwischen den Abbildungen von R und denen von R' besteht erst bei der Frage nach den die Abbildungen definierenden isomorphen Substitutionen der Fundamentalgruppen. Die Abbildungen von R werden definiert durch die Substitutionen $\alpha = a^m b^n$, $\beta = a^{m'} b^{n'}$ der Abelschen Gruppe G_R, bei denen $\begin{vmatrix} m & n \\ m' & n' \end{vmatrix} = \pm 1$ ist. Diese bestimmen zugleich den Abbildungstypus von R'. Die zugehörigen Substitutionen aus der Gruppe $G_{R'}$ sind von der Form der Gleichungen (1) in I und können nach der Methode von I gefunden werden, da das charakteristische Zahlenquadrupel bekannt ist. Es ergibt sich also der Satz:

Die Gruppe der topologisch unterscheidbaren Abbildungstypen der Fläche R' vom Geschlecht 1 *mit einer Randkurve ist isomorph mit der Gruppe der Isomorphismen der allgemeinen, unendlichen, diskontinuierlichen Gruppe $G(a, b)$ mit zwei Erzeugenden, der Fundamentalgruppe von R'.*

Im besonderen entspricht einem Abbildungstypus von R' die invariante Untergruppe aller in eine bestimmte isomorphe Substitution transformierbaren Substitutionen von $G(a, b)$.

Das Erkennungsmerkmal für die Abbildung und die zugehörige Untergruppe von Substitutionen ist das zugehörige charakteristische Zahlenquadrupel.

Für die berandete Fläche vom Geschlecht 1 besteht also der gleiche Zusammenhang zwischen den Abbildungen der Fläche und den Isomorphismen der zugehörigen Fundamentalgruppe, wie ihn M. Dehn für *geschlossene* Flächen beliebigen Geschlechts gefunden hat.**)

Für berandete Flächen von höherem Geschlecht ist das nicht mehr der Fall.

Konstantinopel im Oktober 1917.

---

*) Vgl. meine Schrift „Kurvennetze auf Flächen", Inaug. Diss., Kiel 1913.

**) Nach Mitteilung von M. Dehn.

Sonderabdruck aus Mathematische Annalen. LXXIX Band. Heft 3.
Verlag und Druck von B. G. Teubner in Leipzig.

Über die Isomorphismen unendlicher Gruppen ohne Relation.

Von

J. Nielsen in Konstantinopel.

Eine endliche Anzahl n von erzeugenden Operationen $a_1, a_2, \cdots, a_n$, *zwischen denen keine Relationen bestehen*, definieren die allgemeine unendliche diskontinuierliche Gruppe G_n. Zwei Elemente dieser Gruppe sind nur dann identisch bzw. ineinander transformierbar, wenn die ihnen entsprechenden Ausdrücke in den a_i identisch sind, bzw. in zyklischer Anordnung geschrieben identisch werden. Wählt man n Elemente $\alpha_1, \alpha_2, \cdots, \alpha_n$ der Gruppe so aus, daß man jedes Element der Gruppe als Komposition aus ihnen darstellen kann, daß also hinsichtlich der Erzeugung der Gruppe das System der a_i durch das System der α_i ersetzt werden kann, so definiert ein solches zusammengehöriges System primitiver Elemente α_i eine isomorphe Beziehung von G_n auf sich selbst. Im folgenden wird eine Methode entwickelt, um in einer endlichen Anzahl von Schritten zu entscheiden, ob n vorgegebene Elemente aus G_n in dieser Weise ein isomorphes System bilden.

Die n Elemente $\alpha_1, \alpha_2, \cdots, \alpha_n$ des betrachteten Systems S aus G_n sind als Kompositionen der a_i gegeben, haben also die Form:

$$(1) \qquad \alpha_i = a_{k_1}^{e_1} a_{k_2}^{e_2} \cdots a_{k_l}^{e_l},$$

wobei die Indizes k dem System der Indizes 1 bis n entnommen sind und die Exponenten e positive oder negative ganze Zahlen bedeuten. Notwendige und hinreichende Bedingung dafür, daß man aus den α_i durch Komposition die ganze Gruppe aufbauen kann, ist offenbar, daß man aus ihnen die ursprünglichen Erzeugenden a_i ausdrücken kann. Besteht also jedes α nur aus einer Erzeugenden a, so müssen die α_i eine Permutation der a_i sein, wobei noch jedes a mit dem Exponenten $+1$ oder -1 versehen sein darf. Kommen hingegen unter den α Kompositionen aus den a nach (1) vor, so gibt es Erzeugende a, die Kompositionen aus den α sind:

$$(2) \qquad a_i = \alpha_{\varkappa_1}^{\varepsilon_1} \alpha_{\varkappa_2}^{\varepsilon_2} \cdots \alpha_{\varkappa_\lambda}^{\varepsilon_\lambda}.$$

Schreibt man hierin für die α ihre Ausdrücke nach (1), so nimmt (2) die Gestalt einer identisch erfüllten Gleichung

$$a_i = \Gamma_{i_1} a_i \Gamma_{i_2}$$

an, wobei also die Γ_i Ausdrücke in den a sind, die sich durch Absorption ganz aufzehren. Jedes solche Γ enthält daher notwendig mehr als ein vollständiges Element α, und es muß unter diesen α-Elementen von Γ mindestens eins geben, das von seinen beiden Nachbarelementen (dem vorausgehenden oder dem nachfolgenden oder beiden zugleich, aber nicht mehr als diesen beiden) ganz absorbiert wird; sonst könnte Γ sich nicht selber ganz aufzehren. Sei α_μ dies Element, so muß es also eine Kombination $\alpha_\nu \alpha_\mu$ oder $\alpha_\nu^{-1} \alpha_\mu$ oder $\alpha_\mu \alpha_\nu$ oder $\alpha_\mu \alpha_\nu^{-1}$ geben, die weniger oder höchstens gleichviel Erzeugende enthält wie α_ν. Setzt man in dem System S diese Kombination an Stelle von α_ν, also:

$$\alpha_i' = \alpha_i \qquad (\text{für } i \neq \nu)$$

$$\alpha_\nu' = \alpha_\nu \alpha_\mu \quad (\text{und analog}),$$

so ist S isomorph, wenn sich diese Eigenschaft von dem neu entstehenden System S' nachweisen läßt, und umgekehrt ist wegen

$$\alpha_i = \alpha_i' \qquad (\text{für } i \neq \nu)$$

$$\alpha_\nu = \alpha_\nu' \alpha_\mu'^{-1} \quad (\text{und analog})$$

das System S' isomorph, wenn dies mit S der Fall war. Das Verfahren läßt sich also mit S' fortsetzen, und man erhält in der so entstehenden Systemreihe schrittweise eine Verkleinerung der Erzeugendenzahl, so oft sich eine Kombination $\alpha_\mu^{\pm 1} \alpha_\nu^{\pm 1}$ finden läßt, die *weniger* Erzeugende enthält als der eine ihrer Bestandteile. Gibt es bei einem bestimmten System S keine solche Kombination, so muß es notwendig eine Kombination $\alpha_\mu^{\pm 1} \alpha_\nu \alpha_\varrho^{\pm 1}$ derart geben, daß darin α_ν je zur Hälfte von den Nachbarelementen absorbiert wird. Dann ordne man die Elemente von S derart in eine Reihe $\alpha_1, \alpha_2, \cdots, \alpha_n$, daß kein Element *mehr* Erzeugende umfaßt, als ein späteres der Reihe. Sei nun etwa α_i das erste Element der Reihe, welches die Eigenschaft hat, zwischen zwei anderen je zur Hälfte absorbiert zu werden, so umfaßt α_i eine gerade Anzahl von Erzeugenden, und es sei

$$\alpha_i = A_i B_i$$

die Zerlegung in zwei Hälften. So oft dann ein anderes Element des Systems mit dem Bestandteil B_i endigt bzw. mit dem Bestandteil B_i^{-1} anfängt, bewirke man durch geeignete Kombination mit α_i, daß A_i^{-1} bzw. A_i an die Stelle tritt. Damit hat α_i die betrachtete Eigenschaft verloren, und keines der α_i vorausgehenden Elemente hat sie neu erworben, denn für diese kommen nur Anfangs- und Endbestandteile bis zur Erzeugendenzahl von A_i in Betracht, und solche sind keine neu entstanden, die nicht schon in α_i vertreten waren. Läßt sich nach dieser Umformung des Systems, die die Isomorphieeigenschaft und die Erzeugendenzahl nicht ändert, noch keine Kombination finden, die zu einer Verringerung der Erzeugendenzahl führt, so muß es ein weiteres Element α_k $(k > i)$ der Reihe geben, das zwischen zwei anderen je zur Hälfte absorbiert wird. Mit diesem verfahre man dann wie mit α_i und setze dies Verfahren fort, solange kein Element durch ein anderes *mehr* als zur Hälfte ausgelöscht wird. Da n endlich ist und der Index des zur Komposition verwendeten Elements dauernd wächst, ist dies Verfahren nicht unbegrenzt fortsetzbar. Es muß vielmehr, wenn S ein isomorphes System sein soll, der Fall einmal wieder auftreten, daß ein Element durch ein anderes mehr als zur Hälfte ausgelöscht wird. Also läßt sich durch dies Kompositionsverfahren die Erzeugendenzahl des Systems solange verkleinern, bis jedes der n Elemente nur noch aus einer Erzeugenden besteht. Man kann demnach die Lösung des gestellten Problems so formulieren:

Die n Elemente $\alpha_1, \alpha_2, \cdots, \alpha_n$ aus G_n bilden dann und nur dann ein zusammengehöriges System primitiver Elemente, wenn sich nach bestimmter, oben dargelegter Methode in einer endlichen Zahl von Schritten durch Substitution von Kombinationen $\alpha_\nu^{\pm 1}\alpha_\mu^{\pm 1}$ an Stelle von α_μ die Erzeugendenzahl jedes Elements auf 1 reduzieren läßt und das so entstehende System eine Permutation der mit den Exponenten $+1$ oder -1 versehenen Erzeugenden $a_1, \cdots, a_n$ darstellt.

Bildet man für ein gegebenes System S die Determinante n^{ten} Grades

$$\Delta = \begin{matrix} d_{11} & \cdots & d_{1n} \\ \cdot & \cdot & \cdot \\ d_{n1} & \cdots & d_{nn} \end{matrix},$$

in welcher d_{ik} die algebraische Summe aller Exponenten der Erzeugenden a_k im Ausdruck von α_i bedeutet, so erkennt man nach dem Vorhergehenden unmittelbar, daß Δ für ein *isomorphes* System, dessen Elemente aus je *einer* Erzeugenden bestehen, den Wert $+1$ oder -1 hat. Bei der Substitution von $\alpha_\nu^{\pm 1}\alpha_\mu^{\pm 1}$ an Stelle von α_μ wird die ν^{te} Zeile von Δ mit dem Faktor ± 1 zur μ^{ten} addiert; also ändert sich Δ nicht. Mithin folgt aus obigem Satze im besonderen als notwendige Bedingung für jedes zusammengehörige primitive System, daß $\Delta = \pm 1$ ist.

Bezeichnet man für ein n-zeiliges System die Operation, welche darin besteht, daß man die erste Zeile mit der i^{ten} vertauscht und alles übrige ungeändert läßt, mit $(1i)$, ferner die Operation, welche nur die erste Zeile mit -1 potenziert, mit A, endlich die Operation, welche an die erste Zeile die zweite anfügt und gleichzeitig die zweite mit -1 potenziert, mit B, so erkennt man, daß für das n-zeilige System der α_i die einzelnen auszuführenden Kombinationen wie folgt erhalten werden (in jedem Falle sollen sich die von α_μ verschiedenen Elemente nicht ändern):

$$
\begin{array}{lll}
\alpha_\mu \to \alpha_\mu^{-1} & \text{durch} & (1\mu)A(1\mu) \\
\alpha_\mu \to \alpha_\mu \alpha_\nu & \text{„} & (1\nu)(12)(1\mu)B(1\mu)(12)A(1\nu) \\
\alpha_\mu \to \alpha_\mu \alpha_\nu^{-1} & \text{„} & (1\nu)A(12)(1\mu)B(1\mu)(12)(1\nu) \\
\alpha_\mu \to \alpha_\nu \alpha_\mu & \text{„} & (1\nu)A(12)(1\mu)ABA(1\mu)(12)(1\nu) \\
\alpha_\mu \to \alpha_\nu^{-1} \alpha_\mu & \text{„} & (1\nu)(12)(1\mu)ABA(1\mu)(12)A(1\nu).
\end{array}
$$

Berücksichtigt man noch, daß die $n-1$ Operationen $(1\,i)$ ausreichen, um jede beliebige Permutation der α herzustellen, und daß jede der definierten Operationen, zweimal hintereinander ausgeführt, die Identität ergibt, so folgt der Satz:

Die Gruppe der Isomorphismen der unendlichen Gruppe mit n Erzeugenden ohne Relation läßt sich durch $n+1$ erzeugende Operationen aufbauen; diese lassen sich insbesondere so wählen, daß sie alle von der Ordnung 2 sind.

So lassen sich z. B. alle Isomorphismen der Gruppe $G(a, b, c, d,$ ohne Relation) erzeugen, indem man in geeigneter Reihenfolge die fünf isomorphen Substitutionen ausführt:

$$
\left.\begin{array}{l} a \to b \\ b \to a \\ c \to c \\ d \to d \end{array}\right\},\quad
\left.\begin{array}{l} a \to c \\ b \to b \\ c \to a \\ d \to d \end{array}\right\},\quad
\left.\begin{array}{l} a \to d \\ b \to b \\ c \to c \\ d \to a \end{array}\right\},\quad
\left.\begin{array}{l} a \to a^{-1} \\ b \to b \\ c \to c \\ d \to d \end{array}\right\},\quad
\left.\begin{array}{l} a \to ab \\ b \to b^{-1} \\ c \to c \\ d \to d \end{array}\right\}.
$$

Der Weg zur *Darstellung* eines gegebenen Isomorphismus auf diese Weise ist zugleich durch die obige Methode zur *Auflösung* desselben gegeben.

Konstantinopel, im Juli 1918.

Über fixpunktfreie topologische Abbildungen geschlossener Flächen.

Von

J. Nielsen in Hamburg.

Brouwer hat zu wiederholten Malen[1]) den Satz bewiesen, daß eine eineindeutige und stetige oder, wie Brouwer kurz sagt, „topologische“ Abbildung einer Kugelfläche auf sich selbst mit Erhaltung der Indikatrix mindestens einen Punkt fest läßt. In einer kürzlich publizierten Note[2]) weist Brouwer nach, daß diese Eigenschaft, bei allen topologischen Abbildungen mit Erhaltung der Indikatrix mindestens einen Fixpunkt zu ergeben, unter allen geschlossenen zweiseitigen Flächen allein denjenigen vom Typus der Kugel zukommt, während topologische Abbildungen ohne Fixpunkt mit Umkehrung der Indikatrix bei jeder geschlossenen zweiseitigen Fläche möglich sind.

Im Folgenden soll gezeigt werden, daß die von Brouwer in der letzteren Note angewandte Methode, wenn man statt mit aufgesetzten Henkeln mit aufgesetzten Möbiusschen Bändern operiert, für die einseitigen Flächen ein analoges Resultat in einheitlicher Weise liefert; aus diesem folgt sodann der Brouwersche Satz für die zweiseitigen Flächen, indem man dieselben zwei-eindeutig auf einseitige Flächen abbildet. Dieses Übertragungsprinzip verwendet auch Brouwer, um zu zeigen, daß bei jeder topologischen Abbildung der projektiven Ebene auf sich mindestens ein Punkt fest bleibt[3]).

[1]) Zuerst Amsterdamer Berichte **17** 2, S. 741ff. (holländ. Ausg.). Der einfachste Brouwersche Beweis wurde von Hadamard in der zweiten Auflage der „Introduction à la théorie des fonctions“ von Tannery mitgeteilt. Einen weiteren einfachen Beweis gibt neuerdings B. v. Kérékjártó in den Math. Ann. **80**, S. 30–32.

[2]) Sur les points invariants des transformations topologiques des surfaces. Comptes rendus **168**, S. 1842 (26. Mai 1919).

[3]) Amsterdamer Berichte **17** 2, S. 752.

§ 1.

Zur topologischen Definition einer Fläche dient die Charakteristik

$$K = 2p + k + r - 2.$$

Dabei bedeutet r die Anzahl der Randkurven, ferner k die Anzahl der einrandigen und p die Anzahl der zweirandigen Rückkehrschnitte, welche keinen Punkt untereinander oder mit den Randkurven gemeinsam haben, und nach deren Ausführung die Fläche auf einen schlichten Bereich der Ebene ausgebreitet werden kann (Maximalzahlen nicht zerstückelnder Schnitte.) Bei geschlossenen Flächen, wie sie im Folgenden allein betrachtet werden, ist $r = 0$, bei zweiseitigen $k = 0$ und also bei zweiseitigen geschlossenen K gerade. Bei einseitigen Flächen kann man je zwei einrandige durch einen zweirandigen Rückkehrschnitt ersetzen, so daß die Verteilung der Einheiten von K auf $2p$ und k nicht fest ist. Die Anzahlen r und K und die Eigenschaft der Zweiseitigkeit bzw. Einseitigkeit genügen zur topologischen Charakterisierung der Fläche.

Auf einer Kugelfläche F bedeute $-\frac{\pi}{2} \leqq \psi \leqq +\frac{\pi}{2}$ die geographische Breite und $\varphi \pmod{2\pi}$ die geographische Länge. F hat die Charakteristik $K(F) = -2$. $n \geqq 2$ sei eine ganze Zahl. Um den Punkt $\psi = 0$, $\varphi = 0$ schlage man auf F einen Kreis $\varkappa_0$ mit einem Radius, der einer Winkelöffnung $< \frac{\pi}{n}$ entspricht, und schneide das Innere desselben aus F aus. Das so entstehende Loch schließe man wieder, indem man den Randkreis mit der Randkurve einer gelochten Boyschen Fläche, also eines Möbiusschen Bandes B_0, Punkt für Punkt vereinigt. Die so entstehende geschlossene Fläche F_1 besitzt einen vom Möbiusschen Bande herrührenden einrandigen Rückkehrschnitt und also die Charakteristik $K(F_1) = -1$. Den gleichen Prozeß führe man an den Punkten $\psi = 0$, $\varphi = \frac{2\pi}{n}$, $2 \cdot \frac{2\pi}{n}, \ldots, (n-1) \cdot \frac{2\pi}{n}$ aus. Von den dabei verwendeten Kreisen $\varkappa_0, \varkappa_1, \ldots, \varkappa_{n-1}$ haben nach der Wahl des Radius nie zwei einen gemeinsamen Punkt. Die so entstehende Fläche F_n hat n einrandige Rückkehrschnitte, also $K(F_n) = n - 2$. Nun unterwerfe man zunächst den Teil von F_n, der aus Punkten der ursprünglichen Kugelfläche F besteht, der Abbildung $\psi' = -\psi$, $\varphi' = \varphi + \frac{2\pi}{n}$. Diese hat keinen Fixpunkt, und durch sie wird $\varkappa_0$ auf $\varkappa_1$, $\varkappa_1$ auf $\varkappa_2$ usw., endlich $\varkappa_{n-1}$ auf $\varkappa_0$ abgebildet. Diese Abbildung setze man stetig auf die schließenden Möbiusschen Bänder fort, indem man B_0 auf B_1, B_1 auf $B_2, \ldots, B_{n-1}$ auf B_0 umkehrbar eindeutig abbildet. Die einseitige Fläche F_n, für die $K(F_n) = n - 2 \geqq 0$ ist, ist also einer topologischen Abbildung auf sich ohne Fixpunkt unterworfen worden.

§ 2.

Faßt man die beiden Seiten, die man in der Umgebung jeder Stelle auf der einseitigen Fläche F_n unterscheiden kann, als getrennte, übereinander liegende Blätter auf, so schließen sie sich zu einer geschlossenen zweiseitigen Fläche F_n^* zusammen[4]). Diese hat die Charakteristik $K(F_n^*) = 2K(F_n)$[5]). Jedem Punkte P von F_n entsprechen übereinanderliegende Punkte P_1 und P_2 von F_n^*, dem Bildpunkt P' von P die übereinanderliegenden Punkte P_1' und P_2'. Durch die Festsetzung, daß für einen bestimmten Punkt Q von F_n der Punkt Q_1 von F_n^* dem Punkte Q_1' zugeordnet wird, ist bei stetiger Fortsetzung aus der topologischen Abbildung von F_n auf sich eine ebensolche von F_n^* auf sich bestimmt. Eine zweite aber wird dadurch bestimmt, daß Q_1 dem Punkte Q_2' zugeordnet wird. Und zwar ist die eine dieser Abbildungen von F_n^* auf sich eine solche mit Erhaltung der Indikatrix, die andere eine solche mit Umkehrung der Indikatrix. Denn eine Indikatrix in Q_1 überträgt sich auf zwei sich deckende Indikatrizes in Q_1' und Q_2'. Diese entsprechen aber verschiedenen Indikatrizes von F_n^*, da sich eine Indikatrix auf F_n längs eines Q_1' mit Q_2' verbindenden Weges (von dem einen ins andere Blatt) umkehrt. Ein Fixpunkt bei einer dieser beiden Abbildungen von F_n^* müßte auch ein Fixpunkt bei der Abbildung von F_n sein, kann also nicht existieren. F_n^* besitzt also topologische Abbildungen auf sich ohne Fixpunkte, sowohl mit Erhaltung als auch mit Umkehrung der Indikatrix. Dabei ist für $n = 2, 3, 4, \ldots$ die Charakteristik $K(F_n^*) = 2K(F_n) = 2(n-2) = 0, 2, 4, \ldots$

Es fehlt noch die zweiseitige Fläche der Charakteristik -2. Ersetzt man auf der Kugel jeden Punkt durch den diametral gegenüberliegenden, so hat diese die Indikatrix umkehrende Abbildung keinen Fixpunkt.

Friedrichshof, im August 1919.

---

4) Dies ist die Kleinsche Auffassung der einseitigen Fläche als „Doppelfläche". Vgl. auch W. Boy, Math. Ann. 57.

5) Vgl. Boy, l. c.

(Angenommen September 1919).

Reprinted from
Math. Ann. 81 (1920), 94–96

Über die Minimalzahl der Fixpunkte bei den Abbildungstypen der Ringflächen.

Von

J. Nielsen in Breslau.

Sei O ein Punkt der geschlossenen Fläche Ω, und seien die durch O gehenden geschlossenen Kurven $a_1, a_2, \ldots, a_m$ ein Erzeugendensystem der (die Kompositionseigenschaften der durch O gehenden geschlossenen Kurven von Ω charakterisierenden) „Fundamentalgruppe" F von Ω[1]). Sei t eine topologische Transformation von Ω, d. h. eine umkehrbar eindeutige und stetige Abbildung von Ω auf sich selbst, und t' eine O invariant lassende topologische Transformation von Ω, in welche t sich stetig deformieren läßt, so bestimmt t' einen Isomorphismus

$$a_i' = \varphi_i(a_1, a_2, \ldots, a_m)$$

der Gruppe F. In dieser Weise gehört zu t auf Grund der mehrdeutigen Bestimmung von t' durch t eine *Klasse von Isomorphismen* von F, wenn man den Isomorphismus

$$a_i' = \psi(a_1 \ldots a_m)\varphi_i(a_1 \ldots a_m)\psi^{-1}(a_1 \ldots a_m),$$

in dem ψ ein beliebiges Element aus F bedeutet, mit dem Isomorphismus

$$a_i' = \varphi_i(a_1 \ldots a_m)$$

zur selben Klasse rechnet. Wenn zu zwei topologischen Transformationen t_1 und t_2 von Ω dieselbe Klasse von Isomorphismen von F gehört, — ein Kriterium, das offenbar unabhängig ist sowohl von der Auswahl des Erzeugendensystems a_i als auch des Punktes O, — so werden wir sagen, daß t_1 und t_2 denselben *Typus* besitzen[2]).

---

[1]) Vgl. Poincaré, Journal de l'Ec. Pol. (2) **1**, S. 63. Mit dieser Gruppe arithmetisch äquivalent, obwohl geometrisch anders definiert, ist die unter dem gleichen Namen von Dehn eingeführte Gruppe (vgl. Math. Ann. **71**, S. 131).

[2]) Zweifelsohne sind, wenn die Charakteristik von $\Omega > -2$ ist, zwei Transformationen desselben Typus stetig ineinander deformierbar. Ein strenger Beweis dieser

6*

Fragen wir nun nach der Minimalzahl der Fixpunkte der Transformationstypen geschlossener Flächen, so ist diese Frage für die Flächen der Charakteristik -2 (Kugel) und der Charakteristik -1 (projektive Ebene), die je nur einen einzigen Transformationstypus aufweisen, in bekannter Weise von Brouwer beantwortet worden. Im folgenden soll dasselbe Problem für die Flächen der Charakteristik 0 (Ringflächen) erledigt werden.

§ 1.

Die zweiseitige Ringfläche.

u, v seien rechtwinklige Parallelkoordinaten der Ebene. Man ordne die gegenüberliegenden Seiten des Einheitsquadrates $0 \leqq u \leqq 1$, $0 \leqq v \leqq 1$ so zusammen, daß der Punkt $u, 0$ mit dem Punkte $u, 1$ und der Punkt $0, v$ mit dem Punkte $1, v$ vereinigt wird. Dadurch erhält man eine zweiseitige Ringfläche R (Torus), welche ein überall reguläres rechtwinkliges Koordinatennetz u, v trägt. Faßt man u und v als überall stetig veränderlich auf, so sind sie unendlichvieldeutig und kommen zur Bestimmung eines Punktes nur mod 1, d. h. mit ihrem Bruchbestandteil in Betracht. Die Kurve $v = 0$ sei mit A und die Kurve $u = 0$ mit B bezeichnet, solche Kurven, welche stetig in diese deformiert werden können, sollen „Kurven vom Typus A bzw. B" genannt werden. Dabei seien A und B mit einem Durchlaufungssinn im Sinne der wachsenden u bzw. v behaftet.

Jede geschlossene Kurve auf R ist dann bekanntlich in eine Kombination dieser beiden Fundamentalkurven transformierbar, also durch den Typus $A^x B^y$ zu charakterisieren. A und B bilden ein Erzeugendensystem der Fundamentalgruppe F von R, die in diesem Fall Abelsch ist und die Kompositionseigenschaften geschlossener Kurven hier auch unabhängig von der Bedingung, daß sie durch einen festen Punkt gehen sollen, charakterisiert.

Nun sei eine topologische Abbildung t von R auf sich gegeben. Das Bild A' von A sei vom Typus $A^m B^n$, das Bild B' von B vom Typus $A^p B^q$. Dann sind auch A' und B' ein Erzeugendensystem der Gruppe F, bestimmen also einen Isomorphismus von F, und es ist

$$d = mq - np = \pm 1.$$

Umgekehrt bestimmen vier Zahlen dieser Eigenschaft zwei Kurventypen A' und B', die die Bildkurven von A und B bei einer möglichen topologischen Abbildung charakterisieren können. Da bei einer Abelschen Gruppe

Eigenschaft liegt aber bisher nicht vor. Dagegen verteilt sich im Falle der Charakteristik -2 der *einzige* Transformationstypus über *zwei* Mannigfaltigkeiten von stetig ineinander deformierbaren Transformationen.

eine Klasse von Isomorphismen nur aus einem einzigen Isomorphismus besteht, so entsprechen sich also die Isomorphismen von F und die Abbildungstypen von R eineindeutig. Ist G die Gruppe aller topologischen Abbildungen von R, H die invariante Untergruppe aller mit der Identität zum selben Typus gehörigen Abbildungen, so ist G/H, die „Gruppe der Abbildungstypen", mit der Modulgruppe viereindeutig isomorph.

Wieviel Punkte müssen bei jeder Abbildung t des durch ein solches Zahlenquadrupel m, n, p, q bestimmten Abbildungstypus mindestens fest bleiben?

(u', v') sei der Bildpunkt des Punktes (u, v) bei t. Dann sind auch die Größen $\mathfrak{u} = u' - u$ und $\mathfrak{v} = v' - v$ auf R stetig veränderliche Ortsfunktionen. Zunächst sei angenommen, daß $n \neq 0$ und $p \neq 0$ ist. Durchläuft man A in positivem Sinn, so nimmt dabei u den Summanden 1, u' den Summanden m, also $\mathfrak{u}$ den Summanden $m - 1$ auf. Durchläuft man B in positivem Sinn, so bleibt u konstant, während u' also auch $\mathfrak{u}$ den Summanden p aufnimmt. Ist also Γ eine geschlossene Kurve vom Typus $A^\varrho B^\sigma$, so nimmt $\mathfrak{u}$ beim Durchlaufen von Γ den Summanden $\varrho(m-1) + \sigma p$ auf. Hat nun Γ die Eigenschaft, daß $\mathfrak{u}$ auf Γ einen bestimmten Wert (mod 1) nicht annimmt, so kann $\mathfrak{u}$ längs Γ keinen Summanden aufnehmen, und es folgt

$$\varrho(m-1) + \sigma p = 0.$$

Diese Gleichung ist stets erfüllt, wenn $\varrho = \sigma = 0$ ist, wenn also Γ auf R zusammenziehbar, „reduzibel" ist. Ist Γ nicht zusammenziehbar, „irreduzibel", so ist $\varrho \neq 0$, da aus $\varrho = 0$ wegen $p \neq 0$ auch $\sigma = 0$ folgt. Dann kann die Gleichung geschrieben werden

$$\frac{\sigma}{\varrho} = \frac{1-m}{p} = \frac{\frac{1-m}{e}}{\frac{p}{e}},$$

wenn $e = (1 - m, p)$ ist. Also ist Γ vom Typus

$$\Gamma = \left(A^{\frac{p}{e}} B^{\frac{1-m}{e}}\right)^x,$$

wo x eine ganze Zahl ist. Entsprechend ist eine geschlossene Kurve, auf der ein bestimmter Wert von $\mathfrak{v}$ (mod 1) nicht angenommen wird, vom Typus

$$\Delta = \left(A^{\frac{1-q}{f}} B^{\frac{n}{f}}\right)^y,$$

wenn $f = (1 - q, n)$ und y eine ganze Zahl ist.

Sowohl $\mathfrak{u}$ als $\mathfrak{v}$ nehmen bei unseren Voraussetzungen $n \neq 0$, $p \neq 0$ alle Werte (mod 1) auf R an, da sie längs B bzw. A von Null verschiedene, ganzzahlige Summanden aufnehmen. Die Punkte, für welche $\mathfrak{u} \equiv 0 \pmod 1$ ist, bilden eine nicht leere, abgeschlossene, echte Teilmenge von R, ebenso die Punkte, für welche $\mathfrak{v} \equiv 0 \pmod 1$ ist. Die Menge der Fixpunkte ist der Durchschnitt beider. Es sei δ eine feste Zahl aus dem Intervall $0 < \delta < \frac{1}{4}$. Auch die Punkte $\mathfrak{u} \equiv \pm\delta$ bilden eine nicht leere, abgeschlossene, echte Teilmenge von R. ε_1 sei ihr Abstand von $\mathfrak{u} \equiv 0$. Ebenso sei ε_2 der Abstand der Mengen $\mathfrak{v} \equiv \pm\delta$ und $\mathfrak{v} \equiv 0$. ν sei eine solche positive ganze Zahl, daß $\frac{1}{\nu} < \frac{\varepsilon_1}{\sqrt{2}}$ und $\frac{1}{\nu} < \frac{\varepsilon_2}{\sqrt{2}}$ ist. Dann teile man das Einheitsquadrat durch Parallele zu den Seiten in ν^2 Quadrate von der Kantenlänge $\frac{1}{\nu}$. $M_{\mathfrak{u}}$ sei die Menge der inneren Punkte aller derjenigen Quadrate, in denen oder auf deren Rand mindestens ein Punkt liegt, für den $\mathfrak{u} \equiv 0$ ist, vermehrt um die inneren Punkte derjenigen Quadratseiten, die zwei solche Quadrate trennen, und um diejenigen Quadratecken, an die vier solche Quadrate stoßen. Für alle Punkte von $M_{\mathfrak{u}}$ ist dann $-\delta < \mathfrak{u} < +\delta \pmod 1$, also ist $M_{\mathfrak{u}}$ nur ein Teil von R. $M_{\mathfrak{u}}$ besteht aus einem oder mehreren Gebieten $G_1, G_2, \ldots$ in endlicher Zahl. Jedes G_i wird von einem oder endlich vielen Quadratseitenpolygonen $\Pi_1, \Pi_2, \ldots$ begrenzt. Auf einem solchen Π liegt kein Punkt $\mathfrak{u} \equiv 0$, da diese Punkte alle zu $M_{\mathfrak{u}}$ gehören; auf Π ist also $0 < \mathfrak{u} < \delta \pmod 1$ oder $-\delta < \mathfrak{u} < 0$, so daß wir „positive" und „negative" Polygone unterscheiden können. Da auf den Π z. B. der Wert $\mathfrak{u} \equiv \frac{1}{2}$ nicht angenommen wird, sind sie vom Typus Γ.

Nun durchlaufe man die Kurve B und achte auf ihr Verhalten zu den Gebieten G. Das Betreten eines G an einem negativen Π und das Verlassen eines G an einem positiven Π werde mit dem Index $+1$, das Betreten eines G an einem positiven Π und das Verlassen eines G an einem negativen Π mit dem Index -1 bewertet und die Indexsumme J längs B gebildet. (Diese Vorgänge des Betretens und Verlassens treten nur endlich oft ein). Da $\mathfrak{u}$ längs B den von Null verschiedenen Summanden p aufnimmt, ist $J = 2p$. Der Beitrag J_{Π_1}, den ein bestimmtes Polygon Π_1 vom Typus $\Gamma_{x=x_1} = \left(A^{\frac{p}{e}} B^{\frac{1-m}{e}}\right)^{x_1}$ zu J liefert, ist dem absoluten Betrage nach gleich dem absoluten Betrag der „Charakteristik der beiden Kurven B und Π_1"[3]), d. h. gleich dem Betrag der Exponentendeterminante der beiden Kurventypen, also

$$|J_{\Pi_1}| = \left| \begin{vmatrix} 0 & 1 \\ \frac{p}{e}\cdot x_1 & \frac{1-m}{e}\cdot x_1 \end{vmatrix} \right| = \left|\frac{p}{e}\cdot x_1\right|.$$

---

[3]) Vgl. z. B. H. Weyl: „Strenge Begründung der Charakteristikentheorie auf zweiseitigen Flächen". Jahresbericht d. D. M. V. 25.

Ist also Π_1 reduzibel, also $x_1 = 0$, so liefert es keinen Beitrag zu J. Es muß daher unter den Π auch irreduzible geben. Ist Π_1 irreduzibel, also $x_1 \neq 0$, so muß, da ein Randpolygon eines zusammenhängenden Gebietes sich nicht selbst überschneiden kann, eine Kurve vom Typus $\Gamma_{x=x_1}$ aber, wie man leicht erkennt[4]), mindestens $x_1 - 1$ Selbstüberschneidungen hat, notwendig $x_1 = 1$ sein. Also liefert Π_1 zu J den Beitrag $\pm \frac{p}{e}$, und wir erkennen, daß es mindestens $2e$ irreduzible Polygone Π geben muß, jedes vom Typus $\Gamma_{x=1}$. Da nun ein irreduzibles Polygon (eine irreduzible Kurve) auf R nicht begrenzt, zwei irreduzible Polygone (Kurven) vom selben Typus aber stets begrenzen, so verteilen sich die irreduziblen Randpolygone zu je zweien auf mindestens e aus den Gebieten G_i, die wir kurz die „irreduziblen Gebiete" nennen mögen, im Gegensatz zu „reduziblen Gebieten", die ganz dem Inneren eines reduziblen Randpolygons angehören.

Sei nun G_1 ein irreduzibles Gebiet, Π_1 und Π_2 die beiden zu seiner Begrenzung gehörigen irreduziblen Polygone. Sind diese beide positiv oder beide negativ, so heben sie sich in ihrem Beitrag zu J auf, und wir nennen G_1 ein „unwesentliches irreduzibles Gebiet". Ist aber Π_1 positiv und Π_2 negativ oder umgekehrt, so verdoppeln sie sich in ihrer Wirkung, liefern also zusammen zu J den Beitrag $\frac{+2p}{e}$ oder $\frac{-2p}{e}$. G_1 soll dann ein „wesentliches irreduzibles Gebiet erster Art" bzw. „zweiter Art" heißen. Wir erkennen also:

> Die Anzahl der wesentlichen irreduziblen Gebiete erster Art muß diejenige der zweiten Art um genau e Einheiten überwiegen.

Wir wiederholen nun das ganze Verfahren unter Auszeichnung des Wertes $\mathfrak{u} \equiv \frac{1}{2}$ statt $\mathfrak{u} \equiv 0$. Sei η_1 der Abstand der Mengen $\mathfrak{u} \equiv \frac{1}{2}$ und $\mathfrak{u} \equiv \pm\frac{1}{4}$ und η_2 der Abstand von $\mathfrak{v} \equiv \frac{1}{2}$ und $\mathfrak{v} \equiv \pm\frac{1}{4}$ und ferner μ eine

---

[4]) Es sei $\gamma = A^k B^l$ und $\Gamma = \gamma^x = (A^k B^l)^x = A^K B^L$. Ein Bild von Γ in der u-v-Ebene laufe von u_0, v_0 nach $u_0 + K$, $v_0 + L$. Man betrachte alle diejenigen zu diesem homologen Bilder von Γ, die aus ihm durch alle positiven und negativen Potenzen der Substitution $\bar{u} = u + K$, $\bar{v} = v + L$ entstehen. Diese fügen sich zu einem unendlichen Kurvenzug S aneinander, der die u-v-Ebene teilt, und auf den sich eine Durchlaufungsrichtung von Γ überträgt. Nun sei S_y das Bild von S bei der Substitution $\bar{u} = u + yk$, $\bar{v} = v + yl$, wobei $0 < y < x$ ist. Hätte S_y mit S keinen Punkt gemeinsam, so müßte es z. B ganz links von S liegen; dann läge ebenso S_{2y}, das aus S_y durch nochmalige Anwendung derselben Substitution entsteht, ganz links von S_y, also a fortiori ganz links von S und von diesem durch S_y getrennt, S_{3y} wieder ganz links von S_{2y} usw. Nach spätestens x Schritten ist aber $S_{xy} = S$, woraus ein Widerspruch folgt. Also hat S_y mit S mindestens einen Überkreuzungspunkt; dieser entspricht einer Selbstüberkreuzung von Γ. Da y jeden ganzzahligen Wert aus dem Intervall $1 \leq y \leq x - 1$ annehmen kann, so folgt, daß Γ mindestens $x - 1$ Selbstüberkreuzungen hat, die sich evtl. zu mehrfach zählenden Überkreuzungspunkten vereinigen können.

solche positive ganze Zahl, daß $\frac{1}{\mu} < \frac{\eta_1}{\sqrt{2}}$ und $\frac{1}{\mu} < \frac{\eta_2}{\sqrt{2}}$ ist. Dann konstruieren wir die Quadratteilung mit der Seite $\frac{1}{\mu}$ und zeichnen ebenso wie früher die Menge $N_{\mathfrak{u}}$ aller Quadrate aus, die einen Punkt $\mathfrak{u} \equiv \frac{1}{2}$ besitzen. $N_{\mathfrak{u}}$ hat dann von $M_{\mathfrak{u}}$ einen endlichen Abstand. Dann erschließen wir wie oben die Existenz von $h \geqq e$ irreduziblen Gebieten $H_1, H_2, \ldots$ von $N_{\mathfrak{u}}$, zu deren Berandungen je zwei irreduzible Polygone vom Typus $\Gamma_{x=1}$ gehören.

Die irreduziblen Gebiete H und G haben keine Punkte gegenseitig gemeinsam. Die letzteren werden also durch die ersteren auf R in h Gruppen getrennt, wenn wir gegebenenfalls auch „leere" Gruppen mitzählen. Seien G_1 und G_2 zwei wesentliche irreduzible G-Gebiete derselben Gruppe, die durch kein weiteres wesentliches irreduzibles G-Gebiet innerhalb der Gruppe getrennt werden, und seien Π_1 bzw. Π_2 diejenigen irreduziblen Randpolygone von G_1 bzw. G_2, die diese einander innerhalb der Gruppe zukehren; endlich sei K das von Π_1 und Π_2 begrenzte Zwischengebiet von G_1 und G_2. Dann verbinden wir Π_1 und Π_2 in K durch einen solchen Weg ω, der es vermeidet, etwa in K enthaltene reduzible G- oder H-Gebiete zu betreten. Auf ω wird also $\mathfrak{u} \equiv \frac{1}{2}$ nicht angenommen. $\mathfrak{u} \equiv 0$ kann auf ω nur angenommen werden, wenn unwesentliche irreduzible G-Gebiete zu K gehören. Da aber die irreduziblen Randpolygone eines solchen gleichartig sind, so kann $\mathfrak{u}$ längs ω von Π_1 nach Π_2 auch nicht durch $\mathfrak{u} \equiv 0$ hindurch das Zeichen gewechselt haben. Π_1 und Π_2 sind also beide positiv oder beide negativ und G_1 und G_2 demnach von entgegengesetzter Art, so daß sie sich in ihrem Beitrag zu J aufheben. Wir schließen, daß eine Gruppe günstigstenfalls den Beitrag eines wesentlichen irreduziblen Gebietes erster Art liefern kann, daß es also mindestens e nicht leere Gruppen geben muß. Wir können also e irreduzible Gebiete G_i, die wir mit $G_{\mathfrak{u},1}, G_{\mathfrak{u},2}, \ldots, G_{\mathfrak{u},e}$ bezeichnen, und e irreduzible Gebiete $H_{\mathfrak{u},1}, H_{\mathfrak{u},2}, \ldots, H_{\mathfrak{u},e}$ auf R so auswählen, daß für $e > 1$ die ersteren durch die letzteren getrennt werden und umgekehrt.

Vertauscht man nun die Rollen von $\mathfrak{u}$ und $\mathfrak{v}$, geht also von der Menge $M_{\mathfrak{v}}$ derjenigen ν-Quadrate aus, die einen Punkt $\mathfrak{v} \equiv 0$ besitzen, für die also $-\delta < \mathfrak{v} < \delta \pmod 1$ ist, bzw. von der Menge $N_{\mathfrak{v}}$ derjenigen μ-Quadrate, die einen Punkt $\mathfrak{v} \equiv \frac{1}{2}$ besitzen, so gelangt man durch dieselbe Überlegung zu je f irreduziblen Gebieten $G_{\mathfrak{v},1}, G_{\mathfrak{v},2}, \ldots, G_{\mathfrak{v},f}$ bzw. $H_{\mathfrak{v},1}, H_{\mathfrak{v},2}, \ldots, H_{\mathfrak{v},f}$ von der Eigenschaft, daß für $f > 1$ die ersteren durch die letzteren auf R getrennt werden und umgekehrt. Die je zwei irreduziblen Randpolygone dieser Gebiete sind vom Typus $\Delta_{y=1}$.

Nun betrachte man die sämtlichen Abbilder von $H_{\mathfrak{u},1}$ in der u-v-Ebene.

Diese setzen sich zu Streifen zusammen, die die Ebene in der durch den Typus Γ bestimmten Richtung, der Richtung der Geraden

$$\frac{u}{v} = \frac{p}{1-m}$$

durchziehen und die durch die Parallelverschiebungen des ganzzahligen Koordinatengitters ineinander übergehen. $H_{\mathfrak{v},1}$ erzeugt analoge „Parallelstreifen" in der durch den Typus Δ bestimmten Richtung der Geraden

$$\frac{u}{v} = \frac{1-q}{n}.$$

Derjenige Teil der Ebene, der zwischen zwei benachbarten Streifen der ersten Schar und zwischen zwei benachbarten Streifen der zweiten Schar liegt, möge das „Zwischengebiet" der beiden Streifenpaare heißen. Dann bestimmen die beiden Scharen so viele einander nicht homologe Zwischengebiete, wie der absolute Betrag der Charakteristik der beiden Typen Γ und Δ angibt, also die Zahl

$$Z' = \left| \begin{vmatrix} \frac{p}{e} & \frac{1-m}{e} \\ \frac{1-q}{f} & \frac{n}{f} \end{vmatrix} \right| = \frac{1}{e \cdot f} \cdot |m + q - d - 1|.$$

Nimmt man die durch $H_{\mathfrak{u},2}, \ldots, H_{\mathfrak{u},e}$ erzeugten Streifen hinzu, so wird jedes der bisherigen Zwischengebiete in e neue zerschnitten; und nimmt man dann noch die durch $H_{\mathfrak{v},2}, \ldots, H_{\mathfrak{v},f}$ erzeugten Streifen hinzu, so wird jedes dieser wieder in f neue zerschnitten. Insgesamt bestimmen die $H_{\mathfrak{u},i}$ $(i = 1, \ldots, e)$ und die $H_{\mathfrak{v},k}$ $(k - 1, \ldots, f)$ also

$$Z = e \cdot f \cdot Z' = |m + q - d - 1|$$

einander nicht homologe Zwischengebiete. Da zwischen je zwei benachbarten $H_{\mathfrak{u},i}$- bzw. $H_{\mathfrak{v},k}$-Streifen ein $G_{\mathfrak{u},i}$- bzw. $G_{\mathfrak{v},k}$-Streifen liegt, die sich also im zugehörigen Zwischengebiet überkreuzen, so gibt es also in jedem Zwischengebiet mindestens ein, insgesamt also mindestens Z ν-Quadrate, die sowohl zu $M_{\mathfrak{u}}$ als auch zu $M_{\mathfrak{v}}$ gehören. Man kann also stets Z Durchschnittquadrate von $M_{\mathfrak{u}}$ und $M_{\mathfrak{v}}$ so angeben, daß man von jedem zu einem andern nicht gelangen kann, ohne ein $H_{\mathfrak{u},i}$ oder $H_{\mathfrak{v},k}$ zu durchqueren.

Für jeden Punkt eines solchen Durchschnittquadrates gilt für den Abstand E des Punktes von seinem Bildpunkt

$$E = \sqrt{(u'-u)^2 + (v'-v)^2} = \sqrt{\mathfrak{u}^2 + \mathfrak{v}^2} < \delta\sqrt{2}.$$

Wählt man nun für δ eine gegen 0 konvergierende Reihe von Werten $\delta_1, \delta_2, \ldots$ und wählt man zu jedem $\delta_\varkappa$ die Zahl $\nu_\varkappa$ außer nach den früheren Ungleichungen noch so, daß $\nu_{\varkappa+1}$ ein Vielfaches von $\nu_\varkappa$ ist, so ist die Quadratteilung für den Index $\varkappa + 1$ eine feinere Unterteilung derjenigen für den Index $\varkappa$. Jedes Durchschnittquadrat von $M_{\mathfrak{u}}(\varkappa + 1)$ und $M_{\mathfrak{v}}(\varkappa + 1)$ ist in einem solchen von $M_{\mathfrak{u}}(\varkappa)$ und $M_{\mathfrak{v}}(\varkappa)$ enthalten, da es einen Punkt $\mathfrak{u} \equiv 0$ und einen Punkt $\mathfrak{v} \equiv 0$ besitzt. Eine solche Folge von ineinander geschachtelten Quadraten, für welche jede Quadratseite ein ganzzahliges Vielfaches der folgenden ist, konvergiert gegen einen Punkt, während gleichzeitig der Abstand E der Punkte des einzelnen Quadrates von ihren Bildpunkten mit δ gegen 0 konvergiert. Jeder solche Konvergenzpunkt ist also Fixpunkt.

Die Gebiete $H_{\mathfrak{u},i}$ und $H_{\mathfrak{v},k}$ hängen nur von der festen Zahl μ ab und sind also für alle $\delta_\varkappa$ dieselben, ebenso daher auch die Z Zwischengebiete, die durch Streifen von überall endlicher Breite voneinander getrennt sind. In jedem der Zwischengebiete muß es nun nach dem Obigen mindestens einen Konvergenzpunkt geben, also gibt es mindestens Z voneinander verschiedene Fixpunkte[5]).

Die oben gemachten Voraussetzungen $n \neq 0$ und $p \neq 0$ sind nachträglich aufzuheben. Ist $p = 0$, so ist $mq = \pm 1$, also $m = \pm 1$. Ist $m = -1$, so ist der Gesamtindex längs der Kurve A: $J = -4$. Läßt man dann die Kurve A die Rolle der Kurve B in den obigen Überlegungen übernehmen, so bleiben alle Schlüsse entsprechend in Kraft. Ist $m = +1$, so hat die Abbildung keinen notwendigen Fixpunkt; in diesem Fall ergibt die obige Formel $Z = 0$. Ebenso ist die Beschränkung für n aufzuheben.

Andererseits ist es leicht, Abbildungen von jedem vorgeschriebenen Typus zu konstruieren, welche *nur* diese Minimalzahl Z von Fixpunkten aufweisen; solche werden z. B. durch die lineare Transformation geliefert:

$$u' = mu + pv,$$
$$v' = nu + qv.$$

Damit ist der Satz[6]) bewiesen:

Sind auf einer zweiseitigen geschlossenen Ringfläche A und B zwei geschlossene irreduzible gerichtete Kurven, die sich nur in einem Punkte schneiden, und gehen bei einer Transformation der Fläche

---

[5]) Man findet leicht durch nähere Untersuchung der $H_{\mathfrak{u},i}$ und $H_{\mathfrak{v},k}$, daß man Z Fixpunkte so auswählen kann, daß man von einem zu einem andern nicht gelangen kann, ohne daß $\mathfrak{u}$ oder $\mathfrak{v}$ oder beide einen von Null verschiedenen ganzzahligen Summanden aufnehmen. Ist also ε der kleinere der beiden Abstände zwischen den Mengen $\mathfrak{u} \equiv 0$ und $\mathfrak{u} \equiv \frac{1}{2}$ bzw. $\mathfrak{v} \equiv 0$ und $\mathfrak{v} \equiv \frac{1}{2}$, so daß also ε eine durch die gegebene Abbildung bestimmte endliche Größe ist, so lassen sich unter den Fixpunkten der Abbildung mindestens Z so auswählen, daß keiner einem andern näher als bis auf 2ε kommt.

[6]) Diesen Satz habe ich in meiner Schrift „Kurvennetze auf Flächen" (In.-Diss. Kiel 1913) auf S. 58 ausgesprochen. Es ist dort „Geschlecht 1" statt „Geschlecht 0" zu lesen.

in sich A bzw. B in zwei Kurven vom Typus $A^m B^n$ bzw. $A^p B^q$ über, so hat die Transformation mindestens $|m + q - d - 1|$ verschiedene Fixpunkte, aber nicht notwendig mehr. Dabei ist $d = mq - np = +1$ oder $= -1$, je nachdem die Transformation die Indikatrix erhält oder umkehrt.

§ 2.

Die einseitige Ringfläche.

Im § 1 wurden die Randpunkte des Einheitsquadrates so zusammengeordnet, daß eine geschlossene zweiseitige Ringfläche entstand. Diese deformiere man nun weiter so, daß der Punkt (u, v) in die unmittelbare Nähe des Punktes $(1 - u, v + \frac{1}{2})$ gebracht wird, und mache die so entstehende Doppelfläche zu einer einseitigen Ringfläche R, indem man diese Punkte zur Koinzidenz bringt und als einen auffaßt. Diese Fläche beschreibt Klein als sich selbst durchsetzenden Schlauch[7]). Dehn behandelt sie in seiner Abhandlung „Über unendliche diskontinuierliche Gruppen“[8]). Die Fläche R wird durch die Hälfte des Einheitsquadrates repräsentiert. Als Erzeugendensystem der Fundamentalgruppe F der durch den Punkt $u = v = 0$ gehenden geschlossenen Kurven bieten sich dar die Kurve A ($v = 0$, u von 0 bis 1; bzw. $v = \frac{1}{2}$, u von 1 bis 0) und die Kurve C ($u = 0$, v von 0 bis $\frac{1}{2}$; bzw. $u = 1$, v von 0 bis $\frac{1}{2}$). F ist aber nicht mehr kommutativ. An die Stelle der Relation $ABA^{-1}B^{-1} = 1$ beim zweiseitigen Ring ist vielmehr die Relation $ACAC^{-1} = 1$ getreten.

Um die auf R möglichen topologischen Abbildungstypen aufzusuchen, fragen wir, dem in der Einleitung Gesagten entsprechend, nach allen *Klassen* von Isomorphismen von F.

Nach der definierenden Relation ist $AC = CA^{-1}$ und $CA = A^{-1}C$. Man darf also A mit C vertauschen, wenn nur der Exponent von A das Zeichen wechselt, jedesmal wenn A ein C passiert. Daher ist jedes Element von F in der Form $A^x C^y$ darstellbar. Es sei

$$\begin{cases} A' = A^m C^n \\ C' = A^p C^q \end{cases}$$

ein nicht zur Klasse der Identität gehörender Isomorphismus. Dann muß gelten

$$A'C'A'C'^{-1} = A^m C^n A^p C^q A^m C^n C^{-q} A^{-p} = A^0 C^0 = 1.$$

7) F. Klein, „Über Riemanns Theorie der algebraischen Funktionen usw.“, S. 80.

8) Math. Ann. **71**, S. 126ff.

Daraus folgt $n=0$. Dann ist $q \neq 0$. Da A', C' ein Erzeugendensystem ist, so muß es zwei Zahlen x und y geben, so daß

$$A = A'^x C'^y = A^{mx}(A^p C^q)^y.$$

Daraus folgt $y=0$, weil $q \neq 0$, ferner $xm=1$, also $m=\pm 1$. Ferner muß es zwei Zahlen x' und y' geben, so daß

$$C = A'^{x'} C'^{y'} = A^{\pm x'}(A^p C^q)^{y'}.$$

Daraus folgt $q y'=1$, also $q=\pm 1$, so daß alle Isomorphismen die Form haben:

$$\begin{cases} A' = A^{\pm 1} \\ C' = A^p C^{\pm 1}. \end{cases}$$

Ist hierin $A' = A^{-1}$, so gehört der Isomorphismus

$$\begin{cases} A' = A \\ C' = A^{-p} C^{\pm 1} \end{cases}$$

zur selben Klasse, so daß wir auf jeden Fall $A' = A$ voraussetzen dürfen[9]).

Ist sodann weiter $p = 2r$ bzw. $p = 2r+1$, so gehört zur selben Klasse der Isomorphismus

$$\begin{cases} A' = A \\ C' = A^r C^{\pm 1} A^r \end{cases} \quad \text{bzw.} \quad \begin{cases} A' = A \\ C' = A^{r+1} C^{\pm 1} A^r. \end{cases}$$

Bringt man hierin wieder alle A nach vorne, so heben sich $2r$ Bestandteile A auf, und man erkennt, daß sich alle von der identischen verschiedenen Klassen von Isomorphismen durch die drei folgenden Formen repräsentieren lassen:

$$J_1: \begin{cases} A' = A \\ C' = C^{-1} \end{cases} \qquad J_2: \begin{cases} A' = A \\ C' = AC \end{cases} \qquad J_3: \begin{cases} A' = A \\ C' = AC^{-1}. \end{cases}$$

Diese drei Klassen bilden mit der identischen Klasse $J_0: \begin{cases} A' = A \\ C' = C \end{cases}$ eine Gruppe.

Man erkennt sofort, daß man zu jeder der gefundenen Isomorphismenklassen leicht Abbildungen konstruieren kann. Also folgt:

> Die Gruppe der Abbildungstypen der einseitigen Ringfläche ist endlich; sie hat die Ordnung 4.

Wieviel Fixpunkte müssen bei diesen vier Abbildungstypen mindestens auftreten?

---

[9]) Das besagt geometrisch, daß die einzige doppelpunktlose „Amphidrome" der Fläche (vgl. Dehn, l. c., S. 129) bei jeder topologischen Abbildung in eine homotope Kurve übergeht.

R steht vermöge seiner Entstehung in einer ein-zweideutigen Korrespondenz mit einer zweiseitigen Ringfläche R^*. Der Kurve A auf R entspricht eine Kurve A^* auf R^*. Die Kurve C auf R ist auf R^* erst bei zweimaliger Durchlaufung geschlossen und ergibt dort die im § 1 mit B bezeichnete Kurve. Es ist also $C^2 = B$. Jeder topologischen Abbildung von R auf sich entsprechen nun zwei solche von R^* auf sich[10]), die sich durch die Indikatrix unterscheiden. Ist bei der einen $A^{*\prime} = A^*$, so ist bei der andern $A^{*\prime} = A^{*-1}$. Ferner ist

$$\begin{aligned} &\text{bei } J_0\colon\ B' = C'^2 = C^2 = B,\\ &\text{,, } J_1\colon\ B' = C'^2 = C^{-2} = B^{-1},\\ &\text{,, } J_2\colon\ B' = C'^2 = ACAC = AA^{-1}CC = C^2 = B,\\ &\text{,, } J_3\colon\ B' = C'^2 = AC^{-1}AC^{-1} = AA^{-1}C^{-1}C^{-1} = C^{-2} = B^{-1}. \end{aligned}$$

Aus dem Ergebnis des § 1 läßt sich also bei den Typen J_0 und J_2 für keine der beiden Abbildungen von R^* auf sich die Existenz von Fixpunkten folgern. Bei J_1 und J_3 hat die Abbildung

$$\begin{cases} A^{*\prime} = A^{*-1} \\ B' = B^{-1} \end{cases} \text{ mindestens } Z = |m + q - d - 1| - 1 \text{ Fixpunkte.}$$

Da R^* eindeutig auf R bezogen ist, entspricht einem Fixpunkt auf R^* auch ein solcher auf R. Da aber R zweideutig auf R^* bezogen ist, können einem Fixpunkt auf R zwei verschiedene auf R^* entsprechen. Wir können also nur folgern, daß die Abbildungen von R auf sich von den Typen J_1 und J_3 zwei notwendige Fixpunkte haben. Und es ist auch leicht, Transformationen zu konstruieren, welche im Falle von J_0 und J_2 *keine* und im Falle von J_1 und J_3 *nur zwei* Fixpunkte haben. Wir erhalten also den Satz:

Ist A auf einer einseitigen Ringfläche R eine doppelpunktlose „Amphidrome", d. h. eine nicht auf einen Punkt zusammenziehbare Kurve, die auf R in sich selbst mit umgekehrter Durchlaufungsrichtung stetig deformierbar ist, und C eine geschlossene Kurve, die A nur in einem Punkte schneidet, und ist bei einer topologischen Abbildung von R auf sich das Bild C' von C in C^{-1} oder AC^{-1} stetig überführbar, so hat die Abbildung mindestens 2, aber nicht notwendig mehr als 2 Fixpunkte. Ist aber C' in C oder AC stetig überführbar, — und damit sind alle Möglichkeiten erschöpft, — so hat die Abbildung keinen notwendigen Fixpunkt.

---

[10]) Vgl. meinen Aufsatz „Über fixpunktfreie Abbildungen geschlossener Flächen", Math. Annalen **81**, S. 94–96.

(Eingegangen am 15. Januar 1920.)

Reprinted from
Math. Ann. 82 (1921), 83–93

The Mapping Types of Closed Surfaces and their Relationship with Infinite Groups*

JAKOB NIELSEN

The topological mappings of a closed two-sided surface onto a surface of the same genus, e.g. itself, constitute a mixed group $\mathscr{A}$. Those mappings for which the image of each closed curve is *homotopic* to the original, i.e. convertible to it by continuous transformation on the surface, constitute a normal subgroup $\mathscr{U}$ of $\mathscr{A}$. If one reckons two mappings which result from each other by composition with mappings in $\mathscr{U}$ to belong to the same "type", then the collection of mapping types forms a discrete group $G = \mathscr{A}/\mathscr{U}$. One knows that for the surface of genus $p = 1$ (torus) the group G is isomorphic to the modular group. *Which groups appear as higher analogues of the modular group for* $p > 1$? The case $p = 2$ is treated in what follows.

The closed surface Ω of genus $p = 2$ is most completely characterized topologically by its "fundamental group" $\mathscr{F}$. (Poincaré, J. d. l'Éc. Polyt. II, Série 1. Cahier 1895 and Rendic. Pal. XVIII 1904, Dehn, Math. Ann. **71**, 1911). The closed curves through a fixed point O on Ω correspond to the elements of $\mathscr{F}$ in such a way that curves which are transformable into each other with O fixed correspond to the same element of $\mathscr{F}$, and curves which contract to O are associated with the identity of $\mathscr{F}$. Curves which do not go through O determine, when one deforms them in all possible ways into curves through O, infinitely many "equivalent", conjugate elements of $\mathscr{F}$. If one chooses the fundamental system a, b, c, d of Figure 1 as generators of $\mathscr{F}$, then the only independent relation between them is: $R(abcd) \equiv aba^{-1}b^{-1}cdc^{-1}d^{-1} = 1$. The group diagram is a net of 8-sided cells, 8 of which meet at each vertex, so that it may be constructed in the hyperbolic plane in a metrically regular way. It is a one-to-one image of the universal covering surface of Ω; at the same time it represents $\mathscr{F}$ as a discontinuous group of non-Euclidean motions. An element of the group, and the corresponding motion, is completely determined by the point to which a given point of the net is sent. This motion is always a displacement along a proper line of the hyperbolic plane, the "axis" of the motion.

If Ω is punctured by one hole and the result called Ω^*, and if one takes a, b, c, d again as the fundamental system, then there is no relation between them and hence the fundamental group $\mathscr{F}^*$ of Ω^* is the "free" group with 4 generators.

Under a mapping of Ω onto itself with O fixed the images of the curves a, b, c, d determine four elements $\alpha, \beta, \gamma, \delta$ of $\mathscr{F}$ which constitute an *isomorphic system.* If the mappings A_1 and A_2 correspond to the isomorphisms I_1 and I_2, then the mapping $A_1 \cdot A_2$ corresponds to the isomorphism $I_1 \cdot I_2$ which results from successively making the substitutions of I_1 and I_2. A mapping which does not leave O fixed can be deformed in infinitely many ways into a mapping which leaves O fixed, and defines in this way a "class" of conjugate isomorphisms. Two mappings belong to the same type when they determine the same isomorphism class of $\mathscr{F}$. In this way the problem assumes a pure group theoretic form.

I. *M. Dehn has found the following relationship between mappings and isomorphisms:*

Theorem 1. *Each isomorphism of* $\mathscr{F}$ *is induced by a mapping of* Ω.

*Translation by John Stillwell of Die Abbildungstypen geschlossener Flächen und ihre Beziehungen zu unendlichen Gruppen. [2 Vorträge gehalten in Breslau am. 9 und 11. März 1921 von J. Nielsen.]

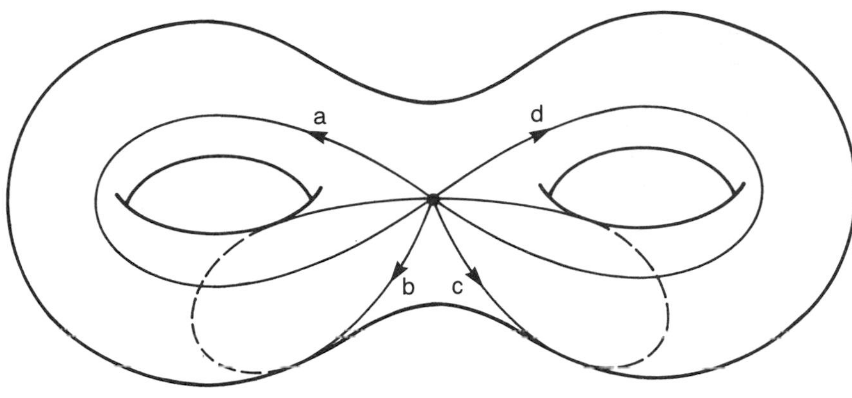

Figure 1

(Thus the above association of mapping types with isomorphism classes is reversible.) The Dehn proof—which is not yet published—uses the theorem that two curves are transformable into disjoint curves if and only if their "axis images" are disjoint (cf. Poincaré, Rend. Pal. XVIII, pp. 79–80). *This "incidence relation" between the axes of two group elements is preserved under isomorphisms.* The group theoretic invariance under isomorphisms corresponds to the topological invariance of the minimum number of intersection points under mappings (cf. also Dehn, Math. Ann. **71**, pp. 132–39). Thus if $[\alpha, \beta, \gamma, \delta]$ is a generator system of $\mathscr{F}$ isomorphic to $[a, b, c, d]$, then the axis images[1] of α and β have one intersection, so do those of γ and δ, while those of α and γ etc. have none. Also, since $R(\alpha, \beta, \gamma, \delta) = 1$, the elements $\delta\gamma\delta^{-1}\gamma^{-1}$ and $\alpha\beta\alpha^{-1}\beta^{-1}$ correspond to homotopic curves, so it follows that Ω can be mapped onto itself in such a way that the axis images of a, b, c, d are mapped respectively onto those of $\alpha, \beta, \gamma, \delta$, with O fixed. This determines an isomorphism of the form $[\mathscr{A}\alpha\mathscr{A}^{-1}, \mathscr{B}\beta\mathscr{B}^{-1}, \mathscr{C}\gamma\mathscr{C}^{-1}, \mathscr{D}\delta\mathscr{D}^{-1}]$.[2] In order to show that there is a mapping which carries this to $[\alpha, \beta, \gamma, \delta]$, it will suffice to show that there is one which carries $[\mathscr{A}a\mathscr{A}^{-1}, \mathscr{B}b\mathscr{B}^{-1}, \mathscr{C}c\mathscr{C}^{-1}, \mathscr{D}d\mathscr{D}^{-1}]$ to $[a, b, c, d]$. Since the axes of the elements $\mathscr{A}a\mathscr{A}^{-1}$ and $\mathscr{B}b\mathscr{B}^{-1}$ relative to a given point of the group diagram intersect, one can multiply $\mathscr{A}$ by a power a^{μ}, and $\mathscr{B}$ by a power b^{ν}, so that $\mathscr{A}a^{\mu} = \mathscr{B}b^{\nu}$. By conjugating the isomorphism with this element—which corresponds to a mapping—one obtains the isomorphism in the form: $[a, b, \phi c\phi^{-1}, \phi d\phi^{-1}]$. Since we must now have $aba^{-1}b^{-1} = \phi(dcd^{-1}c^{-1})\phi^{-1}$, ϕ can only be $(aba^{-1}b^{-1})^{\pm n}$ (n arbitrary); because the endpoints of the net paths $aba^{-1}b^{-1}$ and $dcd^{-1}c^{-1}$ have the same rectilinear distance, and the axis of the motion $aba^{-1}b^{-1}$ is the corresponding cell diagonal. But when $\phi = (aba^{-1}b^{-1})^{\pm n}$, the isomorphism $[a, b, \phi c\phi^{-1}, \phi d\phi^{-1}]$ corresponds to a mapping, and hence the proof of Theorem 1 is complete.

The analogous theorem for bounded surfaces holds only for $p = 1$ (cf. Math. Ann. **78**, p. 397,

[1] I.e. the geodesic representatives of their homotopy classes (Translator's note).

[2] In fact the isomorphism must be of the form $[\mathscr{A}\alpha\mathscr{A}^{-1}, \mathscr{A}\beta\mathscr{A}^{-1}, \mathscr{C}\gamma\mathscr{C}^{-1}, \mathscr{C}\delta\mathscr{C}^{-1}]$, since α, β have a point in common and hence their homotopy classes are determined up to a *common* conjugating factor. Similarly for γ and δ. The reduction to the single conjugating factor ϕ which follows is then immediate by conjugation with $\mathscr{A}$. (Translator's note.)

[N2]). For $p = 2$, $[ac^{-1}, b, c, d]$ is an isomorphism of the fundamental group $\mathscr{F}^*$ of Ω^* which cannot be induced by a mapping of Ω^*, since ac^{-1} has an irreducible double point.

II. For the *isomorphism group itself* we have:

Theorem 2. *The isomorphism group of $\mathscr{F}$ may be generated by 3 elements.*

The proof uses a deep relationship between the isomorphisms of $\mathscr{F}$ and $\mathscr{F}^*$.

Let $I = [\alpha, \beta, \gamma, \delta]$ be a particular isomorphism of $\mathscr{F}$. The elements $\alpha, \beta, \gamma, \delta$ are certain elements of $\mathscr{F}$, however they can obviously be written in terms of a, b, c, d in many ways, since $R(abcd) = 1$. We normalise them as follows. Let A be a mapping of Ω, with fixed point O, which induces I, and let S' be the image of the fundamental system $S = a, b, c, d$. Keeping S' fixed, it is possible to deform A continuously into a mapping A_Q which fixes any point Q not on S or S'. Removal of Q from Ω yields a punctured surface Ω^* which is mapped onto itself by A_Q with O fixed. The expressions for the image curves S' in the generators are uniquely determined on Ω^*, thus $\alpha, \beta, \gamma, \delta$ have unique expressions $\alpha_Q, \beta_Q, \gamma_Q, \delta_Q$ in a, b, c, d as elements of $\mathscr{F}^*$, while retaining their meaning as elements of $\mathscr{F}$. These expressions depend only upon which of the regions on Ω determined by S and S' contains Q. The expression $R(\alpha_Q\beta_Q\gamma_Q\delta_Q)$ is converted to $\phi R(abcd)^{\pm 1}\phi^{-1}$ by cancellation of terms xx^{-1}. The relation $R(\alpha_Q\beta_Q\gamma_Q\delta_Q) = 1$ therefore follows by exactly *one* application of $R(abcd) = 1$, while $R(\alpha\beta\gamma\delta) = 1$ in general requires multiple application of $R(abcd) = 1$. (One then concludes easily from the form of R that $R(\alpha\beta\gamma\delta)$ is always a product of an *odd* number of conjugates of $R(abcd)^{\pm 1}$, in particular, that $R(\alpha\beta\gamma\delta)$ can never be identically 1.)

Definition. 4 elements $\mathrm{A}, \mathrm{B}, \Gamma, \Delta$ of $\mathscr{F}^*$ for which $R(\mathrm{AB}\Gamma\Delta) = \phi R(abcd)^{\pm 1}\phi^{-1}$ are said to constitute a "*basic form.*" As a result of the mapping A_Q, $[\alpha_Q, \beta_Q, \gamma_Q, \delta_Q]$ is an *isomorphism of* $\mathscr{F}^*$, and hence a, b, c, d are expressible in terms of $\alpha_Q, \beta_Q, \gamma_Q, \delta_Q$ *without* application of $R(abcd) = 1$, which again is not generally the case for $\alpha, \beta, \gamma, \delta$. Thus, by writing the isomorphism I of $\mathscr{F}$ in the form $[\alpha_Q, \beta_Q, \gamma_Q, \delta_Q]$ we have made a formal alteration to show it simultaneously as an isomorphism of $\mathscr{F}^*$. Depending on the choice of Q, *I determines different isomorphisms of $\mathscr{F}^*$*. We therefore seek those isomorphisms of $\mathscr{F}^*$ which correspond to mappings of Ω^*; these must be basic forms, as follows topologically from the preservation of the boundary curve $R(abcd)$—a certain number of such isomorphisms of $\mathscr{F}^*$ at a time then corresponds to the same isomorphism of $\mathscr{F}$.

Conjugation of a basic form by any word yields another basic form, and the product of two basic forms (i.e. the result of substituting one in the other) is again a basic form.

We now choose suitable special isomorphisms of $\mathscr{F}$ in basic form (and hence isomorphisms of $\mathscr{F}^*$) and construct the subgroups of the isomorphism group which they generate: $I = [d, c, b, a]$, $I^2 = 1$ (i.e. $= [a, b, c, d]$); $K = [c^{-1}b, a^{-1}, a^{-1}d, c^{-1}]$, $K^6 = 1$; $(IK)^2 = 1$. The group generated by I and K is therefore of order 12. It contains the four group $\mathscr{V}$ as an invariant subgroup: $1, I, K^3 = [c, d, a, b]$ and $IK^3 = [b, a, d, c]$. Also let $\Lambda = [ab, b, c, d]$; $\Lambda^{-1} = [ab^{-1}, b, c, d]$, $\Lambda^{\pm n} = [ab^{\pm n}, b, c, d]$. All the Λ^n are distinct and belong to different classes, hence there are infinitely many different isomorphisms and isomorphism classes. *Let $\mathscr{L}$ be the group generated by I, K and Λ.*

Let $|\alpha|$ be the number of edges in the net path α (sum of the absolute values of the exponents in α), let $|\beta|$ be that of β etc. Call $Z(I) = |\alpha| + |\beta| + |\gamma| + |\delta|$ the "order" of I. We then have the following theorem, whose proof depends upon an intensive investigation of properties of the group diagram: *If $Z(I) > 4$, then $Z(I)$ may be reduced by transformation*[3] *of I or by premultiplication of I by an isomorphism in the group $\mathscr{L}$, when I is given in basic form.* By the properties of the basic form, the process can be continued until $Z = 4$. Then the isomorphism is necessarily one of the 4 isomorphisms in the above group $\mathscr{V}$, which also belong $\mathscr{L}$. Thus $\mathscr{L}$ contains a representative of each isomorphism class. But it also turns out that

$$(K^5\Lambda)^2K\Lambda(K^4\Lambda K\Lambda)^2(K^5\Lambda)^2K\Lambda = [a, aba^{-1}, aca^{-1}, ada^{-1}].$$

One concludes that $\mathscr{L}$ contains all elements of each isomorphism class, and hence *all isomorphisms of $\mathscr{F}$*. One also has: $IK = K^5I$, $I\Lambda = K^2\Lambda^{-1}K^4I$; hence each isomorphism can be written as an expression in K, Λ alone or else as an expression in K, Λ multiplied by I—according as the mapping preserves the orientation of Ω or not. Those isomorphisms of $\mathscr{F}$ for which $R(\alpha\beta\gamma\delta) = \phi R(abcd)^{+1}\phi^{-1}$ when written in basic form therefore constitute a normal subgroup of index 2 in the group of all isomorphisms, and this subgroup may be generated by K and Λ alone. This proves Theorem 2 and shows that $\mathscr{L}$ is the group of isomorphisms of $\mathscr{F}$. The group $\mathscr{G}$ of mapping classes is isomorphic to $\mathscr{L}$ when one regards isomorphisms which result from each other by transformation as equal.

III. In the proof of the above theorem about $Z(I)$—which indeed is the heart of the investigation—it appears possible to do without the assumption that the basic form in question represents an isomorphism of $\mathscr{F}$. It then follows, since each basic form with $Z = 4$ is one of the 4 basic forms in $\mathscr{V}$, that *each basic form is an isomorphism of $\mathscr{F}$*, and at the same time, since I, K, Λ are also isomorphisms of $\mathscr{F}^*$: *a sufficient condition for α, β, γ, δ to define an isomorphism of $\mathscr{F}^*$ is that $R(\alpha\beta\gamma\delta) = \phi R(abcd)^{\pm 1}\phi^{-1}$*. It follows from the irrelevance of order of a, b, c, d in $\mathscr{F}^*$ that the same theroem holds for each permutation of a, b, c, d. Thus, for the isomorphism $[ac^{-1}, b, c, d]$ of $\mathscr{F}^*$ mentioned above we have $R(\alpha\gamma\beta\delta) = R(acbd)$. It now becomes clearer why there are more isomorphisms of $\mathscr{F}^*$ than there are mappings of Ω^*. As a fundamental system on Ω^*, a, b, c, d possess a natural pairing, which they do *not* have in their role as generators of $\mathscr{F}^*$. The foregoing theorem is the generalisation of the theorem for the free group $G(a, b)$ on 2 generators: a *necessary and sufficient* condition for $[\alpha, \beta]$ to be an isomorphism of $G(a, b)$ is that $\alpha\beta\alpha^{-1}\beta^{-1} = \phi(aba^{-1}b^{-1})^{\pm 1}\phi^{-1}$; each isomorphism of $G(a, b)$ is induced by a mapping of the perforated torus (cf. Math. Ann. **78**, p. 397, [N2]).

IV. *Connection with groups of linear transformations.* Each group determines an abelian group, obtained by setting all commutators of generators equal to 1. The element $\alpha(a, b, c, d)$ goes to $a^{\mathrm{A}_\alpha}b^{\mathrm{B}_\alpha}c^{\Gamma_\alpha}d^{\Delta_\alpha}$, where A_α is the exponent sum of a in the expression α. $I = [\alpha, \beta, \gamma, \delta]$ determines the matrix

$$M(I) = \begin{pmatrix} \mathrm{A}_\alpha & \mathrm{B}_\alpha & \Gamma_\alpha & \Delta_\alpha \\ \mathrm{A}_\beta & \mathrm{B}_\beta & \Gamma_\beta & \Delta_\beta \\ \mathrm{A}_\gamma & \mathrm{B}_\gamma & \Gamma_\gamma & \Delta_\gamma \\ \mathrm{A}_\delta & \mathrm{B}_\delta & \Gamma_\delta & \Delta_\delta \end{pmatrix}.$$

If one sets $\{\alpha\beta\} = \begin{vmatrix} \mathrm{A}_\alpha & \mathrm{B}_\alpha \\ \mathrm{A}_\beta & \mathrm{B}_\beta \end{vmatrix} + \begin{vmatrix} \Gamma_\alpha & \Delta_\alpha \\ \Gamma_\beta & \Delta_\beta \end{vmatrix}$, then $M(I)$ is characterised arithmetically by

[3] I.e. by composition with an inner automorphism (Translator's note).

$$\{\alpha\gamma\} = \{\alpha\delta\} = \{\beta\gamma\} = \{\beta\delta\} = 0, \qquad \{\alpha\beta\} = \{\gamma\delta\} = \pm 1,$$

because $M(I)$, $M(K)$ and $M(\Lambda)$ satisfy these equations, and matrices of this kind constitute a group. (Composition of isomorphisms corresponds to multiplication of matrices.) We have $|M(I)| = 1$. The linear transformations which correspond to these matrices (C. Jordan's "substitutions Abéliennes") are known to occur as period transformations in the hyperelliptic case ($p = 2$), and generators of their group $\mathscr{H}$ are known (Jordan, Kronecker, Burkhardt). Their number theoretic characterisation along the above lines—basically an assertion about the minimum number of intersections of certain curve types on Ω—is therefore the topological core of the "Riemann bilinear relation". $\mathscr{H}$ is obviously a quotient group of $\mathscr{G}$, and hence of $\mathscr{L}$. *What is the associated normal subgroup, $\mathscr{K}$, i.e. which are the isomorphisms of $\mathscr{F}$ whose matrix is the identity?* The identity $[a, b, c, d]$ is one such, and so is each isomorphism $[\phi a\phi^{-1}, \phi b\phi^{-1}, \phi c\phi^{-1}, \phi d\phi^{-1}]$ in its class, likewise each isomorphism in the class of $[a, b, k^n ck^{-n}, k^n dk^{-n}]$ for $k = aba^{-1}b^{-1}$ and arbitrary n. Together they constitute an infinite cyclic group $\mathscr{C}$ of classes. The question whether this exhausts $\mathscr{K}$ is, by the proof of Theorem 1, equivalent to the question whether an isomorphism which determines the identity matrix necessarily sends a, b, c, d to elements homotopic to themselves. If α goes to $a^1b^0c^0d^0$ in the abelian group, then α is called "*homologous*" to a (Poincaré) and it then bounds a part of Ω in combination with a. If α is simple, *and if one assumes that α and a are disjoint*, then they divide Ω into two regions Ω_1 and Ω_2, each with two boundary curves. If either of these were to admit a non-separating loop cut, the latter would be a third such cut on Ω, contrary to $p = 2$. Thus, e.g. Ω_1 is planar and α is *homotopic* to a. When α has 2 intersections with a, homology again implies homotopy (because the curves are simple); *however, this is no longer true when one has 4 or more intersections.* Therefore, $\mathscr{C}$ does *not* exhaust the group $\mathscr{K}$. E.g. one constructs the following isomorphism from K and Λ alone:

$$\begin{aligned}
&K^3(\Lambda^{-1}K^5\Lambda^{-1}K)^3\Lambda K^3\Lambda^{-1}K^5\Lambda^{-1}K\Lambda^{-1}K^4\Lambda^{-1}K^5\Lambda^{-1}K^3\Lambda K^3\Lambda^{-1}K^4\Lambda^2K\Lambda^{-1}K^5\Lambda^{-1}\\
&K^4\Lambda^2K\Lambda K^4\Lambda^{-1}K\Lambda^{-1}K\Lambda^{-1}K^4\Lambda K^3\Lambda K^2\Lambda K\Lambda^2K^5\Lambda K\Lambda K^3\\
&= [b^{-1}a^{-1}d^{-1}aba^{-1}cd^{-1}c^{-1}ab^{-1}a^{-1}dabcdc^{-1}ab, \phi b\phi^{-1},\\
&(b^{-2}a^{-1}d^{-1}aba^{-1}c)aba^{-1}cd^{-1}c^{-1}b^{-1}cdc^{-1}ab^{-1}a^{-1}d^{-1}aba^{-1}cd\\
&(b^{-2}a^{-1}d^{-1}aba^{-1}c)^{-1}, \psi d\psi^{-1}],
\end{aligned}$$

$$\text{where} \quad \phi = cd^{-1}c^{-1} \quad \text{and}$$

$$\psi = b^{-2}a^{-1}d^{-1}aba^{-1}cd^{-1}c^{-1}ab^{-1}a^{-1}daba^{-1}cd^{-1}c^{-1}bcdc^{-1}ab^{-1}a^{-1}.$$

It determines the identity matrix; β and δ are homotopic to b and d, but α is not homotopic to a and γ is not homotopic to c. Thus, on a Riemann surface with $p = 2$ one already has *simple period paths which give the same periods without being homotopic.* This shows that, when two canonical cut systems give the same system of elementary periods, it does not follow that they are topologically the same, when $p \geq 2$. When $p > 2$ it still does not follow when a and α are disjoint.

V. *Summary.* We now have a three part answer to the initial question about higher analogues of the modular group:

1. If one views the modular group as a group of linear transformations, then its analogue

is the group $\mathscr{H}$ of linear transformations characterised above. It remains a desideratum for this group, generators of which are known—one need only take the matrices of I, K, Λ—to find a system of independent relations, and hence to characterise the group abstractly.

2. If one views the modular group as the group of mappings of the torus which fix O (or the group of mapping classes, which comes to the same thing for $p = 1$), then its analogue is the above group $\mathscr{L}$, generated by I, K, Λ, and hence the group of mappings of Ω which fix O (or the homomorphic group of mapping classes), which is the same as the group of isomorphisms (resp. isomorphism classes) of $\mathscr{F}$. Its homomorphism onto $\mathscr{H}$ results, briefly, from the "projection" of $\mathscr{F}$ onto its abelianisation, and hence by passing from homotopy to the much less discriminating relation of homology. The desiderata which remain are, above all, (i) presentation of an independent relation system for I, K, Δ; (ii) determination of the normal subgroup $\mathscr{L}/\mathscr{H} = \mathscr{K}$ which corresponds to the identity matrix.
3. If one views the modular group as the group of mappings of the punctured torus which fix O, and hence as the isomorphism group of the free group $G(a, b)$, then its analogue is the isomorphism group of the free group $\mathscr{F}^*(a, b, c, d)$, or more narrowly, the group of isomorphisms of $\mathscr{F}^*$ which correspond to mappings of Ω^*. The concept of a basic form opens the way to the latter group. Passing from it to $\mathscr{L}$ corresponds to "projection" of $\mathscr{F}^*$ onto $\mathscr{F}$ and therefore requires determination of those isomorphisms of $\mathscr{F}^*$ which correspond to the identity isomorphism of $\mathscr{F}$. This is the deepest formulation of the problem, which includes the previous two. Again one comes, as is usually the case in the last analysis of abstract group theory, back to groups without relations.

Math. Scientist, 1981, **6**, 73–85

On Calculation with Non-commutative Factors and its Application to Group Theory

J. Nielsen

(Received: 20 October 1980)

Editor's introduction

Jakob Nielsen's paper 'Om Regning med ikke-kommutative Faktorer og dens Anvendelse i Gruppeteorien', *Mat. Tidskr.* B (1921), 77–94, is fundamental to what is now called combinatorial group theory. In it, the 'Nielsen reduction' of a system of elements of a free group is first described, and used to prove the theorem that a (finitely generated) subgroup of a free group is free. (The condition that the subgroup be finitely generated was later removed, by a different method, by Otto Schreier, 'Die Untergruppen der freien Gruppen', *Abh. math. Sem. Hamburg. Univ.* **5** (1927), 161–183; the theorem is now known as the Nielsen–Schreier theorem.)

Nielsen's paper is a classic, often quoted, but rarely read, because it is in Danish. It is still, 60 years after it was first published, worth reading. To make it accessible to a wider audience, it has now, I believe for the first time, been translated into English. The translation was prepared by Anne W. Neumann with assistance from Walter D. Neumann.

For an obituary of Jakob Nielsen (15 October 1890–3 August 1959) and a list of his publications, see Werner Fenchel, 'Jakob Nielsen in memoriam', *Acta Math.* **103** (1960), vi–xix.

B. H. Neumann

CSIRO Division of Mathematics and Statistics, P.O. Box 1965, Canberra City, ACT 2601, Australia

Let $a_1, a_2, \ldots, a_n$ be a system of n things, which we shall call 'generators'. With each a_i, its 'reciprocal', a_i^{-1}, is also assumed to be given. By 'element' we understand a 'product' of these n generators and their reciprocals, $\prod (a_i^{\pm 1})$, where, however, it is not permitted to reorder the 'factors'; the element $a_4 a_1^{-1} a_7^3 a_1 a_4^{-2}$ is thus different from $a_4 a_1^{-1} a_4^{-2} a_1 a_7^3$. A generator can cancel with its reciprocal if they stand immediately next to each other in a product: $a_2^3 a_5^2 a_5^{-3} = a_2^3 a_5^{-1}$. If all factors in a product are cancellable in this way, the product is denoted by 1; for example, $a_i a_i^{-1} = a_i^{-1} a_i = 1$. For an arbitrary element e, one defines its 'reciprocal', e^{-1}, such that $ee^{-1} = e^{-1}e = 1$; the generators in e^{-1} are thus obtained by reading them backwards in e and changing the signs of their exponents. Example:

$$e = a_4 a_5^{-2} a_3^3 a_4^{-1}; \qquad e^{-1} = a_4 a_3^{-3} a_5^2 a_4^{-1}.$$

1. Reduction of a given system of elements

Now, let $\alpha_1, \alpha_2, \ldots, \alpha_m$ be m arbitrary elements. We can assume that they are written in uncancellable form. Further, let β be any element. We say that the system

$\alpha_1, \alpha_2, \ldots, \alpha_m$ 'generates' the element β if one can find a product $\prod(\alpha_i^{\pm 1})$ which, possibly with cancellation, can be shown to equal β. Thus, given $\alpha_1 = a_1 a_2 a_1^{-2}$, $\alpha_2 = a_2^{-1} a_1^{-2}$, and $\beta = a_1 a_2^2 a_1 a_2^2$, one sees that β is equal to $\alpha_1 \alpha_2^{-1} \alpha_1 \alpha_2^{-1}$. One system of elements is said to generate another if it generates every element in it. Two systems that mutually generate each other are said to be 'equivalent' or to belong to the same 'class'.

We shall denote the system $\alpha_1, \alpha_2, \ldots, \alpha_m$ by S, the system $\alpha_1^{-1}, \alpha_2^{-1}, \ldots, \alpha_m^{-1}$ by S^{-1}, and the system $\alpha_1, \ldots, \alpha_m, \alpha_1^{-1}, \ldots, \alpha_m^{-1}$ by $S + S^{-1}$. By $g(\alpha_1)$, we shall denote the number of generators in α_1, that is the sum of the absolute values of all the exponents that appear in α_1. Then $g(\alpha_1) = g(\alpha_1^{-1})$. Furthermore, $\sum_{i=1}^m g(\alpha_i)$ is denoted by $g(S)$. We shall now seek to replace S by a system S' which is equivalent to S and for which $g(S') < g(S)$. The suffixes in S are assumed to be so arranged that $g(\alpha_1) \leqq g(\alpha_2) \leqq \cdots \leqq g(\alpha_m)$; what a system generates is clearly independent of the order within it. We now seek to achieve a reduction of g-values by multiplying the elements in pairs. Thus we form the elements $\alpha_i\alpha_k, \alpha_i^{-1}\alpha_k, \alpha_k\alpha_i, \alpha_k\alpha_i^{-1}$ for $i = 1, 2, \ldots, m-1$ and $k > i$, and we shall obtain the desired reduction if more than half of α_i can be cancelled in such a product. For example, if $g(\alpha_i\alpha_k) < g(\alpha_k)$, then we define the new system S' by: $\alpha'_\rho = \alpha_\rho$ for $\rho \neq k$, and $\alpha'_k = \alpha_i\alpha_k$. Thus S generates S'. But as $\alpha_\rho = \alpha'_\rho$ for $\rho \neq k$ and $\alpha_k = \alpha_i'^{-1}\alpha'_k$, then S' also generates S; and $g(S') < g(S)$. If $\alpha'_k = 1$, it is omitted. The combinations one would get by considering the case $k = i$ do not produce such an equivalent change of the system; besides, for $\alpha_i \neq 1$ one will always have $g(\alpha_i^2) > g(\alpha_i)$. The same procedure applies in the cases when $\alpha_i^{-1}\alpha_k$ or $\alpha_k\alpha_i$ or $\alpha_k\alpha_i^{-1}$ exhibit the desired reduction. The elements of S' are now reordered according to g-value; then one replaces S' by the same method with an equivalent system S'' such that $g(S'') < g(S')$, and so forth. The sequence of systems thus obtained is finite, as the g-values continually decrease. Let $S^{(p)}$ be the last system in the sequence, so that $g(S^{(p)})$ can not be further reduced by the above procedure. Thus there are no two elements with different suffix in $S^{(p)} + S^{(p)-1}$ such that more than half of one can be cancelled in one of their two products. Next, we try to form products of three elements such that the middle element cancels completely between the other two; this is possible only if the middle element has even g-value and one half of it cancels on each side. Let α_{h_1} be the first element in the system that has this property of being completely cancellable between two others, and let E_1 and E_2 be its two 'half-ends', so that $\alpha_{h_1} = E_1E_2$ and $g(E_1) = g(E_2) = \frac{1}{2}g(\alpha_{h_1})$. We shall now cause α_{h_1} to lose this property by performing equivalent changes of the system such that E_2 is replaced by E_1^{-1} wherever E_2 appears as an 'end' in the system, except in α_{h_1} itself. Thus, if $\alpha_k = E_kE_2$, $(k \neq h_1)$, we replace it by $\alpha'_k = \alpha_k\alpha_{h_1}^{-1} = E_kE_1^{-1}$; and if $\alpha_l = E_2^{-1}E_l$, we replace it by $\alpha'_l = \alpha_{h_1}\alpha_l = E_1E_l$. It is clear that only such elements $\alpha_k, \alpha_l, \ldots$ can come into consideration in these equivalent changes whose g-value is $\geq g(\alpha_{h_1})$, as otherwise more than half of them would cancel in these products. Since their g-values do not change in these equivalent changes, we can keep the old order of the system. None of the elements of the system now begin with E_2^{-1}, and none except α_{h_1} end in E_2; we shall therefore call E_2 an 'isolated end' in the system. None of the elements, if any, which precede α_{h_1} in the order of the system can in these equivalent changes, which for the moment we can call α_{h_1}-changes, have acquired the property of being cancellable between two other elements of the system, half to each side; for, whether such elements have undergone an α_{h_1}-change or not, their half-ends will have g-value $\leqq g(E_1)$; the elements which, in any of the α_{h_1}-changes, have acquired a new form either begin with

E_1 or end in E_1^{-1}; but the end E_1 had already been present in the system before the α_{h_1}-changes, for example in α_{h_1}; and if the element in question has itself undergone an α_{h_1}-change, then it previously had a half-end E_2; but this also had already been present in the system before the α_{h_1}-changes, for example in α_{h_1}. These α_{h_1}-changes have possibly meant that in the new system there is an element (or several) of which more than half can be cancelled in a product with some other element. Then we again apply the first procedure, by which we achieve a reduction of the g-value of the system. If this is not the case, we determine the next element α_{h_2} in the sequence, $(h_2 > h_1)$, with the property of being cancellable between two others—if there is such an element at all—and perform with it the α_{h_2}-changes of the system that correspond to the α_{h_1}-changes described above. Thereafter we once again investigate whether the first procedure can be applied, and so forth. By continuing to alternate the two procedures, in such a way that one always prefers the first whenever possible, one is led to a sequence of equivalent changes of the system S, which include changes of the first kind, by which the g-value of the system is reduced, and which therefore can occur only in finite number, and changes of the second kind, by which the g-value of the system is not changed; but a series of these latter changes, which either come between two changes of the first kind or at the ends of the sequence, consists of a bounded number of α_{h_1}-changes, next a bounded number of α_{h_2}-changes, with $h_2 > h_1$, and so on, with ever-increasing index h_i; these series are, accordingly, likewise finite. In the end one thus arrives at a system S^* which gives room for neither the one procedure nor the other. S^* will be called a 'reduced system corresponding to S'. S^* is equivalent to S. Its number of elements is less than or equal to the number of elements of S. Furthermore, $g(S^*) \leqq g(S)$. Its characteristic property is: *In no product whatsoever of two elements can more than half of one element cancel; in no product whatsoever of three elements can the middle one cancel completely between the other two.* (By this we naturally exclude the case that an element occurs together with its reciprocal.) Or, more explicitly: A product E of generators is called an 'isolated left (respectively, right) end' in a system S if one and only one of the elements of $S + S^{-1}$ begins with E (respectively, ends with E). Then for a reduced system the following holds: Every end of an element that contains more than half the generators of the element is isolated; of the half-ends in an element with even g-value at least one is isolated.

Example. Let the given system S be

$$\alpha_1 = a_4 a_2 a_1^{-1} a_3; \quad \alpha_2 = a_1^2 a_2^{-2} a_4^{-1}; \quad \alpha_3 = (a_2^{-1} a_1^{-1} a_3)^2; \quad \alpha_4 = a_4 a_2^2 a_1^{-1} a_2^{-1} a_1^{-1} a_3.$$

We shall subject this system to the reduction process, and denote the resulting sequence of systems by $S^{(1)}, S^{(2)}, \ldots$. We shall write out only the element that changes in the relevant equivalent change, and think of the others as written down without change; and in this simple example we shall refrain from reordering the elements each time according to g-value.

$$\alpha_4^{(1)} = \alpha_2 \alpha_4 = a_1 a_2^{-1} a_1^{-1} a_3,$$
$$\alpha_3^{(2)} = \alpha_3^{(1)} \alpha_4^{(1)-1} = a_2^{-1} a_1^{-1} a_3 a_1^{-1}.$$

We have thus twice achieved a decrease in the g-value of the system, but this is for the present no longer possible. The present four elements, $\alpha_1^{(2)} = \alpha_1, \alpha_2^{(2)} = \alpha_2, \alpha_3^{(2)}$, and $\alpha_4^{(2)} = \alpha_4^{(1)}$, no longer allow one to form a product of two factors that has smaller g-value than one of its factors. We shall, therefore, look for an element that can be completely cancelled between two others. This possibility arises here in only one way, in that $\alpha_1^{(2)}$ cancels completely in $\alpha_2^{(2)}\alpha_1^{(2)}\alpha_4^{(2)-1}$. We shall thus arrange that, for example, the right half-end of $\alpha_1^{(2)}$ becomes an isolated end in the system. This requires only a single equivalent change, namely:

$$\alpha_4^{(3)} = \alpha_4^{(2)}\alpha_1^{(2)-1} = a_1 a_2^{-1} a_4^{-1}.$$

Now we can again continue with the reduction of the g-value after the first method:

$$\begin{aligned}
\alpha_2^{(4)} &= \alpha_2^{(3)}\alpha_4^{(3)-1} = a_1,\\
\alpha_4^{(5)} &= \alpha_2^{(4)-1}\alpha_4^{(4)} = a_2^{-2}a_4^{-1},\\
\alpha_3^{(6)} &= \alpha_3^{(5)}\alpha_2^{(5)} = a_2^{-1}a_1^{-1}a_3,\\
\alpha_1^{(7)} &= \alpha_1^{(6)}\alpha_3^{(6)-1} = a_4 a_2^2,\\
\alpha_4^{(8)} &= \alpha_4^{(7)}\alpha_1^{(7)} = 1 \text{ and is thus thrown out.}
\end{aligned}$$

The three elements that remain, $\alpha_1^{(8)}, \alpha_2^{(8)}$, and $\alpha_3^{(8)}$, permit no further changes, either by the first or by the second procedure, and thus form the reduced system S^* for S. After arranging the elements by g-value and renaming them, we have:

$$\alpha_1^* = a_1\,(=\alpha_2^{(8)}); \qquad \alpha_2^* = a_2^{-1}a_1^{-1}a_3\,(=\alpha_3^{(8)}); \qquad \alpha_3^* = a_4 a_2^2\,(=\alpha_1^{(8)}).$$

We shall write down the equations that show the equivalence of the systems. How S^* is generated from S can be obtained by following the course of the reduction process:

$$\alpha_1^* = \alpha_2\alpha_1\alpha_4^{-1}\alpha_2^{-1}; \qquad \alpha_2^* = \alpha_3\alpha_4^{-1}\alpha_1\alpha_4^{-1}\alpha_2^{-1}; \qquad \alpha_3^* = \alpha_1\alpha_2\alpha_4\alpha_1^{-1}\alpha_4\alpha_3^{-1}. \tag{1}$$

This generation is, however, not unique; for example, one could replace the second equation by $\alpha_2^* = \alpha_2\alpha_4\alpha_1^{-1}\alpha_4$. How S is generated from S^* is obtained (for example by running through the reduction process in the opposite direction):

$$\alpha_1 = \alpha_3^*\alpha_2^*; \qquad \alpha_2 = \alpha_1^{*2}\alpha_3^{*-1}; \qquad \alpha_3 = \alpha_2^{*2}; \qquad \alpha_4 = \alpha_3^*\alpha_1^{*-1}\alpha_2^*. \tag{2}$$

This generation is unique, as will be shown in Section 2. That the equations (1) are a system of solutions to (2) with respect to the α^*'s follows completely formally, that is to say, without considering the significance of the α's as products of the original generators a_i. On the other hand, (2) is only seen to be a system of solutions of (1) when one inserts the expressions of the α's in terms of the a's.

2. Independent systems of elements

Let $\gamma_1, \gamma_2, \ldots, \gamma_r$ be r given elements in the generators $a_1, a_2, \ldots, a_n$, and $\Gamma = \prod(\gamma_i^{\pm 1})$ an arbitrary product of the γ's. If Γ turns out, after cancellation of the generators a_i, to be equal to 1, then $\Gamma = 1$ is said to be an 'identical relation between $\gamma_1, \gamma_2, \ldots, \gamma_r$ in the generators a_i'. It is called 'trivial' or 'inessential' if it is also 'identical in the elements γ_i', that is to say, if a γ_κ stands immediately next to γ_κ^{-1}, thus can be cancelled as a whole, and if this can be continued until one has cancelled all the γ's in Γ. If there exists no essential relation between $\gamma_1, \gamma_2, \ldots, \gamma_r$, they are said to be 'mutually independent' or to form an 'independent system'. The generators $a_1, a_2, \ldots, a_n$ themselves form an example of such a system, since they are assumed to be mutually distinct. As an example of a dependent system of elements we can use the elements $\alpha_1, \alpha_2, \alpha_3, \alpha_4$ introduced in Section 1, as we can easily set up essential relations between them, for example $(\alpha_2\alpha_4\alpha_1^{-1}\alpha_4)^2\alpha_3^{-1} = 1$. If we here replace S by S^* by means of (2), the relation becomes trivial, namely, identical in the α^*'s. We will now show that the following always holds.

Theorem 1. *A reduced system is independent.*

Proof. Let $\gamma_1, \ldots, \gamma_r$ be a reduced system and $\Gamma = \prod(\gamma_i^{\pm 1}) = 1$. If each factor γ_i in Γ, after cancellation with its preceding and succeeding factors, left a remainder, Γ would consist of all these remainders, between which no further cancellation could take place, contrary to $\Gamma = 1$. Therefore at least one of the factors must cancel completely between its two neighbours. But because of the property of a reduced system, this can occur only if one of the neighbours of that factor is its reciprocal. Their product can thus be eliminated, and, by continuing in the same way, one sees that $\Gamma = 1$ is a trivial relation. A relation between the elements of a reduced system which is an identity in the generators is, consequently, also an identity in the elements of the system.

We now prove the following lemma.

If $\gamma_1, \gamma_2, \ldots, \gamma_r$ *are mutually independent* (*dependent*), *then* $\gamma_1\gamma_2, \gamma_2, \ldots, \gamma_r$ *are also mutually independent* (*dependent*).

Let $S = \gamma_1, \gamma_2, \ldots, \gamma_r$ be an independent system. We set $S' = \gamma_1', \gamma_2', \ldots, \gamma_r'$, where $\gamma_1' = \gamma_1\gamma_2$, and $\gamma_j' = \gamma_j$ for $j = 2, 3, \ldots, r$. Let $\Gamma = \prod(\gamma_i'^{\pm 1}) = 1$ be a valid relation for S'. In order to emphasize γ_1', we shall write it in detail in the following form: $\Gamma = P_1\gamma_1'^{\varepsilon_1}P_2\gamma_1'^{\varepsilon_2}\cdots P_\rho\gamma_1'^{\varepsilon_\rho}P_{\rho+1}$, where all $\varepsilon_i = +1$ and all P_i are products of $\gamma_2', \ldots, \gamma_r'$ or 1. If we introduce S instead of S', we get:

$$\Gamma = P_1(\gamma_1\gamma_2)^{\varepsilon_1}P_2(\gamma_1\gamma_2)^{\varepsilon_2}\cdots P_\rho(\gamma_1\gamma_2)^{\varepsilon_\rho}P_{\rho+1},$$

where each P_i contains only $\gamma_2, \ldots, \gamma_r$. Seeing that $\Gamma = 1$ is an identity in the γ's, the ρ factors γ_1 that occur in Γ must mutually cancel; therefore in at least one place in Γ two factors γ_1 will occur with opposite exponent, separated by a product in $\gamma_2, \ldots, \gamma_r$ which cancels identically in $\gamma_2, \ldots, \gamma_r$. If we assume that the first of these factors γ_1 has, for example, exponent $\varepsilon_{\lambda-1} = +1$, we get:

$$\Gamma = \cdots\gamma_1\gamma_2P_\lambda\gamma_2^{-1}\gamma_1^{-1}\cdots,$$

where $\gamma_2P_\lambda\gamma_2^{-1} = 1$ identically in the γ's, whence $P_\lambda = 1$ identically in the γ's, and thus identically in $\gamma_2', \gamma_3', \ldots, \gamma_r'$. But then $\gamma_1'^{\varepsilon_{\lambda-1}}P_\lambda\gamma_1'^{\varepsilon_\lambda}$ can also be cancelled identically in the γ_i'. And by continuing with the same argument, one sees that $\Gamma = 1$ is a trivial relation in the γ_i'. These are, therefore, independent. The converse theorem follows immediately from the theorem itself by an indirect proof, since one generates S from S' by setting $\gamma_1 = (\gamma_1\gamma_2)\cdot\gamma_2^{-1}$. If S is dependent, and if $\gamma_1\gamma_2 = 1$ and this is eliminated, then $\gamma_2, \gamma_3, \ldots, \gamma_r$ can, of course, be independent.

It is clear that a system which results from a dependent system by the addition of an arbitrary element will also be dependent.

Since the property of a system to be independent (dependent) is not disturbed if one permutes its elements and inverts an arbitrary number of them, one sees immediately that if $\gamma_1, \gamma_2, \ldots, \gamma_r$ are mutually independent (dependent), then the system that results from replacing γ_i by $(\gamma_i^{\pm 1}\gamma_k^{\pm 1})^{\pm 1}$ and leaving all other elements unchanged will also be independent (dependent). By continuing to make this kind of equivalent change in an independent (dependent) system, one thus obtains a whole sequence of independent (dependent) systems which are mutually equivalent. In particular, one sees:

Theorem 2. *A system of elements which has the same number of elements as a reduced system derived from it is independent.*

For in this case the reduction process results only in the above-mentioned equivalent changes; and the reduced system is independent.

Theorem 3. *A system of elements which has a greater number of elements than a reduced system derived from it is dependent.*

For in the course of the reduction process in at least one step an element will be eliminated. The preceding system must then be dependent; and all the systems which precede it, and thus also the given system, must then likewise be dependent.

3. The generation problem

Let a system S of m elements $\alpha_1, \alpha_2, \ldots, \alpha_m$ and an element $\beta \neq 1$ be given. We shall seek a method to decide whether or not S generates β. By means of the method given in Section 1 we construct to S a reduced system $S^* = \alpha_1^*, \alpha_2^*, \ldots, \alpha_p^*$, where, of course, $p \leq m$. Since S^* is equivalent to S, we have only to decide whether S^* generates β. This is clearly the case if there exists a product representation $\beta = \prod_1 (\alpha_i^{\pm 1})$; if there exists simultaneously a product representation $\beta = \prod_2 (\alpha_i^{\pm 1})$, then $\prod_1 \prod_2^{-1} = 1$ is an identity in the generators a. It will then, by Theorem 1, also be an identity in the elements α^*, and both representations will thus be identical, as we can take them to be written without cancellation in the α_i^*. Thus if there exists a generation of β by S^*, then it is also unique. Each factor in this product leaves a remainder after cancellation, and β is the product† of these remainders. Let $\bar{\alpha}$ be the first factor; the remainder that is left over in β is either more than half of $\bar{\alpha}$, or exactly half and then an isolated half. To determine the first factor in a quick and sure manner, we can write down what we can call the 'collection of left large ends' of S^*, where by the 'left large end' of an element we mean the first $r + 1$ generators in the element when its g-value is $2r$ or $2r + 1$; this collection will then consist of the $2p$ left large ends in $S^* + S^{*-1}$. Furthermore we will write down the 'collection of left half-ends' of S^*, into which we admit each left half-end of an element with even g-value in $S^* + S^{*-1}$ whose complementary half-end is not isolated, that is,

† The original has 'Samling', that is 'collection'. [Editor]

occurs in at least two places in S^*. If we now assume that β is generated by S^*, then one of two cases will occur. Either β will begin with one (and only one) large end of the first collection, for example with the left large end in $\alpha_k^{*\varepsilon}$. Then the problem of generating β will be reduced to the problem of generating $\bar{\beta} = \alpha_k^{*-\varepsilon}\beta$, where $g(\bar{\beta}) < g(\beta)$. Or β will not begin with such a large end, but instead with a left half-end that is found in the second collection, for example with the left half-end in $\alpha_l^{*\eta}$. The problem will thus be reduced to the problem of generating $\bar{\beta} = \alpha_l^{*-\eta}\beta$. This will also be a reduction of the problem. For indeed $g(\bar{\beta}) = g(\beta)$, but $\bar{\beta}$ will, if it does not begin with a large end from the first collection, only be able to begin with a half-end of the second collection whose g-value is $>\frac{1}{2}g(\alpha_l^*)$. By continuing with such reductions one can, therefore, very possibly get a sequence of elements $\bar{\beta}, \bar{\bar{\beta}}, \ldots$, which all have the same g-value as β, but yet only a finite sequence, namely in any case no more than there are half-ends in the second collection. It is therefore clear that the generation problem can, by the construction of the two collections described above, be solved in a finite number of steps. One obtains a finite sequence of elements $\beta, \bar{\beta}, \bar{\bar{\beta}}, \ldots, \overset{(q)}{\beta}$, such that each element in the sequence is uniquely determined by those preceding it and by the two collections. β is generated by S^* if $\overset{(q)}{\beta} = 1$, and only in this case. In other words: either we come in this process at some time to an element $\overset{(q)}{\beta}$ which no longer begins with a sequence of generators that is found in one of the two collections, or we get β decomposed into its factors in S^*. From the generation of β by S^*, we naturally can afterwards, without difficulty, find its generation by S. We can say that the generation of β by S, if it is possible, becomes evident when we replace S by S^*.

Example. Let S be the system $\alpha_1, \alpha_2, \alpha_3, \alpha_4$ used in Section 1. From the reduced system S^* constructed there, we find the following collection of left large ends:

$$a_1, a_2^{-1}a_1^{-1}, a_4a_2, a_1^{-1}, a_3^{-1}a_1, a_2^{-2}$$

and no collection of half-ends. Let β be

$$a_1a_4a_2a_1^{-1}a_3a_1^{-1}a_3^{-1}a_1a_2^{-1}a_4^{-1}.$$

The example is so chosen that S generates β. Even if one knows this in advance, it would still, even in this simple and in no way artificially chosen case, be extraordinarily laborious to feel one's way to a representation of β by S. But with the help of the collection of large ends one immediately reads off the representation of S^*. β begins with a_1, $\bar{\beta} = \alpha_1^{*-1}\beta$ begins with a_4a_2, which is a large end in α_3^*. $\bar{\bar{\beta}} = \alpha_3^{*-1}\bar{\beta} = a_2^{-1}a_1^{-1}a_3a_1^{-1}a_3^{-1}a_1a_2^{-1}a_4^{-1}$ begins with $a_2^{-1}a_1^{-1}$, and so forth. One finds in this way that

$$\beta = \alpha_1^*\alpha_3^*\alpha_2^*\alpha_1^{*-1}\alpha_2^{*-1}\alpha_3^{*-1},$$

and can then with the help of the equations (1) write down the representation of β by S, or, more accurately: one of the representations of β by S, as the equations (1) are of course not uniquely determined.

4. Application to group theory. Isomorphism

The collection of all elements generated by $a_1, a_2, \ldots, a_n$ is seen to have the following properties: (1) Two arbitrarily given elements e_1 and e_2 determine 'by composition' (multiplication) a new element of the collection: e_1e_2, and this multiplication is associative, that is to say, $(e_1e_2)e_3 = e_1(e_2e_3)$; (rule of composition and associativity). (2) There exists an element in the collection which, upon composition with an arbitrary element, does not alter it (unit element, 1). (3) To each element e there exists one element —we denote it by e^{-1}—which satisfies the equations $ee^{-1} = e^{-1}e = 1$ (inverse). A collection of things which satisfies these three conditions is said, as is well known, to form a group. In our case we will call it the 'free group G_n' generated by the generators $a_1, \ldots, a_n$ and write it $[a_1, a_2, \ldots; a_n]$. A collection of the elements that are generated by a system of elements $S = \alpha_1, \alpha_2, \ldots, \alpha_m$ likewise fulfils the three conditions mentioned. The group $[\alpha_1, \alpha_2, \ldots, \alpha_m]$ that they form is a subgroup of G_n, since all its elements are clearly elements of G_n. (It must be said from the outset, however, that one can not expect to get all subgroups of G_n in this way, as a collection of elements of G_n that fulfils the group conditions can not always be generated by a finite system of elements; compare Dehn [2].) With respect to the group $[\alpha_1, \alpha_2, \ldots, \alpha_m]$, we also use the name 'generators' for $\alpha_1, \alpha_2, \ldots, \alpha_m$. This system of generators is by no means determined by the group; we can always replace it by an equivalent system. All systems of the same class accordingly determine the same subgroup of elements of G_n.

Here a difference must be noticed in the identification of the elements of the groups in the different cases. In G_n, two distinct products of the a_i, irreducible in the a_i, always describe two different elements. It is for this reason that G_n is called a free group. In $[\alpha_1, \alpha_2, \ldots, \alpha_m]$, two distinct products of the a_i, irreducible in the α_i, can determine the same element, namely if they are seen to be identical as soon as one writes for the α_i their expressions in the a_i. However, one can, of course, always replace $\alpha_1, \alpha_2, \ldots, \alpha_m$ by a reduced system $\alpha_1^*, \alpha_2^*, \ldots, \alpha_p^*$, and two distinct products in these will always produce different elements of the group, since the α_i^* are, by Theorem 1, independent. The group $[\alpha_1, \alpha_2, \ldots, \alpha_m]$, represented, therefore, in the form $[\alpha_1^*, \alpha_2^*, \ldots, \alpha_p^*]$, is a free group, and all subgroups of G_n that are determined by a finite system of elements can accordingly be represented in this way.

It will now be of special interest to know whether the group $[\alpha_1, \ldots, \alpha_m]$ includes all the elements of G_n, and therefore coincides with G_n. In this case we call $S = \alpha_1, \alpha_2, \ldots, \alpha_m$ a 'primitive system'. A necessary and sufficient condition for this will be that S generates $a_1, a_2, \ldots, a_n$; but this is the same as a reduced system $S^* = \alpha_1^*, \ldots, \alpha_p^*$ of S generating $a_1, \ldots, a_n$, and that will only occur, by the solution of the generation problem, if each a_i occurs among the α_j^* with exponent either $+1$ or -1. Thus S^* can not contain more elements, as otherwise it would not be independent. The condition therefore is that S^* is a permutation of the original generators $a_1, \ldots, a_n$ provided with exponents $+1$ or -1. We shall thus have $p = n$ and $m \geqq n$.

Let us consider the case that $m = n$, thus that there is no element of S eliminated in the reduction process. In this case the system S is independent, by Theorem 2. One can thus express the relations between S and the original system of generators $A = a_1, \ldots, a_n$ as follows: (1) S generates A, and A generates S. (2) A and S have the same number of elements. (3) If one pairs the elements of S in a one-to-one way with those of

A, then every valid relation in A will give rise to a valid relation in S by replacing each a by the corresponding α, and conversely. (This is simply because the relations in both cases will be identities.) Such a pairing of the original generators in a particular order with the n elements $\alpha_1, \ldots, \alpha_n$ is called an 'isomorphic transformation' or 'autoisomorphism'† of the group G_n. The group is 'mapped' onto itself. For a system of elements to determine such an autoisomorphism of G_n, it must accordingly be an independent primitive system, hence have n elements and generate A. One can, without great difficulty, give an account of those properties of independent primitive systems that are of particular interest in the free group G_2 with only two generators; see Nielsen [4], [5]. If one wishes to trace the deeper properties of the group, then one must deprive the system A of its special status, as it is of course in no way distinguished as against other independent primitive systems. What is interesting about a group is its 'intrinsic' properties, which do not depend upon the fortuitous representation of the group, but are invariant under isomorphic transformation.‡

Here only one further property of such an independent primitive system $S = \alpha_1, \ldots, \alpha_n$ will be mentioned. In α_i the generators a_k will in general occur several times and with both positive and negative exponents. We shall denote the algebraic sum of the exponents of a_k in α_i by d_{ik} and form the square array of numbers (a 'matrix') from these n^2 numbers d_{ik}. For the system A itself, this matrix is seen to be the 'unit matrix':

$$\begin{matrix} 1 & 0 & \cdots & 0 \\ 0 & 1 & \cdots & 0 \\ \cdots & \cdots & \cdots & \cdots \\ 0 & 0 & \cdots & 1 \end{matrix}$$

Its determinant is thus in this case $|d_{ik}| = +1$. We can now pass from A to S by a sequence of the following operations: (1) Change the sign in a single element; this means in the matrix: change the sign of one of the rows of the matrix. (2) Permute the elements: permute the rows of the matrix. (3) Replace α_i by $\alpha_i\alpha_k^{\pm 1}$: add the kth row with the factor ± 1 to the ith row. The matrix can thereby assume a complicated form, but its determinant will remain ± 1. A necessary condition for an isomorphic transformation is, therefore, that the determinant of the transformation $|d_{ik}|$ is equal to ± 1; but this condition is naturally not sufficient.

Clearly one here sees a connection with the usual linear equations. We assumed the system of n equations:

$$\alpha_i = \prod^{(i)} (a_j^{\pm 1}) \tag{3}$$

and defined the array of numbers d_{ij} mentioned above.

Let us simultaneously write down the n equations that express the a_i as a product of the α_k:

$$a_i = \Psi^{(i)}(\alpha_k^{\pm 1}) \tag{4}$$

† The term 'automorphism' was not yet generally accepted in 1921. [Editor]
‡ Such properties are now called 'characteristic'. [Editor]

and denote the algebraic sum of the exponents of α_k in a_i by δ_{ik}. If we substitute (3) in (4), we get the n equations

$$a_i = \Psi^{(i)}(\prod^{(k)} (a_j^{\pm 1})^{\pm 1}), \tag{5}$$

which will, of course, be identities in the a_i, as these are independent. A particular generator must therefore have, in a particular one of these equations (5), the same sum of exponents on both sides of the equality sign. We thereby get the n^2 equations:

$$\delta_{i1}d_{1l} + \delta_{i2}d_{2l} + \cdots + \delta_{in}d_{nl} = \begin{cases} 0 & \text{for } i \neq l, \\ 1 & \text{for } i = l. \end{cases} \tag{6}$$

And clearly what these equations (6) express is that n linear inhomogeneous equations in n indeterminates with the integers d_{ik} as coefficients have a system of solutions with the integers δ_{ik} as coefficients; but then one clearly must have $|d_{ik}| = \pm 1$. This connection can be even more palpably expressed when one recalls that the a_i can stand for anything whatever, and one so chooses them as such objects whose composition is commutative. Let us, for example, choose the a_i as positive numbers. We get thus by rearranging factors:

$$\alpha_i = a_1^{d_{i1}} \alpha_2^{d_{i2}} \cdots a_n^{d_{in}} \tag{7}$$

$$a_i = \alpha_1^{\delta_{i1}} \alpha_2^{\delta_{i2}} \cdots \alpha_n^{\delta_{in}} \tag{8}$$

and, taking logarithms and denoting them by the corresponding capital letters:

$$\mathrm{A}_i = d_{i1}A_1 + d_{i2}A_2 + \cdots + d_{in}A_n \tag{7a}$$

$$A_i = \delta_{i1}\mathrm{A}_1 + \delta_{i2}\mathrm{A}_2 + \cdots + \delta_{in}\mathrm{A}_n. \tag{8a}$$

This last system of n linear equations and its inverse, which hold for arbitrary values $A_1, A_2, \ldots, A_n$, show that $|d_{ik}| = \pm 1$. Within the commutative domain, (6) is a necessary and sufficient condition for the simultaneous validity of (7) and (8), or (7a) and (8a), within the non-commutative domain, on the other hand, it is only a necessary condition for the simultaneous validity of (3) and (4). This observation shows that the proposition $|d_{ik}| = \pm 1$ is not a profound proposition in calculating with non-commutative elements, as it has after all nothing to do with the non-commutative character of the calculation. We could have proved this simply by using the fact that the non-commutative domain covers the commutative one, a fact one can always use to advantage as control, and to find necessary conditions in calculating with non-commutative elements.

5. Equivalent reduced systems

A class of systems, that is, the collection of all systems equivalent to a given system and therefore to each other, will in general contain several reduced systems. Let $S = \alpha_1, \alpha_2, \ldots, \alpha_m$ and $T = \beta_1, \beta_2, \ldots, \beta_p$ be two equivalent reduced systems. We have then

$$\beta_i = \prod^{(i)} (\alpha_j^{\pm 1}) \tag{9}$$

but also

$$\alpha_i = \Psi^{(i)}(\beta_j^{\pm 1}) = \Psi^{(i)}(\textstyle\prod^{(j)}/(\alpha_k^{\pm 1})^{\pm 1}). \tag{10}$$

(10) is an identity in the α_j for all values of i. If we ignore for the moment that the α_i and the β_i are products of the original generators a_i and regard the α_i as generators of a free group G_m, then T is seen to be a primitive system in the α_i. From this it follows that $p \geqq m$. By reversing the roles of the systems, one gets $m \geqq p$, thus $m = p$.

Theorem 4. *All reduced systems of the same class have the same number of elements.* We shall call it the 'basic number of the class'.

In combination with Theorems 2 and 3 one sees now immediately:

Theorem 5. *The number of elements of an independent system is minimal within its class, and equal to the basic number of the class; and every system in the class with this number of elements is independent.*

Thus if a system is dependent, then every system equivalent to it with the same or a greater number of elements will likewise be dependent.

We will look more closely at the relations between the above two reduced systems S and T. Thus the transition from S to T is an isomorphic transformation in the α_i, and vice versa. Let d_{ik} be the algebraic sum of the exponents of α_k in β_i. We know that $|d_{ik}| = \pm 1$. Therefore at least one term of this determinant is $\neq 0$. We choose a particular term of this kind. This has m factors, all $\neq 0$, which are distributed over all the rows of the determinant and also over all its columns. This term thus defines a one-to-one pairing between the rows and the columns of the determinant, and therefore between the elements of T and of S in such a way that each element of S occurs as a factor in its corresponding element of T. Since the properties of the system clearly are independent of the order of its elements, we can take the subscripts in S and in T so arranged that α_i corresponds to β_i, that is, is a factor in β_i. It is easy to see that an arbitrary product of elements of a reduced system has a g-value at least as great as that of any of its factors. We have, therefore,

$$g(\beta_i) \geqq g(\alpha_i). \tag{11}$$

Here we have now not yet the right to reverse the roles of the systems, since we do not know whether the same pairing can be used for the generation of S by T. But we can deduce:

$$\sum_{i=1}^{m} g(\beta_i) \geqq \sum_{i=1}^{m} g(\alpha_i).$$

And as here we have the right to reverse the roles of the systems, it follows that

$$\sum_{i=1}^{m} g(\beta_i) = \sum_{i=1}^{m} g(\alpha_i),$$

hence in combination with (11)

$$g(\beta_i) = g(\alpha_i).$$

Theorem 6. *Equivalent reduced systems have the same g-value and the same distribution of this between the g-values of their individual elements.*

As one can find only a finite number of systems with the same g-value, one has with this

Theorem 7. *Every class of systems contains only a finite number of reduced systems.*

With the help of the basic properties of reduced systems, the reader will easily be able to construct the reduced systems equivalent to a given reduced system in a finite (and in general small) number of steps.

The reduced systems in a class distinguish themselves from the other independent systems in the class really only by their practical importance in solving a given generation problem, as in Section 3. In respect to the generation of a subgroup, we shall regard all independent systems as equally good. Of the above-mentioned class invariants, $m, \sum_{i=1}^{m} g(\alpha_i)$, and the system of numbers $g(\alpha_1), g(\alpha_2), \ldots, g(\alpha_m)$, it is therefore the first, the basic number of the class, which has real methodical significance. In establishing the isomorphic relationship between S and T in this section, we have used only that both were independent and equivalent, and none of their special properties as reduced systems. We have then this result:

Theorem 8. *If one knows that a given group is generated by an independent system of elements, one finds all other ways of generating the group by independent systems of elements by means of all possible isomorphic transformations applied to the elements of the first system.*

The representation of a given group as a free group, when this is possible, is therefore all but determined by an isomorphic transformation. The number of these is infinite, but one can determine them all in a systematic way by starting from a suitable finite set from among them and combining these (see [5], pp. 271–272).

General remarks on the application to the theory of groups

To characterise correctly the significance for the theory of groups of the preceding investigations, one must stress that it has its limitations, in that these methods are only adequate for the treatment of groups that can be expressed as free groups. As concerns the general discontinuous groups, this is only the simplest case. The theory of general discontinuous groups in the particularly interesting case that they are infinite but can be generated by a finite number of elements has recently been treated especially by M. Dehn.† The characteristic difficulty that these investigations present, in comparison with those carried out here, lies in the following fact:

Let $A = a_1, a_2, \ldots, a_n$ again be a system of generators of a group G, whose elements are thus given as products of the a's. The difference now lies in the identification of the

† Of a series of papers by M. Dehn on this subject, [2] is to be specially underlined here. [Author's footnote]

elements. Suppose a system of elements $R_1, R_2, \ldots, R_m$ is given, which are not 1 in the group G_n previously considered, that is to say, do not become equal to 1 by cancellation in the a's, but which by definition are stipulated to be 1 in G. These relations $R_i = 1$, which thus are not identities in the a's, are called 'defining relations'. It will now be permissible to delete the complex R_i of generators, wherever they occur in that order in an element, and likewise to insert them at pleasure. The expression of the elements in terms of the generators thereby becomes highly indeterminate. In this way the element 1 of the group can not only be written $\prod (a_i^{\pm 1})$, where this product is identically equal to 1 in the a's, but also, for example $\prod (R_i^{\pm 1})$, where this product can not be completely cancelled in the R's, nor in the a's either. But the real difficulty lies in the following fact: $\Phi R_i \Phi^{-1}$ will clearly also be 1, where Φ is an arbitrary element; therefore also any product $\prod (\Phi R_i^{\pm 1} \Phi^{-1})$ with arbitrarily many factors and arbitrary choice of Φ in each factor will be 1. The group can now also be defined by postulating that an element shall be equal to 1 if, and only if, it can be written in the form $\prod (\Phi R_i^{\pm 1} \Phi^{-1})$. Dehn appears to have been the first who has clearly emphasised that this belongs logically to the definition of the group. With that it is in principle decided whether two products P and Q of the generators stand for the same element in G: this is the case if, and only if, there exists an equation, identical in the a's,

$$PQ^{-1} = \prod (\Phi R_i^{\pm 1} \Phi^{-1}).$$

But at the same time there arises the problem of deciding in a finite number of steps whether or not such an equation exists, when P, Q and $R_1, \ldots, R_m$ are given by their expression in the a's. This 'identity problem', as Dehn has called it, is thus an extension of the 'generation problem' dealt with in Section 3, but its general solution has so far presented insuperable difficulties. The relationship between these two problems can be properly perceived if one compares the method set forth here in Section 3 with the solution of the identity problem that one knows in special cases (compare, for example, M. Dehn [3]), and the graphic treatment of the identity problem one arrives at by applying the 'group graphs' introduced by Dehn (see [1], especially p. 140 ff.). While in the solution of the generation problem one arranges elements in sequence in all possible ways—one could speak of a 'linear' problem—one represents the elements R_i in the group graph by polygons and in the treatment of the identity problem one puts these together in a map. This latter problem has, therefore, in an expression of Dehn's, 'one more dimension' than the first. Whether this is a real or only an apparent difference in level, in other words, whether one can succeed in constructing from the system $R_1, R_2, \ldots, R_m$ another finite system which allows one to reduce the given identity problem to a generation problem, remains undecided.

References

[1] Dehn, M. (1910) Über die Topologie des dreidimensionalen Raumes. *Math. Ann.* **69,** 137–168.
[2] Dehn, M. (1912) Über unendliche diskontinuierliche Gruppen. *Math. Ann.* **71,** 116–144.
[3] Dehn, M. (1912) Transformation der Kurven auf zweiseitigen Flächen. *Math. Ann.* **72,** 413–421.
[4] Nielsen, J. (1918) Die Isomorphismen der allgemeinen, unendlichen Gruppe mit zwei Erzeugenden. *Math. Ann.* **78,** 385–397.
[5] Nielsen, J. (1919) Über die Isomorphismen unendlicher Gruppen ohne Relation. *Math. Ann.* **79,** 269–272.

The Torus and the Plane*

JAKOB NIELSEN

In this paper we study point sets, especially curves, on the ring surface (torus), with particular attention to properties which depend on the connectivity of the surface. An example of a natural problem of this kind is that of finding the minimum number of intersection points of two given closed curves on a closed surface of higher connectivity, which in the simplest case is given by v. Staudt's theorem that two "curves of odd order" in the projective plane always intersect each other.[1] We shall begin by proving this theorem as a pointer towards the direction we should take in the investigation of the torus which follows.

We first consider a spherical surface K. If P is a point on K, we denote the diametrically opposite point by $\bar{P}$. If M is a point set on K we denote by $\bar{M}$ the set of "diametric images", i.e. the set of all points opposite those of M. If $\bar{M}$ is the same as M we call M "self-diametric" (a great circle, for example). If O is a region (i.e. a point set whose members are all interior points, any two of which can be connected by a polygonal path in O consisting of great circle segments), then $\bar{O}$ is also a region, and if P is a boundary point of O then $\bar{P}$ is a boundary point of $\bar{O}$. If O is bounded by a simple closed curve $\mathscr{F}$ (Jordan curve) on K then O cannot be self-diametric, because it follows from $O = \bar{O}$ that $\mathscr{F} = \bar{\mathscr{F}}$, and one could then give a one-to-one continuous (even mirror-image congruent in fact) map of $O + \mathscr{F}$ onto itself by sending each point P of $O + \mathscr{F}$ to $\bar{P}$; under this mapping the orientation of $O + \mathscr{F}$ is reversed, in particular, the sense of $\mathscr{F}$ is reversed, so the mapping will have a fixed point on $\mathscr{F}$—contrary to the fact that each point is mapped to its opposite. If $\mathscr{F}$ is a self-diametric Jordan curve it must therefore "halve" K into a region and its diametric image.

Let us consider a continuum Q on K (i.e. a closed point set which cannot be divided into closed subsets). The "complementary set" $K—Q$, i.e. the set of all points of K which are not in Q (we assume that Q is not all of K) then consists of a finite or countable number of regions. Let O be one of these, let P be a point of O, and let us test the assertion that O is self-diametric. We connect P to $\bar{P}$ by a simple polygonal great circle path Π in O. $\bar{\Pi}$ then lies in O also. We let the variable point X traverse Π from P to $\bar{P}$ and at the same time we pursue $\bar{X}$'s path on $\bar{\Pi}$. If $\bar{X}$ does not set foot on Π until P, then $\Pi + \bar{\Pi}$ is a closed Jordan polygon. If $\bar{X}$ first meets Π at $P_1 \neq P$, we consider the subpath $P_1\bar{P}_1$ of Π in place of Π. Finally, if $\bar{X}$ begins to follow Π in the reverse direction from $\bar{P}$, and eventually leaves Π at the point P_2, we shall take $P_2\bar{P}_2$ in place of Π. In both the latter cases we obtain a path connecting opposite points in O, but with at least one side less than Π. Hence by repeated applications of this reduction we arrive, in a finite number of steps, at a reduced path Π such that $\Pi + \bar{\Pi}$ is a closed Jordan polygon in O, which determines regions Ω and $\bar{\Omega}$ on K. Then $\Pi + \bar{\Pi}$ has no point in common with Q, but Q

*Translation by John Stillwell of Ringfladen og Planen. [Mat. Tidsskr. B 1924, 1–22.]

[1] I am indebted to Prof. Juel for the information that this theorem which was often used but not properly proved by v. Staudt, neither was it proved later. Cf. C. Juel. "Eine Studie über v. Staudt's Geometrie der Lage", Matematiker-kongressen in Helsingfors 1922, p. 23. After this paper was sent to press I was informed that E. Stenfors has recently proved this theorem for projective Jordan curves: E. Stenfors, Der Jordansche Kurvensatz in der projektiven Ebene und die v. Staudt'schen Schnittpunktssätze. Soc. Scient. Fenn., Comm. Phys.-Math. II, 5 (1923).

is entirely contained in one of these regions, for example $\bar{\Omega}$ (otherwise we would have a division of Q into closed subsets). Ω is then entirely contained in $K—Q$ and hence in O, but $\bar{\Omega}$ not entirely, contrary to the assumption that O is self-diametric. None of the complementary regions determined by a continuum on K can therefore be self-diametric.

Now let Q_1 be a self-diametric continuum and let O_1 be one of the regions which make up the complementary set $K - Q_1$. $\bar{O}_1$ is also one of these regions, and we have established that $O_1 \neq \bar{O}_1$. Therefore $K - Q_1$ consists of paired regions,

$$O_1 + \bar{O}_1 + O_2 + \bar{O}_2 + \cdots,$$

finite or countable in number. Now if Q_2 is also a self-diametric continuum and if P is a point of Q_2 but not of Q_1, then P lies in one of the complementary regions in $K - Q_1$, say in O_1. $\bar{P}$, which also belongs to Q_2, then lies in $\bar{O}_1$. Thus Q_2 has points outside O_1 as well as inside O_1, and hence also points on the boundary of O_1, i.e. on Q_1. We have therefore shown: two self-diametric continua on K have at least one point in common. (The same is true for self-diametric regions, since we can use these as above to construct self-diametric polygons, i.e. continua.)

This theorem contains v. Staudt's theorem. In the projective plane p we use homogeneous coordinates x, y, z which we can normalize by demanding that $x^2 + y^2 + z^2 = 1$. Each point P in p therefore has coordinates (x, y, z) and $(-x, -y, -z)$. We can interpret these as coordinates of the points P_1 and $P_2 = \bar{P}_1$ on the sphere $x^2 + y^2 + z^2 = 1$ in x-y-z-space. This two-to-one mapping of the sphere K onto p can be realized, e.g. by projecting the sphere's surface onto its tangent plane from the midpoint of the sphere. A continuous curve κ in p is defined by 3 arbitrary continuous functions of one parameter which runs through a finite interval:

$$\kappa\colon \begin{cases} x = x(t) \\ y = y(t) \qquad 0 \leq t \leq 1 \\ z = z(t) \end{cases}$$

and κ is called a closed curve when $(x(0), y(0), z(0))$ is the same point on p as $(x(1), y(1), z(1))$. A curve is now of even or odd order, by v. Staudt's definition, according as $x(0) = x(1)$, $y(0) = y(1)$, $z(0) = z(1)$ or $x(0) = -x(1)$, $y(0) = -y(1)$, $z(0) = -z(1)$. In the first case the point sets $(x(t), y(t), z(t))$ and $(-x(t), -y(t), -z(t))$ for $0 \leq t \leq 1$ make up two closed curves (and therefore continua) on K which are diametric images of each other and may have no common point (if they have a common point, the two together constitute a self-diametric continuum). In the latter case we always have a self-diametric continuum. Two closed curves of odd order must therefore always have a common point by the theorem proved above.

As we now move on to deal with the torus, we are guided by the principles of the above method, namely:

1. Instead of the surface itself we consider a "covering surface" which is many-to-one related to the given surface (just as K can be said to doubly cover p).
2. We consider special point sets on the covering surface, corresponding to general point sets on the surface itself (such as the self-diametric sets on K which correspond to arbitrary sets on p).

§1.
A Topological Lemma

Let $ABCD$ be a parallelogram in the plane and let Π denote the set of all points on $ABCD$ and in its interior. We say that a point set on Π separates the opposite sides AB and CD when each polygonal path connecting AB to CD contains at least one point of the set. Then the following result holds:

(I) *Let M be a closed point set on Π which separates AB from CD, and let N be a closed point set on Π which separates AD from BC. Then M and N have at least one common point.*

Proof. We suppose that M and N have no common point; then, being closed sets, they have a minimal distance $\varepsilon > 0$. We divide Π into small parallelograms (cells) by means of lines parallel to the sides, in such a way that the greater diagonal of each cell is $<\varepsilon$. A cell will be called an M-cell if it contains a point of M in its interior or on its boundary. No M-cell contains a point of N. Let S denote the collection of all sides of M-cells. S contains at least one side on AD and at least one on BC. On AB as well as CD there is at least one cell side which does not belong to S, since AB and CD contain at least one point of N. Then there must be a path v of sides in the collection S which connects AD to BC; if not, one could connect AB and CD by a path (not necessarily of cell sides) which avoided S and therefore did not enter any M-cells, thus containing no points of M. Then since S is free of N-points by construction, v is a path which connects AD to BC and contains no N-points, contrary to hypothesis.

§2.
Functions which are Linear at Infinity

A real continuous function $f(x, y)$ defined over the whole x-y-plane will be called "linear at infinity" if there are real numbers a and b such that

$$\lim_{r\to\infty} \frac{f(x, y) - (ax + by)}{r} = 0, \qquad \text{where } r = \sqrt{x^2 + y^2}.$$

The linear "approximation function" $ax + by$ is uniquely determined by f, since two different linear homogeneous functions have a difference of order r as $r \to \infty$ for suitable values of the variables. A linear combination with constant coefficients of functions which are linear at infinity is again linear at infinity, with approximation function determined by the combination. If one performs an affine transformation of the variables:

$$x = \alpha\xi + \beta\eta, \qquad \begin{vmatrix} \alpha & \beta \\ \gamma & \delta \end{vmatrix} \neq 0,$$
$$y = \gamma\xi + \delta\eta,$$

and sets $\xi^2 + \eta^2 = \rho^2$, then the ratio ρ/r lies between two fixed positive constants for all points in the plane. The remainder function

$$f_1(x, y) = f(x, y) - (ax + by)$$

which by definition $= o(r)$, is therefore $o(\rho)$ when expressed as a function of ξ and η. The property of being linear at infinity is therefore invariant under linear homogeneous transformations of the variables (this is clearly the case for parallel displacements also, and hence for arbitrary affine transformations).

We shall now prove the following theorem:

(II) *Two functions linear at infinity, whose approximation functions have a non-zero determinant, assume all pairs of values.*

Let the functions be

$$\begin{aligned} f(x, y) &= ax + by + f_1(x, y), \qquad \text{where } f_1(x, y) = o(r) \\ g(x, y) &= cx + dy + g_1(x, y), \qquad \text{where } g_1(x, y) = o(r) \end{aligned} \Bigg\} \quad \text{as } r \to \infty$$

and suppose

$$D = \begin{vmatrix} a & b \\ c & d \end{vmatrix} \neq 0.$$

We shall prove that for any two real numbers α and β there is at least one point satisfying the pair of equations

$$f(x, y) = \alpha,$$

$$g(x, y) = \beta.$$

We can set the constants on the right hand side equal to 0, 0 by taking $f_1 - \alpha$ and $g_1 - \beta$ in place of f_1 and g_1, respectively. Since $D \neq 0$ we can evidently introduce new variables by

$$\xi = ax + by,$$

$$\eta = cx + dy$$

and thereby reduce the problem to solving the following equations (rewritten in the original notation):

$$\begin{aligned} f(x, y) &= x + f_1(x, y) = 0 \qquad & f_1 &= o(r) \\ g(x, y) &= y + g_1(x, y) = 0, \qquad & g_1 &= o(r) \end{aligned} \Bigg\} \quad r \to \infty.$$

For this purpose we determine a positive value R such that

$$\left| \frac{f_1(x, y)}{r} \right| < \frac{1}{\sqrt{2}},$$

$$\left| \frac{g_1(x, y)}{r} \right| < \frac{1}{\sqrt{2}}$$

for $r \geq R$, draw a circle of radius R about $x = y = 0$ and the lines $y = x$ and $y = -x$ (see Figure 1).

Then $f > 0$ in the infinite sector $AA_0E_1B_0B$, since

$$|f_1| < \frac{r}{\sqrt{2}} \leq x$$

everywhere in this sector. Similarly, $f < 0$ in $DD_0E_2C_0C$, $g > 0$ in $AA_0E_3D_0D$ and $g < 0$ in

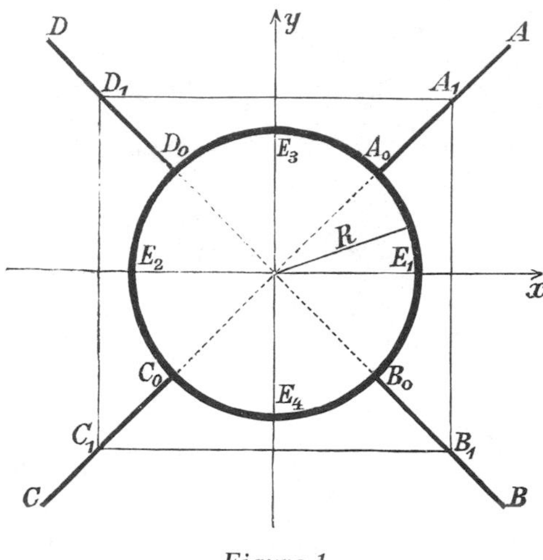

Figure 1

$BB_0E_4C_0C$. Because of the continuity of the functions, there must then be a (closed) set of points in the square $A_1B_1C_1D_1$ where $f = 0$, separating the side A_1B_1 from the side C_1D_1; and a set of points at which $g = 0$, separating A_1D_1 from B_1C_1. Then by (I) these two point sets have at least one point in common. Q.E.D.

Theorem (II) can be viewed as a generalisation of theorems on the solution of linear equations to the case of functions linear at infinity, though of course the solution is no longer necessarily unique. There can be arbitrarily many solutions; the set of solution points can even have the cardinality of the continuum; however, we can see that the set must always be bounded: in our case all solutions lie inside the circle of radius R.

The definition of linearity at infinity transfers immediately to functions of an arbitrary number of variables, there is likewise a theorem analogous to (II) on solutions of n equations, linear at infinity, in n unknowns. A proof of this fact, though not quite as elementary as that given above for the two-dimensional case, can be given by means of *J. W. Alexander's*[2] extension of the concept of "angle variation" to n dimensions. *A. Ostrowski* has communicated to me a shorter proof which transforms the n-dimensional space by reciprocal radii and thereby obtains continuous functions which are differentiable at the origin.

A special class of functions linear at infinity, which we shall use in what follows, is determined by the following theorem:

(III) *An additive-periodic function is linear at infinity.*

Let $f(x, y)$ be a continuous function which satisfies the following functional equations

$$f(x + \phi, y) = f(x, y) + \Phi,$$

$$f(x, y + \psi) = f(x, y) + \Psi,$$

[2] *J. W. Alexander*, On transformations with invariant points. Transactions of the American Math. Society, vol. **23** (1922).

where ϕ, ψ, Φ, Ψ are real constants; $\phi \neq 0$, $\psi \neq 0$. We can assume $\phi > 0$, $\psi > 0$. If one sets

$$x = h\phi + x_0,$$

$$y = k\psi + y_0,$$

where h and k are integers and $0 \leq x_0 < \phi$, $0 \leq y_0 < \psi$, then

$$f(x, y) - \left(\frac{\Phi}{\phi}x + \frac{\Psi}{\psi}y\right) = f(x_0, y_0) - \left(\frac{\Phi}{\phi}x_0 + \frac{\Psi}{\psi}y_0\right).$$

Thus if F is the highest value assumed by the right hand side of this equation for $0 \leq x_0 \leq \phi$, $0 \leq y_0 \leq \psi$ we have

$$\left|f(x, y) - \left(\frac{\Phi}{\phi}x + \frac{\Psi}{\psi}y\right)\right| \leq F$$

for all values of x and y. The remainder function is therefore even bounded.

§3.
Mappings

1. Mappings of the plane by two functions which are linear at infinity. If one sets

$$\left.\begin{aligned} x' &= f(x, y) = ax + by + o(r) \\ y' &= g(x, y) = cx + dy + o(r) \end{aligned}\right\} \quad r \to \infty$$

and lets $P' = (x', y')$ be the image of $P = (x, y)$, then one obtains a continuous (though not in general one-to-one) mapping of the x-y-plane onto itself. The coordinates of P' relative to P are then

$$\xi = x' - x,$$

$$\eta = y' - y$$

so that the vector PP' is likewise a function of x and y which is linear at infinity, with approximation functions $(a - 1)x + by$ and $cx + (d - 1)y$. If we assume

$$\Delta = \begin{vmatrix} a - 1 & b \\ c & d - 1 \end{vmatrix} \neq 0,$$

then it follows from (II) that the equations $\xi = \alpha$, $\eta = \beta$ can always be solved. The vector therefore assumes all sizes and directions; in particular, such a mapping has at least one fixed point. Moreover, the set of all points for which the vector takes a given value, in particular the set of fixed points, is bounded.

Example. Choose $a = b = c = d = 0$, so that the image point is of smaller order than the original point as $r \to \infty$, and $\Delta = 1$; then there is always at least one fixed point. This includes the case where the whole plane is mapped continuously into a bounded region.

2. Modulo 1 periodic mappings of the plane. We shall now consider a "modulo 1 periodic mapping" of the x-y-plane given by transformations

$$x' = f(x, y),$$
$$y' = g(x, y),$$

where f and g satisfy the functional equations

$$f(x + 1, y) = f(x, y) + m, \qquad f(x, y + 1) = f(x, y) + n,$$
$$g(x + 1, y) = g(x, y) + p, \qquad g(x, y + 1) = g(x, y) + q,$$

where m, n, p, q are integers: The functions f and g are then linear at infinity because of (III), with approximation functions $mx + ny$ and $px + qy$, respectively. The vector defined by

$$\xi = x' - x,$$
$$\eta = y' - y$$

is likewise linear at infinity with approximation functions

$$(m - 1)x + ny \qquad \text{for } \xi$$
$$px + (q - 1)y \qquad \text{for } \eta.$$

Let M denote the matrix of these functions

$$M = \begin{pmatrix} m - 1 & n \\ p & q - 1 \end{pmatrix}$$

and let Δ be its determinant:

$$\Delta = \begin{vmatrix} m - 1 & n \\ p & q - 1 \end{vmatrix}.$$

We assume $\Delta \neq 0$.

Two number pairs (x_1, y_1) and (x_2, y_2) will be called "congruent modulo 1" or "equivalent" when

$$(x_1, y_1) \equiv (x_2, y_2) \qquad (\text{mod } 1),$$

i.e. when $x_2 - x_1$ and $y_2 - y_1$ are integers. Our mapping therefore sends equivalent points to equivalent points and hence is called "periodic mod 1". Two number pairs will be called "congruent modulo M" when

$$(x_1, y_1) \equiv (x_2, y_2) \qquad (\text{mod } M),$$

i.e. when there are integers h and k such that

$$x_2 - x_1 = h(m - 1) + kp,$$
$$y_2 - y_1 = hn + k(q - 1).$$

Two mod M congruent points are always equivalent, but not conversely, because a complete set of equivalent points divides into $|\Delta|$ residue classes mod M.

When the vectors (ξ_1, η_1) and (ξ_2, η_2) corresponding to two points P_1 and P_2 have integral differences $\xi_2 - \xi_1$ and $\eta_2 - \eta_1$, then we can say the points are "of equal status under the mapping" and write

$$P_1 P_1' \equiv P_2 P_2' \qquad (\text{mod } 1).$$

Two equivalent points are therefore of equal status, because

$$P_1 P_1' \equiv P_2 P_2' \qquad (\text{mod } M),$$

but the converse does not hold in general, not even if $\Delta = \pm 1$. We shall examine a complete system of points of equal status a little more closely and choose those for which the vectors have integral components, i.e. a system of "fixed points modulo 1":

$$(\xi, \eta) \equiv (0, 0) \qquad (\text{mod } 1),$$

This system does not occupy a special position among all systems of points of equal status, because application of an arbitrary parallel displacement to the given mapping yields another mapping of the same type. Let $\alpha_1 = 0$, $\beta_1 = 0$; $\alpha_2, \beta_2; \ldots; \alpha_{|\Delta|}, \beta_{|\Delta|}$ be $|\Delta|$ pairs of integers which represent the residue classes mod M. It follows from (II) that the set N_i of all points for which $\xi = \alpha_i$, $\eta = \beta_i$ is non-empty and bounded. N_1 is the set of actual fixed points of the mapping. N_i and N_k have no common point. The whole set

$$N - N_1 + N_2 + \cdots + N_{|\Delta|}$$

contains no two equivalent points, but an arbitrary fixed point mod 1 is equivalent to some point in N. The bounded set N is therefore a "fundamental system" for fixed points mod 1.

In particular, we conclude that the mapping has at least $|\Delta|$ mutually inequivalent fixed points mod 1; and one can attain this minimal number by mapping with the usual linear functions

$$x' = mx + ny,$$

$$y' = px + qy$$

with integer coefficients m, n, p, q. Finally, one can see that when $\Delta = 0$ one can construct mappings of the required kind without fixed points mod 1.

3. Mappings of the torus. When one takes (Figure 2) a point on the circle determined by centre angle v from a fixed radius, then this point corresponds to infinitely many values of v which are congruent mod 2π, and if the circle is then rotated about the axis AA we obtain a torus T. If the rotation angle is w mod 2π, a point on T can be given by coordinates $x = v/2\pi$, $y = w/2\pi$, so that mod 1 congruent number pairs determine the same point on T. If x and y are interpreted as rectilinear coordinates in a plane Π, then Π is called a "covering surface" of T, and equivalent points of Π are said to "lie over" the same point of T.

We now consider a continuous mapping t of T onto itself. Each point P of T has a single image P' (but in general not conversely), and when $P_1, P_2, \ldots \to P$ we have $P_1', P_2', \ldots \to P'$. We now choose a point P_0 on T and a particular pair x_0, y_0 among the infinite set of its coordinates, then subsequently (and independently) a particular pair x_0', y_0', among the coordinates of P_0'. Then there is exactly one pair of functions

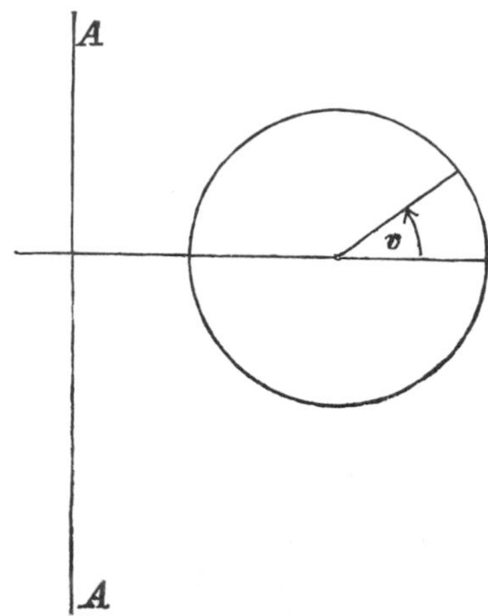

Figure 2

$$x' = f(x, y),$$
$$y' = g(x, y)$$

for which (x', y') is always the image of the point (x, y) on T and for which $f(x_0, y_0) = x'_0$, $g(x_0, y_0) = y'_0$. This pair of functions is a uniquely determined, continuous and mod 1 periodic mapping τ of Π onto itself; (and one obtains infinitely many different mappings of Π according to the choice of x_0, y_0 and the choice of coordinates for P'_0). The numbers m, n, p, q (defined in 2) depend only on t, and we call

$$\begin{pmatrix} m & p \\ n & q \end{pmatrix}$$

the mapping type matrix of t. Then fixed points mod 1 of τ lie over real fixed points of t, and it follows from 2) that t has at least $|\Delta|$ different fixed points; in fact the fixed points on T can be divided into $|\Delta|$ different classes according to the residue character mod M of their image vectors.

Example. Let $m = n = p = q = 0$ so that Π is mapped onto a bounded region, because the remainder functions for an additive periodic function are certainly bounded. The image of each closed curve can then be "contracted to a point" on T. Consequently, such a mapping always has a fixed point.

In the paper "Über die Minimalzahl der Fixpunkte bei den Abbildungstypen der Ringflächen" in Math. Annalen **82**, [N7], the author derived this minimal number $|\Delta|$ of T's fixed points by operating directly on T, but only for one-to-one mappings. In another paper in the same volume Brouwer extended this result to arbitrary continuous mappings. However, the division of T's fixed points into finitely many classes as above seems to be the simplest and most natural way of proving minimality.

§4.
Point Sets on the Torus

A point set M in the x-y-plane Π will be called a "torus set" when each point equivalent to a point of M itself belongs to M; there is then a point set m on T over which M lies, and m and M determine each other. A "subcontinuum" of M is a subset of M which is closed, bounded (i.e. situated entirely in the finite) and not divisible into closed subsets. M is called "rationally connected" when there is a subcontinuum of M which contains two equivalent points, otherwise "simple". If a, b, p, q are integers, and if one has a subcontinuum k of M which contains points (x_0, y_0) and $(x_0 + a, y_0 + b)$, then a displacement (p, q) gives a subcontinuum which contains $(x_0 + p, y_0 + q)$ and $(x_0 + p + a, y_0 + q + b)$. If l_1 and l_2 are two lines with slope b/a supporting k such that k lies on the parallel strip σ bounded by l_1 and l_2, and if one subjects k to displacements (ma, mb) where m runs through all the integers, and if we use the fact that two subcontinua which have a point in common together constitute a continuum, then it follows that (x_0, y_0) is connected to $(x_0 + ma, y_0 + mb)$ in a subcontinuum; the collection of all subcontinua which appear in k as a result of these displacements gives an infinite continuum K of M situated on σ. This separates l_1 from l_2 in the sense defined in §1.

We shall assume that a and b are not relatively prime, say $a = ea_1$, $b = eb_1$ where $e > 1$, and construct a subcontinuum which contains (x_0, y_0) and $(x_0 + a_1, y_0 + b_1)$. Let k go to k' under the displacement (a_1, b_1); so k' lies on σ and contains at least one point on l_1 and at least one on l_2. Then it follows from (I) that k' has at least one point S in common with K. Let K' be a subcontinuum of K which contains S and (x_0, y_0). Then $K' + k'$ is a subcontinuum of M which contains (x_0, y_0) and $(x_0 + a_1, y_0 + b_1)$. Thus (x_0, y_0) is connected to each equivalent point on the line through (x_0, y_0) in direction (a, b). We then say that M is rationally connected in the direction (a, b) and prove the following theorem:

(IV) *A torus set which is rationally connected in two different directions is rationally connected in every (rational) direction.*

Proof. Let M be connected in the directions (a, b) and (c, d) where $ad - bc \neq 0$. As above, we construct parallel-sided strips σ_1 and σ_2 containing the infinite continua K_1 and K_2, the former of which is mapped into itself by the displacement (a, b) and the latter by the displacement (c, d). By (I), these have at least one common point S, and this point is then connected to all its images under the two displacements and any of their integral linear combinations; e.g. with the points generated from S by the displacements $(da - bc \neq 0, dc - cd = 0)$ and $(0, bc - ad \neq 0)$. S is therefore connected to points generated from itself by displacements $(1, 0)$ and $(0, 1)$ and therefore with all its equivalent points; M is therefore rationally connected in all rational directions. Q.E.D.

If M is rationally connected in two directions, then its complement $M' = \Pi - M$ (which is likewise a torus set and which lies over $m' = T - m$) is simple. Since each rationally connected torus set contains an infinite continuum, it suffices to prove that the distance between points of M' belonging to a subcontinuum is bounded by a fixed constant (this is a stronger theorem). We construct two strips containing infinite continua of M in the directions $(0, 1)$ and $(1, 0)$ respectively and displace these parallel to the axes an integral number of times so that the sides of rectangle R enclosed by the 4 strips are >1. Then R contains at least one point equivalent to each point of M'. A point of M' inside R and a point of M' outside the outer rectangle determined by the 4 strips cannot belong to the same subcontinuum of M', since such a subcontinuum would "cut" at least one strip, so that M' and M would have a common point, which is impossible.

If both M and $\Pi - M$ are rationally connected then obviously the direction must be the same for both.

It is possible to construct a torus set M such that neither M nor $\Pi - M$ is rationally connected. Let M be the collection of all straight lines with slope $\sqrt{2}$ which contain rational points (i.e. points whose coordinates are both rational). This is obviously a torus set. It cannot be rationally connected since, e.g., the line through $(0, \sqrt{3})$ with slope $\sqrt{2}$, which is contained in $\Pi - M$, will cut each parallel-sided strip with a rational slope. For the same reason, $\Pi - M$ cannot be rationally connected.

§5.

One Curve on the Torus

A curve k on the torus T is understood to be a continuous (but not necessarily one-to-one) image of the unit interval:

$$x = f(t), \qquad y = g(t), \qquad (0 \leq t \leq 1).$$

A "point on the curve" is a value of the parameter t, and a "point on the surface" is a pair of numbers (x, y) (mod 1). Each point on the curve coincides with a certain point on the surface, but there can be several, perhaps infinitely many, curve points at the same point on the surface. A curve is called "closed" when $t = 0$ and $t = 1$ coincide with the same point on the surface, in other words, when

$$a = f(1) - f(0),$$

$$b = g(1) - g(0)$$

are integers. The pair (a, b) is then said to determine the "type" of the curve. In this case it is convenient to let t vary freely through all values, so that the curve point is determined by t's residue mod 1 and f and g are defined outside the original interval $0 \leq t \leq 1$ by the functional equations

$$f(t + 1) = f(t) + a,$$

$$g(t + 1) = g(t) + b,$$

making them additive-periodic with summands a and b respectively for period 1. If $(a, b) \neq (0, 0)$ then the torus set K over k in the x-y-plane Π is rationally connected in the direction (a, b).

A "double point" (t_1, t_2) $(t_1 \not\equiv t_2 \bmod 1)$ is a pair of curve points which coincide with the same surface point. The "class" (p, q) of the double point is the pair of integers

$$p = f(t_2) - f(t_1),$$

$$q = g(t_2) - g(t_1).$$

If one changes t_1 or t_2 by an integer, (p, q) changes by a multiple of (a, b). We shall therefore only compute class pairs mod (a, b), so that (p, q) and $(p + na, q + nb)$ will be the same double point class. We then have the following theorem:

(V) *A closed curve k has only a finite number of double point classes.*

Proof. We first assume that the curve is of type $(a, b) = (0, 0)$. The continuous curve K_0 in Π determined by $x = f(t)$, $y = g(t)$ is then also closed in Π and determines a bounded continuum. The torus set K in Π which lies over k consists of K_0 and distinct point sets $K_1, K_2, \ldots$ resulting from K_0 by integral parallel displacements, K_i being determined by $x = f(t) + \alpha_i$, $y = g(t) + \beta_i$ where α_i, β_i are the components of the displacement vector. Now let (t_1, t_2) be a double point on k of class (p, q), so that $f(t_2) = f(t_1) + p$, $g(t_2) = g(t_1) + q$. This corresponds to a point of intersection of K_0 and K_i, where $\alpha_i = p$, $\beta_i = q$. Since only a finite number (possibly none at all) of the sets K_i can have a point in common with the bounded set K_0, there can be only a finite number of classes; double points of K_0 itself correspond to double points of class $(0, 0)$ on k.

We now assume $(a, b) \neq (0, 0)$. We let t vary without restriction, then the curve K_0: $x = f(t)$, $y = g(t)$ becomes an infinite continuum in Π which is contained in a parallel-sided strip σ_0, bounded by supporting lines l_1 and l_2 with slope determined by (a, b); K_0 is mapped onto itself by the displacements (na, nb) for all integers n. Let m be the greatest common divisor of a and b, $a = ma_1$, $b = mb_1$, and let r and s be integers chosen so that

$$\begin{vmatrix} a_1 & b_1 \\ r & s \end{vmatrix} = 1.$$

We obtain the whole torus set K by parallel displacements of K_0 of the form $(\alpha a_1 + \beta r, \alpha b_1 + \beta s)$, where α runs through the values $0, 1, \ldots, m - 1$ and β runs through all the integers. The infinitely many continua K_i generated from K_0 in this way are represented by

$$x = f(t) + \alpha_i a_1 + \beta_i r,$$

$$y = g(t) + \alpha_i b_1 + \beta_i s.$$

Now if (p, q) is the class of a double point, we represent this pair in the form $p = \alpha a_1 + \beta r$, $q = \alpha b_1 + \beta s$, where we can choose α among the values $0, \ldots, m - 1$, since classes are only determined mod (a, b). The double point therefore corresponds to the intersection of K_0 and the K_i in question. Under the displacement $(\alpha a_1, \alpha\beta_1)$, the strip σ_0 goes into itself, under $(\beta r, \beta s)$ it goes into a parallel strip; there can be only a finite number of these (possibly none) which do not lie entirely outside σ_0, and therefore only a finite number of corresponding values of β. Thus there can be only a finite number of classes. Q.E.D.

For $m > 1$ there are always double points on k, namely, at least one in each of the classes determined by $\beta = 0$, $\alpha = 1, 2, \ldots, m - 1$. As shown in the previous paragraph, the continua generated from K_0 by the corresponding displacements have points in common with K_0. We call these $m - 1$ classes "irreducible", and the possible classes $\beta \neq 0$ and the null class $\alpha = \beta = 0$ are called "reducible". We can formulate the following theorem:

(VI) *A closed curve the type of which is the mth power of a prime type has at least $m - 1$ double points, because $m - 1$ different irreducible classes are represented.*

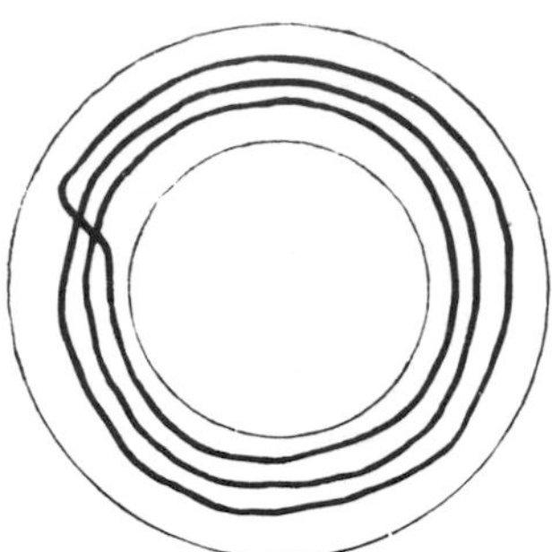

Figure 3

The terms "irreducible" and "reducible" classes indicate that the first type of intersection never vanishes, whereas the second one does, when one "straightens the curve out". By that we mean taking a system of closed curves, all of type (a, b) and the same initial point, with form depending continuously on a parameter λ according to the equations

$$\begin{aligned} x(t, \lambda) &= f(t) + \lambda(f(0) + at - f(t)), \\ y(t, \lambda) &= g(t) + \lambda(g(0) + bt - g(t)). \end{aligned} \qquad (0 \leq \lambda \leq 1)$$

For $\lambda = 0$ one has the original curve, for $\lambda = 1$ the image of the curve in the plane is a straight line. All reducible double point classes vanish for $\lambda = 1$, but the irreducible ones are still present (each point on the curve is m-tuple). The minimal number, $m - 1$, of double points mentioned in (VI) can be realized by placing the curve inside a small flat strip whose image in the plane is an infinite strip. Such a strip can be mapped onto an annulus, and one can then draw a curve as shown for $m = 3$ in Figure 3.

One may of course allow these $m - 1$ double points to coincide with the same surface point.

§6.

Two Curves on the Torus

We consider two closed curves k and l on T, given by

$$\begin{aligned} &k: x = f(t), \quad y = g(t), \qquad 0 \leq t \leq 1, \\ &l: x = \phi(u), \quad y = \psi(u), \qquad 0 \leq u \leq 1 \end{aligned}$$

and set:

$$\begin{aligned} f(1) - f(0) &= a, \qquad g(1) - g(0) = b, \\ \phi(1) - \phi(0) &= c, \qquad \psi(1) - \psi(0) = d, \end{aligned}$$

$$\begin{pmatrix} a & b \\ c & d \end{pmatrix} = M, \qquad \begin{vmatrix} a & b \\ c & d \end{vmatrix} = D.$$

An "intersection" of k and l is a pair (t_1, u_1) of values such that the curve points t_1 and u_1 coincide with the same surface point. The "class of the intersection pair (t_1, u_1, t_2, u_2)" is the pair (p, q) defined by

$$p = f(t_2) - f(t_1) + \phi(u_1) - \phi(u_2),$$

$$q = g(t_2) - g(t_1) + \psi(u_1) - \psi(u_2).$$

(In visual terms, this means that one runs along k from t_1 to t_2 and then back along l from u_2 to u_1, and determines the type of the closed curve which results.) Evidently, if $(t_2, u_2; t_3, u_3)$ is of class (p', q') then it follows from the definition that $(t_1, u_1; t_3, u_3)$ is of class $(p + p', q + q')$. If one allows k or l to be traversed an additional integral number of times, then t_2 is replaced by $t_2 + r$ and u_1 by $u_1 + s$, where r and s are integers, and the class pair changes to

$$(p + ra + sc, q + rb + sd).$$

We shall therefore let this pair denote the same intersection class as (p, q), and thus only consider class pairs as residue classes mod M.

We now assume $D \neq 0$. Then there are $|D|$ residue classes mod M. Thus in any case there are no more than $|D|$ intersection pair classes. We shall now show that all $|D|$ residue classes are really represented by intersection pairs. We construct the following continuous functions defined over the whole t-u-plane:

$$F(t, u) = f(t) - \phi(u),$$

$$G(t, u) = g(t) - \psi(u).$$

Since f, g, ϕ and ψ are additive-periodic, these are likewise additive-periodic, and linear at infinity by (III), with approximation functions

$$at - cu \qquad \text{for } F,$$

$$bt - du \qquad \text{for } G,$$

and the determinant $-D \neq 0$. We then use (II) and determine a solution (t_1, u_1) for $F = 0$, $G = 0$ and a solution (t_2, u_2) for $F = p$, $G = q$, where p and q are arbitrary integers and therefore representative of an arbitrary residue class mod M. Then the intersection pair $(t_1, u_1; t_2, u_2)$ is of class (p, q), and all residue classes are represented. Further we can use the same solution (t_1, u_1) to $F = 0$, $G = 0$ for all pairs (p, q) in this construction. We can therefore choose intersections (t_2, u_2), (t_3, u_3), $\ldots$, $(t_{|D|}, u_{|D|})$ so that these give, in conjunction with (t_1, u_1), all intersection pair classes except the null class. It is clear that $(t_i, u_i) \not\equiv (t_k, u_k) \pmod 1$ for $i \neq k$; $i, k = 1, \ldots, |D|$. Since (t_1, u_1) does not occupy a distinguished position, we can call this a set of "representative intersections". We then have the theorem:

(VII) *Two closed curves of types* (a, b) *and* (c, d), *respectively, with* $D = ad - bc \neq 0$, *have at least* $|D|$ *intersections, because we can take* $|D|$ *intersections as a representative set of intersection pair classes.*

As before, different intersections, even among representative intersections, can coincide with the same surface point. The minimal number $|D|$ can be obtained by really distinct surface points, e.g., when one "straightens out" two primitive curve types as in the previous paragraph.

§7.
Connection with the Kronecker Characteristic

Let k and l again be two continuous closed curves of types (a, b) and (c, d) respectively, where $D = ad - bc \neq 0$. We shall be content to look at the very simple case[3] in which the parameter intervals $0 \leq t \leq 1$ and $0 \leq u \leq 1$ are divided into finitely many subintervals in which x and y are linear functions. The images in the plane are then polygons. Evidently we can suppose that k and l have no line segment in common, and hence only a finite number of common points. The positive direction of each curve is defined as the direction in which the parameter increases. Let P be a common point of k and l, corresponding to certain values of t and u, and let C_P be a circle with centre P and radius sufficiently small that

1. there are no other points of intersection inside C_P, and
2. k, if it has a double point at P, so that several intersections of k and l occur at P, does not return to P without having been outside C_P, and
3. the same holds for l.

Let k^+, l^+, k^-, l^- be the four points on C_P which one first encounters following k and l in the positive and negative directions from P, respectively. (We can always choose C_P so that these 4 points are different).

If the pair (k^+, k^-) does not separate the pair (l^+, l^-) on C_P we give the intersection in question "index" 0. If k^+, l^+, k^-, l^- lie on C_P like $(1, 0), (0, 1), (-1, 0), (0, -1)$ on the unit circle we give the intersection index 1, if they lie like $(0, 1), (1, 0), (0, -1), (-1, 0)$ we give the intersection index -1. In other words: intersections have index $+1$, -1 or 0 according as k cuts l from left to right, from right to left, or just "touches". Following Kronecker, we shall take the index sum over all intersections and call it the "characteristic" of the curves in question. For that purpose we divide the intersections into classes with the help of the results in §6, reckoning two intersections to be in the same class when their intersection pair is of class $(0, 0)$. Intersections between the infinite continua $K_0 : x = f(t),\ y = g(t),\ -\infty < t < \infty$ and $L_0 : x = \phi(u),\ y = \psi(u),\ -\infty < u < \infty$ in the plane fully represent all intersections of a particular class. We can therefore find the index sum for this class by finding the index sum for intersections between K_0 and L_0.

K_0 and L_0 run along within parallel-sided strips which intersect in a parallelogram H. All intersections lie in H. Let P^- and P^+ be two points on K_0 outside H and on opposite sides (Figure 4). We connect these by a simple polygonal path P^+P^- running around H. As a result we obtain a closed polygon $\bar{K}_0$, consisting of a piece of K_0 between P^- and P^+ and the polygonal path just constructed. K_0's orientation is carried over to $\bar{K}_0$. We choose Q^- and Q^+ in the same way on L_0 and similarly form a closed polygon $\bar{L}_0$. This can be done in such a way that the supplementary polygonal paths have only one common point S (see Figure 4). $\bar{K}_0$ and $\bar{L}_0$ have index sum zero. This is obvious when one of them is a simple closed polygon (without double points); but then it will also be true for an arbitrary closed polygon, since one can split it into simple loops such that the conditions which determine index values are preserved. Now one sees that the point S, which should not be reckoned as an intersection of K_0 and L_0, has

[3] One can generalise this to far less special assumptions; but the above assumption simplifies the presentation and illustrates the method sufficiently well.

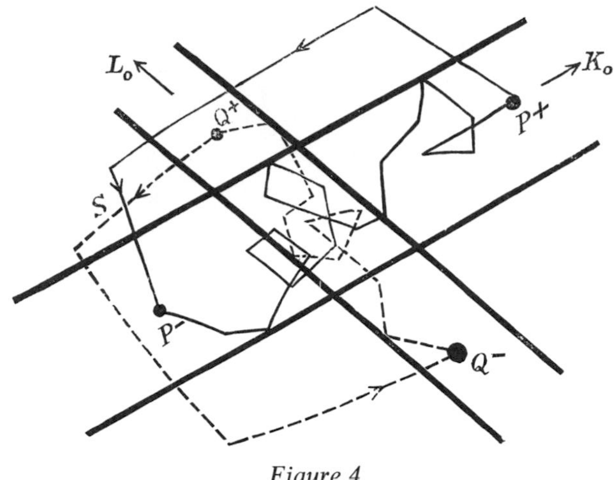

Figure 4

index -1 or $+1$ according as $D > 0$ or $D < 0$. Thus the intersection in H, i.e. those between K_0 and L_0, have index sum $+1$ or -1 according as $D > 0$ or $D < 0$. This holds for all classes, and there are $|D|$ classes. We therefore have the following theorem.

(VIII) *The index sum over each simple intersection class is $+1$ or -1 according as the curve type determinant is positive or negative. The index sum over all intersections is therefore equal to this determinant.*

DIE GRUPPE DER DREIDIMENSIONALEN GITTER-TRANSFORMATIONEN

JAKOB NIELSEN

Es ist bekannt, wie man die Transformationsgruppe des zweidimensionalen Gitters, also die (durch Spiegelung erweiterte, homogene) Modulgruppe, durch Angabe eines Systems erzeugender Operationen und definierender Relationen abstrakt charakterisieren kann. Dasselbe geschieht im Folgenden für das räumliche Gitter.

Um die im Folgenden benutzte, für den Beweis der Vollständigkeit eines Relationensystems für eine unendliche diskontinuierliche Gruppe oft anwendbare Methode am einfachsten Fall zu illustrieren, wird im § 1 das ebene Gitter ausführlich behandelt.

§ 1.

Es sei

$$M = \begin{pmatrix} e_{11} & e_{12} \\ e_{21} & e_{22} \end{pmatrix}$$

die Matrix einer homogenen linearen ganzzahligen Substitution

$$x_i' = e_{i1}\,x_1 + e_{i2}\,x_2 \qquad (i = 1, 2)$$

mit der Determinante

$$D = |e_{ik}| = \pm 1,$$

und es sei bezeichnet

$$s_{ik} = e_{i1}\,e_{k1} + e_{i2}\,e_{k2} = s_{ki} \qquad (i, k = 1, 2).$$

Dann ist

$$D^2 = \begin{vmatrix} s_{11} & s_{12} \\ s_{21} & s_{22} \end{vmatrix} = s_{11}\,s_{22} - s_{12}^2 = 1.$$

1*

Nimmt man an, es sei gleichzeitig

$$2\,|s_{12}| \leq s_{11}$$
$$2\,|s_{12}| \leq s_{22},$$

so folgt wegen der Ganzzahligkeit der s_{ik} aus der letzten Gleichung

$$s_{11} = s_{22} = 1, \quad s_{12} = 0,$$

also dass M orthogonal ist. Bezeichnet $\mathfrak{v}_i$ den Vektor mit den Komponenten e_{i1}, e_{i2} und $|\mathfrak{v}_i| = \sqrt{s_{ii}}$ seine Länge, so erfüllt also ein transformiertes, nicht orthogonales Gitter mindestens eine der folgenden vier Ungleichungen:

$$|\mathfrak{v}_1 + \mathfrak{v}_2| < |\mathfrak{v}_1|$$
$$|\mathfrak{v}_1 - \mathfrak{v}_2| < |\mathfrak{v}_1|$$
$$|\mathfrak{v}_1 + \mathfrak{v}_2| < |\mathfrak{v}_2|$$
$$|\mathfrak{v}_1 - \mathfrak{v}_2| < |\mathfrak{v}_2|.$$

Im Einheitsparallelogramm eines nicht orthogonalen Gitters ist also die kürzeste Diagonale stets kürzer als die längste Seite. Als Erzeugende der gesuchten Transformationsgruppe G wählen wir daher zunächst die Substitutionen:

$$P = \begin{pmatrix} 0 & 1 \\ 1 & 0 \end{pmatrix}$$
$$O_1 = \begin{pmatrix} -1 & 0 \\ 0 & 1 \end{pmatrix}$$
$$O_2 = \begin{pmatrix} 1 & 0 \\ 0 & -1 \end{pmatrix}$$
$$12 = \begin{pmatrix} 1 & 1 \\ 0 & 1 \end{pmatrix} \qquad \left(12^{-1} = \begin{pmatrix} 1 & -1 \\ 0 & 1 \end{pmatrix}\right)$$
$$21 = \begin{pmatrix} 1 & 0 \\ 1 & 1 \end{pmatrix}.$$

Denn mittels 12 und 21 kann man eine beliebige Matrix M nach dem Vorstehenden auf eine orthogonale reduzieren, und jede orthogonale wird durch P, O_1, O_2 erzeugt. — Die Zahl $\sigma(M) = |\mathfrak{v}_1|^2 + |\mathfrak{v}_2|^2 = s_{11} + s_{22} = \Sigma e_{ik}^2$ kann als Mass für die »Schiefe« des Gitters dienen.

Wir benutzen folgende Bezeichnungsweise: Sind M und N die Matrizen zweier Substitutionen, so ist die Substitution MN mittels der Matrizenmultiplikation (Zeile von $M \times$ Spalte von N) erklärt. M_N bedeutet $N^{-1}MN$, und $M \rightleftarrows N$ bedeutet die Relation $MN = NM$, oder anders geschrieben $M_N = M$. Endlich bezeichnet $\{M, N\}$ die aus M und N erzeugte Untergruppe, und $\{M, N\} \rightleftarrows \{M', N'\}$ bedeutet, dass man zu irgend zwei Elementen E bezw. E' aus diesen beiden Gruppen stets zwei andere F und F' so finden kann, dass $EE' = F'F$ ist.

Aus der Definition der Erzeugenden bestätigt man unmittelbar die Richtigkeit folgender Relationen:

(1)	$P^2 = 1$	(5)	$12_P = 21$
(2)	$O_1^2 = 1$	(6)	$12_{O_1} = 12^{-1}$
(3)	$O_{1P} = O_2$	(7)	$12_{O_2} = 12^{-1}$
(4)	$O_1 \rightleftarrows O_2$	(8)	$O_1 \cdot P \cdot 12 \cdot 21^{-1} \cdot 12 = 1.$

Es soll gezeigt werden, dass dies Relationensystem für G vollständig ist, d. h. dass jede in G gültige Relation zwischen den Erzeugenden eine formale Folge aus (1)—(8) ist.

Aus (2) bezw. (3) folgt formal durch Transformation mit P wegen (1):

(9) $$O_2^2 = 1$$

(10) $$O_{2P} = O_1.$$

(3) und (10) besagen, dass $P \rightleftarrows \{O_1, O_2\}$, und wegen (4) kann dann jedes Element aus $\{P, O_1, O_2\} = \Omega$ in der Form

(11) $$O_1^\alpha O_2^\beta P^\gamma$$

geschrieben werden, wo α, β, γ wegen (2), (9) und (1) auf die Werte 0 und 1 beschränkt werden können. Die Normalform (11) stellt also jedenfalls nicht mehr als 8 Elemente dar; da aber Ω seiner Bedeutung nach 8 orthogonale Transformationen enthält, ist (1)—(4) für Ω vollständig.

Aus (5), (6) und (7) folgt durch Transformation mit P wegen (3), (10) und (1):

$$21_P = 12 \tag{12}$$

$$21_{O_2} = 21^{-1} \tag{13}$$

$$21_{O_1} = 21^{-1}. \tag{14}$$

(5)—(7) und (12)—(14) sagen aus, dass $\Omega \rightleftarrows \{12, 21\} = Z$; aber (8) zeigt, dass G nicht das »direkte Produkt« von Ω und Z ist. ω bezw. ζ sei das allgemeine Zeichen für eine Operation aus Ω bezw. Z.

Aus (8) folgt durch Transformation mit P:

$$O_2 \cdot P \cdot 21 \cdot 12^{-1} \cdot 21 = 1. \tag{15}$$

Ein $\zeta = \Pi(12, 21)$, das Exponenten beiderlei Zeichens enthält, enthält zwei benachbarte Faktoren 12 und 21 mit entgegengesetzten Exponentenzeichen. Aus (8) oder (15) und danach (5)—(7), (12)—(14) bekommt man dann eine Umformung

$$\zeta = \omega \cdot \zeta'$$

wo ζ' mindestens einen Faktor weniger enthält. In endlich vielen Schritten kann man daher jedes Element g aus G in die »Normalform«

$$g = O_1^{\alpha} O_2^{\beta} P^{\gamma} \cdot \Pi(12, 21) \tag{16}$$

bringen, wo die Exponenten in Π entweder alle positiv oder alle negativ sind und $\alpha, \beta, \gamma = 0, 1$. Treten in Π überhaupt von 0 verschiedene Exponenten auf, so ist $\sigma(\Pi) > 3$, da σ mit jedem neuen Faktor 12 oder 21 wächst, daher auch $\sigma(g) > 3$. Das Einheitselement wird daher in (16) **eindeutig** dargestellt, indem α, β, γ und alle Exponenten in Π verschwinden müssen.

Ist nun

(17) $\Pi_1 (P, O_1, O_2, 12, 21) = \Pi_2 (P, O_1, O_2, 12, 21)$

eine in G gültige Relation, so forme man die linke Seite von

(18) $$\Pi_1 \Pi_2^{-1} = 1$$

in die Normalform (16) um, was einer formalen Anwendung von (1)—(8) entspricht. Wie gezeigt, wird (18) dadurch zu einer Identität, also folgt (17) formal aus (1)—(8), w. z. b. w.

Nachträglich kann man dann die Erzeugendenzahl reduzieren und z. B. P und 12, das weiterhin mit U bezeichnet sei, als Erzeugende wählen.[1] Denn aus (5), (8) und (3) ergeben sich

$$21 = U_P$$
$$O_1 = P U U_P^{-1} U$$
$$O_2 = U U_P^{-1} U P.$$

Das vollständige Relationensystem zwischen P und U erhält man nun, indem man in den übrigen fünf Relationen die gefundenen Ausdrücke für 21, O_1, O_2 einsetzt: (1) bleibt ungeändert, und man erhält aus (2)

(19) $$(PUPU^{-1}PU)^2 = 1$$

und aus (4), in der Form $(O_1 O_2)^2 = 1$ geschrieben:

(20) $$(UPU^{-1}PU)^4 = 1\,,$$

während (6) und (7) keine neuen Bindungen zwischen P und U mehr liefern. So führt z. B. (6), in der Form $(O_1 \cdot 12)^2 = 1$ geschrieben, zu der Relation

[1] Dass auch die **erweiterte** Modulgruppe durch nur **zwei** Substitutionen erzeugbar ist, findet sich übrigens m. W. in der Literatur nicht bemerkt.

$$(PUPU^{-1}PU^2)^2 = 1;$$

die linke Seite dieser Relation kann aber, identisch in $P = P^{-1}$ und U, in

$$(PUPU^{-1}PU)^2\,(PUPU^{-1}PU)^2_U$$

umgeformt werden, die Relation folgt also aus (1) und (19). Entsprechendes gilt für (7). Wir haben also dieses abstrakte Schema für G:

Erzeugende: P, U;

Definierende Relationen $\Big\}$ $P^2 = (PUPU^{-1}PU)^2 = (PU^{-1}PU^2)^4 = 1.$

Will man die invariante Untergruppe $\overline{G}$ von G haben, die zu der Determinante $D = +1$ gehört, die also die Gitterorientierung erhält, so weiss man, wegen $D(U) = +1$, $D(P) = -1$, dass die Erzeugende P in den Elementen von $\overline{G}$ eine gerade Anzahl von Malen auftritt. Man kann also

$$V = U_P \quad (= 21)$$

einführen und $\overline{G} = \{U, V\}$ schreiben. (19) und (20) ergeben die Relationen

(21) $$VU^{-1}VUV^{-1}U = 1,$$

(22) $$(UV^{-1}U)^4 = 1.$$

Um die Vollständigkeit eines Relationensystems für $\overline{G}$ auf die bekannte Vollständigkeit für G stützen zu können, ist zu berücksichtigen, dass die Transformation von (19) und (20) mit P zwar in G nichts Neues liefern kann, wohl aber möglicherweise in $\overline{G}$, das ja P nicht enthält. Die Ausrechnung ergibt zwei Relationen, die leicht aus (21) und (22) gefolgert werden, sodass in der Tat (21) und (22) für $\overline{G}$ vollständig sind. — Führt man statt V als neue Erzeugende neben U

$$T = UV^{-1}U$$

ein, was wegen $V = UT^{-1}U$ möglich ist, so erhält man für $\overline{G}$ das Schema:

Erzeugende: T, U;
Definierende Relationen: $UT^{-1}UT^{-1}UT = T^4 = 1$.

Will man endlich von hier zu der Gruppe F der inhomogenen Modulsubstitutionen hinabsteigen, die in $\overline{G}$ als Faktorgruppe enthalten ist, so geschieht dies durch Hinzufügung der Relation

$$\begin{pmatrix} -1 & 0 \\ 0 & -1 \end{pmatrix} = T^2 = 1,$$

wonach man für F das übliche[1] Schema hat:

Erzeugende: T, U;
Definierende Relationen: $(UT)^3 = T^2 = 1$.

Diese Methode suchen wir nun im § 2 für das räumliche Gitter anzuwenden, indem wir zum Zwecke des Vollständigkeitsbeweises wieder die Konstruktion einer für das Einheitselement eindeutigen Normalform anstreben. Die Bezeichnungen werden im Wesentlichen ungeändert übernommen, um einen kurzen Ausdruck zu ermöglichen.

§ 2.

Es sei $M = (e_{ik})$ die Matrix einer ganzzahligen Substitution

$$x'_i = e_{i1}x_1 + e_{i2}x_2 + e_{i3}x_3 \qquad (i = 1, 2, 3)$$

[1] Vgl. F. Klein, Vorl. über die Theorie der ell. Modulfunktionen. Ausgearbeitet von R. Fricke. Teubner 1890. Bd. I pg. 452—454. U ist dort mit S bezeichnet.

mit der Determinante

$$D = |e_{ik}| = \pm 1.$$

Es bezeichne $\mathfrak{v}_i$ den Vektor mit den Komponenten e_{i1}, e_{i2}, e_{i3} und $|\mathfrak{v}_i|$ seine Länge, ferner

$$s_{ik} = e_{i1}e_{k1} + e_{i2}e_{k2} + e_{i3}e_{k3} = s_{ki},$$

$$\sigma(M) = |\mathfrak{v}_1|^2 + |\mathfrak{v}_2|^2 + |\mathfrak{v}_3|^2 = \sum_i s_{ii} = \sum_{i,k} e_{ik}^2.$$

Hülfssatz: Gilt für jedes i und $k \neq i$ die Ungleichung

$$2\,|s_{ik}| \leqq s_{ii},$$

so ist M orthogonal.

Beweis: Wir können annehmen, es sei $s_{11} \leqq s_{22} \leqq s_{33}$, und setzen

$$s_{22} = s_{11} + h, \qquad h \geqq 0$$

$$s_{33} = s_{22} + k, \qquad k \geqq 0.$$

Dann ist

$$1 = D^2$$

$$= |s_{ik}| = s_{11}s_{22}s_{33} - (s_{11}s_{23}^2 + s_{22}s_{13}^2 + s_{33}s_{12}^2) + 2s_{12}s_{13}s_{23}$$

$$\geqq s_{11}(s_{11}+h)(s_{11}+h+k)$$

$$-\left[s_{11}\frac{(s_{11}+h)^2}{4} + (s_{11}+h)\frac{s_{11}^2}{4} + (s_{11}+h+k)\frac{s_{11}^2}{4}\right] - \frac{s_{11}^2(s_{11}+h)}{4}$$

$$= \frac{s_{11}}{4}(3s_{11}h + 3h^2 + 4hk + 3s_{11}k).$$

Daraus folgt $h = 0$ und (zunächst für $s_{11} > 1$, dann aber auch für $s_{11} = s_{22} = 1$) $k = 0$.

Die Zahl $s_{11} = s_{22} = s_{33}$ ist dann wegen

$$1 = s_{11}^3 - s_{11}(s_{12}^2 + s_{13}^2 + s_{23}^2) + 2s_{12}s_{13}s_{23}$$

ungerade, etwa $s_{11} = 2n+1$ und also $|s_{ik}| \leqq n$ $(i \neq k)$, also

$$1 \geq (2n+1)^3 - (2n+1)\,3n^2 - 2n^3 = 9n^2 + 6n + 1,$$

also $n = 0$, also M orthogonal, w. z. b. w.

Für eine nicht orthogonale Matrix lässt sich also stets ein i und k so finden, dass eine der beiden Ungleichungen

$$|\mathfrak{v}_i + \mathfrak{v}_k| < |\mathfrak{v}_i|$$
$$|\mathfrak{v}_i - \mathfrak{v}_k| < |\mathfrak{v}_i|$$

erfüllt ist. Das Grundparallelepiped eines schiefen Gitters ist also immer so schief, dass in mindestens einer Seitenfläche die kürzere Diagonale kürzer ist, als die längere Seite.

Mit ik sei die Substitution

$$\begin{aligned} x_i' &= x_i + x_k \\ x_h' &= x_h \qquad (h \neq i) \end{aligned}$$

bezeichnet; dann lässt sich also jede Substitution durch fortgesetzte Prämultiplikation mit Faktoren $ik^{\pm 1}$ auf eine orthogonale reduzieren, und zwar so, dass dabei das »Schiefemass« $\sigma(M)$ monoton abnimmt.

Als Erzeugende der Gruppe Ω orthogonaler Substitutionen nehmen wir zunächst die folgenden:

$$P_{ik} = P_{ki}: \begin{cases} x_i' = x_k \\ x_k' = x_i \\ x_h' = x_h \end{cases} \qquad \begin{pmatrix} i \neq k \\ h \neq i,\ k \end{pmatrix}$$

$$O_i: \begin{cases} x_i' = -x_i \\ x_h' = x_h \qquad (h \neq i). \end{cases}$$

Zwischen diesen bestehen die Relationen[1]

[1] Verschieden bezeichnete Indices sind verschieden.

$$
(A) \left\{
\begin{array}{ll}
(1) & P_{ik}^2 = 1 \\
(2) & P_{ik}^{P_{kl}} = P_{il} \\
(3) & O_i^2 = 1 \\
(4) & O_i \rightleftarrows O_k \\
(5) & O_i^{P_{ik}} = O_k \\
(6) & O_i^{P_{kl}} = O_i .
\end{array}
\right.
$$

(A) ist für Ω vollständig. Denn (5) und (6) besagen, dass

$$\{O_i\} \rightleftarrows \{P_{ik}\}.$$

Ist also wieder ω das Zeichen für ein Element aus Ω, so kann man umformen:

$$\omega = \Pi(O_i, P_{ik}) = \Pi_1(O_i) \cdot \Pi_2(P_{ik}).$$

$\Pi_1(O_i)$ kann wegen (3) und (4) in die Form

$$O_1^{\alpha_1} O_2^{\alpha_2} O_3^{\alpha_3} \qquad (\alpha_1, \alpha_2, \alpha_3 = 0, 1)$$

gebracht werden. In $\Pi_2(P_{ik})$ kann die Faktorenzahl mittels (1) und (2) verkleinert werden, solange ein Faktor zweimal auftritt, und auch noch, wenn P_{12}, P_{13}, P_{23} je einmal auftreten; es bleibt also entweder kein Faktor übrig (1 Möglichkeit), oder ein Faktor (3 Möglichkeiten), oder 2 Faktoren; in diesem letzteren Fall kann man mittels (2) erreichen, dass der erste Faktor P_{12} ist. Jedes ω kann also

$$\omega_n = O_1^{\alpha_1} O_2^{\alpha_2} O_3^{\alpha_3} P_{12}^{\beta_1} P_{13}^{\beta_2} P_{23}^{\beta_3} \qquad (\alpha_i, \beta_i = 0, 1; \beta_2 \cdot \beta_3 = 0)$$

geschrieben werden. Diese »Normalform« stellt 48 Elemente dar, ist also für Ω eindeutig, womit die Vollständigkeit von (A) erkannt ist.

Wir erzeugen nun die Gesamtgruppe G durch die ik, P_{ik} und O_i. Dann haben wir zunächst folgende weitere Relationen:

$$
(B)\begin{cases}
(7) & ik_{P_{ik}} = ki \\
(8) & ik_{P_{il}} = lk \\
(9) & ik_{P_{kl}} = il \\
(10) & ik_{O_l} = ik, \quad \text{d. h. } ik \rightleftarrows O_l \\
(11) & ik_{O_i} = ik^{-1} \\
(12) & ik_{O_k} = ik^{-1}.
\end{cases}
$$

(B) besagt, dass $\Omega \rightleftarrows \{ik\}$. Jedes Element g aus G kann also geschrieben werden:

$$g = \omega \cdot \Pi(ik)$$

(G ist aber nicht das direkte Produkt von Ω und $\{ik\}$, wie (13) zeigt). Wir haben nun noch die Relationen

$$
(C)\begin{cases}
(13) & O_i P_{ik} \cdot ik \cdot ki^{-1} \cdot ik = 1 \\
(14) & ik \rightleftarrows il \\
(15) & ik \rightleftarrows lk \\
(16) & ik \cdot kl \cdot ik^{-1} \cdot kl^{-1} \cdot il^{-1} = 1.
\end{cases}
$$

(16) sagt aus, dass il der »Kommutator« von ik und kl ist. Wegen (14) und (15) ist er mit beiden vertauschbar. Man sieht daher die Tragweite der Relation (16) am besten, wenn man sie in der Form schreibt:

(16a) $\quad ik \cdot kl = il \cdot kl \cdot ik = kl \cdot il \cdot ik = kl \cdot ik \cdot il.$

Der Rest dieses § bringt den Nachweis dafür, dass (A), (B), (C) für G vollständig ist. Wir benutzen dazu, dass jedes Element g von G in der Form

$$g = \omega \cdot \Pi(ik)$$

geschrieben werden kann, und betrachten ein solches Produkt $p = \Pi(ik)$. Bezeichnen wir einen Augenblick die Faktoren desselben mit

$$p = F_1 \cdot F_2 \ldots F_r,$$

so heisse die Folge positiver Zahlen ≥ 3:

$$\begin{aligned} \sigma_1 &= \sigma(F_1 \cdot F_2 \ldots F_r) \\ \sigma_2 &= \sigma(F_2 \cdot F_3 \ldots F_r) \\ &\ldots\ldots\ldots\ldots \\ \sigma_r &= \sigma(F_r) \\ \sigma_{r+1} &= \sigma(1) = 3 \end{aligned}$$

das »Diagramm« von p. Es hängt von der besonderen Wahl der Produktdarstellung von p ab. Aus dem Hülfssatz folgt, dass es für p eine Produktdarstellung

$$p = F_1' \cdot F_2' \ldots F_s' \cdot \omega$$

gibt, in der das Diagramm von $F_1' \cdot F_2' \ldots F_s' = p_n$ monoton abnimmt. Die Vertauschung eines ω mit einem $\Pi(ik)$ mittels (B) lässt, da sie nur einer Umbenennung der Variablen entspricht, die Faktorenzahl und das Diagramm von $\Pi(ik)$ ungeändert. Also lässt sich jedes g in der Form schreiben:

$$g_n = \omega_n \cdot p_n,$$

wo ω_n die obige Normalform für ω ist. In dieser Normalform g_n ist das Einheitselement von G nur auf eine Weise darstellbar, denn wegen $\sigma(1) = 3$ kann p_n, dessen Diagramm monoton ist, überhaupt keinen Faktor enthalten, und in ω_n ist die 1 eindeutig. Unsere Aufgabe ist also darauf zurückgeführt, zu zeigen, dass sich jedes p unter alleiniger Verwendung von (A), (B) und (C) in die Form $p = \omega \cdot p_n$ umformen lässt. Und hierzu genügt es, zu zeigen, dass, wenn das Diagramm von p nicht monoton ist und λ die grösste der vorkommenden Diagrammzahlen ist, eine aus (A), (B), (C) folgende Umformung $p = \omega \cdot p'$

gefunden werden kann, wo keine Diagrammzahl von p' grösser als λ ist und λ im Diagramm von p' einmal weniger vorkommt.

Ist das Diagramm von p nicht monoton, so gibt es einen Index h, sodass $\sigma_h < \lambda$, $\sigma_{h+1} = \lambda$ ist. Wir setzen $I = F_{h+2} F_{h+3} \ldots F_r$ und betrachten den Endbestandteil $\pi = F_h F_{h+1} I$ von p. Wir können $F_h = 12$ annehmen, da wir dies durch Umbenennung der Variablen (d. h. Transformation mit einem ω) stets erreichen können; enthält ferner F_{h+1} den Index 3, so können wir den Exponenten von F_{h+1} als $+1$ annehmen, da wir das durch Zeichenwechsel in x_3 (d. h. Transformation mit O_3) erreichen können, ohne $F_h = 12$ zu ändern; endlich können wir F_{h+1} als von 12^{-1} verschieden annehmen, da wir sonst einfach $F_h F_{h+1}$ streichen können. Somit haben wir für F_{h+1} folgende Möglichkeiten, die wir einzeln erledigen müssen:

1) 32 2) 12 3) 21 4) 21^{-1} 5) 13 6) 23 7) 31.

Wir setzen dabei $I = (e_{ik})$ und haben $\sigma(I) \leq \lambda$ wegen der Maximaleigenschaft von λ im Diagramm von p.

Fall 1) $\pi = 12 \cdot 32 \cdot I$.

Die beiden Faktoren wirken unabhängig von einander auf die Matrix $M(I)$ ein. Also ist $\sigma(12 \cdot I) < \sigma(I) \leq \lambda$. In

$$\pi_1 = 32 \cdot 12 \cdot I$$

tritt also λ einmal weniger auf. $\pi = \pi_1$ folgt aus (15).

Fall 2) $\pi = 12 \cdot 12 \cdot I$.

Hier soll also sein:

$$\sigma(12 \cdot I) \geq \sigma(I)$$

d. h.

$$\sum_i (e_{1i} + e_{2i})^2 \geq \sum_i (e_{1i})^2$$

oder

$$2 s_{12} + s_{22} \geq 0 \qquad (\alpha)$$

und ferner

$$\sigma(12\cdot 12\cdot I) < \sigma(12\cdot I)$$

d. h.

$$\sum_i (e_{1i} + 2e_{2i})^2 < \sum_i (e_{1i} + e_{2i})^2$$

oder

$$2s_{12} + 3s_{22} < 0. \qquad (\beta)$$

Die Ungleichung (β) steht im Widerspruch zu (α), wegen $s_{ii} > 0$, also ist der Fall 2) unmöglich.

Fall 3) $\pi = 12\cdot 21\cdot I$.

Hier soll sein:

$$2s_{12} + s_{11} \geqq 0 \qquad (\alpha)$$

und:

$$4s_{12} + 3s_{11} + s_{22} < 0 \qquad (\beta)$$

(α) und (β) stehen im Widerspruch, der Fall ist also unmöglich.

Fall 4) $\pi = 12\cdot 21^{-1}\cdot I$.

Wir setzen $\pi_1 = P_{12}O_1\cdot 12^{-1}I$, wobei $\pi = \pi_1$ aus (13) folgt. Dann ist $\sigma(12^{-1}\cdot I) = \sigma(\pi) < \lambda$. Den Faktor $P_{12}O_1$ schafft man mittels (B) über den Anfangsbestandteil $F_1\cdot F_2 \ldots F_{h-1}$ von p hinweg nach vorne, ohne dass dabei an dem Diagramm dieses Teils etwas geändert wird. λ tritt dann einmal weniger auf. — Analog denke man sich stets im Folgenden bei Umformungen auftretende Faktoren ω über $F_1 \ldots F_{h-1}$ hinweg an den Anfang des Produkts geschafft.

Fall 5) $\pi = 12\cdot 13\cdot I$.

Die Voraussetzungen: $\sigma(13\cdot I) \geqq \sigma(I)$, bezw. $\sigma(12\cdot 13\cdot I) < \sigma(13\cdot I)$ besagen, dass

$$\sum_i (e_{1i}+e_{3i})^2 \geqq \sum_i e_{1i}^2 \quad \text{bezw.} \quad \sum_i (e_{1i}+e_{3i}+e_{2i})^2 < \sum_i (e_{1i}+e_{3i})^2,$$

sodass wir für die s_{ik} folgende Ungleichungen haben:

$$2s_{13} + s_{33} \geqq 0 \qquad (\alpha)$$
$$2s_{12} + 2s_{23} + s_{22} < 0. \qquad (\beta)$$

Es sei $\pi_1 = 13 \cdot 12 \cdot I$. Die Umformung von π in π_1, deren Zulässigkeit aus (14) folgt, genügt unserer Forderung, λ einmal weniger hervorzurufen, falls $\sigma(12 \cdot I) < \sigma(13 \cdot I)$, also $2s_{12} + s_{22} < 2s_{13} + s_{33}$ ist. Wir nehmen also das Gegenteil an und notieren die Ungleichungen:

$$2s_{12} + s_{22} - 2s_{13} - s_{33} \geqq 0 \qquad (\gamma)$$
$$2s_{12} + s_{22} \geqq 0 \qquad (\delta) \text{ aus } (\gamma) + (\alpha)$$
$$s_{23} < 0. \qquad (\varepsilon) \text{ » } (\beta) - (\delta)$$

Die Umformung $\pi \rightarrow \pi_2 = 32^{-1} \cdot 13 \cdot 32 \cdot I$ mittels (16 a) genügt, falls $\sigma(32 \cdot I) < \sigma(I)$. Wir nehmen also das Gegenteil an und notieren:

$$2s_{23} + s_{22} \geqq 0 \qquad (\zeta)$$
$$s_{12} < 0. \qquad (\eta) \text{ aus } (\beta) - (\zeta)$$

$\pi \rightarrow \pi_3 = 23^{-1} \cdot 12 \cdot 23 \cdot I$ mittels (16 a) genügt, falls $\sigma(23 \cdot I) < \sigma(I)$; also:

$$s_{23} + s_{33} \geqq 0 \qquad (\vartheta)$$
$$s_{13} < 0. \qquad (\iota) \text{ aus } (\beta) - (\gamma) - (\vartheta)$$

Wir wenden nun den Hülfssatz auf die (z. B. wegen (ε)) nicht orthogonale Matrix I an. Die Differenz zweier Vektoren ist wegen (ε), (η), (ι) stets grösser als die einzelnen Summanden. Wir prüfen die Vektorsummen und finden:

$$|\mathfrak{v}_1 + \mathfrak{v}_3| \geqq |\mathfrak{v}_1| \text{ wegen } (\alpha)$$
$$|\mathfrak{v}_1 + \mathfrak{v}_2| \geqq |\mathfrak{v}_1| \text{ » } (\delta)$$
$$|\mathfrak{v}_2 + \mathfrak{v}_3| \geqq |\mathfrak{v}_2| \text{ » } (\vartheta)$$
$$|\mathfrak{v}_3 + \mathfrak{v}_2| \geqq |\mathfrak{v}_3| \text{ » } (\zeta).$$

Also ist nach dem Hülfssatz entweder $|\mathfrak{v}_2 + \mathfrak{v}_1| < |\mathfrak{v}_2|$ oder $|\mathfrak{v}_3 + \mathfrak{v}_1| < |\mathfrak{v}_3|$.

Ist $|\mathfrak{v}_2+\mathfrak{v}_1| < |\mathfrak{v}_2|$, also $2s_{12}+s_{11} < 0$, so genügt die Umformung:

$$\pi \twoheadrightarrow \pi_4 = O_2 P_{12} \cdot 12^{-1} \cdot 13 \cdot 23 \cdot 21 \cdot I,$$

denn nach Annahme ist $\sigma(21 \cdot I) < \sigma(I)$, ferner aber $\sigma(23 \cdot 21 \cdot I) < \sigma(21 \cdot I)$, denn $|\mathfrak{v}_2+\mathfrak{v}_1+\mathfrak{v}_3| < |\mathfrak{v}_2+\mathfrak{v}_1|$ folgt aus

$$2s_{23}+2s_{13}+s_{33} < 0 \quad \text{aus } (\beta)-(\gamma);$$

wir haben also $\sigma(23 \cdot 21 \cdot I) < \sigma(I)$, und da die erste und dritte Zeile der Matrix noch ungeändert sind, folgt $\sigma(13 \cdot 23 \cdot 21 \cdot I) < \sigma(13 \cdot I) = \lambda$; und endlich ist $\sigma(12^{-1} \cdot 13 \cdot 23 \cdot 21 \cdot I) = \sigma(\pi_4) = \sigma(\pi) < \lambda$. Es ist noch zu zeigen, dass $\pi = \pi_4$ aus (C) folgt. Es ist

$$\pi_4 \underset{(14)}{=} O_2 P_{12} \cdot 12^{-1} \cdot 13 \cdot 21 \cdot 23 \cdot I \underset{(16\,a)}{=} O_2 P_{12} \cdot 12^{-1} \cdot 21 \cdot 13 \cdot I$$
$$\underset{(13)}{=} 12 \cdot 13 \cdot I = \pi,$$

wobei unter dem Gleichheitszeichen jedesmal die verwendete Relation angegeben ist; (13) ist dabei in der Form

$$(13\,a) \qquad O_k P_{ik} \cdot ik^{-1} \cdot ki \cdot ik^{-1} = 1$$

verwendet, die aus (13) durch Transformation mit O_i unter Benutzung von (A) und (B) hervorgeht.

Also sei $|\mathfrak{v}_2+\mathfrak{v}_1| \geqq |\mathfrak{v}_2|$ angenommen, und also $|\mathfrak{v}_3+\mathfrak{v}_1| < |\mathfrak{v}_3|$. Wir notieren:

$$2s_{12}+s_{11} \geqq 0 \qquad (\varkappa)$$
$$2s_{13}+s_{11} < 0 \qquad (\mu)$$
$$s_{33} > s_{11}. \qquad (\nu) \text{ aus } (\alpha)-(\mu)$$

$\pi_5 = O_3 P_{13} O_2 P_{23} \cdot 32^{-1} \cdot 13^{-1} \cdot 23 \cdot 31 \cdot I$ genügt, falls $\sigma(13^{-1} \cdot 23 \cdot 31 \cdot I) < \lambda$, d. h. falls $|\mathfrak{v}_2+\mathfrak{v}_3+\mathfrak{v}_1| < |\mathfrak{v}_2|$; denn es ist $\sigma(31 \cdot I) < \sigma(I)$ nach (μ) und $\sigma(23 \cdot 31 \cdot I) < \sigma(13^{-1} \cdot 23 \cdot 31 \cdot I)$ nach (ν). $\pi = \pi_5$ folgt so:

$$\pi_5 \underset{(15)}{=} O_3 P_{13} O_2 P_{23} \cdot 32^{-1} \cdot 23 \cdot 13^{-1} \cdot 31 \cdot I \underset{(13\,a)}{=} O_3 P_{13} \cdot 32 \cdot 13^{-1} \cdot 31 \cdot I$$
$$\underset{(8),\,(10)}{=} 12 \cdot O_3 P_{13} \cdot 13^{-1} \cdot 31 \cdot I \underset{(13\,a)}{=} 12 \cdot 13 \cdot I = \pi.$$

Wir nehmen also $|\mathfrak{v}_2+\mathfrak{v}_3+\mathfrak{v}_1| \geqq |\mathfrak{v}_2|$ an, also

$$2s_{23}+2s_{12}+2s_{13}+s_{33}+s_{11} \geqq 0. \qquad (\xi)$$

Dann genügt $\pi_6 = O_3 P_{13} \cdot 13^{-1} \cdot 12 \cdot 32 \cdot 31 \cdot I$, denn $\sigma(31 \cdot I) < \sigma(I)$ nach (μ), $\sigma(32 \cdot 31 \cdot I) < \sigma(31 \cdot I)$ nach (β). Es ist also nur noch zu zeigen, dass $\sigma(12 \cdot 32 \cdot 31 \cdot I) < \lambda$, also dass gilt:

$$\text{(U)} \qquad |\mathfrak{v}_1+\mathfrak{v}_2|+|\mathfrak{v}_3+\mathfrak{v}_1+\mathfrak{v}_2| < |\mathfrak{v}_3|+|\mathfrak{v}_1+\mathfrak{v}_3|.$$

Dazu wenden wir den Hülfssatz auf die Matrix mit den Vektoren $\mathfrak{v}_1$, $\mathfrak{v}_2$ und $\mathfrak{v}_3' = \mathfrak{v}_3+\mathfrak{v}_1+\mathfrak{v}_2$ an. Es ist

$\lvert\mathfrak{v}_1 \pm \mathfrak{v}_2\rvert \geqq \lvert\mathfrak{v}_1\rvert$	nach	(δ)	und	(η)
$\lvert\mathfrak{v}_2 \pm \mathfrak{v}_1\rvert \geqq \lvert\mathfrak{v}_2\rvert$	»	$(\varkappa)$	»	(η)
$\lvert\mathfrak{v}_3' - \mathfrak{v}_2\rvert > \lvert\mathfrak{v}_3'\rvert$	»	(β)		
$\lvert\mathfrak{v}_3' + \mathfrak{v}_2\rvert > \lvert\mathfrak{v}_3'\rvert$	»	(δ)	»	(ζ)
$\lvert\mathfrak{v}_2 + \mathfrak{v}_3'\rvert > \lvert\mathfrak{v}_2\rvert$	»	(δ)	»	(ζ) und (ξ)
$\lvert\mathfrak{v}_2 - \mathfrak{v}_3'\rvert > \lvert\mathfrak{v}_2\rvert$	»	(ξ)	»	(β)
$\lvert\mathfrak{v}_3' - \mathfrak{v}_1\rvert > \lvert\mathfrak{v}_3'\rvert$	»	(μ)	»	(η)
$\lvert\mathfrak{v}_1 - \mathfrak{v}_3'\rvert > \lvert\mathfrak{v}_1\rvert$,	»	(ζ)	»	(ν)

also ist nach dem Hülfssatz

entweder $|\mathfrak{v}_1+\mathfrak{v}_3'| < |\mathfrak{v}_1|$

oder $|\mathfrak{v}_3'+\mathfrak{v}_1| < |\mathfrak{v}_3'|$.

Die erste Ungleichung besagt:

$$\begin{aligned} 0 &\geqq 3s_{11}+s_{22}+s_{33}+4s_{12}+4s_{13}+2s_{23} \\ &= [s_{11}+2s_{22}-s_{33}+4s_{12}+2s_{23}]+2\,[2s_{23}+2s_{12}+2s_{13}+s_{33}+s_{11}] \\ &\quad -2\,[s_{22}+2s_{12}+2s_{23}]+s_{22}. \end{aligned}$$

Die zweite Ungleichung besagt:

$$\begin{aligned} 0 &\geqq 3s_{11}+2s_{12}+2s_{13} \\ &= [s_{11}+2s_{22}-s_{33}+4s_{12}+2s_{23}]+[2s_{23}+2s_{12}+2s_{13}+s_{33}+s_{11}] \\ &\quad -2\,[2s_{22}+2s_{12}+2s_{23}]+s_{11}. \end{aligned}$$

In beiden Fällen folgt wegen (ξ) und (β)

$$0 > s_{11}+2s_{22}-s_{33}+4s_{12}+2s_{23},$$

2*

welches mit der zu beweisenden Ungleichung (U) gleichbedeutend ist.

$\pi = \pi_6$ folgt so:

$$\pi_6 \underset{(16\,a)}{=} O_3 P_{13} \cdot 13^{-1} \cdot 31 \cdot 12 \cdot I \underset{(13\,a)}{=} 13 \cdot 12 \cdot I \underset{(14)}{=} 12 \cdot 13 \cdot I = \pi.$$

Damit ist der Fall 5 erledigt.

Fall 6) $\pi = 12 \cdot 23 \cdot I$. — Nach Voraussetzung ist:

$$\begin{array}{ll} 2s_{23} + s_{33} \geqq 0 & (\alpha) \\ s_{22} + s_{33} + 2s_{23} + 2s_{12} + 2s_{13} < 0 & (\beta) \\ s_{22} + 2s_{12} + 2s_{13} < 0. & (\gamma) \text{ aus } (\beta) - (\alpha) \end{array}$$

$\pi \twoheadrightarrow \pi_1 = 23 \cdot 13 \cdot 12 \cdot I$ mittels (16 a). Hier ist $\sigma(13 \cdot 12 \cdot I) < \sigma(I)$ nach (β). Also genügt π_1, falls $\sigma(12 \cdot I) < \sigma(23 \cdot I) = \lambda$. Wir nehmen also das Gegenteil an:

$$\begin{array}{ll} 2s_{12} + s_{22} - 2s_{23} - s_{33} > 0 & (\delta) \\ 2s_{12} + s_{22} \geqq 0 & (\varepsilon) \text{ aus } (\delta) + (\alpha) \\ s_{13} < 0. & (\zeta) \text{ » } (\gamma) - (\varepsilon) \end{array}$$

$\pi \twoheadrightarrow \pi_2 = 23 \cdot 12 \cdot 13 \cdot I$ mittels (16 a) genügt analog, falls nicht

$$\begin{array}{ll} s_{13} - s_{23} \geqq 0 & (\eta) \\ 2s_{13} + s_{33} \geqq 0 & (\vartheta) \text{ aus } (\eta) \text{ und } (\alpha) \\ s_{23} < 0. & (\iota) \text{ » } (\zeta) \text{ » } (\eta) \end{array}$$

$\pi \twoheadrightarrow \pi_3 = O_3 P_{23} \cdot 23^{-1} \cdot 13 \cdot 32 \cdot I$ genügt, falls nicht $\sigma(32 \cdot I) \geqq \sigma(23 \cdot I)$, d. h.

$$\begin{array}{ll} s_{22} \geqq s_{33} & (\varkappa) \\ 2s_{23} + s_{22} \geqq 0 & (\mu) \text{ aus } (\varkappa) + (\alpha) \\ s_{12} < 0. & (\nu) \text{ » } (\beta) - (\mu) - (\vartheta) \end{array}$$

$\pi = \pi_3$ folgt so: $\pi_3 \underset{(15)}{=} O_3 P_{23} \cdot 13 \cdot 23^{-1} \cdot 32 \cdot I \underset{(9)\,(10)}{=} 12 \cdot O_3 P_{23} \cdot 23^{-1} \cdot 32 \cdot I \underset{(13\,a)}{=} 12 \cdot 23 \cdot I = \pi$. Wegen ($\zeta$), ($\iota$), ($\nu$) ist die Differenz zweier Vektoren stets grösser, als der einzelne Summand. Ferner ist

$$|\mathfrak{v}_2+\mathfrak{v}_3| \geqq |\mathfrak{v}_2| \quad \text{nach } (\alpha)$$
$$|\mathfrak{v}_3+\mathfrak{v}_2| \geqq |\mathfrak{v}_3| \quad \text{» } (\mu)$$
$$|\mathfrak{v}_1+\mathfrak{v}_2| \geqq |\mathfrak{v}_1| \quad \text{» } (\varepsilon)$$
$$|\mathfrak{v}_1+\mathfrak{v}_3| \geqq |\mathfrak{v}_1| \quad \text{» } (\vartheta),$$

also ist nach dem Hülfssatz

$$\text{entweder } |\mathfrak{v}_2+\mathfrak{v}_1| < |\mathfrak{v}_2|$$
$$\text{oder } |\mathfrak{v}_3+\mathfrak{v}_1| < |\mathfrak{v}_3|.$$

Ist $|\mathfrak{v}_2+\mathfrak{v}_1| < |\mathfrak{v}_2|$, also $|\mathfrak{v}_2| > |\mathfrak{v}_1|$, so genügt $\pi_4 = O_2 P_{12} \cdot 12^{-1} \cdot 23 \cdot 21 \cdot I \underset{(14)\,(13\,a)}{=} \pi$; denn $\sigma(21 \cdot I) < \sigma(I)$; und $\sigma(23 \cdot 21 \cdot I) < \sigma(\pi)$ wegen $|\mathfrak{v}_1| < |\mathfrak{v}_2| \underset{(\alpha)}{\leqq} |\mathfrak{v}_2+\mathfrak{v}_3|$. Ist $|\mathfrak{v}_3+\mathfrak{v}_1| < |\mathfrak{v}_3|$, also $|\mathfrak{v}_3| > |\mathfrak{v}_1|$, so genügt $\pi_5 = O_2 P_{12} \cdot 12^{-1} \cdot 31^{-1} \cdot 23 \cdot 31 \cdot I \underset{(16\,a)}{=} \pi_4$; denn $\sigma(31 \cdot I) < \sigma(I)$, $\sigma(23 \cdot 31 \cdot I) < \sigma(31^{-1} \cdot 23 \cdot 31 \cdot I) < \sigma(\pi)$, weil

$$|\mathfrak{v}_1| < |\mathfrak{v}_3| \underset{(\varkappa)}{\leqq} |\mathfrak{v}_2| \underset{(\alpha)}{\leqq} |\mathfrak{v}_2+\mathfrak{v}_3|.$$

Fall 7) $\pi = 12 \cdot 31 \cdot I$. — Nach Voraussetzung ist

$$2s_{13}+s_{11} \geqq 0 \quad (\alpha)$$
$$2s_{12}+s_{22} < 0 \quad (\beta)$$
$$s_{12} < 0. \quad (\gamma) \text{ aus } (\beta)$$

$\pi \rightarrow \pi_1 = O_2 P_{12} \cdot 12^{-1} \cdot 31 \cdot 21 \cdot I \; (\underset{(15)}{=} O_2 P_{12} \cdot 12^{-1} \cdot 21 \cdot 31 \cdot I \underset{(13\,a)}{=} 12 \cdot 31 \cdot I)$ genügt, ausser wenn

$$2s_{12}+s_{11} \geqq 0 \quad (\delta)$$
$$s_{11} > s_{22}. \quad (\varepsilon) \text{ aus } (\delta)-(\beta)$$

$\pi \rightarrow \pi_2 = 31 \cdot 32^{-1} \cdot 12 \cdot I \; (\underset{(16\,a)}{=} \pi)$ genügt, ausser wenn $\sigma(32^{-1} \cdot 12 \cdot I) \geqq \lambda$, d. h.

$$2s_{22}-s_{11}+2s_{12}-2s_{13}-2s_{23} \geqq 0 \quad (\zeta)$$
$$s_{22}-2s_{23} > 0. \quad (\eta) \text{ aus } (\zeta)+(\alpha)-(\beta)$$

$\pi \rightarrow \pi_3 = 32^{-1} \cdot 31 \cdot 12 \cdot I \; (\underset{(16\,a)}{=} \pi)$ genügt, ausser wenn $\sigma(31 \cdot 12 \cdot I) \geqq \lambda$, d. h.

$$s_{22}+2s_{12}+s_{23} \geqq 0 \qquad (\vartheta)$$
$$s_{23} > 0 \qquad (\iota) \text{ aus } (\vartheta)-(\beta)$$
$$s_{13} < 0. \qquad (\nu) \text{ aus } (\beta)-(\varepsilon)-(\zeta)-2\times(\iota)$$

$\pi \twoheadrightarrow \pi_4 = O_1 P_{13} \cdot 31^{-1} \cdot 32^{-1} \cdot 13 \cdot I \underset{(14)}{(=} O_1 P_{13} \cdot 32^{-1} \cdot 31^{-1} \cdot 13 \cdot I \underset{(8)(11)}{=} 12 \cdot O_1 P_{13} \cdot 31^{-1} \cdot 13 \cdot I \underset{(13a)}{=} \pi)$ genügt, ausser wenn $\sigma(32^{-1} \cdot 13 \cdot I) \geqq \lambda$; denn $\sigma(32^{-1} \cdot 13 \cdot I) > \sigma(13 \cdot I)$ wegen (η). Also:

$$s_{33}+s_{22}-s_{11}-2s_{23} \geqq 0 \qquad (\varkappa)$$
$$s_{33}-2s_{23} > 0. \qquad (\mu) \text{ aus } (\varkappa)+(\varepsilon)$$

Nun wende man den Hülfssatz auf die Matrix $12 \cdot I$ mit den Vektoren $\mathfrak{v}_1' = \mathfrak{v}_1 + \mathfrak{v}_2$, $\mathfrak{v}_2$ und $\mathfrak{v}_3$ an. Es ist

$$|\mathfrak{v}_1'-\mathfrak{v}_2| > |\mathfrak{v}_1'| \quad \text{nach } (\beta)$$
$$|\mathfrak{v}_1'+\mathfrak{v}_2| > |\mathfrak{v}_1'| \quad \text{aus } (\zeta)+(\alpha)+2\times(\iota)$$
$$|\mathfrak{v}_2-\mathfrak{v}_1'| > |\mathfrak{v}_2| \quad \text{nach } (\varepsilon)$$
$$|\mathfrak{v}_2+\mathfrak{v}_1'| > |\mathfrak{v}_2| \quad \text{aus } (\delta)+(\zeta)+(\alpha)+2\times(\iota)$$
$$|\mathfrak{v}_2 \pm \mathfrak{v}_3| > |\mathfrak{v}_2| \quad \text{nach } (\iota) \text{ und } (\mu)$$
$$|\mathfrak{v}_3 \pm \mathfrak{v}_2| > |\mathfrak{v}_3| \quad \text{» } (\iota) \text{ » } (\eta)$$
$$|\mathfrak{v}_3+\mathfrak{v}_1'| > |\mathfrak{v}_3| \quad \text{aus } (\vartheta)+(\alpha)+(\iota)$$
$$|\mathfrak{v}_3-\mathfrak{v}_1'| > |\mathfrak{v}_3| \quad \text{» } (\zeta)+(\varepsilon)$$
$$|\mathfrak{v}_1'-\mathfrak{v}_3| > |\mathfrak{v}_1'|. \quad \text{» } (\mu)-2\times(\nu)$$

Also ist nach dem Hülfssatz $|\mathfrak{v}_1'+\mathfrak{v}_3| < |\mathfrak{v}_1'|$, d. h.

$$s_{33}+2s_{13}+2s_{23} < 0 \qquad (\xi)$$
$$s_{33}+2s_{13} < 0. \qquad (\varrho) \text{ aus } (\xi)-2\times(\iota)$$

Wegen (ϱ) ist $|\mathfrak{v}_1+\mathfrak{v}_3| < |\mathfrak{v}_1|$. Wir setzen $\mathfrak{v}_1+\mathfrak{v}_3 = \mathfrak{v}_1''$ und wenden den Hülfssatz auf die Matrix mit den Vektoren $\mathfrak{v}_1''$, $\mathfrak{v}_2$, $\mathfrak{v}_3$ an. Da $|\mathfrak{v}_2 \pm \mathfrak{v}_3| > |\mathfrak{v}_2|$ und $> |\mathfrak{v}_3|$ ist, wie früher gezeigt, und $|\mathfrak{v}_1''-\mathfrak{v}_3| = |\mathfrak{v}_1| > |\mathfrak{v}_1''|$ ist, prüfen wir:

$$|\mathfrak{v}_1''+\mathfrak{v}_2| > |\mathfrak{v}_1''| \quad \text{aus } (\vartheta)+(\iota)$$
$$|\mathfrak{v}_1''-\mathfrak{v}_2| > |\mathfrak{v}_1''| \quad \text{» } (\eta)-2\times(\gamma)$$
$$|\mathfrak{v}_2-\mathfrak{v}_1''| > |\mathfrak{v}_2|. \quad \text{» } (\alpha)+(\mu)-2\times(\gamma)$$

Wenn nun $|\mathfrak{v}_2 + \mathfrak{v}_1''| \leq |\mathfrak{v}_2|$, also $|\mathfrak{v}_1 + \mathfrak{v}_2 + \mathfrak{v}_3| \leq |\mathfrak{v}_2|$ ist, so genügt:

$\pi \Longrightarrow \pi_5 = O_1 P_{13} \cdot 21^{-1} \cdot 32^{-1} \cdot 21 \cdot 13 \cdot I \underset{(16\,a)}{(=} O_1 P_{13} \cdot 32^{-1} \cdot 31^{-1} \cdot 13 \cdot I = \pi)$; denn dann ist $\sigma(21 \cdot 13 \cdot I) \underset{(8)\,(11)\,(13\,a)}{\leq} \sigma(13 \cdot I) < \sigma(I)$ und $\sigma(32^{-1} \cdot 21 \cdot 13 \cdot I) \leq \sigma(\pi_5) = \sigma(\pi)$; also sei $|\mathfrak{v}_2 + \mathfrak{v}_1''| > |\mathfrak{v}_2|$; ferner ist $|\mathfrak{v}_3 - \mathfrak{v}_1''| > |\mathfrak{v}_3|$ aus $(\varrho) - (\alpha)$. Also ist nach dem Hülfssatz

entweder $|\mathfrak{v}_1'' + \mathfrak{v}_3| < |\mathfrak{v}_1''|$ d. h. $3s_{33} + 2s_{13} < 0$ (ψ)

oder $|\mathfrak{v}_3 + \mathfrak{v}_1''| < |\mathfrak{v}_3|$ d. h. $s_{11} + 4s_{13} + 3s_{33} < 0$. (χ)

Da (ψ) aus $(\chi) - (\alpha)$ folgt, gilt ψ in beiden Fällen. Dann genügt π_5.

Denn es ist $\sigma(13 \cdot I) < \sigma(I)$, ferner $\sigma(21 \cdot 13 \cdot I) < \sigma(32^{-1} \cdot 21 \cdot 13 \cdot I)$, weil $|\mathfrak{v}_3| < |\mathfrak{v}_1 + \mathfrak{v}_2|$, wie aus $(\alpha) + (\vartheta) + (\mu) - (\psi) + (\iota)$ folgt. Es ist also nur noch zu zeigen, dass $\sigma(32^{-1} \cdot 21 \cdot 13 \cdot I) < \lambda$, also $|\mathfrak{v}_1 + \mathfrak{v}_2 + \mathfrak{v}_3| + |\mathfrak{v}_1 + \mathfrak{v}_2| < |\mathfrak{v}_1| + |\mathfrak{v}_2|$, also

$$s_{11} + s_{22} + s_{33} + 4s_{12} + 2s_{23} + 2s_{13} < 0$$

ist; und das folgt aus $2 \times (\beta) + (\psi) - (\varkappa)$.

— Damit ist die Vollständigkeit des Relationensystems (A), (B), (C) für die Gruppe G bewiesen. Im nächsten § wird dies Gruppenschema auf eine kürzere — dafür aber nicht symmetrische — Form gebracht.

§ 3.

Wir haben bisher als Erzeugende der Gruppe G die 12 Operationen P_{ik}, O_i und ik benutzt. Es ist nun leicht, die Erzeugendenzahl zunächst auf 4, dann weiter auf 3 Operationen zu verkleinern. Wir bezeichnen:

$$P = P_{12}, \quad O = O_1,$$
$$Q = P_{23} \cdot P_{12}, \quad U = 12.$$

Dann ist

$$\begin{array}{lllll} P_{23} & = QP & \text{wegen} & (1) & \\ P_{13} & = PQ & » & (2), & (\text{nämlich} = P_{23\,P_{12}} = (QP)_P = PQ), \\ O_2 & = O_P & \left.\right\} & » & (5) \\ O_3 & = O_{PQ} & & & \\ 21 & = U_P & » & (7) & \\ 32 & = U_{PQ} & » & (8) & \\ 13 & = U_{QP} & » & (9) & \\ 23 & = 13_{P_{12}} = U_Q & \text{wegen} & (8) & \\ 31 & = 32_{P_{12}} = U_{PQP} & » & (9). & \end{array}$$

Also ist $G = \{P, Q, O, U\}$. Ein vollständiges Relationensystem zwischen diesen erhält man, wenn man die eben gefundenen Ausdrücke für die P_{ik}, O_i, ik in (A), (B), (C) einsetzt. Vereinfacht man das sich ergebende System nach Möglichkeit dadurch, dass man Relationen, die eine Folge der übrigen sind, fortlässt, so lässt sich das Ergebnis auf die folgende Form bringen[4]:

$$(D)\quad G = \{P, Q, O, U\} \left\{ \begin{array}{ll} (D_1): P^2 = 1 & (D_6): U \rightleftarrows O_{Q^2} \\ (D_2): (QP)^2 = 1 & (D_7): U_O = U^{-1} \\ (D_3): Q^3 = 1 & (D_8): U_{POP} = U^{-1} \\ & (D_9): OPUU_P^{-1}U = 1 \\ (D_4): O^2 = 1 & \\ (D_5): O \rightleftarrows QP & \\ \multicolumn{2}{l}{(D_{10}): U \rightleftarrows U_{QP}} \\ \multicolumn{2}{l}{(D_{11}): U \rightleftarrows U_{PQ}} \\ \multicolumn{2}{l}{(D_{12}): UU_QU^{-1}U_Q^{-1} = U_{QP}.} \end{array} \right.$$

Die Relation (D_9) zeigt, dass man O durch P und U ausdrücken kann:

[4] Ich unterdrücke die Zwischenrechnung und verweise auf die weitgehend analoge entsprechende Durchrechnung in meinem Aufsatz: »Die Isomorphismengruppe der freien Gruppen«. Math. Ann. (Im Erscheinen).

$$(D_9): \quad O = PUU_P^{-1}U$$

unter Benutzung von (D_4). Substituiert man diesen Ausdruck für O in den übrigen Gleichungen, so erhält man, indem ausser (D_9) auch (D_7) und (D_8) entbehrlich werden, das folgende Schema:

$$(E) \quad G = \{P, Q, U\} \left\{ \begin{array}{ll} (E_1):\ P^2 = 1 & (E_4):\ (PUU_P^{-1}U)^2 = 1 \\ (E_2):\ (QP)^2 = 1 & (E_5):\ PUU_P^{-1}U \rightleftarrows QP \\ (E_3):\ Q^3 = 1 & (E_6):\ PUU_P^{-1}U \rightleftarrows U_Q \\ \multicolumn{2}{c}{(E_7):\ U \rightleftarrows U_{QP}} \\ \multicolumn{2}{c}{(E_8):\ U \rightleftarrows U_{PQ}} \\ \multicolumn{2}{c}{(E_9):\ UU_QU^{-1}U_Q^{-1} = U_{QP}} \end{array} \right.$$

Hier ist $D(P) = -1$, $D(Q) = +1$, $D(U) = +1$. Gittertransformationen mit Erhaltung der Orientierung, deren Gruppe $\overline{G}$ heissen möge, enthalten also die Erzeugende P eine gerade Anzahl von Malen. Also ist $\overline{G} = \{Q, U, Q_P, U_P\}$. Wegen (E_2) ist $Q_P = Q^{-1}$, also ist $\overline{G} = \{Q, U, U_P\}$. Zur Erzeugung dieser die Orientierung erhaltenden Gruppe $\overline{G}$, die in G invariant vom Index 2 enthalten ist, müssen wir also neben Q und U noch eine Operation mit Determinante $+1$ einführen. Es ist dazu am bequemsten, auf das Schema (D) zurückzugreifen, und dort O, für welches $D(O) = -1$ ist, durch die neue Erzeugende

$$T = PO, \qquad D(T) = +1,$$

zu ersetzen, was wegen $O = PT$ möglich ist. Wegen (D_4) ist dann $(PT)^2 = 1$, also $T_P = T^{-1}$. Wegen (D_9) ist $T = UU_P^{-1}U$, also $U_P = UT^{-1}U$. Also ist $\overline{G} = \{Q, T, U\}$, da man Q_P, T_P und U_P durch Q, T, U ausdrücken kann, und $G = \{Q, T, U, P\}$ ist. (E_4) nimmt die Form

$$UT^{-1}UT^{-1}UT = 1$$

an, sodass wir $UT^{-1}U = U_T^{-1}$ und also $U_P = U_T^{-1}$ haben.

Das Schema für $G = \{Q, T, U, P\}$ zerfällt nun in zwei Teile, einen P enthaltenden Teil:

$$(F) \quad P^2 = 1\,, \quad Q_P = Q^{-1}\,, \quad T_P = T^{-1}\,, \quad U_P = U_T^{-1}\,,$$

mittels dessen die Stellung der invarianten Untergruppe $\overline{G}$ innerhalb G gekennzeichnet wird, und das Schema für $\overline{G}$, das die übrigen aus (D) erhaltenen Relationen enthält. Man kann es hier auf folgende Form bringen:

$$\begin{array}{c} (H) \\ \overline{G} = \{Q, T, U\} \end{array} \left\{ \begin{array}{ll} (H_1):\ Q^3 = 1 & (H_4):\ (UT^{-1})^3 = T^2 \\ (H_2):\ T^4 = 1 & (H_5):\ U_{T_Q^2} = U^{-1} \\ (H_3):\ (QT)^2 = 1 & \\ \quad (H_6):\ U_T \rightleftarrows U_Q & \\ \quad (H_7):\ U_T \rightleftarrows U_{Q^{-1}} & \\ \quad (H_8):\ U U_Q U^{-1} U_Q^{-1} = U_{T_Q}. & \end{array} \right.$$

Man darf dies Schema einer schon in § 1 gemachten Bemerkung zufolge erst dann als das Schema für $\overline{G}$ ansprechen, wenn die Transformation der einzelnen Relationen mit P keine neuen liefert; (H) geht durch Transformation mit P mittels (F) in sich über.

Die Aeqvivalenz des Schemas $(F) + (H)$ für G mit dem in § 2 benutzten Schema (A) + (B) + (C), erkennt man übrigens, anstatt durch (D) zu gehen, leichter direkt, indem man formal definiert:

$$\begin{aligned} 12 &= U \\ 21 &= U_P \\ P_{12} &= P \\ O_1 &= PT \end{aligned}$$

und dann, ebenfalls per definitionem, der Transformation mit Q eine zyklische Permutation der Indizes in diesen Ausdrücken entsprechen lässt, z. B. $23 = U_Q$, $O_2 = (PT)_Q$

u. s. w. Dann ergeben sich die Relationen (1)—(16) durch leichte Rechnung aus (F) und (H).

Für das geometrische Studium der Gittertransformationen ist es natürlich ausreichend, sich mit der die Orientierung erhaltenden Gruppe $\overline{G}$ zu beschäftigen. Die geometrische Bedeutung der Erzeugenden geht aus den Matrizen hervor:

$$Q = \begin{pmatrix} 0 & 1 & 0 \\ 0 & 0 & 1 \\ 1 & 0 & 0 \end{pmatrix}, \quad T = \begin{pmatrix} 0 & 1 & 0 \\ -1 & 0 & 0 \\ 0 & 0 & 1 \end{pmatrix}, \quad U = \begin{pmatrix} 1 & 1 & 0 \\ 0 & 1 & 0 \\ 0 & 0 & 1 \end{pmatrix}.$$

Q ist also eine Drehung durch $\frac{2\pi}{3}$ um die Gerade $x = y = z$ und T eine Drehung durch $\frac{\pi}{2}$ um die z-Achse. Diese beiden erzeugen die orthogonale Gruppe $\overline{\Omega}$, die Oktaedergruppe; und (H_1), (H_2), (H_3) ist ein bekanntes Schema für diese Gruppe. U endlich ist die Affinität, welche durch die Substitution

$$\begin{aligned} x' &= x + y \\ y' &= y \\ z' &= z \end{aligned}$$

bestimmt wird. Die Potenzen von U sind also alle verschieden, und das Mass und die Art der Verknüpfung von U mit den Drehungen Q und T wird durch die Gleichungen (H_4)—(H_8) vollständig beschrieben.

Zum Schluss sei eine Bemerkung über den Zusammenhang mit freien Gruppen angefügt. Seien a_1, a_2, a_3 die erzeugenden Operationen einer freien Gruppe Φ (ohne verbindende Relationen zwischen den a_i). Wir bezeichnen vier Isomorphismen von Φ mit sich selbst, wie folgt:

$$P: \begin{cases} a_1' = a_2 \\ a_2' = a_1 \\ a_3' = a_3 \end{cases} \qquad Q: \begin{cases} a_1' = a_2 \\ a_2' = a_3 \\ a_3' = a_1 \end{cases}$$

$$O: \begin{cases} a_1' = a_1^{-1} \\ a_2' = a_2 \\ a_3' = a_3 \end{cases} \qquad U: \begin{cases} a_1' = a_1 a_2 \\ a_2' = a_2 \\ a_3' = a_3 \end{cases}$$

Dann erzeugen diese die gesamte Gruppe der Selbstisomorphismen von Φ, $I_\Phi = \{P, Q, O, U\}$, und das Gruppenschema ist[5]:

$$\begin{matrix} I_\Phi: \\ = \{P, Q, O, U\} \end{matrix} \begin{cases} P^2 = 1 & U \rightleftarrows O_{Q^2} & U_{O_P} = U^{-1} \\ (QP)^2 = 1 & U \rightleftarrows U_O & O P U_O^{-1} U_P^{-1} U = 1 \\ Q^3 = 1 & U \rightleftarrows U_{OQP} & \\ O^2 = 1 & U \rightleftarrows U_{PQ} & \\ O \rightleftarrows QP & U U_Q U^{-1} U_Q^{-1} = U_{QP} & \\ O \rightleftarrows O_Q & & \end{cases}$$

Ist a_1, a_2, a_3 ein Isomorphismus von Φ, und bezeichnet e_{ik} die Exponentensumme von a_k in a_i, so ist (e_{ik}) eine Matrix mit Determinante ± 1, also in G. G ist also Faktorgruppe von I_Φ, und die zugehörige invariante Untergruppe I_Φ° von I_Φ ist die Gruppe derjenigen Isomorphismen, die zur Einheitsmatrix gehören. Nun geht das Schema (D) von G aus dem Schema von I_Φ durch Hinzufügung der einzigen Relation

$$U_O U = 1$$

hervor. $U_O U$ ist der Isomorphismus

$$U_O U: \begin{cases} a_1' = a_2^{-1} a_1 a_2 \\ a_2' = a_2 \\ a_3' = a_3 . \end{cases}$$

Dass $U_O U$ in G, nicht aber in I_Φ der Einheit entspricht, ist der Ausdruck dafür, dass die Variabeln unserer linearen

[5] Siehe die unter [4] zitierte Abhandlung.

Substitutionen, nicht aber die Erzeugenden a_i von Φ vertauschbar sind. Die obige Tatsache besagt nun, dass die zur Einheitsmatrix gehörige invariante Untergruppe I_{Φ}° durch alle Transformierten von $U_0 U$ mit beliebigen Elementen von I_{Φ} erzeugt wird.

Es scheint mir eine für die Grundaufgaben der allgemeinen Gruppentheorie bedeutungsvolle Frage zu sein, ob diese Tatsache unabhängig von der Beschränkung auf die Variablenzahl 3, für die sie hier allein bewiesen ist, ihre Gültigkeit bewahrt.

Kopenhagen, Oktober 1923.

Reprinted from
Math.-fys. Medd. Danske Vid. Selsk. 5, No. 12 (1924)

Færdig fra Trykkeriet d. 13. Marts 1924.

The Isomorphism Group of a Free Group*

JAKOB NIELSEN

A set of elements of a group is called *independent* when no relations other than identities hold between elements of the set. A group is called *free* when it may be engendered by an independent set of *generators*. An abstract theory of discontinuous groups, which must deal with *infinite* groups from the outset, is where the study of free groups has its natural starting point.[1] A glance at previous investigations shows the importance of the systematic change of generating sets in this field.[2] The passage between two sets of independent generators corresponds to a one-one mapping of the group onto itself, an "automorphism". The automorphisms of a group themselves constitute a group, and its study is one of the fundamental problems of abstract group theory.

The goal of the present work is to determine the structure of the (auto-)isomorphism group of a finitely generated free group. The approach involves the following steps:

§1. Generators of the isomorphism group.
§2. The relation system for the symmetric group.
§3. The relation system for the symmetric group extended by sign changes.
§4. The relation system for the whole group.
§5. Completeness proof for the whole group: reduction to a lemma.
§6. Proof of the lemma, Part I.
§7. Proof of the lemma, Part II.

§1.
Generators of the Isomorphism Group

Let F_n be the free group, with free generators $a_1, a_2, \ldots, a_n$, Γ_n the group of its isomorphisms.[3] The set of elements[4]

$$\alpha_i = \Pi^{(i)}(a_1, \ldots, a_n) \qquad (i = 1, 2, \ldots, n) \tag{1}$$

is a "primitive system", i.e. able to serve as a new generating system for F_n, when the original generators can be expressed in terms of the α's, and thus when there are solution equations

$$a_i = \bar{\Pi}^{(i)}(\alpha_1, \ldots, \alpha_n) \qquad (i = 1, 2, \ldots, n) \tag{2}$$

which are identities in the a's. The correspondence between a_i and α_i $(i = 1, 2, \ldots, n)$ deter-

* Translation by John Stillwell of Die Isomorphismengruppe der freien Gruppen. [Math. Ann., 91 (1924), 169–209.]

[1] See, e.g., W. Dyck's "Gruppentheoretische Studien", Math. Ann. 20 & 22.

[2] See, e.g., the Dehn method for the determination of quotient groups (Report of the lecture by M. Dehn to the Jahresversammlung of the D.M.V., Leipzig 1922).

[3] The index n does *not*, as is otherwise usual in group theory, denote the order of the group. F_n and Γ_n are infinite groups.

[4] The product sign Π without an index range stands for a general non-commutative product in the arguments and their inverses.

mines an isomorphism I of F_n, which will be denoted by

$$I = [\alpha_1, \alpha_2, \ldots, \alpha_n] = [\alpha].$$

Let $I' = [\alpha']$ be a second isomorphism of F_n. Then if one replaces each a_i in the α's by α'_i one obtains a new primitive system; we denote the new isomorphism by II'. The construction of a "product" $II'I'' \ldots I^{(r)}$ of isomorphisms can therefore be carried out *either* by beginning with the leftmost factor and *substituting* for the a's according to successive factors, *or* by beginning with the rightmost factor and *composing* with successive factors to the left. This "multiplication" of isomorphisms obviously satisfies the associative law. The inverse I^{-1} of I is given by (2) when one exchanges the symbols a and α. If one denotes the identity isomorphism $[a]$ by 1, then the relation $I^{-1}I = 1$ is a consequence of equations (1) and (2). The group property of Γ_n follows from the composition rule, associative law and the existence of "left" inverses. For II^{-1} is likewise an isomorphism, say I', and it has a left inverse I'^{-1}. If one premultiplies the correct equation $I^{-1}I \cdot I^{-1} = I^{-1}$ by $I'^{-1}I$ then one gets $I'^{-1}I' \cdot I' = I'^{-1}I'$ by application of the associative law, and hence $I' = 1$. This says that the equations (1) are solutions of (2). One can also see this directly by substituting the expressions (2) for the a in (1) and using the fact that the resulting relations between the α's must be identities.[5]

In order to obtain a set of generators for Γ_n, we first recall that an arbitrary permutation of the a's represents the simplest form of an isomorphism of F_n. One therefore has the symmetric group as a subgroup of Γ_n. We introduce the following notation:

$$P = P_{12} = [a_2, a_1, a_3, \ldots, a_n] = (a_1 a_2), \tag{3}$$

$$Q = [a_2, a_3, \ldots, a_n, a_1] = (a_1 a_2 \ldots a_n). \tag{4}$$

As usual, the round brackets denote that each element should be replaced by its successor in the cycle, and that elements not mentioned remain unchanged. If one sets

$$T_i = (a_1 a_2 \ldots a_i) \qquad (i = 2, 3, \ldots, n),$$

then each permutation may be written in the form

$$T_2^{\varepsilon_2} T_3^{\varepsilon_3} \ldots T_n^{\varepsilon_n}, \qquad 0 \leq \varepsilon_i < i.$$

But since

$$Q^{-(i-1)} P Q^{i-1} = P_{i,i+1} = (a_i a_{i+1}) \qquad (i = 1, 2, \ldots, n-1),$$
$$T_i = P_{i-1,i} P_{i-2,i-1} \ldots P_{23} P_{12},$$

it follows that P and Q alone generate Σ_n. We write $\Sigma_n = \{P, Q\}$.[6]

Equally simple isomorphisms are obtained when one replaces a generator by its inverse. We let

$$O = O_1 = [a_1^{-1}, a_2, \ldots, a_n] = [a_1 \mapsto a_1^{-1}], \tag{5}$$

[5] More detailed investigations of the properties of independent systems may be found in my essay: "Om Regning med ikke-kommutative Faktorer og dens Anvendelse i Gruppeteorien", Matematisk Tidsskrift 1921, [N8].

[6] The generation of the symmetric group by two operations is used, among others, by A. Capelli, Giorn. mat. (2) **4** (1897), p. 354; E. H. Moore, Proc. Lond. Math. Soc. (1) **28** (1896/7), p. 363.

where the arrow in the last term serves as a *replacement* sign, and elements not mentioned are replaced by themselves. It follows from

$$Q^{-(i-1)}OQ^{i-1} = O_i = [a_i \mapsto a_i^{-1}] \qquad (i = 1, \ldots, n)$$

that O, together with P and Q, generates the group which extends Σ_n by sign change

$$\Omega_n = \{P, Q, O\}$$

and which has order $2^n \cdot n!$.

If α is any product of generators and their inverses in "freely reduced form", i.e. containing no subproduct $a_i a_i^{-1}$, then let $L(\alpha)$ be the "length of α", i.e. the number of symbols in α, and thus the sum of the absolute values of the exponents of a's in α. If $I = [\alpha_1, \ldots, \alpha_n]$ we set

$$L(I) = L(\alpha_1) + L(\alpha_2) + \cdots + L(\alpha_n).$$

Ω_n includes all isomorphisms of length n. In order to obtain isomorphisms of greater length, we introduce a generator of length $n + 1$:

$$U = U_{12} = [a_1 a_2, a_2, \ldots, a_n] = [a_1 \mapsto a_1 a_2]. \tag{6}$$

Then $U^{-1} = [a_1 \mapsto a_1 a_2^{-1}]$ and also

$$OU^{\mp 1}O = V^{\pm 1} = V_{12}^{\pm 1} = [a_1 \to a_2^{\pm 1} a_1]. \tag{7}$$

The permutation

$$N_{ik} = \begin{cases} (QP)^{k-i-1}Q^{i-1} & \text{for } i < k, \\ (QP)^{k-i} \ \ Q^{i-1} & \text{for } i > k \end{cases} \tag{8}$$

brings a_i to the place of a_1 and a_k to the place of a_2, since

$$QP = (a_2 a_3 \ldots a_n) \tag{9}$$

as one sees immediately. In particular, (8) says that $N_{12} = 1$ and $N_{21} = P$. One then has

$$N_{ik}^{-1} P N_{ik} = P_{ik} = P_{ki} = (a_i a_k), \tag{10}$$

$$N_{ik}^{-1} U^{\pm 1} N_{ik} = U_{ik}^{\pm 1} = [a_i \mapsto a_i a_k^{\pm 1}], \tag{11}$$

$$N_{ik}^{-1} V^{\pm 1} N_{ik} = V_{ik}^{\pm 1} = [a_i \mapsto a_k^{\pm 1} a_i]. \tag{12}$$

In an earlier short work[7] I have shown that one can proceed from an arbitrary isomorphism I of length $L(I) > n$, by repeated premultiplications with isomorphisms of the form $U_{ik}^{\pm 1}$ and $V_{ik}^{\pm 1}$, to isomorphisms with steadily decreasing length, finally reaching an isomorphism of length n, and therefore in Ω_n. This "dismantling" of I corresponds to a "construction" of I from Ω_n by premultiplication by the inverses of the factors in opposite order. Since, by (7), (11) and (12), all these factors result from $U^{\pm 1}$ by conjugation with isomorphisms from Ω_n, it follows that $\Gamma_n = \{P, Q, O, U\}$, i.e.:

The isomorphism group Γ_n of the free group F_n is generated by the four isomorphisms P, Q, O and U, which are given by the defining equations (3) *to* (6).

For $n = 2$, Q and P are identical, and hence the number of generators reduces to 3.

[7] "Über die Isomorphismen unendlicher Gruppen ohne Relation", Math. Ann. 79, [N3]. An extended treatment is found in the work cited in [5].

§2.

The Relation System for the Symmetric Group

Now that generators whose number is independent of n have been given for Γ_n, we come to the problem of finding a complete relation system. It is a question of finding a system of relations between P, Q, O and U—which we shall call the system S—with the following two properties:

(a) The relations of the system S are consequences of the defining equations (3) to (6) of P, Q, O, U.
(b) Each relation between P, Q, O, U which is a consequence of the defining equations (3) to (6), follows *formally* from S.

Following the path laid out in §1, in the next paragraphs we shall derive such a relation system stepwise for Σ_n, Ω_n and finally Γ_n. Since conjugation and commutation play a commanding role in the system, we shalll use the following abbreviated notation for the sake of clarity: if X and Y are two group elements, the conjugate of X by Y will be denoted X_Y:

$$X_Y = Y^{-1}XY.$$

The commutativity of X and Y will be denoted by a double arrow: thus

$$X \rightleftarrows Y$$

denotes

$$XY = YX$$

or

$$XYX^{-1}Y^{-1} = 1.$$

The most immediate relations between P and Q result from the orders of P, Q and QP, which are immediate from (3), (4) and (9), and also the commutativity of P with permutations of $a_3, \ldots, a_n$. Since $P_{Q^i} = (a_{i+1}a_{i+2})$, the latter are completely expressed by $P \rightleftarrows P_{Q^i}$ for $2 \leq i \leq n-2$; and $P_{Q^{n-i}} \rightleftarrows P$ follows from $P_{Q^i} \rightleftarrows P$. We therefore note the following relation system.

$$\begin{aligned} &\text{(a)} \qquad P^2 = 1,\\ &\text{(b)} \qquad Q^n = 1,\\ &\text{(c)} \qquad (QP)^{n-1} = 1,\\ &\text{(d)} \qquad P \rightleftarrows P_{Q^i} \quad \left(i = 2, 3, \ldots, \left[\frac{n}{2}\right]\right). \end{aligned} \tag{13}$$

For small values of n one may note the following: for $n = 4$, relation (d) follows from (a) to (c), and for $n = 3$ it drops out, while for $n = 2$, (c) implies $Q = P$ and thus only (a) remains.

It is now claimed that (13) is *complete* for Σ_n. To prove this it suffices to show that the order of the abstract group defined by (13) cannot be greater than $n!$. Let

$$q = Q^2PQ^{-1} = (QP)_{Q^{-1}}.$$

Then it follows from (13c) that

$$q^{n-1} = 1. \tag{β}$$

Using (13a and d) one has

$$P_q = QPQ^{-2} \cdot P \cdot Q^2PQ^{-1} = Q \cdot P \cdot P_{Q^2} \cdot P \cdot Q^{-1} = Q \cdot P_{Q^2} \cdot Q^{-1} = P_Q.$$

We assume $P_{q^{i-1}} = P_{Q^{i-1}}$, as is true for $i = 2$, and find

$$P_{q^i} = q^{-1} \cdot P_{q^{i-1}} \cdot q = QPQ^{-2} \cdot P_{Q^{i-1}} \cdot Q^2PQ^{-1} = Q \cdot P \cdot P_{Q^{i+1}} \cdot P \cdot Q^{-1}.$$

Now if $i + 1 < n - 2$, so $i \leq n - 3$, then by (13d)

$$P_{q^i} = Q \cdot P_{Q^{i+1}} \cdot Q^{-1} = P_{Q^i}.$$

Thus these relations hold for $i = 1, 2, \ldots, n - 3$. Then it follows from (13d) that

$$P \rightleftarrows P_{q^i} \qquad \left(i = 2, 3, \ldots, \left[\frac{n-1}{2}\right]\right). \tag{δ}$$

Finally, the last but one equation together with (13a–c) and (β) gives

$$\begin{aligned}(Pq^{-1})^{n-2} &= P \cdot q^{-1}Pq \cdot q^{-2}Pq^2 \cdot \ldots \cdot q^{-(n-3)}Pq^{n-3} \cdot q^{-(n-2)} \\ &= P \cdot P_q \cdot P_{q^2} \cdot \ldots \cdot P_{q^{n-3}} \cdot q = P \cdot P_Q \cdot \ldots \cdot P_{Q^{n-3}} \cdot q \\ &= (PQ^{-1})^{n-2} \cdot Q^{n-2} \cdot q = QPQ^{-2} \cdot q = 1.\end{aligned}$$

Thus

$$(qP)^{n-2} = 1. \tag{γ}$$

The relations (β), (γ), and (δ), show that the subgroup $\{P, q\}$ of $\{P, Q\}$ satisfies the system (13) when one replaces Q by q and n by $n - 1$. This subgroup commutes with the subgroup $\{Q\}$. For one has

$$Q^iP = Q^{-(n-i)}P = P_{Q^{n-i}}Q^i = P_{q^{n-i}}Q^i,$$

valid for $n - i = 1, 2, \ldots, n - 3$, and thus $i = n - 1, n - 2, \ldots, 3$. However, one can read the relations for the missing exponents directly from the definition of q:

$$Q^2P = qQ,$$
$$QP = q^{-1}Q^2.$$

Similarly one has

$$Q^iq = Q^{i+2}PQ^{-1} = P_{Q^{n-i-2}}Q^{i+1} = P_{q^{n-i-2}}Q^{i+1} = P_{q^{-i-1}}Q^{i+1},$$

valid for $n - i - 2 = 0, 1, \ldots, n - 3$, thus for $i = n - 2, n - 3, \ldots, 1$, and from the definition of q one reads the relation for the missing exponent:

$$Q^{-1}q = q^{-1}Q.$$

Each element in $\{P, Q\}$ can therefore be written as an element in $\{P, q\}$, postmultiplied by a power of Q. Since $Q^n = 1$, (13) can therefore represent, for a particular value of n, at most n

times as many elements as for the value $n-1$. And for $n = 2$, the two elements 1 and P are represented. This ends the completeness proof.

Complete relation systems for the symmetric group have been presented many times, first by E. H. Moore[8] and W. Burnside.[9] In particular, Moore has given a relation system between two generators, which consists of the above system (13) with an additional relation, $(P_Q P)^3 = 1$ in our notation. The proof above guarantees that this must follow from (13); one can also confirm this directly from (13d):

$$P \rightleftarrows P_{Q^{n-2}} \cdot P_{Q^{n-3}} \cdot \ldots \cdot P_{Q^2},$$

the right-hand side transforms into QPP_Q by means of (13c) and one can immediately write down the relation from which Moore's emerges.[10,11]

§3.

The Relation System for the Symmetric Group Extended by Sign Changes

If one extends Σ_n to Ω_n by adding the generator O, then obvious relations result from the order of O, the commutativity of O with the permutations of $a_2, \ldots, a_n$, and the commutativity of sign changes in the different elements. These properties are expressed in the following relations:

$$\begin{aligned} &\text{a)} \quad O^2 = 1, \\ &\text{b)} \quad O \rightleftarrows P_Q, \\ &\text{c)} \quad O \rightleftarrows QP, \\ &\text{d)} \quad O \rightleftarrows O_Q, \end{aligned} \tag{14}$$

and (14), together with (13), is complete for Ω_n. Again one only has to show that the order of the abstract group defined by (13), (14) is no greater than the order $n!\,2^n$ of Ω_n. Let

$$O_i = O_{Q^{i-1}} \qquad (i = 1, \ldots, n), \tag{15}$$

so that $O_1 = O$. Then it follows from (14a) that $O_i^2 = 1$. Now for each value of h

$$Q(QP)^h = Q^{-(n-2)}P(QP)^{h-1} = P_{Q^{n-2}}P_{Q^{n-3}} \ldots P_{Q^{n-h-1}} \cdot Q^{h+1}.$$

Thus for $h = 0, 1, \ldots, n-3$

$$P_{Q(QP)^h} = P_{Q^{h+1}}.$$

---

[8] Proc. Lond. Math. Soc. (1) **28** (1896/97).

[9] Same as [8]; see also the same author's "Theory of Groups of Finite Order", Note C.

[10] In Comptes rend. Acad. sc. Paris **132** (1901), p. 1031, J. Séguier gives a relation system for the symmetric group, which he describes as that of Moore; however the relation (13c) is missing, and a glance at the exponent sum of Q in (13), and the Moore relations, shows that (13c) cannot follow from the others. The system of Séguier is therefore incomplete.

[11] Cf. also L. E. Dickson in Proc. Lond. Math. Soc. (1) **31**.

When one conjugates (14b) with $(QP)^{i-1}$, it follows from (14c) that

$$O \rightleftarrows P_{Q^i} \qquad (i = 1, 2, \ldots, n-2). \tag{ε}$$

When one conjugates (14d) with $(QP)^h$ the result is

$$O \rightleftarrows O_{Q(QP)^h} = O_{Q^{h+1}}$$

for $h + 1 \leq n - 1$, therefore $O \rightleftarrows O_i$ for all i, and by conjugation with all powers of Q

$$O_i \rightleftarrows O_k$$

for all i and k. The O_i therefore constitute an abelian group whose order is at most 2^n. It commutes with $\{P, Q\}$, because

$$QO_i = O_{i-1}Q \qquad (i \bmod n),$$
$$PO_i = Q^{-(i-1)}P_{Q^{-(i-1)}}OQ^{i-1} = Q^{-(i-1)}OP_{Q^{-(i-1)}}Q^{i-1} = Q_{Q^{i-1}}P = O_iP$$

are valid for $i \geq 3$ because of (ε). For the missing exponents we have

$$PO_2 = O_1P,$$
$$PO_1 = O_2P,$$

because $POP = O_P = O_{QP \cdot P} = O_Q = O_2$. This ends the completeness proof.

In what follows it is often useful to generate Ω_n by a larger set of elements, consisting of all P_{ik} and all $O_i = [a_i \mapsto a_i^{-1}]$. We note the following relation system (16) between these, symmetric in the indices, and which is guaranteed to be a consequence of (13), (14) as well as the *formal* defining equations (8), (10) and (15), because of the above proof. In using the indices the following remark holds: indices of the letters P, O, as well as for the later letters U, V, W, which are denoted differently, are actually different; if the particular case of two equal indices has to be included, it will each time be added with the same notation for them

$$\left\{\begin{array}{ll} \text{a)} & P_{ik}^2 = 1 \qquad (P_{ik} = P_{ki}), \\ \text{b)} & P_{ik} \rightleftarrows P_{lm}, \\ \text{c)} & \underline{P_{ik}P_{kl} = P_{il}P_{ik} = P_{kl}P_{il}}, \\ \text{d)} & O_i^2 = 1, \\ \text{e)} & \underline{O_i \rightleftarrows O_k}, \\ \text{f)} & P_{ik} \rightleftarrows O_l, \\ \text{g)} & P_{ik}O_i = O_kP_{ik}. \end{array}\right. \tag{16}$$

§4.
The Relation System for the Whole Group

We now add the generator U. The powers of U are all different. Turning to the relations between U and P, Q, O, a large number of these, especially commutativity relations. are again immediate. Table (17) gives a suitable choice of such relations, whose correctness is confirmed without difficulty from the meaning of the generating isomorphisms:

$$
\begin{array}{ll}
\text{(a)} & \\
\text{(b)} & \\
\text{(c)} & \\
\text{(d)} & U \rightleftarrows \left\{ \begin{array}{l} P_{Q^2}, \\ QPP_Q, \\ O_{Q^2}, \\ U_{Q^2}, \\ U_O, \\ U_{OQP}, \\ U_{P_{QP}}, \end{array} \right. \\
\text{(e)} & \\
\text{(f)} & \\
\text{(g)} & \\
\text{(h)} & UU_Q = U_{QP}U_QU, \\
\text{(i)} & U_{O_P} = U^{-1}, \\
\text{(j)} & U^{-1}U_PU_O = OP.
\end{array}
\tag{17}
$$

The task of the following paragraphs is to show that (17), together with (13) and (14), is complete for the whole isomorphism group Γ_n. (13), (14) and (17) will be called the "relation system S."

It is useful to extend the generator system with the isomorphisms U_{ik} and V_{ik} defined in (11) and (12) and to add the relations which these satisfy between themselves and with the P_{ik} and O_i. This will be done in the present paragraph. The symbol W_{ik} will be introduced as a joint name for both the U_{ik} and V_{ik}. The new relations will be derived as consequences of S. The meaning of the W_{ik} as isomorphisms will nowhere be used, the W_{ik} will only be defined formally by (7), (8), (11) and (12) as abbreviations for certain products in P, Q, O, U. When products of P, Q, O alone are involved it is permissible to make use of the meaning of these symbols, since the derivability from S of every relation between them is guaranteed.

$P_{Q^2} = (a_3a_4)$ and $QPP_Q = (a_3a_4 \dots a_n)$ generate all permutations of $a_3, \dots, a_n$. Let σ be the symbol for *any* such permutation, so that

$$\sigma = \Pi(P_{Q^2}, QPP_Q)$$

for an arbitrary product Π. Then (17a) and (17b) together say

$$U \rightleftarrows \sigma.$$

Let N_{iklm} be the symbol for a permutation which brings a_i, a_k, a_l, a_m into the places of a_1, a_2, a_3, a_4. Here, i, k, l, m are any four different numbers among 1, 2, …, n. Then $N_{iklm} = \sigma \cdot N_{ik}$. Thus if one conjugates (17a) with N_{iklm} the result is

$$U_{N_{iklm}} = U_{\sigma \cdot N_{ik}} = U_{N_{ik}} = U_{ik} \rightleftarrows P_{Q^2 \cdot N_{iklm}} = P_{lm}.$$

It follows from $U \rightleftarrows \sigma$, by conjugation with O, that

$$V \rightleftarrows \sigma,$$

hence from (17a), by conjugation with ON_{iklm}, that

$$V_{ik} \rightleftarrows P_{lm},$$

and hence in summary

$$W_{ik} \rightleftarrows P_{lm}. \tag{18a}$$

We have $PN_{ik} = \sigma \cdot N_{ki}$. If one conjugates $W_{12P}^{\pm 1}$ with N_{ik}, then on the one side

$$W_{12PN_{ik}}^{\pm 1} = W_{12N_{ik}P_{ik}}^{\pm 1} = W_{ikP_{ik}}^{\pm 1},$$

and on the other side

$$W_{12PN_{ik}}^{\pm 1} = W_{12\sigma\cdot N_{ki}}^{\pm 1} = W_{12N_{ki}}^{\pm 1} = W_{ki}^{\pm 1},$$

and thus

$$P_{ik}W_{ik}^{\pm 1} = W_{ki}^{\pm 1}P_{ik}. \tag{18b}$$

Let N_{ikl} be a permutation which brings a_i, a_k, a_l to the places of a_1, a_2, a_3. Then

$$N_{ikl} = \sigma \cdot N_{ik},$$

$$P_{13}N_{ikl} = \sigma \cdot N_{lk},$$

$$P_{23}N_{ikl} = \sigma \cdot N_{il}.$$

If one conjugates $W_{12P_{13}}^{\pm 1}$ with N_{ikl}, then on the one side

$$W_{12P_{13}N_{ikl}}^{\pm 1} = W_{12N_{ikl}P_{il}}^{\pm 1} = W_{12N_{ik}P_{il}}^{\pm 1} = W_{ikP_{il}}^{\pm 1},$$

and on the other side

$$W_{12P_{13}N_{ikl}}^{\pm 1} = W_{12\sigma\cdot N_{lk}}^{\pm 1} = W_{lk}^{\pm 1},$$

hence

$$P_{il}W_{ik}^{\pm 1} = W_{lk}^{\pm 1}P_{il}. \tag{18c}$$

If one conjugates $W_{12P_{23}}^{\pm 1}$ with N_{ikl}, then

$$P_{kl}W_{ik}^{\pm 1} = W_{il}^{\pm 1}P_{kl}. \tag{18d}$$

follows in exactly the same way. Relations (18a–d) say that the groups $\{P_{ik}\}$ and $\{W_{ik}\}$ commute. If one conjugates (17c) with O then the result is

$$V \rightleftarrows O_{Q^2}.$$

Under conjugation with N_{ikl}, this relation and (17c) gives

$$W_{ik} \rightleftarrows O_i. \tag{18e}$$

Since $O_P = O_2$, (17i) says:

$$U_{O_2} = U^{-1},$$

whence it follows, by conjugation with O, that

$$V_{O_2}^{-1} = V,$$

and hence in summary:

$$W_{12O_2}^{\pm 1} = W_{12}^{\mp 1},$$

and then, by conjugation with N_{ik}:

$$O_k W_{ik}^{\pm 1} = W_{ik}^{\mp 1}O_k. \tag{18f}$$

Finally, conjugation by N_{ik} in (7) gives

$$O_i U_{ik}^{\pm 1} = V_{ik}^{\mp 1} O_i, \tag{18g}$$

which one can also write

$$O_i V_{ik}^{\pm 1} = U_{ik}^{\mp 1} O_i.$$

Relations (18e–g) say that $\{O_i\}$ commutes with $\{W_{ik}\}$, and hence (18a–g) say that Ω_n commutes with $\{W_{ik}\}$.

However, Γ_n is not the "direct pruduct" of Ω_n and $\{W_{ik}\}$, since these two subgroups, as (17j) shows, have more than the identity element in common. Since $PN_{ik} = \sigma \cdot N_{ki}$ and (7) hold, it follows from (17j) by conjugation with N_{ik} that

$$U_{ik}^{-1} U_{ki} V_{ik}^{-1} = O_i P_{ik}, \tag{18h}$$

and from this, by conjugation with O_k, and (18f,g), that

$$U_{ik} V_{ki}^{-1} V_{ik} = O_k P_{ik}. \tag{18i}$$

The relations (17d–h) still remaining yield relations inside the group $\{W_{ik}\}$. Since $N_{iklm} = \sigma \cdot N_{ik}$ and $Q^2 N_{iklm} = \sigma \cdot N_{lm}$, (17d) yields, under conjugation with N_{iklm},

$$U_{ik} \rightleftarrows U_{lm},$$

which then extends to

$$W_{ik} \rightleftarrows W_{lm} \tag{18k}$$

by conjugation with O_i and $O_i O_l$, by (18e–g). We have $P_{QP} = (a_1 a_3)$, hence $P_{QP} N_{ikl} = \sigma \cdot N_{lk}$; and since $N_{ikl} = \sigma \cdot N_{ik}$, (17g) yields, under conjugation with N_{ikl}:

$$U_{ik} \rightleftarrows U_{lk},$$

which extends to

$$W_{ik} \rightleftarrows W_{lk} \tag{18l}$$

by conjugation with O_i and $O_i O_l$. Relation (17e) says $U \rightleftarrows V$, and under conjugation with N_{ik} it yields

$$U_{lk} \rightleftarrows V_{ik}. \tag{18m}$$

Relation (17f) says $U \rightleftarrows V_{QP}$ and since $QPN_{ikl} = \sigma \cdot N_{il}$ it yields

$$U_{ik} \rightleftarrows V_{il} \tag{18n}$$

by conjugation with N_{ikl}. Finally we have to derive the consequences of the important relation (17h), which says that U_{QP} is the commutator of U and U_Q. Since

$$N_{ikl} = \sigma \cdot N_{ik},$$

$$QN_{ikl} = \sigma \cdot N_{kl},$$

$$QPN_{ikl} = \sigma \cdot N_{il},$$

it follows from (17h), by conjugation with N_{ikl}, that

$$U_{ik} U_{kl} = U_{il} U_{kl} U_{ik},$$

which yields, by (18e,f) and conjugation with O_i:

$$U_{ik}U_{kl}^{-1} = U_{il}^{-1}U_{kl}^{-1}U_{ik},$$

and in summary, bearing in mind that U_{il} commutes with U_{kl} by (18 l):

$$U_{ik}U_{kl}^{\pm 1}U_{ik}^{-1}U_{kl}^{\mp 1}U_{il}^{\mp 1} = U_{ik}U_{kl}^{\pm 1}U_{ik}^{-1}U_{il}^{\mp 1}U_{kl}^{\mp 1} = 1, \tag{18o}$$

which yields, under conjugation with O_k:

$$U_{ik}^{-1}V_{kl}^{\mp 1}U_{ik}V_{kl}^{\pm 1}U_{il}^{\mp 1} = U_{ik}^{-1}V_{kl}^{\mp 1}U_{ik}U_{il}^{\mp 1}V_{kl}^{\pm 1} = 1, \tag{18p}$$

and with O_i:

$$V_{ik}^{-1}U_{kl}^{\pm 1}V_{ik}U_{kl}^{\mp 1}V_{il}^{\pm 1} = V_{ik}^{-1}U_{kl}^{\pm 1}V_{ik}V_{il}^{\pm 1}U_{kl}^{\mp 1} = 1, \tag{18q}$$

and with O_iO_k:

$$V_{ik}V_{kl}^{\mp 1}V_{ik}^{-1}V_{kl}^{\pm 1}V_{il}^{\pm 1} = V_{ik}V_{kl}^{\mp 1}V_{ik}^{-1}V_{il}^{\pm 1}V_{kl}^{\pm 1} = 1. \tag{18r}$$

The passage from (17) to (18) shows the significance of the relations in (17) individually. Use of the properties of the subgroup Ω_n of Γ_n makes it possible to reduce the relations which go beyond Ω_n to the concise form (17).

§5.

Completeness Proof for the Whole Group. Reduction to a Lemma

In place of the presentation $\Gamma_n = \{P, Q, O, U\}$ we shall use the presentation $\Gamma_n = \{P_{ik}, O_i, W_{ik}\}$ from now on. By the last paragraph, $\Omega_n = \{P_{ik}, O_i\}$ commutes with $\{W_{ik}\}$. Thus if Ω is the general symbol for an operation in Ω_n, then each isomorphism can be brought into each of the following forms:

$$\Omega \cdot \Pi(W_{ik}),$$

$$\Pi(W_{ik}) \cdot \Omega,$$

$$\Omega \cdot \Pi(W_{ik}) \cdot \Omega.$$

Now we again take up the concept of the *length* $L(I)$ of an isomorphism I defined in §1. We have $L(\Omega) = n$, in particular $L(1) = n$ and

$$L(I\Omega) = L(\Omega I) = L(\Omega I \Omega) = L(I).$$

The length of an isomorphism is therefore determined by the *core* $\Pi(W_{ik})$ alone. Let

$$K = W^{(1)}W^{(2)}\dots W^{(r)}$$

be such a core, where each $W^{(k)}$ denotes any W_{ik} and K is a freely reduced word in these terms. We define

$$L(W^{(1)}W^{(2)}\dots W^{(r)}) = \lambda_1 = L(K),$$

$$L(W^{(2)}\dots W^{(r)}) = \lambda_2,$$

$$\dots$$

$$L(W^{(r)}) = \lambda_r = n + 1$$

and call the sequence

$$\Delta(K) = \lambda_1, \lambda_2, \ldots, \lambda_r$$

of integers $\geq n$ the *diagram* of K. If one conjugates K with an Ω and uses the commutation relations (18a–g) of the previous paragraph, then one obtains a new freely reduced core of r factors with the same diagram. If one also allows factors O_i and P_{ik} next to the $W^{(h)}$, then the definition of the diagram $\Delta(I)$ extends to a representation of an arbitrary isomorphism as a product. A diagram is called *normal* when the λ constitute a nowhere increasing sequence, and the corresponding product representation is a *normal form* of the isomorphism being represented. As we already know from §1, each isomorphism can be represented as a product in the P_{ik}, O_i, W_{ik} *in normal form.* We are going to show the following

Main Theorem. *For each isomorphism given in the form* $I = \Pi(P_{ik}, O_i, W_{ik})$, *a representation in the normal form,* $I = \Pi'(P_{ik}, O_i, W_{ik})$, *may be found so that the relation*

$$\Pi(P_{ik}, O_i, W_{ik}) = \Pi'(P_{ik}, O_i, W_{ik})$$

is a formal consequence of the relations (16) *and* (18) *between the* P_{ik}, O_i, W_{ik}, *derived above from* S.

With the proof of this main theorem, the completeness of the relation system S for Γ_n will be shown. Because, suppose

$$R(P, Q, O, U) = 1 \tag{19a}$$

is an arbitrary relation in Γ_n which is correct according to the meaning of P, Q, O, U. One introduces the symbols P_{ik}, O_i, U_{ik}, V_{ik} merely as formal abbreviations for certain products of P, Q, O, U, given by (7), (8), (10), (11), (12) and (15). Then

$$P_{n-1,n}P_{n-2,n-1}\cdots P_{23}P_{12} = P_{Q^{n-2}}P_{Q^{n-3}}\cdots P_Q P = Q^{-(n-1)}(QP)^{n-1},$$

and by S this is equal to Q. If one now substitutes P_{12} for P, O_1 for O, U_{12} for U and $P_{n-1,n}P_{n-2,n-1}\cdots P_{12}$ for Q in (19a), then (19a) has the form

$$R'(P_{ik}, O_i, W_{ik}) = 1. \tag{19b}$$

Now the Main Theorem says that one can convert the left-hand side of this relation, by application of the consequence relations (16) and (18) of S alone, to a normal form

$$R''(P_{ik}, O_i, W_{ik}) = 1. \tag{19c}$$

Since $R'' = 1$, $L(R'') = n$, since R'', as a normal form, has a nowhere increasing diagram, $\Delta(R'')$ consists simply of the number n. It follows that *no* W_{ik} *at all can appear in* (19c), because the appearance of the first factor W_{ik} would give rise to a number of $n + 1$ in the diagram. Thus in reality (19c) reads

$$R''(P_{ik}, O_i) = 1 \tag{19d}$$

and when P_{ik}, O_i are replaced by their expressions in P, Q, O:

$$R'''(P, Q, O) = 1. \tag{19e}$$

The passage from (19a) to (19e) means that the generator U is removed from the previous relation by application of S alone. Also, (19e) can be shown to be an identity by means of S, by the completeness proof given in §2 and §3. Thus each relation between P, Q, O and U can be reduced to an identity by S. Q.E.D.

Thus it all comes down to a proof of the Main Theorem. We seek to reduce this to a theorem of more limited extent. First, as we said above,

$$\Pi(P_{ik}, O_i, W_{ik}) = \Omega \cdot K(W_{ik}) \cdot \Omega,$$

and the diagram of the right-hand side is monotonic when that of $K(W_{ik})$ is monotonic. Thus we can confine ourselves to the core K. Let the factors of K and the numbers λ of the diagram $\Delta(K)$ be denoted as above. If Δ is not monotonic, let ρ be the greatest index such that $\lambda_\rho < \lambda_{\rho+1}$; thus ρ denotes the position of the last non-monotonicity in the diagram of K. Then we consider the subcore $W^{(\rho)}W^{(\rho+1)}\dots W^{(r)}$ and show that this can be transformed by S until none of the numbers in the new diagram of this subcore reaches the value $\lambda_{\rho+1}$. Obviously one can make $\Delta(K)$ monotonic in a finite number of steps by continued application of such transformations. Possible factors Ω, which appear between the W_{ik} in the process, can always be moved to the beginning or end without otherwise disturbing the process. The Main Theorem is thereby reduced to the following

Lemma. *If*

$$K(W_{ik}) = W^{(1)}W^{(2)}\dots W^{(r)},$$
$$L(W^{(i)}W^{(i+1)}\dots W^{(r)}) = \lambda_i \qquad (i = 1, 2, \dots, r),$$
$$\lambda_1 < \lambda_2 \geq \lambda_3 \geq \lambda_4 \geq \dots \geq \lambda_r;$$

then S implies an equation

$$K(W_{ik}) = \Omega \cdot K'(W_{ik}) \cdot \Omega$$

such that no number in the diagram $\Delta(K')$ reaches the value λ_2.

In order to emphasize the special position of λ_2 as the greatest number in $\Delta(K)$, we denote it by Λ from now on. In §6 the Lemma is proved in the case that $\lambda_2 > \lambda_3$, i.e. when Λ appears only once in $\Delta(K)$. However, this condition is not made an assumption in §6, since the latter is also used as preparation for the general case in §7.

§6.
Proof of the Lemma. Part 1

By conjugation of K with a suitable Ω one can arrange that $W^{(1)} = U_{12}$: because if $W^{(1)}$ has the indices i, k one converts these into 1, 2 by conjugation with $N_{i,k}^{-1}$, by (11) and (12); $V_{12}^{\pm 1}$ is converted to $U_{12}^{\mp 1}$ by conjugation with O_1, by (18g); U_{12}^{-1} is converted to U_{12} by conjugation with O_2, by (18f). Suppose then that $W^{(1)} = U_{12}$. Also let

$$W^{(3)}W^{(4)}\dots W^{(r)} = I = [\alpha_1, \alpha_2, \dots, \alpha_n].$$

In this paragraph we do not have to worry about the factor sequence of I and therefore we have only the following hypotheses

$$K = U_{12}W^{(2)}I,$$
$$L(K) = \lambda_1 < \Lambda,$$
$$L(W^{(2)}I) = \lambda_2 = \Lambda,$$
$$L(I) = \lambda_3 \leq \Lambda.$$

Now it all comes down to the factor $W^{(2)}$. We first suppose that it has no index in common with U_{12}:

$$W^{(2)} = W_{ik}^{\pm 1} \qquad (i, k = 3, 4, \ldots, n).$$

If one then puts

$$K' = W_{ik}^{\pm 1}U_{12}I,$$

then $K = K'$ is a consequence of (18k). The effect of the factor U_{12} on the isomorphism here is independent of whether $W_{ik}^{\pm 1}$ is applied beforehand or afterwards. Hence $L(U_{12}I) < L(I)$. Thus if $\lambda_3 < \Lambda$, Λ does not appear in $\Delta(K')$ at all, and the lemma holds; if $\lambda_3 = \Lambda$, then we are reduced to the investigation of the shorter core $U_{12}I$, which again satisfies the hypotheses of the lemma.

A similar viewpoint guides us in the more difficult cases below, where $W^{(2)}$ has one or both indices in common with U_{12}. Let t denote an arbitrary number among 3, 4, ..., n. Then it is still easy to settle the case $W^{(2)} = W_{t2}^{\pm 1}$. Because the latter commutes with U_{12} by (18 l), and the transformed core $K' = W_{t2}^{\pm 1}U_{12}I$ satisfies the same conditions as $W_{ik}^{\pm 1}U_{12}I$ did above. We therefore still have the following individual cases for $W^{(2)}$ to consider:

I. $W^{(2)} = W_{12}^{\pm 1}$,

II. $W^{(2)} = W_{21}^{\pm 1}$,

III. $W^{(2)} = W_{t1}^{\pm 1}$,

IV. $W^{(2)} = W_{1t}^{\pm 1}$,

V. $W^{(2)} = W_{2t}^{\pm 1}$.

By (18a), P_{3t} commutes with U_{12}. Thus by using conjugation by P_{3t} in III, IV, V, and (18c,d) we can arrive at the special index 3 for t.

I. $K = U_{12}W_{12}^{\pm 1}I$. By the assumption that K is freely reduced, $W_{12}^{\pm 1}$ cannot be U_{12}^{-1}. It also cannot be U_{12}, because we would have

$$U_{12}I = [\alpha_1\alpha_2, \alpha_2, \ldots],$$
$$U_{12}^2 I = [\alpha_1\alpha_2\alpha_2, \alpha_2, \ldots]$$

with $L(\alpha_1\alpha_2) \geq L(\alpha_1)$ and $L(\alpha_1\alpha_2\alpha_2) < L(\alpha_1\alpha_2)$. Thus with the introduction of the first α_2 at most half of α_1 is lost by cancellation, while with the introduction of the second α_2 more than half is lost. Thus the two factors of the product $\alpha_2\alpha_2$ must cancel at least half each other, which is impossible for a freely reduced word α_2. We therefore have only the subcases V_{12} and V_{12}^{-1} for $W_{12}^{\pm 1}$.

(a) $K = U_{12}V_{12}I$.

$$V_{12}I = [\alpha_2\alpha_1, \alpha_2, \ldots],$$

$$U_{12}V_{12}I = [\alpha_2\alpha_1\alpha_2, \alpha_2, \ldots].$$

Here, $L(\alpha_2\alpha_1) \geq L(\alpha_1)$ and $L(\alpha_2\alpha_1\alpha_2) < L(\alpha_2\alpha_1)$. In what follows we shall often have to compute with length numbers of the α as well as their parts and combinations. It will not lead to misunderstanding if we omit the length sign $L(\quad)$ and, e.g., write the above inequalities

$$\alpha_2\alpha_1 \geq \alpha_1,$$

$$\alpha_2\alpha_1\alpha_2 < \alpha_2\alpha_1.$$

It is now entirely a question of how the absorption of the generators a occurs in the word $\alpha_2\alpha_1\alpha_2$. In order to obtain an overview of the possible cases more easily, and to be as brief as possible, we use the intuitive aid of an infinite graph N with the following properties:

1. N is a "tree", i.e. any two points A and B in N determine a unique connecting path AB in N, when one disallows successive traversals of the same edge in opposite directions.

2. There are $2n$ edges issuing from each vertex of N. The topological possibility of such a graph in the plane follows, e.g., from the fact that it can be constructed in a metrically regular way in the hyperbolic plane, with all edges of equal length and the full angle around each vertex divided into $2n$ equal parts. Obviously one has only to choose the edge length so large that the vertices of a polygonal path with angles π/n follow a distance curve.

3. Each edge carries the name a_i of a generator and an orientation, in such a way that each generator issues from each vertex with both the positive and negative orientation. With these properties, N is the Dehn "group diagram" of F_n. If in addition one chooses the labelling of the hyperbolically regular complex so that the order of labels around each vertex is the same, then the group of motions of the hyperbolic plane which bring the labelled net into coincidence with itself is isomorphic to F_n. However, in what follows the hyperbolically regular construction of N, or even the whole of N, will not be used—only the topological properties of finite subcomplexes of N.

If AB and CD are two reduced paths in N, then $AB = CD$ shall mean that they have "equal length", i.e. the same number of edges. By $AB \equiv CD$ we mean in addition that the paths correspond to the same word in the generators (a_1 in the negative direction corresponds to a_1^{-1}). Each word in the generators can be described as a path in N in a unique way from a given vertex. For two different points A and B we always have $AC \not\equiv BC$ and $AB \not\equiv BA$. If A, B and C are three different points and if the reduced path AB contains the point C, then we say that C separates A and B on N.

In the case we have above we now describe the word $\alpha_2\alpha_1$ from a point A of N, and likewise

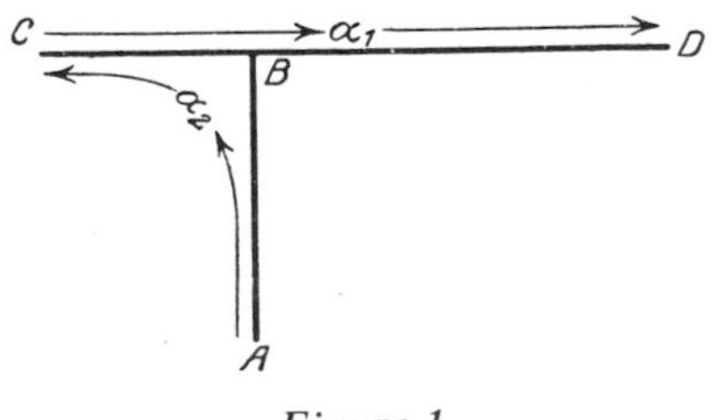

Figure 1

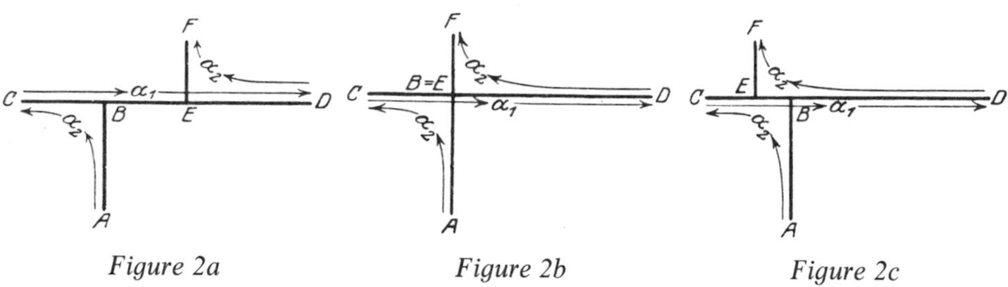

Figure 2a Figure 2b Figure 2c

α_2 on its own. Suppose $AC \equiv \alpha_2$ and $CD \equiv \alpha_1$, and hence that $AD \equiv \alpha_2\alpha_1$, (see Figure 1). It can happen that CA and CD have a piece CB in common, but it is also possible that $CB = 0$. In either case we know, from the hypothesis $\alpha_2\alpha_1 \geq \alpha_1$, that $AD \geq CD$, and hence that $BA \geq BC$. Now in order to come to $\alpha_2\alpha_1\alpha_2$, one describes α_2 from D, $\alpha_2 \equiv DF$. Let E be the point at which DF branches off the previous figure. E can coincide with F in a special case, namely, when DF runs entirely on the previous figure.

We suppose first that E lies on DC, and hence does not separate A from B; see Figures 2(a–c). In the first two cases DE, and in the last case DB, is the piece cancelled when $\alpha_2\alpha_1$ is multiplied by α_2, and hence it is greater than $\alpha_2/2$ because $\alpha_2\alpha_1\alpha_2 < \alpha_2\alpha_1$. Consequently $\alpha_1\alpha_2 < \alpha_1$. If one then puts

$$K' = V_{12}U_{12}I,$$

then $K = K'$ follows from (18m). We thereby get $L(U_{12}I) < L(I)$ and hence $< \Lambda$. Thus for $L(I) < \Lambda$ the lemma holds, and for $L(I) = \Lambda$ we are reduced to investigating the shorter core $U_{12}I$.

We therefore suppose that E separates the points A and B; see Figure 3. Let M_1 be the midpoint of AC, and hence a vertex if $L(AC)$ is even and the midpoint of an edge if $L(AC)$ is odd. Similarly, M_2 is the midpoint of DF. Since $AM_1 \equiv DM_2$ (both are initial halves of α_2), M_1 and M_2 cannot coincide. Since $\alpha_2\alpha_1 \geq \alpha_1$, M_1 lies on AB, and since $\alpha_2\alpha_1\alpha_2 < \alpha_2\alpha_1$, M_2 lies on DE and not at E. $BC \neq BD$ because $M_1 \neq M_2$. If we had $BC > BD$, then M_1M_2 would lie on BE with M_1 nearer to B. It would then follow from $M_1A \equiv M_2D$ that $M_1M_2 = M_2M_1$, which is impossible. Thus $BC < BD$, and hence $\alpha_1\alpha_2 < \alpha_2\alpha_1$, because a larger constituent is cancelled in $\alpha_1\alpha_2$; thus in K' we again have $L(U_{12}I) < \Lambda$.

(b) $K = U_{12}V_{12}^{-1}I$.

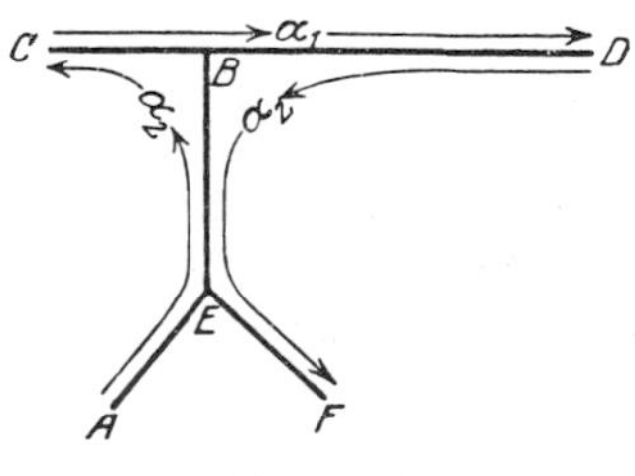

Figure 3

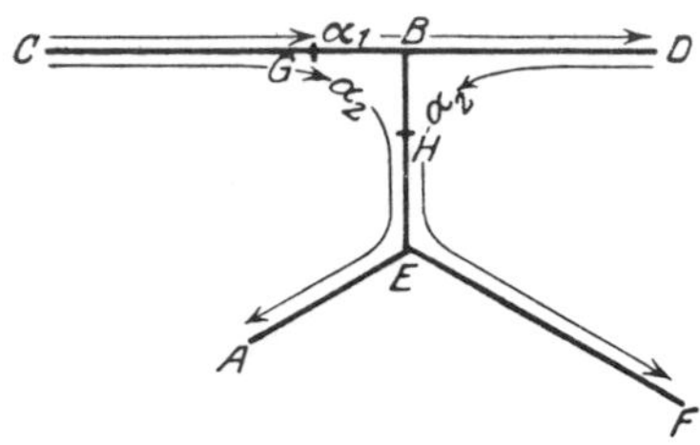

Figure 4

$$V_{12}^{-1}I = [\alpha_2^{-1}\alpha_2, \alpha_2, \ldots],$$

$$U_{12}V_{12}^{-1}I = [\alpha_2^{-1}\alpha_1\alpha_2, \alpha_2, \ldots].$$

Here $\alpha_2^{-1}\alpha_1 \geq \alpha_1$ and $\alpha_2^{-1}\alpha_1\alpha_2 < \alpha_2^{-1}\alpha_1$. One traces $\alpha_2^{-1}\alpha_1\alpha_2$ and obtains four figures corresponding to 2(a–c) and 3, but with the direction of the arrow on AC reversed. The relation which again follows from (18m),

$$K' = V_{12}^{-1}U_{12}I$$

also suffices in the first three cases, and likewise in the fourth case when $CB < BD$. $CB = BD$ is impossible because $BC \not\equiv BD$. Suppose therefore that $BC > BD$ (see Figure 4). Let $CG = DB$, so that $CG \equiv DB$ is an initial part of α_2. Since $\alpha_2^{-1}\alpha_1 \geq \alpha_1$, $CD \leq \alpha_2/2$, and since $\alpha_2^{-1}\alpha_1\alpha_2 < \alpha_2^{-1}\alpha_1$ we have $DE > \alpha_2/2$, hence $DE > CB$ and therefore $BE > GB$. Let H be the point on BE for which $BH = GB$. Then $BH \equiv GB$. Therefore α_1 and α_2 can be written as the following products of words:

$$\alpha_1 \equiv (CG)(GB)(GC),$$

$$\alpha_2 \equiv (CG)(GB)(GB)(HA).$$

We construct the new core

$$K' = U_{12}^{-1}V_{21}^{-1}U_{12}V_{21}^{-1}V_{21}^{-1}I,$$

from the successive subcores which look as follows:

$$V_{21}^{-1}I = [\alpha_1, \alpha_1^{-1}\alpha_2 \equiv (CG)(GB)(HA), \ldots],$$

$$V_{21}^{-2}I = [\alpha_1, \alpha_1^{-2}\alpha_2 \equiv (CG)(HA), \ldots],$$

$$U_{12}V_{21}^{-2}I = [\alpha_1^{-1}\alpha_2, \alpha_1^{-2}\alpha_2, \ldots],$$

$$V_{21}^{-1}U_{12}V_{21}^{-2}I = [\alpha_1^{-1}\alpha_2, \alpha_2^{-1}\alpha_1^{-1}\alpha_2, \ldots],$$

$$K' = U_{12}^{-1}V_{21}^{-1}U_{12}V_{21}^{-2}I = [\alpha_2, \alpha_2^{-1}\alpha_1^{-1}\alpha_2, \ldots].$$

The lengths of all these isomorphisms remain below $\Lambda = L([\alpha_2^{-1}\alpha_1, \alpha_2, \ldots])$, because: the first two are $< L(I)$; $L(K') = L(K) < \Lambda$; the fourth is $< L(K')$ because $\alpha_1^{-1}\alpha_2 < \alpha_2$; the third is $< \Lambda$ because $\alpha_1^{-1}\alpha_2 = \alpha_2^{-1}\alpha_1$ and $\alpha_1^{-2}\alpha_2 < \alpha_2$. We have $K = P_{12}O_2K'$ and it remains to show that this relation follows from S. The relation to be proved is

$$U_{12}V_{12}^{-1}V_{21}^{2}U_{12}^{-1}V_{21}U_{12}O_2P_{12} = 1.$$

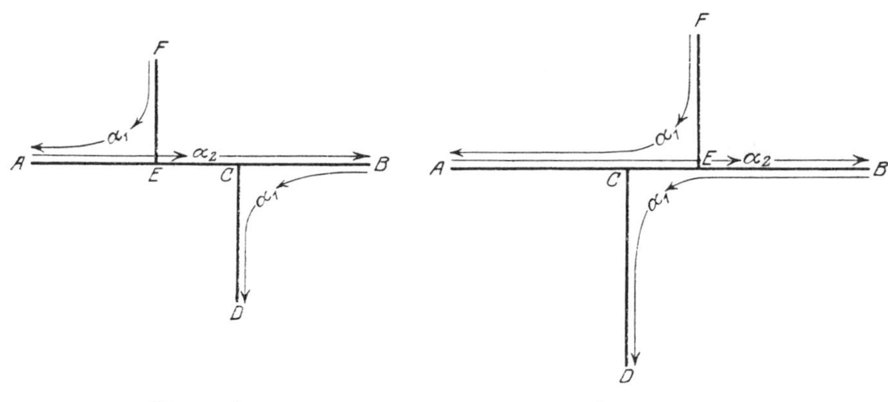

Figure 5a *Figure 5b*

By means of (18i,h) the left-hand side becomes

$$U_{12} \cdot U_{21}^{-1} O_1 P_{12} \cdot P_{12} O_2 U_{21}^{-1} \cdot V_{21} U_{12} O_2 P_{12},$$

and hence, exchanging U_{21}^{-1} and V_{21} by (18m) and applying (18h) afresh:

$$U_{12} U_{21}^{-1} O_1 O_2 V_{21} \cdot O_2 P_{12} V_{21} \cdot O_2 P_{12},$$

which reduces to 1 by (18f,g).

II. $K = U_{12} W_{21}^{\pm 1} I$. We have to replace $W_{21}^{\pm 1}$ by U_{21}, U_{21}^{-1}, V_{21}, V_{21}^{-1} in turn.
(a) $K = U_{12} U_{21} I$.

$$U_{21} I = [\alpha_1, \alpha_2\alpha_1, \ldots] \qquad (\text{with } \alpha_2\alpha_1 \geq \alpha_2),$$

$$K = U_{12} U_{21} I = [\alpha_1\alpha_2\alpha_1, \alpha_2\alpha_1, \ldots] \qquad (\text{with } \alpha_1\alpha_2\alpha_1 < \alpha_1).$$

One traces $AB = \alpha_2$, $BD = \alpha_1$ with branching point C and finally $AF = \alpha_1{}^1$ with branching point E. If one first takes E on AB, then one has one of the Figures 5(a) or (b) where, e.g. one includes the case $E = C$ in Figure 5(a), in which case $EC = 0$. One constructs the core

$$K' = U_{21} V_{21}^{-1} V_{12} V_{21} I,$$

which corresponds to the following build-up from I with the attached length numbers:

$\alpha_1\alpha_2\alpha_1$	$\alpha_1\alpha_2\alpha_1$	$\alpha_1\alpha_2\alpha_1$	α_1	α_1
$\alpha_2\alpha_1$	α_1^{-1}	$\alpha_1\alpha_2$	$\alpha_1\alpha_2$	α_2
$\lambda_1 < \Lambda$	μ_3	μ_2	μ_1	$\lambda_3 \leq \Lambda$

where the unaltered elements $\alpha_3, \ldots, \alpha_n$ are not written. In Figure 5(a), $CD \geq CB$, $EC + CD < EA$, hence $CB < EA$ and therefore $\mu_1 < \Lambda$, $\mu_2 < \lambda_1$, $\mu_3 < \lambda_1$. But the same inequalities also hold in case 5b because $CD \geq CB$, $CA > CD$, hence $CA > CB > EB$. We still have to derive from S that $K = K'$, i.e. that

$$U_{21} V_{21}^{-1} V_{12} V_{21} U_{21}^{-1} U_{12}^{-1} = 1;$$

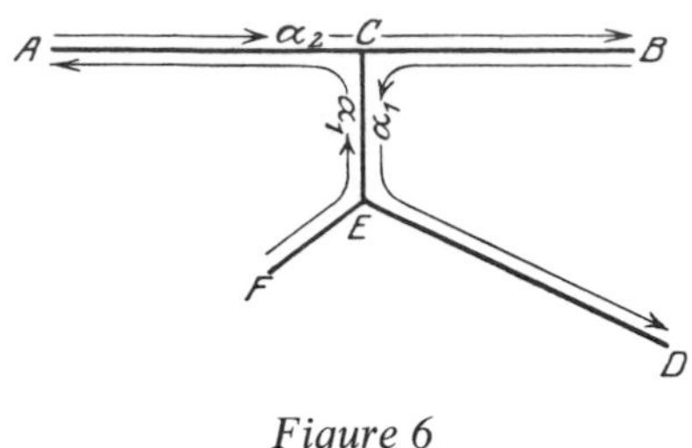

Figure 6

its left-hand side becomes

$$U_{21}\cdot U_{12}^{-1}O_2P_{12}\cdot U_{21}^{-1}\cdot P_{12}O_2U_{21}^{-1},$$

by means of (18m,i,h), and this reduces to 1 by (18b,f).

The remaining case is where E separates the points C and D (Figure 6). Here $BC \leq CD$ and $ED < EA$. Thus the midpoint M of AD lies on AE. M cannot lie on CE, because $MA = MD$ would imply $MA \equiv MD$ as a final segment of α_1. Thus M separates A and C; hence $AC > CD$ and therefore $AC > CB$.

$$K' = V_{12}^{-1}U_{21}V_{21}I$$

$\alpha_1^{-1}\alpha_2^{-1}$	α_1	α_1	α_1
$\alpha_1\alpha_2\alpha_1$	$\alpha_1\alpha_2\alpha_1$	$\alpha_1\alpha_2$	α_2
$\lambda_1 < \Lambda$	μ_2	μ_1	$\lambda_3 \leq \Lambda$

Here, $\mu_1 < \lambda_3$ because $AC > CB$ implies $EF < ED$, hence $FC < DC < AC$ and therefore $\alpha_1\alpha_2 < \alpha_2$. Also, $\mu_2 < \lambda_1$, since $\alpha_1 < \alpha_2\alpha_1$ because $CB < CA$. Thus $K = P_{12}O_1K'$, and this is a consequence of (18m,i).

(b) $K = U_{12}U_{21}^{-1}I$. By tracing $\alpha_1\alpha_2\alpha_1^{-1}$ one obtains three figures corresponding to 5(a,b) and 6, but with the oppositely directed arrow on DB. In cases 5(a,b) one has the same magnitude relations as above and

$$K' = V_{21}V_{21}U_{12}^{-1}V_{21}I$$

satisfies on the same grounds. We have $K = O_1O_2K'$. It follows from (18m,h) that

$$\begin{aligned} O_1O_2V_{21}^2U_{12}^{-1}V_{21}U_{21}U_{12}^{-1} &= O_1O_2V_{21}\cdot V_{21}U_{12}^{-1}U_{21}\cdot V_{21}U_{12}^{-1} \\ &= O_1O_2V_{21}\cdot P_{12}O_2\cdot P_{12}O_2U_{21}^{-1} \end{aligned}$$

and this becomes 1 by means of (18f,g).

In the case of Figure 6, $CA \neq CB$ because $CA \not\equiv CB$. Assume first that $CA > CB$. Then $EF < ED$. Let $K' = V_{12}^{-1}U_{21}^{-1}V_{21}I$

$\alpha_1\alpha_2^{-1}$	α_1	α_1	α_1
$\alpha_1\alpha_2\alpha_1^{-1}$	$\alpha_1\alpha_2\alpha_1^{-1}$	$\alpha_1\alpha_2$	α_2
$\lambda_1 < \Lambda$	μ_2	μ_1	$\lambda_3 \leq \Lambda$

Then $\mu_1 < \Lambda$ since $FB < AD$; $\mu_2 < \lambda_1$ since $\alpha_1 < \alpha_1\alpha_2^{-1}$ because $CB < CA$. We have $K = P_{12}O_1K'$ and this is a consequence of (18m,i).

Finally, suppose that $CA < CB$. Then let G be the point on BC for which $BG = AC$ and hence $BG \equiv AC$ is an initial segment of α_1^{-1}. By hypothesis

$$CE + ED \geq CB = CG + CA,$$

$$CA + CE > ED,$$

whence, by addition,

$$2CE > CG.$$

The word GC therefore repeats (as a part of α_1^{-1}) on CE from C as far as CE extends, i.e. for more than half of GC. Let

$$K' = U_{12}^{-1}V_{21}U_{12}^{-1}U_{12}^{-1}I$$

$\alpha_1\alpha_2^{-1}\alpha_1^{-1}$	$\alpha_1\alpha_2^{-2}$	$\alpha_1\alpha_2^{-2}$	$\alpha_1\alpha_2^{-1}$	α_1
$\alpha_1\alpha_2^{-1}$	$\alpha_1\alpha_2^{-1}$	α_2	α_2	α_2
$\lambda_1 < \Lambda$	μ_3	μ_2	μ_1	$\lambda_3 \leq \Lambda$

Here $\mu_1 < \lambda_3$, because $CA < CB$; also, $\mu_2 < \mu_1$, because if one describes α_2^{-1} from A it first runs to C because $AC = BG$ and thereafter describes more than half the word GC on CE, as found above. However, $L(\alpha_2)$ is $2BG + GC$. Since $\alpha_1\alpha_2^{-2} < \alpha_1$ as a result, we have $\mu_3 < \Lambda$. $K = O_1O_2K'$, and by (18h) one has:

$$O_1O_2U_{12}^{-1}\cdot V_{21}U_{12}^{-1}\cdot U_{12}^{-1}U_{21}\cdot U_{12}^{-1} = O_1O_2U_{12}^{-1}\cdot P_{12}O_2U_{21}^{-1}\cdot O_1P_{12}V_{12}\cdot U_{12}^{-1}$$

and this becomes 1 by (18f,g,m).

(c) $K = U_{12}V_{21}I$.

$\alpha_1\alpha_1\alpha_2$	α_1	α_1
$\alpha_1\alpha_2$	$\alpha_1\alpha_2$	α_2
$\lambda_1 < \Lambda$	Λ	$\lambda_3 \leq \Lambda$

In $\alpha_1\alpha_1\alpha_2$, the middle factor α_1 loses less than $\alpha_1/2$ in cancellation by the first factor, and at most $\alpha_1/2$ with the factor α_2, because $\alpha_1\alpha_2 \geq \alpha_2$. The middle factor is therefore not entirely absorbed, and we have only one possible form for $\alpha_1\alpha_1\alpha_2$ (Figure 7), with $CD > 0$. Let

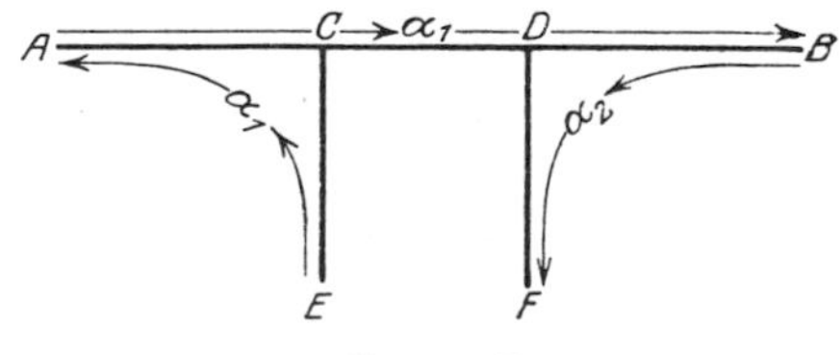

Figure 7

$$K' = V_{21}U_{12}^{-1}V_{21}^{-1}U_{12}I$$

$\alpha_1\alpha_1\alpha_2$	$\alpha_1\alpha_1\alpha_2$	$\alpha_1\alpha_2$	$\alpha_1\alpha_2$	α_1
$\alpha_1\alpha_2$	$\alpha_2^{-1}\alpha_1^{-1}\alpha_2$	$\alpha_2^{-1}\alpha_1^{-1}\alpha_2$	α_2	α_2
$\lambda_1 < \Lambda$	μ_3	μ_2	μ_1	$\lambda_3 \leq \Lambda$

By hypothesis, $AC + CD \geq DB$ and $AC > CD + DF$, hence by addition: $2AC > DB + DF = \alpha_2$, and since $AC < \alpha_1/2$ (because $\alpha_1^2 > \alpha_1$), it follows that $\alpha_1 > \alpha_2$, hence $\mu_1 < \Lambda$. Since $AC < CD + DB$ (because $<\alpha_1/2$), it follows that $DB > DF$, hence $\alpha_1\alpha_2 < \alpha_1$. Thus $BD > \alpha_2/2$ and $AC > \alpha_2/2$; the first $[\alpha_2/2] + 1$ generators of AC therefore coincide with the first $[\alpha_2/2] + 1$ of BD (as the beginning of α_1^{-1}). Therefore $\alpha_2^{-1}\alpha_1^{-1}\alpha_2 < \alpha_2^{-1}\alpha_1^{-1}$ and it follows from $\alpha_1\alpha_2 < \alpha_1$ that $\mu_2 < \Lambda$. At the same time it follows that $\mu_3 < \lambda_1$.

We have $K = K'$, and this is a consequence of a double application of (18h), together with (18m,f,g,b).

(d) $K = U_{12}V_{21}^{-1}I$. One has $K' = V_{12}^{-1}I$. The relation $K = P_{12}O_1K'$ follows from (18i). The cases III, IV, V still missing assume $n > 2$. Taking our previous results into account, this concludes the completeness proof for S when $n = 2$. (Of course, this can be done more simply when one assumes $n = 2$ from the beginning because one can then make quite special assumptions about the form of α.) In what follows we have to draw a third element of I, α_3, into the investigation.

III. $K = U_{12}W_{31}^{\pm 1}I$. Since $O_3 \rightleftarrows U_{12}$ by (18e) we can use (18g), with possible conjugation by O_3, arrange for W_{31} to receive the exponent $+1$ (analogously below in IV and V). We therefore have to replace W_{31} first by U_{31}, then by V_{31}.

(a) $K = U_{12}U_{31}I$.

$\alpha_1\alpha_2$	α_1	α_1
α_2	α_2	α_2
$\alpha_3\alpha_1$	$\alpha_3\alpha_1$	α_3
$\lambda_1 < \Lambda$	Λ	$\lambda_3 \leq \Lambda$

One draws the following figure: $AB \equiv \alpha_1$; $AD \equiv \alpha_3^{-1}$ with the branching point C of AB; $BF \equiv \alpha_2$ with the branching point E of the previously drawn figure. If E does not separate A and C, then one constructs

$$K' = U_{32}^{-1}U_{31}U_{12}I$$

$\alpha_1\alpha_2$	$\alpha_1\alpha_2$	$\alpha_1\alpha_2$	α_1
α_2	α_2	α_2	α_2
$\alpha_3\alpha_1$	$\alpha_3\alpha_1\alpha_2$	α_3	α_3
$\lambda_1 < \Lambda$	μ_2	μ_1	$\lambda_3 \leq \Lambda$

It follows from $\alpha_1\alpha_2 < \alpha_1$ that $\mu_1 < \lambda_3$, $\mu_2 < \lambda_1$, by the above hypothesis on E. $K = K'$ is a consequence of (18o). But if E separates A and C, then $CB \geq CA$ because $\alpha_3\alpha_1 \geq \alpha_3$, hence $EB > EA$ and therefore $\alpha_1\alpha_2 < \alpha_2$. Then Λ is not attained in

$$K' = V_{12}^{-1}U_{31}V_{21}I.$$

We have $K = P_{12}O_1K'$. This is a consequence of (18 l,i).

(b) $K = U_{12}V_{31}I$. One draws $AB \equiv \alpha_1$; $BD \equiv \alpha_3$ with the branching point C of AB; $BF \equiv \alpha_2$ with the branching point E of the figure previously drawn. Then one constructs

$$K' = V_{31}V_{32}^{-1}U_{12}I$$

$\alpha_1\alpha_2$	$\alpha_1\alpha_2$	$\alpha_1\alpha_2$	α_1
α_2	α_2	α_2	α_2
$\alpha_1\alpha_3$	$\alpha_2^{-1}\alpha_3$	α_3	α_3
$\lambda_1 < \Lambda$	μ_2	μ_1	$\lambda_3 \leq \Lambda$

Here, $\mu_1 < \lambda_3$ because $\alpha_1\alpha_2 < \alpha_1$ by hypothesis. If E does not separate A and C, then $BE > \alpha_2/2$ (because $\alpha_1\alpha_2 < \alpha_1$) and hence $\alpha_2^{-1}\alpha_3 < \alpha_3$ also, so $\mu_2 < \mu_1$. Let E therefore separate A and C. We set up the condition for $\mu_2 < \Lambda$. We have

$$\Lambda = \lambda_3 + AC - CB,$$

$$\mu_2 = \lambda_3 + (EF - EB) + (CF - CB),$$

therefore $\mu_2 < \Lambda$ as long as $AC - EF + EB - CF > 0$,

$$\alpha_1 - 2EF > 0,$$

and since $EF < EB$, because $\alpha_1\alpha_2 < \alpha_1$, this is satisfied as long as

$$\alpha_1 - 2EB \geq 0,$$

i.e. $AE \geq EB$. Thus only $AE < EB$ is still to be considered. In this case $\alpha_1\alpha_2 < \alpha_2$. Then $K'' = V_{12}^{-1}V_{31}V_{21}I$ suffices. $K = K'$ is a consequence of (18q), $K = P_{12}O_1K''$ is a consequence of (18 l,i).

IV. $K = U_{12}W_{13}I$.

(a) $K = U_{12}U_{13}I$.

$\alpha_1\alpha_3\alpha_2$	$\alpha_1\alpha_3$	α_1
α_2	α_2	α_2
α_3	α_3	α_3
$\lambda_1 < \Lambda$	Λ	$\lambda_3 \leq \Lambda$

One draws $AB \equiv \alpha_3$ and $AD \equiv \alpha_1^{-1}$ with the branching point C, and finally $BF \equiv \alpha_2$ with the branching point E. If E does not separate C and D, then one constructs

$$K' = U_{32}^{-1}U_{13}U_{32}I$$

$\alpha_1\alpha_3\alpha_2$	$\alpha_1\alpha_3\alpha_2$	α_1	α_1
α_2	α_2	α_2	α_2
α_3	$\alpha_3\alpha_2$	$\alpha_3\alpha_2$	α_3
$\lambda_1 < \Lambda$	μ_2	μ_1	$\lambda_3 \leq \Lambda$

It follows from $\alpha_1\alpha_3\alpha_2 < \alpha_1\alpha_3$ that $BE > \alpha_2/2$, hence $\alpha_3\alpha_2 < \alpha_3$ and therefore $\mu_1 < \lambda_3$, $\mu_2 < \lambda_1$. $K = K'$ is a consequence of (18o). The remaining case is where C and D are separated by E. If $CF < CB$ then K' still satisfies. Suppose therefore that $CF \geq CB$. Since $\alpha_1\alpha_3 \geq \alpha_1$ we have $CB \geq CA$. If $CB > CA$ then $K'' = V_{23}^{-1}U_{12}V_{23}I$ satisfies, because $\alpha_3\alpha_2 < \alpha_2$. $K = K''$ is a consequence of (18p). If $CB = CA$, and hence $\lambda_3 = \Lambda$, then we can still settle the case where $\alpha_1 < \alpha_3$. We then have the hypothesis $CD < CA = CB \leq CF$ and hence $\alpha_1 < \alpha_3 \leq \alpha_2$. Let

$$K''' = V_{32}^{-1}V_{13}^{-1}V_{23}V_{31}I$$

α_3^{-1}	α_3^{-1}	α_1	α_1	α_1
$\alpha_1\alpha_3\alpha_2$	$\alpha_1\alpha_3\alpha_2$	$\alpha_1\alpha_3\alpha_2$	α_2	α_2
α_2^{-1}	$\alpha_1\alpha_3$	$\alpha_1\alpha_3$	$\alpha_1\alpha_3$	α_3
$\lambda_1 < \Lambda$	μ_3	μ_2	μ_1	$\lambda_3 = \Lambda$

Here $\mu_1 < \Lambda$, $\mu_2 < \lambda_1$ because $\alpha_1\alpha_3 < \alpha_3$, and $\alpha_1 < \alpha_2$, $\mu_3 < \lambda_1$ because $\alpha_1\alpha_3 < \alpha_3 \leq \alpha_2$. We have $K = P_{23}P_{12}O_1O_3K'''$ and this is derivable, since by (18 l,i):

$$\begin{aligned} P_{23}P_{12}O_1O_3V_{32}^{-1}V_{13}^{-1}V_{23}V_{31}U_{13}^{-1}U_{12}^{-1} &= P_{23}P_{12}O_1O_3(V_{32}^{-1}V_{23})(V_{13}^{-1}V_{31}U_{13}^{-1})U_{12}^{-1} \\ &= P_{23}P_{12}O_1O_3(P_{23}O_2U_{32})(P_{13}O_3)U_{12}^{-1} \\ &= 1 \end{aligned}$$

by (18c,e).

With this, case IV(a) is settled apart from the following:

Remaining case: $CA = CB \begin{cases} \leq CD \\ \leq CF \end{cases}$

For the further treatment of this remaining case, in which $\lambda_3 = \Lambda$ and equal amounts of α_3 are cancelled by α_1 and α_2, and the latter are at least as long as α_3, the frontmost factor of the subcore I must be brought into consideration. For this, we refer to the next paragraph.

(b) $U_{12}V_{13}I$. One draws $AB \equiv \alpha_1$, $AD \equiv \alpha_3^{-1}$ with the branching point C, and then $BF \equiv \alpha_2$ with the branching point E. If E does not separate C and D, then $K' = V_{13}U_{12}I$ satisfies. $K = K'$ is a consequence of (18n). Suppose that E separates C and D. K' still satisfies as long as $\alpha_1\alpha_2 < \alpha_3\alpha_1$, i.e. as long as $AF < DB$. We therefore suppose $AF \geq DB$.

Let $CB > CA$. Then $CD < CF$, hence $\alpha_3\alpha_1 < \alpha_2$. And since $CD \geq CA$, because $\alpha_3\alpha_1 \geq \alpha_1$, we have $CF > CA$, and hence $\alpha_2 > \alpha_1$. Then $K'' = V_{12}^{-1}V_{13}V_{23}V_{21}I$ satisfies. We have $K = P_{12}O_1K''$. By (18r),

$$P_{12}O_1V_{12}^{-1}(V_{13}V_{23}V_{21}V_{13}^{-1})U_{12}^{-1} = P_{12}O_1V_{12}^{-1}V_{21}U_{12}^{-1},$$

and this is 1 because of (18i).

Let $CB < CA$. Since $\alpha_3\alpha_1 \geq \alpha_1$, and therefore $CD \geq CA$, we then have $\alpha_3 > \alpha_1$. One constructs

$$K''' = U_{13}^{-1}U_{12}U_{32}U_{31}I$$

α_3^{-1}	$\alpha_1\alpha_2$	α_1	α_1	α_1
α_2	α_2	α_2	α_2	α_2
$\alpha_3\alpha_1\alpha_2$	$\alpha_3\alpha_1\alpha_2$	$\alpha_3\alpha_1\alpha_2$	$\alpha_3\alpha_1$	α_3
$\lambda_1 < \Lambda$	μ_3	μ_2	μ_1	$\lambda_3 \leq \Lambda$

Here $\mu_1 < \lambda_3$, $\mu_2 < \lambda_1$; also $\mu_3 \leq \lambda_1$ when $\alpha_1\alpha_2 \leq \alpha_3$, and this is the case when $EF \leq ED$. For $EF > ED$ we use

$$K'''' = U_{23}U_{12}U_{23}^{-1}U_{32}U_{31}I$$

α_3^{-1}	α_3^{-1}	α_1	α_1	α_1	α_1
α_2	$\alpha_1^{-1}\alpha_3^{-1}$	$\alpha_1^{-1}\alpha_3^{-1}$	α_2	α_2	α_2
$\alpha_3\alpha_1\alpha_2$	$\alpha_3\alpha_1\alpha_2$	$\alpha_3\alpha_1\alpha_2$	$\alpha_3\alpha_1\alpha_2$	$\alpha_3\alpha_1$	α_3
$\lambda_1 < \Lambda$	$\mu_4^{(1)}$	$\mu_3^{(1)}$	μ_2	μ_1	$\lambda_3 \leq \Lambda$

where $\mu_3^{(1)}$ and $\mu_4^{(1)}$ are still to be investigated. We have $\mu_3^{(1)} < \mu_4^{(1)}$ because $\alpha_3 > \alpha_1$, and $\mu_4^{(1)} < \lambda_1$ because $ED < EF$, hence $\alpha_3\alpha_1 < \alpha_2$. $K'''' = K'''$ is a consequence of (18o). $K = P_{13}O_1K'''$ is derived as follows:

$$\begin{aligned} P_{13}O_1U_{13}^{-1}U_{12}U_{32}\cdot U_{31}V_{13}^{-1}\cdot U_{12}^{-1} &= P_{13}O_1U_{13}^{-1}U_{12}U_{32}\cdot O_1P_{13}V_{31}^{-1}\cdot U_{12}^{-1} \qquad \text{(by (18i))}\\ &= V_{31}V_{32}^{-1}U_{12}V_{31}^{-1}U_{12}^{-1} \qquad \text{(by (18e,g,c,b))}\\ &= 1 \qquad \text{(by (18q)).}\end{aligned}$$

The case $CB = CA$ still remains. If $CD > AC$, then $\lambda_3 < \Lambda$, $\alpha_1 < \alpha_3 \leq \alpha_2$, since $ED \leq EF$ because $AF \geq DB$; also, $\alpha_3\alpha_1 \leq \alpha_2$ and $\alpha_1\alpha_2 = \alpha_2$. Then the above core K'' satisfies. Since $\alpha_3\alpha_1 > \alpha_1$, and hence $CD \geq AC$, we therefore still have the

Remaining case: $CA = CB = CD \leq CF$

with $\lambda_3 = \Lambda$, $\alpha_1 = \alpha_3 \leq \alpha_2$.

V. $K = U_{12}W_{23}I$.

(a) $K = U_{12}U_{23}I$. One draws $AB \equiv \alpha_2$, $AD \equiv \alpha_1^{-1}$ with branching point C and $BF \equiv \alpha_3$ with branching point E. The core $K' = U_{23}U_{13}U_{12}I$ satisfies, in case E does not separate the points C and D. Suppose then that E does separate C and D. Then K' still satisfies in case $CA > CB$, and also when $CA = CB$ and $\alpha_3 > \alpha_2$ simultaneously. If $CA < CB$, then $K'' = V_{23}^{-1}U_{13}V_{32}I$ satisfies because $\alpha_2 \leq \alpha_2\alpha_3 < \alpha_3$. The case $CA = CB = CF$ remains. If, at the same time, $CD < CA$, then $K''' = V_{12}^{-1}U_{23}V_{21}I$ satisfies.

Remaining case: $CA = CB = CF \leq CD$

We have $K = K'$, and this is a consequence of (18o); $K = P_{23}O_2K''$, and this is a consequence of (18 l,i,d,e); $K = P_{12}O_1K'''$, and this is a consequence of (18n,i).

(b) $K = U_{12}V_{23}I$. One draws $AB \equiv \alpha_3$, $AD \equiv \alpha_1^{-1}$ with the branching point C and $BF \equiv \alpha_2$ with the branching point E. For each position of E, $EA \geq EB$ because $\alpha_3\alpha_2 \geq \alpha_2$. If A and C are separated by E, then $K' = V_{23}U_{12}U_{13}$ satisfies, because $CA > EA \geq EB > CB$ and hence $\alpha_1\alpha_3 < \alpha_1$. K' also satisfies for $E = C$, in case $EA > EB$. If $E = C$ and $EA = EB$, then $EF < EA = EB$ because $\alpha_1\alpha_3\alpha_2 < \alpha_1$; thus $\alpha_3\alpha_2 < \alpha_3$ and $\alpha_2 < \alpha_3$. Then $K'' = U_{23}^{-1}U_{13}U_{32}I$ satisfies. Suppose now that E separates C and B. Since $\alpha_1\alpha_3\alpha_2 < \alpha_1$, we have $CF < CA$ and hence either $CB < CA$, in which case K' satisfies, or else $EF < EB$, and then K'' satisfies. Finally, suppose E separates C and D. Then $CA \geq CB$ because $\alpha_3\alpha_2 \geq \alpha_2$. For $CA > CB$, K' satisfies. When $CA = CB$ and $\alpha_2 < \alpha_3$ simultaneously then K'' satisfies. When $CA = CB \leq CF$ we can still settle the case $\alpha_1 < \alpha_3$. Namely, if $CD < CA$ then $K''' = V_{12}^{-1}V_{31}^{-1}V_{23}V_{31}I$ satisfies.

$$\text{Remaining case: } CA = CB \begin{cases} \leq CF \\ \leq CD \end{cases}$$

$K = K'$ is a consequence of (18p), $K = P_{23}O_2K''$ is a consequence of (18 l,h,e,d), $K = P_{12}O_1K'''$ is a consequence of (18r,i).

§7.

Proof of the Lemma. Part II

The proof of the lemma still lacks settlement of the four cases remaining under IV(a), IV(b), V(a), V(b). Common to them all is the condition $\lambda_3 = \Lambda$; the maximal diagram number is "stationary", i.e. the diagram number takes the same value at least twice in succession immediately before the introduction of the shortening factor. In all cases one is dealing with a product of three elements in which the middle factor is cancelled in equal amounts by the two others. In no case does the length of this middle factor exceed that of either of the others. In cases IV(b) and V(a) the length of the first cancelling factor equals the length of the factor being cancelled. This property makes it possible to reduce the last three cases to the first by an easy transformation, thereby unifying the treatment which follows.

In the remaining case under IV(b) one sets

$$U_{12}V_{13}I = P_{23}O_2O_1U_{31}^{-1}U_{32}U_{12}U_{13}P_{23}P_{12}O_3O_2O_1I,$$

which is a consequence relation of (18a–h,r). Here one exchanges $P_{23}P_{12}O_3O_2O_1$ and I. Beginning with the new core $I^{(1)} = [\alpha_2^{-1}, \alpha_3^{-1}, \alpha_1^{-1}, \alpha_4, \ldots, \alpha_n]$, one has the following isomorphism sequence with the attached length numbers:

$\alpha_2^{-1}\alpha_1^{-1}\alpha_3^{-1}$	$\alpha_2^{-1}\alpha_1^{-1}\alpha_3^{-1}$	$\alpha_2^{-1}\alpha_1^{-1}\alpha_3^{-1}$	$\alpha_2^{-1}\alpha_1^{-1}$	α_2^{-1}
α_3^{-1}	α_3^{-1}	α_3^{-1}	α_3^{-1}	α_3^{-1}
α_2	$\alpha_1^{-1}\alpha_3^{-1}$	α_1^{-1}	α_1^{-1}	α_1^{-1}
λ_1	μ_2	μ_1	Λ	Λ

with $\mu_1 \leq \lambda_1$ because $\alpha_1 \leq \alpha_2$ and $\mu_2 = \mu_1$ because $\alpha_3\alpha_1 = \alpha_1$. The subcore $U_{12}U_{13}I^{(1)}$ now corresponds exactly to the remaining case under IV(a), and we can confine ourselves to this, since the two preceding factors U_{32} and U_{31}^{-1} do not come into consideration for the reduction of Λ. Similarly, in the remaining case under V(a) one puts:

$$U_{12}U_{23}I = U_{23}U_{13}U_{12}I = P_{23}U_{32}U_{12}U_{13}P_{23}I$$

by (18o) and (18b,d), and in the remaining case under V(b)

$$U_{12}V_{23}I = V_{23}U_{12}U_{13}I$$

by (18p). One can restrict the remaining case under IV(a) still further by the condition $\alpha_2 \leq \alpha_1$. Namely, if $\alpha_2 > \alpha_1$ one interchanges the rôles of α_2 and α_1 by means of the equation

$$U_{12}U_{13}I = O_3O_1V_{23}U_{21}^{-1}U_{12}U_{13}P_{12}O_3O_2O_1I,$$

(which is a consequence of (18e–g,i,p)), and within this brings $P_{12}O_3O_2O_1$ to the right of I.

Thus we have to deal with the remaining case under IV(a) (with the additional condition $\alpha_2 \leq \alpha_1$). A direct investigation of this remaining case would lead to further remaining cases of higher type, in which the maximal diagram number Λ remained stationary for at least two steps. We therefore view our remaining case IV(a) as a special case of a more general case, in which Λ is stationary for a number of steps, and for this purpose take the following form of the core, which we shall assume to be freely reduced:

$$K = U_{12}U_{1r_\nu}^{\varepsilon_\nu}U_{1r_{\nu-1}}^{\varepsilon_{\nu-1}}\cdots U_{1r_2}^{\varepsilon_2}U_{1r_1}^{\varepsilon_1}W_{xy}^{\pm 1}I.$$

The r_i are any indices, not necessarily different, among 3, 4, …, *n; the* $\varepsilon_i = \pm 1$. *The diagram numbers are*

$$L(I) = \lambda_0,$$
$$L(W_{xy}^{\pm 1}I) = L(U_{1r_1}^{\varepsilon_1}W_{xy}^{\pm 1}I) = \cdots = L(U_{1r_\nu}^{\varepsilon_\nu}\cdots U_{1r_1}^{\varepsilon_1}W_{xy}^{\pm 1}I) = \Lambda \geq \lambda_0,$$
$$L(K) = \lambda_1 < \Lambda.$$

We let

$$I = [\alpha_1, \alpha_2, \ldots, \alpha_n],$$
$$W_{xy}^{\pm 1}I = [\beta_1, \beta_2, \ldots, \beta_n],$$

and we have the condition

$$\beta_i \leq \beta_2 \leq \beta_1 \qquad (i = r_1, r_2, \ldots, r_\nu).$$

A core of this form is called "*normal* ν-tuply stationary" with respect to the subcore $U_{12}U_{1r_\nu}^{\varepsilon_\nu}\cdots U_{1r_1}^{\varepsilon_1}$.

In order to visualise the construction and express ourselves briefly in what follows, we represent as before the first element of K in the group diagram, beginning from the β. Let $AB_0 \equiv \beta_1$ and let M_1 be the midpoint of AB_0, so that M_1 is a vertex if $L(\beta_1)$ is even and the midpoint of an edge if $L(\beta_1)$ is odd. Also, let $B_0B_1 \equiv \beta_{r_1}^{\varepsilon_1}$; the branching point C_1 of B_0A lies on B_0M_1 and is the midpoint of B_0B_1 ($L(\beta_{r_i})$ is even). Further, let $B_1B_2 \equiv \beta_{r_2}^{\varepsilon_2}$ be described, with midpoint C_2; D_2 is then the branching point of the figure previously drawn. If $\beta_{r_2} < \beta_{r_1}$, then $D_2 = C_2$ separates the points C_1 and B_1; if $\beta_{r_2} > \beta_{r_1}$ then $D_2 = C_2$ lies on C_1M_1, and in fact not at C_1; if $\beta_{r_2} = \beta_{r_1}$ then $C_2 = C_1$ and either $D_2 = C_2 = C_1$ or else D_2 separates C_1 and

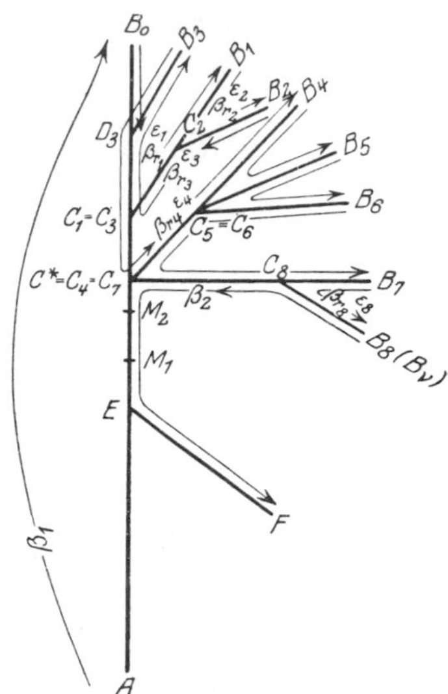

Figure 8

B_0. Next one draws $B_2B_3 \equiv \beta_{r_3}^{\varepsilon_3}$ with midpoint C_3 and the branching point D_3 from the previous figure, etc. The points $B_0, B_1, \ldots, B_v$ obtained in this way are all different; namely, if $B_h = B_l$ $(l > h)$ then $\beta_{r_{h+1}}^{\varepsilon_{h+1}}\beta_{r_{h+2}}^{\varepsilon_{h+2}}\ldots\beta_{r_l}^{\varepsilon_l} = 1$ would be a relation, which would have to be satisfied *identically* in the β, and it would therefore correspond to a cancellation in K, contrary to hypothesis. Also, $AB_i = AB_0$ because $\beta_1\beta_{r_1}^{\varepsilon_1}\ldots\beta_{r_i}^{\varepsilon_i} = \beta_1$. And each B_i is separated from A by M_1 because $\beta_{r_i} \leq \beta_1$. Thus M_1 is at an equal distance from all B_i. Let C^* denote the C-point on AB_0 which lies next to A; C^* lies on B_0M_1 and can in particular lie at M_1. C^* is the midpoint of the (or the) β_i of greatest length and $C^* = M_1$ when this length equals $L(\beta_1)$. C^* is also at equal distance from all B_i, and it separates each B_i from A. Finally, draw $B_vF \equiv \beta_2$ with midpoint M_2 and branching point E of the previous figure. Since $\beta_2 \geq \beta_{r_i}$, $B_vM_2 \geq B_vC^*$.

Since $\beta_1\beta_{r_1}^{\varepsilon_1}\ldots\beta_{r_v}^{\varepsilon_v}\beta_2 < \beta_1\beta_{r_1}^{\varepsilon_1}\ldots\beta_{r_v}^{\varepsilon_v}$, B_vF has *more* than B_vM_2 in common with B_vA. Thus C^* lies on B_vM_2, and in particular $C^* = M_2$ in case a $\beta_{r_i} = \beta_2$. Also, E separates the points M_2 and A. Since $\beta_1 \geq \beta_2$ we have $B_0M_1 \geq B_vM_2 = B_0M_2$, and hence M_1 lies on AM_2.

Figure 8 shows an example of such a word $\beta_1\beta_{r_1}^{\varepsilon_1}\ldots\beta_{r_v}^{\varepsilon_v}\beta_2$ for $v = 8$. $C_3 \neq D_3$, but elsewhere C_i has been chosen equal to D_i. Also, we have chosen $\beta_{r_i} < \beta_2 < \beta_1$, hence $B_0C^* < B_0M_2 < B_0M_1$.

We shall now draw the factor $W_{xy}^{\pm 1}$ into the investigation, and prove the following two theorems.

Theorem A. *If $\Lambda > \lambda_0$, so that the factor $W_{xy}^{\pm 1}$ effects an increase of the diagram number in K, then it is possible to find a consequence relation $K_1 = \Omega \cdot K_1'$ of S to apply to the subcore $K_1 = U_{12}U_{1r_v}^{\varepsilon_v}\ldots U_{1r_1}^{\varepsilon_1}W_{xy}^{\pm 1}$ so that Λ appears in $K_1'I$ less often than in K_1I and, if its appearance is multiple, it is again normal stationary.*

It follows from this theorem that Λ, if $\Lambda > \lambda_0$, can be entirely removed from the diagram by repeated application of such relations.

Theorem B. *If $\Lambda = \lambda_0$, then it is possible to find a consequence relation $K_1 = \Omega \cdot K_1'$ of S to apply to the subcore K_1 so that either the subcore K_1' of $K_1'I$ does not give rise to the diagram number Λ, or else Λ appears normal stationary with respect to a final segment of this subcore.*

In the latter case one draws the frontmost factor of I into the investigation and continues the process. Since the numbers in the diagram $\Delta(I)$ must decrease once, the condition $\Lambda > \lambda_0$ of Theorem A necessarily occurs once, and with it the possibility of removing Λ completely. Thus the following joint proof of Theorems A and B concludes the proof of the lemma. The proof that the relations used are consequences of the relations (18) offers no essential difficulties, and is therefore left to the reader.

In analogy with the previous paragraph, we now have to consider a series of distinct cases for the indices x and y separately.

I. $x \neq 1, 2, r_1, \ldots, r_v;\ y \neq 1.$

$$K' = W_{xy}^{\pm 1} U_{12} U_{1r_v}^{\varepsilon_v} \ldots U_{1r_1}^{\varepsilon_1} I$$

satisfies in Theorems A and B; because in this new position each factor alters the diagram number by the same value as in the old position. Thus if $\Lambda > \lambda_0$, Λ does not appear in K' at all, and hence Theorem A holds; if $\Lambda = \lambda_0$, then Λ appears in K' once less often than in K, and the subcore $U_{12} \ldots I$ of K' is normal stationary with respect to $U_{12} \ldots U_{1r_1}^{\varepsilon_1}$, hence Theorem B is satisfied. (In the latter case the frontmost factor of I can become drawn into the investigation, and one thereby has the same hypotheses as for K.)

II. $x \neq 1, 2, r_1, \ldots, r_v;\ y = 1.$

(a) $K = U_{12} U_{1\varepsilon_v}^{\varepsilon_v} \ldots U_{1r_1}^{\varepsilon_1} U_{x1} I$. Thus $\beta_t \equiv \alpha_t$ for $t \neq x$ and $\beta_x \equiv \alpha_x \alpha_1$.

$$K' = U_{xr_1}^{-\varepsilon_1} \ldots U_{xr_v}^{-\varepsilon_v} U_{x2}^{-1} U_{x1} U_{12} U_{1r_v}^{\varepsilon_v} \ldots U_{1r_1}^{\varepsilon_1} I$$

satisfies in Theorems A and B.

Proof. $L(U_{12} U_{1r_v}^{\varepsilon_v} \ldots U_{1r_1}^{\varepsilon_1} I) < L(I) \leq \Lambda$. In order to ascertain the influence of the factor U_{x1} on $\Delta(K')$, one draws $AH = \alpha_x^{-1}$ with the branching point G on the Figure 8 already present. We shall show that U_{x1} causes no more enlargement in its new position than in the old, so that $\alpha_x \alpha_1 \alpha_{r_1}^{\varepsilon_1} \ldots \alpha_{r_v}^{\varepsilon_v} \alpha_2 \leq \alpha_x \alpha_1$ and hence $HF \leq HB_0$, or what is the same, $GF \leq GB_0$. If G lies on AF, then this follows from $EF < EB_0$. If G lies on EB_0, then it must lie on EM_1, since $GB_0 \geq GA$ because $\alpha_x \alpha_1 \geq \alpha_x$. Then $GF \leq GA \leq GB_0$. On the same grounds, it follows more generally that AH can branch away from AB_0 only on AM_1. Thus one obtains other possible positions for G only when $C^* = M_2 = M_1$ and G is separated from A as well as B_0 by C^*. Then $GF = GB_0$. Thus $L(U_{x1} \ldots I) < \Lambda$ in all cases. Then Λ can also not be reached by the other factors, since in all cases G lies at least as close to the points $B_v, \ldots, B_1$ as to the point B_0.

(b) $K = U_{12} U_{1r_v}^{\varepsilon_v} \ldots U_{1r_1}^{\varepsilon_1} V_{x1} I$; $\beta_x \equiv \alpha_1 \alpha_x$, or else $\beta_t \equiv \alpha_t$.

$$K' = V_{x1} V_{x2}^{-1} V_{xr_v}^{-\varepsilon_v} \ldots V_{xr_1}^{-\varepsilon_1} U_{12} U_{1r_v}^{\varepsilon_v} \ldots U_{1r_1}^{\varepsilon_1} I$$

satisfies in Theorems A and B.

The proof is similar to that under (a): $L(U_{12} \ldots U_{1r_1}^{\varepsilon_1} I) < L(I) \leq \Lambda$. One draws $B_0 H \equiv \alpha_x$ with branching point G. If G lies on the part of the figure different from $M_1 A + EF$ then $B_i G < AG$ and $FG \leq AG$; if G lies on $M_1 F$ then $B_i G = B_0 G$ and $FG \leq AG$; if G lies on EA then

$B_iG = B_0G$, $FG < B_0G$. Recalling that $\alpha_1\alpha_x \geq \alpha_x$, and hence $AG \geq B_0G$, it follows in all cases that none of the factors introduced after U_{12} can give rise to Λ in $\Delta(K')$.

The cases U_{x1}^{-1} and V_{x1}^{-1} are reduced to the present one by conjugation with O_x. Thus from now on we have only to consider values for x from the series $1, 2, r_1, \ldots, r_v$.

III. $x = 1$; $y \neq 2$. If $\Lambda > \lambda_0$ one chooses

$$K' = V_{2r_v}^{-\varepsilon_v} \ldots V_{2r_1}^{-\varepsilon_1} \cdot U_{12} W_{1y}^{\pm 1} \cdot V_{2r_1}^{\varepsilon_1} \ldots V_{2r_v}^{\varepsilon_v} I;$$

because Λ appears in $\Delta(K')$ only once, namely, on application of $W_{1y}^{\pm 1}$. We therefore take $\Lambda = \lambda_0$.

(a) $W_{1y}^{\pm 1} = U_{1y}$. The case U_{1y}^{-1} reduces to this by conjugation with O_y, with possible sign changes in some of the ε_i. If $\alpha_y \leq \alpha_2$, then K itself is already normal $(v + 1)$-tuply stationary with respect to $U_{12} U_{1r_v}^{\varepsilon_v} \ldots U_{1r_1}^{\varepsilon_1} U_{1y}$. If $\alpha_y > \alpha_2$, then $y \neq r_i$; one then constructs

$$K'' = U_{yr_1}^{-\varepsilon_1} \ldots U_{yr_v}^{-\varepsilon_v} U_{y2}^{-1} \cdot U_{1y} \cdot U_{y2} U_{yr_v}^{\varepsilon_v} \ldots U_{yr_1}^{\varepsilon_1} I,$$

where the factor U_{y2} already brings the diagram number ultimately under the value Λ, and K'' is normal stationary with respect to $U_{y2} U_{yr_v}^{\varepsilon_v} \ldots U_{yr_1}^{\varepsilon_1}$.

(b) $W_{1y}^{\pm 1} = V_{1y}$. One draws $AH \equiv \alpha_y$ with the branching point G in Figure 8. If M_2 does not lie on AG, then one chooses

$$K' = V_{1y} U_{12} U_{1r_v}^{\varepsilon_v} \ldots U_{1r_1}^{\varepsilon_1} I,$$

because K' is normal stationary with respect to $U_{12} U_{1r_v}^{\varepsilon_v} \ldots U_{1r_1}^{\varepsilon_1}$. If the part of AG which lies on AB_0 contains M_1 as an interior point, then $\alpha_y\alpha_1 < \alpha_y$ and $y \neq r_i$. Then one takes

$$K'' = U_{1y}^{-1} U_{12} U_{1r_v}^{\varepsilon_v} \ldots U_{1r_1}^{\varepsilon_1} \cdot U_{y2} U_{yr_v}^{\varepsilon_v} \ldots U_{yr_1}^{\varepsilon_1} U_{y1} I.$$

Because already $L(U_{y1}I) < L(I) = \Lambda$ here, and since $HB_i < HA$ and $HF \leq HA$, Λ is not attained at all in $\Delta(K'')$ outside $\Delta(I)$. Since $AM_2 \geq AM_1$, this case always occurs when the part of AG which lies on AB_0 contains M_2 as an interior point. Thus the only two possibilites remaining are:

1. $G = M_1 = M_2$.
2. $M_1 = M_2 = C^*$ and G lies on the part of the figure which is separated from A and B_0 by C^*.

In both cases K'' is normal $(v + 1)$-tuply stationary with respect to $U_{y2} U_{yr_v}^{\varepsilon_v} \ldots U_{yr_1}^{\varepsilon_1} U_{y1}$, and the remaining factors do not evoke any more Λ.

IV. $x = 1$; $y = 2$. The case $W_{xy}^{\pm 1} = U_{12}^{\pm 1}$ can be excluded; because $\alpha_1\alpha_2^{\pm 1} \geq \alpha_1$ implies that the midpoint of word $\alpha_2^{\pm 1}$ within the word $\alpha_1\alpha_2^{\pm 1}$ represented by AB_0 must be a point on AB_0, and then since $C^*B_0 = C^*B_v$ it must coincide with M_2; however, $M_2B_0 \not\equiv M_2B_v$ and $M_2B_0 \not\equiv M_2F$. Thus $W_{xy}^{\pm 1} = V_{12}^{\pm 1}$. The midpoint M_3 of the word $\alpha_2^{\pm 1}$ in $AB_0 \equiv \alpha_2^{\pm 1}\alpha_1$ again lies on AB_0, but not at M_2. Since $EF \leq EA$ and $AM_3 = M_2F$ we then have $AM_3 < AM_2$. Let $AH \equiv \alpha_2^{\pm 1}$ with the branching point G. If A is separated from M_2 by G, one chooses

$$K' = V_{12}^{\pm 1} U_{12} U_{1r_v}^{\varepsilon_v} \ldots U_{1r_1}^{\varepsilon_1} I.$$

Thus M_2 lies on AG; it follows that $\alpha_2^{\pm 1}\alpha_1 > \alpha_1$ and hence $\Lambda > \lambda_0$. We have $GA > GB_i$

because $M_2A > \alpha_2/2 = M_2B_i$, and hence one can take

$$K'' = U_{12}V_{12}^{\pm 1}U_{1r_v}^{\varepsilon_v}\dots U_{1r_1}^{\varepsilon_1}I.$$

V. $x = 2$; $y \neq 1$. The exponent of W_{2y} can be assumed to be $+1$. For $\Lambda > \lambda_0$,

$$K' = U_{12}W_{2y}U_{1r_v}^{\varepsilon_v}\dots U_{1r_1}^{\varepsilon_1}I$$

satisfies, so we assume $\Lambda = \lambda_0$.

(a) $W_{2y} = V_{2y}$. Let $B_vH \equiv \alpha_y$, with branching point G. If M_2 lies on AG, then

$$K'' = V_{2y}U_{12}U_{1y}U_{1r_v}^{\varepsilon_v}\dots U_{1r_1}^{\varepsilon_1}I$$

is normal $(v + 1)$-tuply stationary with respect to $U_{12}U_{1y}\dots U_{1r_1}^{\varepsilon_1}$. Thus M_2 does not lie on AG. Then $\alpha_y\alpha_2 < \alpha_y$ and $y \neq r_i$. One chooses

$$K''' = U_{2y}^{-1}U_{1y}U_{1r_v}^{\varepsilon_v}\dots U_{1r_1}^{\varepsilon_1}U_{y2}I.$$

(b) $W_{2y} = U_{2y}$. Let $FH \equiv \alpha_y^{-1}$ with branching point G. If M_2 does not lie on FG, then

$$K' = U_{2y}U_{1y}U_{12}U_{1r_v}^{\varepsilon_v}\dots U_{1r_1}^{\varepsilon_1}I$$

is normal v-tuply stationary with respect to $U_{12}\dots U_{1r_1}^{\varepsilon_1}$, and U_{1y} does not evoke Λ. If $M_2 = G$, then K' is normal $(v + 1)$-tuply stationary with respect to $U_{1y}\dots U_{1r_1}^{\varepsilon_1}$. Thus assume that M_2 separates the points F and G. If M_2G has more than the point M_2 in common with M_2B_v, then $\alpha_2\alpha_y < \alpha_y$ and $y \neq r_i$; one then takes

$$K'' = V_{2y}^{-1}U_{1y}U_{1r_v}^{\varepsilon_v}\dots U_{1r_1}^{\varepsilon_1}V_{y2}I.$$

If M_2G has only the point M_2 in common with M_2B_v, then we must have $M_2 = C^*$, and G lies on the part of the figure which is separated from A and B_v by C^*. Then K' is normal $(v + 1)$-tuply stationary.

VI. $x = 2$; $y = 1$.

(a) $W_{xy}^{\pm 1} = V_{21}^{\pm 1}$. One draws $B_vH \equiv \alpha_1^{\pm 1}$ in Figure 8, with midpoint M_3 and branching point G of the figure originally present. Since $\alpha_1^{\pm 1}\alpha_2 \geq \alpha_2$, M_3 lies on B_vF, and in fact, since $M_3B_v = \alpha_1/2$, $C^*B_v \leq \alpha_1/2$, it lies on C^*F. Since $C^*B_0 \not\equiv C^*B_v$, the case V_{21}^{-1} is then excluded, and hence $W_{xy}^{\pm 1} = V_{21}$. If M_3 lay on C^*E we would have $M_3B_v = M_3B_0 = M_3A$; this is excluded because $M_3B_v \not\equiv M_3A$. Thus M_3 lies on EF and in fact not at E. Hence $M_3F < M_3B_v = M_3H$ and then $GF < GH$ also, and hence $\alpha_1\alpha_2 < \alpha_1$ and $\alpha_2 < \alpha_1$. One then constructs

$$K' = U_{2r_v}^{\varepsilon_v}\dots U_{2r_1}^{\varepsilon_1}\cdot V_{1r_v}^{-\varepsilon_v}\dots V_{1r_1}^{-\varepsilon_1}\cdot V_{21}^{-1}\cdot U_{21}^{-1}\cdot V_{1r_1}^{\varepsilon_1}\dots V_{1r_v}^{\varepsilon_v}U_{12}I$$

$\alpha_1\alpha_2$ $\alpha_2^{-1}\alpha_1^{-1}\alpha_{r_v}^{-\varepsilon_v}\dots\alpha_{r_1}^{-\varepsilon_1}\alpha_1^{-1}$	$\alpha_1\alpha_2$ $\alpha_2^{-1}\alpha_1^{-1}\alpha_{r_v}^{-\varepsilon_v}\dots\alpha_{r_1}^{-\varepsilon_1}\cdot\alpha_1^{-1}\cdot\alpha_{r_v}^{-\varepsilon_v}\dots\alpha_{r_1}^{-\varepsilon_1}$
$\lambda_1 < \Lambda$	μ_5

$\alpha_{r_1}^{\varepsilon_1}\dots\alpha_{r_v}^{\varepsilon_v}\alpha_1\alpha_2$ $\alpha_2^{-1}\alpha_1^{-1}\alpha_{r_v}^{-\varepsilon_v}\dots\alpha_{r_1}^{-\varepsilon_1}\cdot\alpha_1^{-1}\cdot\alpha_{r_v}^{-\varepsilon_v}\dots\alpha_{r_1}^{-\varepsilon_1}$	$\alpha_{r_1}^{\varepsilon_1}\dots\alpha_{r_v}^{\varepsilon_v}\alpha_1\alpha_2$ $\alpha_1^{-1}\alpha_{r_v}^{-\varepsilon_v}\dots\alpha_{r_1}^{-\varepsilon_1}$
μ_4	μ_3

$\alpha_{r_1}^{\varepsilon_1}\ldots\alpha_{r_v}^{\varepsilon_v}\alpha_1\alpha_2$	$\alpha_1\alpha_2$	α_1
α_2	α_2	α_2
μ_2	μ_1	$\lambda_0 \leq \Lambda$

where $\mu_2 = \mu_1 < \lambda_0$, $\mu_4 = \mu_5 = \lambda_1$ and $\mu_3 = \Lambda$ since $\alpha_{r_1}^{\varepsilon_1}\ldots\alpha_{r_v}^{\varepsilon_v}\alpha_1\alpha_2 = \alpha_1\alpha_2 \equiv \beta_2$ because $B_0F = B_vF$, and $\alpha_1^{-1}\alpha_{r_v}^{-\varepsilon_v}\ldots\alpha_{r_1}^{-\varepsilon_1} = \alpha_1$ because $HB_0 = HB_v$. Thus Λ appears in K' only once, namely, with the application of the factor U_{21}^{-1}. We have $K = P_{12}O_2K'$.

(b) $W_{xy}^{\pm1} = U_{21}^{\pm1}$. Let $FH \equiv \alpha_1^{\pm1}$ with the midpoint M_3. Since $\alpha_2\alpha_1^{\pm1} \geq \alpha_2$, M_3 lies on FB_v, and since $FM_2 = \beta_2/2 \leq \alpha_1/2 = FM_3$, M_2 lies on FM_3. The case U_{21} is then excluded because $M_2F \not\equiv M_2B_0$. Thus $W_{xy}^{\pm1} = U_{21}^{-1}$. Since $EF \not\equiv EA$, we then have $EF < EA$, and hence $FB_v < AB_0$, i.e. $\alpha_1\alpha_2^{-1} < \alpha_1$, so $B_0M_1 > B_0C^*$ and $\alpha_1 > \alpha_2$.

If $\alpha_2\alpha_1^{-1} > \alpha_2$, so that $\Lambda > \lambda_0$, then one constructs

$$K' = U_{12}U_{21}^{-1}\cdot U_{2r_v}^{\varepsilon_v}\ldots U_{2r_1}^{\varepsilon_1}\cdot U_{1r_v}^{\varepsilon_v}\ldots U_{1r_1}^{\varepsilon_1}I.$$

Here $L(U_{1r_i}^{\varepsilon_i}\ldots U_{1r_1}^{\varepsilon_1}I) = \lambda_0 < \Lambda$. The factors $U_{2r_i}^{\varepsilon_i}$ can indeed increase the diagram number, but not up to the value Λ; because if one draws $B_0G \equiv \alpha_2^{-1}$, then it does not contain M_1 because $\alpha_2\alpha_1^{-1} > \alpha_2$; thus $GB_i < GA$ and hence $\alpha_2\alpha_{r_1}^{\varepsilon_1}\ldots\alpha_{r_i}^{\varepsilon_i} < \alpha_2\alpha_1^{-1}$. Thus Λ appears only once, namely, with the application of U_{21}^{-1}. The case remaining is $\alpha_2\alpha_1^{-1} = \alpha_2$, so that $\Lambda = \lambda_0$. B_0G branches off B_0A at $M_1 \neq C^*$. Then one constructs

$$K'' = U_{12}^{-1}V_{21}\cdot U_{2r_v}^{\varepsilon_v}\ldots U_{2r_1}^{\varepsilon_1}\cdot U_{12}^{-1}\cdot U_{1r_1}^{-\varepsilon_1}\ldots U_{1r_v}^{-\varepsilon_v}\cdot U_{12}^{-1}I.$$

Here

$$L(U_{1r_i}^{-\varepsilon_i}\ldots U_{12}^{-1}I) = L(U_{12}^{-1}I) < L(I),$$

$$L(U_{12}^{-1}U_{1r_1}^{-\varepsilon_1}\ldots U_{1r_v}^{-\varepsilon_v}U_{12}^{-1}I) < L(K),$$

because $\alpha_2 = \alpha_2\alpha_1^{-1}$, $GF < AF$, not only when M_1 lies on EA but also when M_1 lies on EC^* because $GM_1 < \alpha_1/2$. Also, $L(U_{2r_i}^{\varepsilon_i}\ldots I) = L(U_{12}^{-1}\ldots I)$ because $GB_i = GB_0$, and finally $L(V_{21}\ldots I) < L(K)$ because $GF < AF$. $K = O_1O_2K''$.

VII. $x = r_{h_1} = r_{h_2} = \cdots = r_{h_s}$; $y = 1$. It follows from $\beta_x \geq \alpha_x$ that at least one final half of the generators of α_1 in β_x stands at an end, and it follows from $\beta_x \leq \alpha_1$ that it makes up at least half of β_x.

Let $\beta_x = \alpha_1$. Then $C^* = M_1$, and this point coincides with the midpoints C_{h_i} $(i = 1, \ldots, s)$; one of the segments C^*A or C^*B_0 is thereby identical with one of the segments $C^*B_{h_i-1}$ or $C^*B_{h_i}$. This is only possible for $h_i = 1$, hence for $x = r_1 \neq r_2, \ldots, r_v$. We can then assume $\varepsilon_1 = +1$, since this can always be arranged by conjugation with O_{r_1}, and we then have $W_{xy}^{\pm1} = V_{r_11}^{-1}$, $\beta_{r_1} \equiv B_0B_1$, $\alpha_{r_1} \equiv AB_1$, $C_1 = C^* = M_1$. The core

$$K' = V_{1r_2}^{\varepsilon_2}\ldots V_{1r_v}^{\varepsilon_v}V_{12}V_{1r_1}^{-1}U_{r_12}U_{r_1r_v}^{\varepsilon_v}\ldots U_{r_1r_2}^{\varepsilon_2}I$$

is then normal stationary with respect to $U_{r_12}\ldots U_{r_1r_2}^{\varepsilon_2}$ and the other factors do not evoke Λ, since $FB_0, B_vB_0, \ldots, B_1B_0 \leq AB_0$.

Let $\beta_x < \alpha_1$.

(a) $W_{xy}^{\pm1} = U_{x1}$.

$$K'' = (V_{x2}^{-1}V_{x1}^{-1})_{T_{2r_1}\ldots T_{2r_v}}\cdot V_{1x}I$$

satisfies, where $T_{2r_i} = V_{2r_i}^{\varepsilon_i}$ for $i \neq h_k$ and $T_{2r_i} = V_{21}^{\varepsilon_i}$ for $i = h_k$ $(k = 1, 2, \ldots, s)$. Here $L(V_{1x}I) < L(I)$, the factors T leave the diagram number unaltered, and only the factor V_{x1}^{-1} evokes Λ once.

(b) $W_{xy}^{\pm 1} = V_{x1}$.

$$K''' = (V_{x2}^{-1}U_{x1}^{-1})_{T_{2r_1}\ldots T_{2r_v}} \cdot U_{1x}I$$

analogously to (a), with the same meaning of the T.

(c) $W_{xy} = U_{x1}^{-1}$.

$$K'''' = (U_{x2}V_{x1})_{T_{2r_1}\ldots T_{2r_v}} \cdot U_{1x}^{-1}I,$$

where $T_{2r_i} = V_{2r_i}^{\varepsilon_i}$ for $i \neq h_k$ and $T_{2r_i} = V_{21}^{-\varepsilon_i}$ for $i = h_k$.

(d) $W_{xy}^{\pm 1} = V_{x1}^{-1}$.

$$K''''' = (U_{x2}U_{x1})_{T_{2r_1}\ldots T_{2r_v}} \cdot V_{1x}^{-1}I$$

with the same meaning for the T as in (c).

VIII. $x = r_{h_1} = r_{h_2} = \cdots = r_{h_s}$; $y = 2$. It follows from $\beta_x \geq \alpha_x$ that at least one final half of the generators of α_2 in β_x stands at an end, and it follows from $\beta_x \leq \alpha_2$ that it makes up at least half of β_x.

Let $\beta_x = \alpha_2$. Then $C^* = M_2$ and this point coincides with the midpoints C_{h_i} $(i = 1, \ldots, s)$; one of the segments C^*F or C^*B_v is identical with one of the segments $C^*B_{h_i-1}$ or $C^*B_{h_i}$. This is only possible for $h_i = v$, hence $x = r_v \neq r_1, \ldots, r_{v-1}$. We can then assume $\varepsilon_v = +1$ and we have $W_{xy}^{\pm 1} = U_{r_v 2}^{-1}$, $\beta_{r_v} \equiv B_{v-1}B_v$, $\alpha_{r_v} \equiv B_{v-1}F$, $C_v = C^* = M_2$. Then the core

$$K' = U_{r_v 2}^{-1}U_{1r_v}U_{1r_{v-1}}^{\varepsilon_{v-1}}\cdots U_{1r_1}^{\varepsilon_1}I$$

is normal stationary with respect to $U_{1r_v}U_{1r_{v-1}}^{\varepsilon_{v-1}}\cdots U_{1r_1}^{\varepsilon_1}$.

Let $\beta_x < \alpha_2$. Then one takes

$$K'' = Z \cdot Y \cdot T_{1r_v}\ldots T_{1r_1} \cdot X \cdot I,$$

where $T_{1r_i} = U_{1r_i}^{\varepsilon_1}$ for $i \neq h_k$ and the meaning of the remaining symbols depends on the form of $W_{xy}^{\pm 1}$ according to the following table.

$W_{xy}^{\pm 1}$	X	$T_{1r_{h_k}}$	Y	Z
U_{x2}	V_{2x}	$U_{12}^{\varepsilon_{h_k}}$	V_{x2}^{-1}	U_{1x}^{-1}
U_{x2}^{-1}	U_{2x}^{-1}	$U_{12}^{-\varepsilon_{h_k}}$	V_{x2}	U_{1x}
V_{x2}	U_{2x}	$U_{12}^{\varepsilon_{h_k}}$	U_{x2}^{-1}	U_{1x}^{-1}
V_{x2}^{-1}	V_{2x}^{-1}	$U_{12}^{-\varepsilon_{h_k}}$	U_{x2}	U_{1x}

In all cases here, $L(XI) < L(I)$ because $\beta_x < \alpha_2$, the factors T leave the diagram number unaltered, and the value Λ appears only once, namely, with the application of Y.

IX. $x = r_{h_1} = r_{h_2} = \cdots = r_{h_s}$; $y \neq 1, 2$. We can assume $W_{xy}^{\pm 1} = U_{xy}$, since we can arrive at this by conjugation by one or both of O_x or O_y if necessary (with possible sign changes in some of the ε_i).

(a) $\Lambda > \lambda_0$, so that $\alpha_x\alpha_y > \alpha_x$. If $x \neq r_1$ then

$$K' = U_{12} U_{1r_v}^{\varepsilon_v} \dots U_{1r_2}^{\varepsilon_2} U_{xy} U_{1r_1}^{\varepsilon_1} I$$

satisfies Theorem A, because Λ occurs once less and K' is normal stationary with respect to $U_{12} U_{1r_v}^{\varepsilon_v} \dots U_{1r_2}^{\varepsilon_2}$.

Suppose then that $x = r_1$. Also let $\varepsilon_1 = +1$.

One draws $B_0 H \equiv \alpha_{r_1}$. Let G be the point at which $B_0 H$ branches off $B_0 B_1$. If G is a point of $B_0 C_1$ different from C_1, and at the same time $GH \leq GB_0$, or if G lies on $C_1 B_1$, in which case $GH < GB_1$ because $\alpha_{r_1}\alpha_y > \alpha_{r_1}$, then one puts

$$K'' = U_{12} U_{1r_v}^{\varepsilon_v} \dots U_{1r_2}^{\varepsilon_2} U_{r_1 y} U_{1y} U_{1r_1} I.$$

If G lies on $B_0 C_1$, not at C_1, and if at the same time $GH > GB_0$, then $\alpha_{r_1}\alpha_y < \alpha_y$. Then one takes

$$\begin{aligned} K''' &= U_{12} U_{1r_v}^{\varepsilon_v} \dots U_{1r_2}^{\varepsilon_2} P_{r_1 y} O_{r_1} V_{r_1 y}^{-1} U_{1y} V_{yr_1} I \\ &= U_{12} U_{1r_v}^{\varepsilon_v} \dots U_{1r_2}^{\varepsilon_2} U_{yr_1} U_{1r_1} V_{r_1 y}^{-1} P_{r_1 y} O_{r_1} I. \end{aligned}$$

The case $\varepsilon_1 = -1$ is settled completely analogously.

(b) $\Lambda = \lambda_0$.

(b_1) $\alpha_x = \alpha_x\alpha_y < \alpha_y$. If $r_1 \neq x$ and $\neq y$ then one constructs

$$\begin{aligned} K' &= U_{12} U_{1r_v}^{\varepsilon_v} \dots U_{1r_2}^{\varepsilon_2} P_{xy} O_x V_{xy}^{-1} U_{1r_1}^{\varepsilon_1} V_{yx} I \\ &= U_{12} U_{1r_v}^{\varepsilon_v} \dots U_{1r_2}^{\varepsilon_2} U_{yx} U_{1r_1}^{\varepsilon_1} V_{xy}^{-1} I^{(1)}, \end{aligned}$$

where $I^{(1)} = P_{xy} O_x I$ and the factors $P_{xy} O_x$ herein can be carried to the right of I. We now have

$$L(U_{1r_1}^{\varepsilon_1} V_{xy}^{-1} I^{(1)}) = L(V_{xy}^{-1} I^{(1)}) < L(I^{(1)}) = \Lambda,$$

the factor U_{yx} again evokes Λ, and Λ is normal stationary with respect to $U_{12} U_{1r_v}^{\varepsilon_v} \dots U_{1r_2}^{\varepsilon_2}$. Thus one can continue by case (a).

If $x = r_1$ then one constructs

$$\begin{aligned} K'' &= U_{12} U_{1r_v}^{\varepsilon_v} \dots U_{1r_2}^{\varepsilon_2} \cdot P_{r_1 y} O_x V_{xy}^{-1} U_{1r_1}^{\varepsilon_1} V_{yx} I \\ &= U_{12} U_{1r_v}^{\varepsilon_v} \dots U_{1r_2}^{\varepsilon_2} \cdot U_{yr_1} U_{1r_1}^{\varepsilon_1} V_{r_1 y}^{-1} (P_{r_1 y} O_{r_1} I), \end{aligned}$$

which satisfies just as before.

Finally, if $y = r_1$ and, e.g., $\varepsilon_1 = +1$, then one draws $B_0 H \equiv \alpha_x^{-1}$; this must branch off $B_0 B_1 \equiv \alpha_y \equiv \alpha_{r_1}$ at C_1, because $\alpha_x\alpha_{r_1} = \alpha_x$, and so we must have $C_1 H < C_1 B_0$ because $\alpha_x\alpha_{r_1} < \alpha_{r_1}$. Then one takes

$$\begin{aligned} K''' &= U_{12} U_{1r_v}^{\varepsilon_v} \dots U_{1r_2}^{\varepsilon_2} \cdot P_{r_1 x} O_x V_{xr_1}^{-1} U_{1r_1} U_{1x}^{-1} V_{r_1 x} I \\ &= U_{12} U_{1r_v}^{\varepsilon_v} \dots U_{1r_2}^{\varepsilon_2} \cdot U_{r_1 x} U_{1x} U_{1r_1} V_{xr_1}^{-1} (P_{r_1 x} O_x I). \end{aligned}$$

The case $\varepsilon_1 = -1$ is settled completely analogously, by drawing $\alpha_x^{-1} \equiv B_1 H$. It again branches off $B_1 B_0$ at C_1 and $C_1 H < C_1 B_1$.

(b_2) $\alpha_x = \alpha_x\alpha_y \geq \alpha_y$. Figure 9 represents $\alpha_x\alpha_y$. Here we have $P_4 P_2 = P_4 P_3$ because $\alpha_x\alpha_y = \alpha_x$, and $P_4 P_1 \geq P_4 P_2$ because $\alpha_x\alpha_y \geq \alpha_y$. The midpoint P_5 of $P_1 P_3$ therefore lies on $P_1 P_4$ and it can in particular coincide with P_4, in case $\alpha_x\alpha_y = \alpha_y$. Then one constructs

$$K' = U_{xy} U_{12} T_{1r_v} \dots T_{1r_1} I,$$

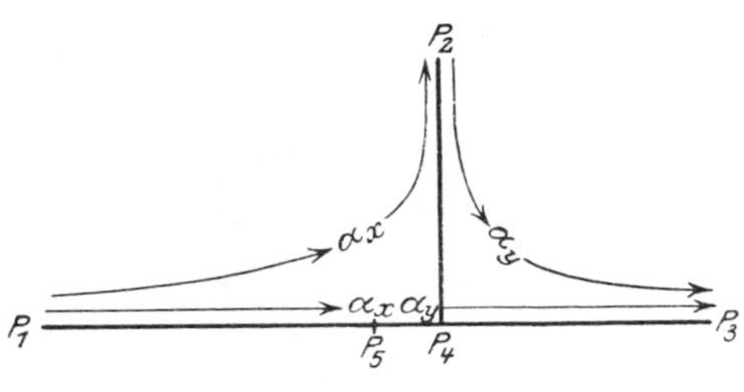

Figure 9

where $T_{1r_i} = U_{1r_i}^{\varepsilon_i}$ if $r_i \neq x = r_{h_k}$, $T_{1r_{h_k}} = U_{1y}U_{1r_{h_k}}$ for $\varepsilon_{h_k} = +1$ and $T_{1r_{h_k}} = U_{1r_{h_k}}^{-1}U_{1y}^{-1}$ for $\varepsilon_{h_k} = -1$. The meaning of this relation is obviously that one attaches in K' first α_x and then α_y where and only where in K the element $\beta_x = \alpha_x\alpha_y$ (e.g. with exponent $+1$) has to be attached to the first element. Now if $P_4 \neq P_5$ then K' is normal stationary with respect to $U_{12}\dots T_{1r_1}$. In order to see this, one considers an r_{h_k} $(=x)$, with $\varepsilon_{h_k} = +1$, say; then $\beta_{r_{h_k}} \equiv \alpha_x\alpha_y \equiv B_{h_k-1}B_{h_k} \equiv P_1P_3$ of Figure 9. If one brings P_1P_3 to $B_{h_k-1}B_{h_k}$, then P_5 coincides with C_{h_k} and P_4 separates C_{h_k} from B_{h_k}. Thus $AP_2 = AP_5 + P_5P_4 + P_4P_2 = AP_4 + P_4P_3 = AB_{h_k}$, and if one attaches α_x to AB_{h_k-1} the diagram number does not alter. All places at which the point P_2 can occur therefore have the same distance from A_1 as the B_i, and since $\alpha_x = \alpha_x\alpha_y \leq \alpha_2$, K' is normal stationary with respect to $U_{12}\dots T_{1r_1}$.

If $P_4 = P_5$ then it is still conceivable that all positions at which P_2 can occur have the same distance from A as B_0. However, other positions can also occur. In fact $P_4 = P_5 = C_{h_k}$ and now when P_4P_2 starts out on $C_{h_k}A$, a position arises for P_2 at which P_2 and A are not separated by C_{h_k}, so that $AP_2 < AB_0$. Let $h_1 < h_2 < \cdots < h_s$ and let h_m be the smallest of these values for which one such position of P_2 occurs. Suppose $\varepsilon_{h_m} = +1$. (The case of -1 is settled quite analogously.) The first applied factor $U_{1r_{h_m}}$ of $T_{1r_{h_m}}$ then evokes a diagram number $<\Lambda$ in K'. We therefore still have to show how K', which is therefore not normal stationary in this case, can be changed to another relation so as to enable the process to continue. Let l be the greatest index $<h_m$ such that $\beta_{r_l} > \beta_{r_{h_m}}$ $(=\beta_x)$, if such exists. Then r_l is different from $x, y, r_{l+1}, \dots, r_{h_m-1}$. Also, none of the indices r_j $(j = l+1, l+2, \dots, h_{m-1})$ equals x or y because for $r_j = x$ or y we would have C_j coincident with C_{h_m}, and at this point, which would correspond to $P_4 = P_5$ in Figure 9, like-labelled segments depart differently. We can assume $\varepsilon_l = +1$, since we can arrive at this by conjugation with O_{r_l} if necessary, because $r_l \neq x, y, 1, 2$. Then we first construct the subcore

$$K_1I = U_{r_lr_{h_m}}U_{r_lr_{h_m-1}}^{\varepsilon_{h_m-1}}\dots U_{r_lr_{l+1}}^{\varepsilon_{l+1}}I.$$

We have $L(K_1I) < L(I) = \Lambda$ and K_1I is a normal stationary core with respect to K_1. We now change K to

$$K'' = U_{xy}(U_{12}T_{1r_v}\dots T_{1r_1})_{K_1}I,$$

where $T_{1r_i} = U_{1r_i}^{\varepsilon_i}$ for $r_i \neq x$ and $\neq r_l$, $T_{1r_i} = (U_{1y}U_{1x})^{\varepsilon_i}$ for $i = h_k$ (and hence for $r_i = x$), and finally $T_{1r_i} = (U_{1r_{l+1}}^{-\varepsilon_{l+1}}\dots U_{1r_{h_m-1}}^{-\varepsilon_{h_m-1}}U_{1r_{h_m}}^{-1}U_{1r_l})^{\varepsilon_i}$ for $r_i = r_l$. Then none of the factors $U_{12}T_{1r_v}\dots T_{1r_1}$ in K'' produces a greater length for the first element in the isomorphism than $L(\alpha_1) = AB_0$, but the length AB_0 of the first element corresponds in K'' to a diagram number $<\Lambda$, since the length of α_{r_l} is reduced by the previous application of K_1. Thus in K'', Λ appears only normal stationary inside K_1, and not at all elsewhere, which establishes that the process can be continued by bringing in the frontmost factor of I.

It was assumed above that such an index l could be found. If this is not the case then it follows, first, that $h_m = h_1$, and then that the subcore (for $\varepsilon_{h_1} = +1$)

$$K_2 I = U_{1r_{h_1}} U_{1r_{h_1-1}}^{\varepsilon_{h_1-1}} \cdots U_{1r_1}^{\varepsilon_1} I$$

is normal stationary with respect to K_2. Then one constructs

$$K'''(U_{12})_{V_{2y} V_{2r_{h_1+1}}^{\varepsilon_{h_1+1}} \cdots V_{2r_v}^{\varepsilon_v}} \cdot U_{xy} \cdot K_2 I.$$

Here K''' is normal stationary with respect to K_2, $L(K_2 I) < L(I) = \Lambda$, and the remaining factors no longer evoke the value Λ, whence the possibility of continuation by bringing in the frontmost factor of I is established.

This brings the proof of the lemma to a conclusion, and with it the proof of completeness of the relation system (13), (14), (17) for the isomorphism group Γ_n.

ZUR TOPOLOGIE DER GESCHLOSSENEN FLÄCHEN

VON

JAKOB NIELSEN

Die in diesem Vortrag besprochenen topologischen Probleme beziehen sich auf geschlossene zweiseitige Flächen vom Geschlechte $p > 1$. Wir charakterisieren die Punkte der Fläche durch eine komplexe Koordinate x in folgender Weise:

Sei E der Einheitskreis der x-Ebene, Φ das Innere von E, B ein reguläres $4p$-seitiges Polygon mit $x = 0$ als Zentrum und dem Polygonwinkel $\frac{\pi}{2p}$, gebildet aus Bögen von auf E senkrechten Kreisen. Die Seiten von B seien fortlaufend numeriert. Mit $a_1, a_2, \ldots, a_p, b_1, b_2, \ldots, b_p$ seien lineare Substitutionen von x bezeichnet, die E in sich transformieren und durch folgende Festsetzung bestimmt sind: a_1 führt die dritte Polygonseite in die erste über, a_2 die siebente in die fünfte u. s. w. Ebenso führt b_1 die zweite Polygonseite in die vierte, b_2 die sechste in die achte u. s. w. über. Und zwar ist jedesmal nicht die Drehung um $x = 0$ gemeint, sondern die andere Substitution, die das Geforderte leistet. Jede der Substitutionen lässt dann zwei Punkte auf E invariant; für a_1 z. B. sind das die beiden Punkte, in denen der zur ersten und dritten Polygonseite und E senkrechte Kreis K auf E trifft. Im Sinne einer auf E als unendlich fernes Gebilde gegründeten hyperbolischen Metrik, bei der Φ das konforme Abbild der hyperbolischen Ebene ist, ist a_1 also eine »Verschiebung« mit der »Geraden« K als »Schiebungsachse«.

Die a_i, b_i sind erzeugende Operationen einer unendlichen Gruppe F. Sie genügen dabei der Beziehung

$$a_1\, b_1\, a_1^{-1}\, b_1^{-1}\, a_2\, b_2\, a_2^{-1}\, b_2^{-1} \ldots a_p\, b_p\, a_p^{-1}\, b_p^{-1}(x) = x$$

als einziger unabhängiger Relation. Die Bilder von B bei allen Operationen von F bilden ein unendliches nichteuklidischreguläres Netz, dessen Maschen sich gegen E häufen. Punkte von Φ, die sich bei einer Operation aus F entsprechen, sollen äquivalent heissen. Die Gesamtheit der Systeme äquivalenter Punkte bildet eine geschlossene zweidimensionale Mannigfaltigkeit φ vom Geschlecht p, eine »geschlossene Fläche«. Diese hat Φ als universelle Überlagerungsfläche und F als Fundamentalgruppe. — Man kann diese geschlossene Fläche φ in abstracto auffassen oder auch an die Einführung einer komplexen Koordinate auf einer im Raum gelegenen geschlossenen Fläche φ vom Geschlecht p denken, indem man diese nach Anbringung eines kanonischen Schnittsystems mit einem gemeinsamen Punkt zu dem Polygon B in eine bis auf den Rand eineindeutige und stetige Beziehung setzt.

Einem stetigen Kurvenstück

$$x = k(t),\ 0 \leqq t \leqq 1,\ |k(t)| < 1,$$

in Φ entspricht ein stetiges Kurvenstück auf φ; demselben Kurvenstück auf φ entspricht jedes mit dem ersteren äquivalente Kurvenstück in Φ. Sind die Endpunkte in Φ äquivalent, z. B.

$$k(1) = f(k(0)),\quad f \text{ ein Element von } F,$$

so wird eine geschlossene Kurve auf φ dargestellt. Dieser Kurve ist also das Element f, oder mit demselben Recht jedes aus f durch Transformation innerhalb F hervorgehende Element, also eine ganze »Elementklasse«, zugeordnet. Zwei geschlossene Kurven sind homotop, d. h. auf φ in einander deformierbar, wenn sie zu derselben Elementklasse von F gehören *(Poincaré)*. Das Homotopieproblem führt also auf das gruppentheoretische Problem, zu entscheiden, ob zwei Elemente von F ineinander transformierbar sind. Dieses Problem ist rein kombinatorisch gelöst *(Dehn)*.

Ist es möglich, auch offene, unendliche Kurven auf der Fläche ihrem topologischen Verlauf nach in ähnlicher Weise mittels der Fundamentalgruppe zu charakterisieren? Um eine Antwort hierauf zu finden, wollen wir als Zwischenglied ein Normalbild einer Kurve betrachten, das in den Untersuchungen von *Poincaré* und *Dehn* eine wesentliche Rolle spielt.

Jedes Element von F hat zwei getrennte Fixpunkte auf E. Der zu E senkrechte Kreisbogen in Φ, der sie verbindet, heisse die »Achse« des Elements. Die abzählbare Menge der Achsenendpunkte erfüllt E überall dicht. Jeder Achsenendpunkt gehört nur einer bestimmten Achse und damit einer unendlichen zyklischen Untergruppe von F an. Nun ist jeder zu E senkrechte Kreis Bild einer geodätischen Linie in der mit hyperbolischer Metrik ausgestatteten geschlossenen Mannigfaltigkeit φ. Die Achsen der F-Elemente, und nur diese, sind Bilder geschlossener geodätischer Linien, indem die Grundpotenz der zu einer Achse gehörigen unendlichen zyklischen Untergruppe auf der Achse ein »Verschiebungsstück« bestimmt, das einer einmaligen Durchlaufung der geschlossenen geodätischen Linie auf φ entspricht. Ein auf E senkrechter Kreisbogen, der in einem Achsenendpunkt endigt, ohne selbst Achse zu sein, stellt auf φ eine geodätische Linie dar, die asymptotisch gegen eine geschlossene geodätische Linie verläuft. Wie ist der Verlauf einer geodätischen Linie auf φ zu charakterisieren, wenn ihr Bild in Φ gegen einen Punkt von E strebt, der nicht Achsenendpunkt ist? Hier erweist sich, wie zu erwarten ist, eine Bestimmung der Lage der einzelnen Punkte auf E mittels der Gruppe F erforderlich, die wie folgt vorgenommen werden kann:

Man teile E in $4p$ gleich grosse Segmente, die der Reihe nach mit A_1, B_1^{-1}, A_1^{-1}, B_1, A_2, B_2^{-1}, A_2^{-1}, B_2, ..., B_p bezeichnet werden, und zwar so, dass die Mittelpunkte von A_1, B_1^{-1},... beziehungsweise mit den Punkten $a_1(0)$, $b_1^{-1}(0)$,.... auf demselben Radius liegen. Nun sei Q ein beliebiger Punkt von E. Er liege etwa in A_1. Dann sucht man $a_1^{-1}(Q)$ auf. Fällt dieser Punkt z. B. in B_2, so sucht man $b_2^{-1}\, a_1^{-1}(Q)$ auf. Fällt dieser etwa in B_1^{-1}, so fährt man mit $b_1\, b_2^{-1}\, a_1^{-1}(Q)$ fort u. s. w. So bestimmt Q eindeutig eine unendliche Erzeugendenfolge:

$$a_1\; b_2\; b_1^{-1} \ldots\ldots$$

(Für die abzählbar vielen Punkte, die bei diesem Prozess einmal in einen der benutzten Teilungspunkte auf E fallen, kann man, wenn man will, das Verfahren leicht durch besondere Verabredungen eindeutig machen). Die Bedeutung dieser »Entwickelung des Randpunktes Q nach der Gruppe F« beruht darauf, dass für jedes $|x| < 1$ die Folge

$$a_1(x), \quad a_1 b_2(x), \quad a_1 b_2 b_1^{-1}(x), \ldots.$$

gegen Q konvergiert.

Diese Entwickelung nach F wird für alle Achsenendpunkte, und nur für diese, periodisch. Die Periode entspricht der zu der Achse gehörigen Grundpotenz. So wie hiermit die geschlossenen und die gegen solche asymptotisch verlaufenden geodätischen Linien auf φ durch die Gruppenentwickelung ihres »Unendlichkeitspunktes« charakterisiert sind, gilt das Gleiche auch im allgemeinen Fall. Hieraus fliessen zunächst metrische Resultate, von denen ich die beiden folgenden anführe: [1])

1^0. *Es gibt offene geodätische Linien auf φ, die jedem Punkt von φ beliebig nahe kommen. Dabei liegen überdies noch die Richtungen der geodätischen Normalen beliebig kleiner Länge vom Punkte auf die Linie überall dicht um den Punkt. Dieser Fall tritt ein, wenn in der F-Entwickelung des Randpunktes allen Elementen von F entsprechende Ausschnitte vorkommen.*

2^0. *Zu jeder vorgegebenen Länge ε gibt es geschlossene geodätische Linien auf φ, die jedem Punkte von φ näher als bis auf den Abstand ε kommen.*

Der Begriff des durch eine F-Entwickelung bestimmten »Unendlichkeitspunktes« einer gerichteten Kurve hat topologische Bedeutung für eine ausgedehnte Klasse von Kurven:

Unter einer »offenen Halbkurve« verstehen wir eine durch eine stetige Funktion

$$x = k(t), \ 0 \leqq t < \infty, \ |k(t)| < 1,$$

dargestellte Kurve. Zwei offene Halbkurven wird man auf φ homotop nennen, wenn man Darstellungen $x = k(t)$ und $x = l(t)$ für sie so wählen kann, dass der nichteuklidische Abstand zwischen zwei zum selben t-Wert gehörigen Punkten auf Φ beschränkt bleibt, da man alsdann, wie unmittelbar ersichtlich, einen stetigen Übergang von der einen zur anderen mittels einer einparametrigen Schar von offenen Halbkurven bewerkstelligen kann. Das wird zunächst immer dann eintreten, wenn die Bilder der beiden Kurven ganz innerhalb eines Kreises um den Nullpunkt mit einem Radius < 1 verlaufen. Wir denken

[1]) Elementargeometrische Beweise dieser Sätze ohne Benutzung der Randpunktentwickelung findet man in meinem Aufsatz in der Juel-Festschrift, Mat. Tidsskrift B, 1925.

daher im Folgenden an solche Halbkurven, deren Bilder in Φ Grenzpunkte auf E haben, speziell an eine Halbkurve K, deren Bild $x = k(t)$ nur einen Grenzpunkt Q auf E hat, so dass dabei $k(t) \rightarrow Q$ für $t \rightarrow \infty$ und $k(t)$ ganz innerhalb eines Kreisbogenzweiecks mit Spitzen auf E (die eine in Q) verläuft. Eine solche Halbkurve K nennen wir »richtungsbestimmt« auf φ und Q ihren »Richtungspunkt« für die gewählte Darstellung. (Wählt man $x = f(k(t))$ anstatt $x = k(t)$ als Darstellung für K, so wird Q durch $f(Q)$ ersetzt). Da die Seiten des Kreiszweiecks nichteuklidische Abstandslinien sind, ist eine solche Halbkurve homotop mit ihrer nichteuklidischen Projektion auf irgend eine in Q endigende geodätische Linie. Notwendige und hinreichende Bedingung für die Homotopie zweier richtungsbestimmter Halbkurven ist daher zunächst die Äquivalenz ihrer Richtungspunkte und danach die Homotopie ihrer Projektionen auf irgend eine geodätische Linie »gleicher Richtung« (ein eindimensionales Problem). Hier hat uns nur die erste Bedingung zu beschäftigen, und die Antwort ist diese: Zwei Randpunkte sind dann und nur dann äquivalent vermittelst F, wenn ihre F-Entwickelungen gleich endigen, d. h. bis auf endliche Anfangsabschnitte (nicht notwendig gleicher Länge) übereinstimmen.

In dieser Form umfasst die Homotopiebedingung auch die auf φ geschlossenen Kurven, und in diesem speziellen Fall ist die erste Bedingung allein wegen des periodischen Verlaufes der Bilder in Φ ausreichend. *Dehns* kombinatorische Lösung des Transformationsproblems spielt dabei dadurch hinein, dass alle F-Entwickelungen »reduzierte« Erzeugendenausdrücke im *Dehn*schen Sinne sind[1]).

Man dehnt diese Betrachtungen leicht auf nach beiden Seiten offene und richtungsbestimmte Kurven aus.

Wir wollen nun topologische Abbildungen der Fläche φ auf sich betrachten. Solche werden dargestellt durch Abbildungsfunktionen $y = g(x)$, welche Φ topologisch auf sich abbilden und dabei in äquivalenten Punkten äquivalente Werte haben. Dies Letztere spricht sich dadurch aus, dass eine solche Abbil-

[1]) M. Dehn, Transformation der Kurven auf zweiseitigen Flächen. Math. Annalen 72 (1912).

dungsfunktion $g(x)$ unendlich vielen Funktionalgleichungen der Form

$$g(f(x)) = f'(g(x))$$

genügt, wo f und f' Elemente von F sind, deren Zuordnung zu einander, wenn f die Gruppe F durchläuft, einen Isomorphismus von F mit sich selbst konstituiert. Die durch $g(x)$ dargestellte Abbildung von φ auf sich wird auch durch jede »mit g konjugierte« Abbildungsfunktion $g_1(x) = f_1(g(x))$ dargestellt, wo f_1 irgend ein Element aus F ist. g_1 »induziert« dabei den Isomorphismus

$$f \rightarrow f_1 f' f_1^{-1},$$

den wir mit dem ersteren zur selben »Klasse« rechnen. Jede Abbildung von φ auf sich gehört so mit einer bestimmten Isomorphismenklasse von F zusammen. Zwei Abbildungen, die in diesem Sinn zur selben Klasse gehören, sind in einander deformierbar, wie neuerdings *R. Baer* bewiesen hat.

Es ist nun von entscheidender Bedeutung für die hier behandelten Fragen, dass die durch eine Abbildungsfunktion bewirkte topologische Abbildung von Φ auf sich (wobei die obigen Funktionalgleichungen gültig sind) auf den Rand E erweitert werden kann.[1]) Die topologische Abbildung von E auf sich, die sich so stetig an die gegebene von Φ anschliesst, hängt dabei nur von dem induzierten Isomorphismus ab, ist also ein Charakteristikum der Klasse von Flächenabbildungen. Diese Abbildung von E ist dadurch bestimmbar, dass die Achsenendpunkte des Elements f in diejenigen von f' übergehen, wodurch die zyklische Anordnung der Achsenendpunkte auf E nicht geändert wird.

Wir benutzen im Folgenden von den Abbildungsfunktionen nur die Stetigkeit und die Gültigkeit der Funktionalgleichungen, nicht aber die eindeutige Umkehrbarkeit der topologischen Abbildungen. Das Folgende gilt also für beliebige stetige Abbildungen, wenn nur die zugehörigen Abbildungsfunktionen den Funktionalgleichungen genügen. Auch in diesem allgemeineren Fall ist nämlich die Erweiterung auf den Rand möglich, und die Randabbildung ist dieselbe, allein durch den Isomorphismus bestimmte wie im topologischen Spezialfall. Man sieht das un-

[1]) Siehe des Vf. Aufsatz in den »Abhandl. aus dem Math. Seminar der Hamburgischen Univ.« III, 1924.

mittelbar im einfachsten Fall, wo $g(x)$ den identischen Isomorphismus induziert:

$$g(f(x)) = f(g(x)), \tag{1}$$

wo also g mit allen Elementen von F vertauschbar ist. Bildet man nämlich die Funktion

$$\frac{|x - g(x)|^2}{(1 - |x|^2)(1 - |g(x)|^2)},$$

so bestätigt man leicht, dass sie sich bei gleichzeitiger linearer Transformation von x und $g(x)$ nicht ändert; sie hat also wegen (1) in äquivalenten Punkten den gleichen Wert; und da ihr Wert in einem Fundamentalbereich von F beschränkt ist, ist er in ganz Φ beschränkt. Daraus folgt $g(x) \to x$ für $|x| \to 1$. Die stetige Abbildung $g\Phi$ wird also durch die identische Abbildung von E abgeschlossen.

Wir fragen jetzt bei einer gegebenen stetigen Abbildung von φ auf sich selbst nach Punkten, die mit ihrem Bildpunkt zusammenfallen. Dieses »Fixpunkteproblem« hat in neuerer Zeit in der Mechanik bei der Bestimmung periodischer Bewegungen in einem dynamischen System mit 2 Freiheitsgraden eine Rolle gespielt. Wir wollen die Annahme machen, dass unter den Gebieten, die auf φ bei Fortnahme der Fixpunktmenge übrigbleiben, eines das ursprüngliche Geschlecht p hat. Es heisse ω und das Teilgebiet von Φ, das die entsprechenden Koordinatenwerte enthält, Ω.

Nun sei x ein Punkt von Ω, $y = g(x)$ der (von x verschiedene) Bildpunkt. Wir legen einen zu E senkrechten Kreis durch x und y. Durchläuft man diesen in der Richtung von x nach y, so sei $z = z(x)$ der Punkt, in dem man auf E stösst, also

$$z(x) = \frac{x\bar{y} - \bar{x}y + |(x - y)(1 - x\bar{y})|}{\bar{x}y(x - y) - (\bar{x} - \bar{y})}.$$

Diese Funktion ist, wie unmittelbar ersichtlich, auf solche Teile des Randes E fortsetzbar, wo $y \neq x$, und in solchen Randpunkten wird $z = y$. Wir wollen daher die fixpunktefreien Intervalle auf E mit zur Punktmenge Ω rechnen. Man kann sagen, dass wir es hier mit einer »Arcusvariation« zu tun haben, insofern wir die Richtung von Punkt zu Bildpunkt durch einen Punkt des unendlichfernen Kegelschnitts repräsentieren. Bei simultaner nichteuklidischer Bewegung von Punkt und Bildpunkt nimmt der »Richtungspunkt« an der Bewegung Teil. Beschreibt x einen geschlossenen Weg in Ω, so beschreibt z einen geschlossenen Weg auf E. Ist der erstere in Ω zusammenziehbar, so ist der letztere auf E zusammenziehbar. Lässt sich der letztere stetig in eine von Null verschiedene Anzahl von Umläufen um E überführen, so hat der erstere Fixpunkte der Abbildung $g\Phi$ umschlossen. Wir messen daher die Variation von $z(x)$ mittels der in ganz Ω definierten reellen Funktion

$$\xi(x) = \frac{1}{2\pi i} \log z(x).$$

Jedem stetigen Weg in Ω wird das $\int d\xi$ längs des Weges und jedem geschlossenen gerichteten Weg in Ω also eine ganze Zahl, der »Index« des geschlossenen Weges, zugeordnet. Zur Bestimmung von Fixpunkten für $g\Phi$ gilt es also, geschlossene Wege mit einem von Null verschiedenen Index zu finden. Da nun die Gültigkeit der Funktionalgleichungen das Einzige ist, was man ausser der Stetigkeit von der Abbildung $g\Phi$ weiss, ist man darauf angewiesen, eine Bestimmung von $\int d\xi$ durch gleichzeitige Betrachtung äquivalenter Wegstücke zu ermöglichen. Man sieht aber sofort aus den Funktionalgleichungen, dass dabei als äquivalenzvermittelnde Funktionen nur solche Elemente von F in Betracht kommen, die mit g vertauschbar sind. Aus der obigen Bemerkung über simultane Bewegung von Punkt und Bildpunkt folgert man:

Ist f ein Element aus F, das die Gleichung (1) erfüllt, und bildet man die ebenfalls in ganz Ω definierte Funktion

$$\delta_f(x) = \xi(f(x)) - \xi(x),$$

so ist $\int d\delta_f$ unabhängig vom Integrationsweg in Ω. Der Index eines geschlossenen Weges lässt sich auf Integrale dieser Art zurückführen, wenn der Weg aus Stücken zusammengesetzt ist, die paarweise mittels solcher mit g vertauschbaren Gruppenelemente äquivalent sind. Wir wählen als solchen Weg daher den Rand eines Fundamentalbereichs für die Untergruppe der beim Isomorphismus ungeänderten Elemente.

Die analoge Bestimmung ist dann für die mit $g(x)$ konjugierten Abbildungsfunktionen, d. h. für die anderen Isomorphismen der Isomorphismenklasse auszuführen. Hierbei ist aber Folgendes zu beachten[1]):

Es sei hg eine zu g konjugierte Funktion (also h ein Element aus F). Diese habe in x_1 einen Fixpunkt:

$$hg\,x_1 = x_1 .$$

Dann ist für beliebiges f aus F:

$$gf x_1 = f'g\,x_1 = f'h^{-1}x_1 = f'h^{-1}f^{-1}\cdot fx_1 ,$$

also hat die zu g konjugierte Funktion $fhf'^{-1}g$ eine Fixpunktmenge, die aus derjenigen von hg durch f hervorgeht. Wir nennen die Gruppenelemente h und fhf'^{-1} und ebenso die durch sie hervorgerufenen zu g konjugierten Funktionen »isogredient« mit Bezug auf den gegebenen Isomorphismus. Für festes h und variables f erhalten wir eine Isogredienzklasse, der wir somit eine »Fixpunktklasse« der gegebenen Abbildung von φ zuordnen können. Nun gilt: Jede Abbildung von φ hat nur endlich viele Fixpunktklassen, d. h. die oben formal bestimmten Fixpunktklassen sind bis auf endlich viele leer.[2])

Damit ist folgender Weg zur Lösung des allgemeinen Fixpunktproblems vorgezeichnet:

> *Gegeben ist eine stetige Abbildung von φ auf sich, etwa durch Angabe einer Abbildungsfunktion $g(x)$. I: $(f \rightarrow f')$ sei der durch g induzierte Isomorphismus von F, U die Untergruppe von F, deren Elemente bei I ungeändert bleiben. Der Index des Randes eines Fundamentalbereichs von U ist eine allein von I abhängige ganze Zahl, die den Beitrag einer Fixpunktklasse zur Indexsumme misst. Man wiederhole das Verfahren für die anderen Isomorphismen $I_{(h^{-1})}$: $(f \rightarrow hf'h^{-1})$ der Isomorphismenklasse von I, wo h je einen Repräsentanten aller Isogredienzklassen bezüglich I durchläuft. Dabei führen nur endlich viele h auf einen von Null verschiedenen Beitrag zur Indexsumme (topologisch notwendige Fixpunktklassen).*

Ich bespreche kurz einige einfache Beispiele:

Beispiel 1: Die Deformationen. Hier gibt es unter den konjugierten Abbildungsfunktionen eine, $g(x)$, die den identischen

[1]) Ich unterdrücke im Folgenden die Klammern um die Funktionsargumente.

[2]) Den Beweis siehe in der auf Seite 268 zitierten Arbeit.

Isomorphismus induziert. Die stetige Abbildung $g(\Phi + E)$ ist also auf dem Rande die Identität. E besteht ganz aus Fixpunkten.

Hier ist die Untergruppe U mit F identisch. Bringt man in ω ein kanonisches Schnittsystem an, so überträgt es sich auf Ω als Rand eines Fundamentalbereiches von F mit Seiten, die paarweise mittels eines Fixelements äquivalent sind. Das obige Verfahren der Zurückführung auf Integrale der Form $\int d\,\delta_f$, die durch einfache Kreisbögen auf E gemessen werden, liefert vermittels einer einfachen Abzählung den Index $2-2p$.

Die konjugierten Abbildungsfunktionen hg induzieren Isomorphismen $I_{\langle h^{-1}\rangle} : (f \rightarrow hfh^{-1})$. U ist dabei die zu der Achse von h gehörige unendliche zyklische Untergruppe von F, und die Endpunkte dieser Achse sind die einzigen Fixpunkte auf E. Ein Fundamentalbereich für U wird also abgegrenzt durch einen beliebigen Querschnitt von E zu E in Ω, der die beiden Fixpunkte trennt, und sein Bild bei der Grundpotenz von U, und der Index bestimmt sich sofort zu Null. (Das gilt allgemein immer dann, wenn E bis auf die Endpunkte einer Achse von Fixpunkten frei ist). Wir haben also:

Bei den Deformationen einer Fläche vom Geschlecht p ist die Indexsumme über alle Fixpunktklassen $2-2p$. Für $p \neq 1$ gibt es also immer Fixpunkte. Für $p > 1$ rührt die Indexsumme vom Beitrag einer Fixpunktklasse her, und eventuelle andere Fixpunktklassen (die in endlicher Zahl auftreten können) haben den Index Null. Diese letzteren lassen sich zum Verschwinden bringen, und die eine topologisch notwendige Klasse lässt sich, wie man leicht sieht, durch einen einzigen Punkt repräsentieren.

Dies ist der topologische Teil der in der *Birkhoff*schen Stabilitätsformel enthaltenen Aussage. Dieser topologische Satz ergibt sich auf diesem Wege, ohne dass man einschränkende Voraussetzungen irgendwelcher Art einzuführen braucht ausser der obigen Voraussetzung der Existenz eines fixpunktefreien Gebietes ω vom Geschlecht p auf φ. Und diese Voraussetzung ist natürlich und notwendig. Wenn nämlich durch die Fortnahme der Fixpunktmenge von φ der topologische Charakter der Fläche vollkommen zerstört wird, — im extremen Fall kann ja die Abbildung die Identität als Spezialfall einer Deformation sein, — so entfällt damit natürlich die Möglichkeit, die Rolle, die die einzelnen Fixpunkteklassen spielen, durch Indizes zu messen.

Beispiel 2: Das Gegenstück des identischen Isomorphismus ist der Isomorphismus ohne von der Identität verschiedene Fixelemente, z. B. Isomorphismen endlicher Ordnung. Wir nehmen als Beispiel:

$$I: \begin{pmatrix} a_i \to a_{i+1} \\ b_i \to b_{i+1} \end{pmatrix}, \ (i \bmod p).$$

Hier ist $I^p = 1$. Dieser Isomorphismus wird z. B. durch eine Drehung um den Nullpunkt durch den Winkel $\frac{2\pi}{p}$ induziert. Aus der Periodizität von I folgt, dass E fixpunktefrei, also I fixelementefrei ist. Hier ist U also die Identität und Φ der Fundamentalbereich von U. Wir haben somit den Index von E selbst zu bilden (E gehört ja hier ganz zu Ω), und dieser ergibt sich bei positivem Umlauf zu 1. (Dies ist mit anderen Worten einfach eine Anwendung des bekannten *Brouwer*schen Satzes). Wir betrachten dann weiter die anderen Isomorphismen der Klasse von I:

$$I_{\langle h^{-1}\rangle}: \begin{pmatrix} a_i \to h\, a_{i+1}\, h^{-1} \\ b_i \to h\, b_{i+1}\, h^{-1} \end{pmatrix}, \ (i \bmod p).$$

$I_{\langle h^{-1}\rangle}$ hat das Fixelement

$$k = h \cdot I(h) \cdot I^2(h) \ldots\ldots I^{p-1}(h),$$

wo $I(h)$, $I^2(h)$, u. s. w. die Elemente bedeuten, in die h bei I, I^2, u. s. w. übergeht. Hier wissen wir nun, dass die Randabbildung hgE als Superposition zweier linearen Transformationen (g ist nämlich auf E einfach eine p-tel Drehung) höchstens zwei Fixpunkte auf E hat. Ist also $k \neq 1$, so ist die zur Achse von k gehörige unendliche zyklische Untergruppe von F die gesuchte Gruppe U, und analog wie im Beispiel 1 führen alle solche h auf die Indexsumme Null. Es bleiben also nur solche h noch zu untersuchen, für die $k = 1$ ist, also Lösungen der Isomorphiegleichung

$$h \cdot I(h) \cdot I^2(h) \ldots\ldots I^{p-1}(h) = 1\,.$$

Nun zeigt eine einfache gruppentheoretische Analyse, dass die Lösungen dieser Isomorphiegleichung entweder die Form $l\, l'^{-1}$ oder die Form $l \cdot a_1\, b_1\, a_1^{-1}\, b_1^{-1} \cdot l'^{-1}$ haben, wo l ein willkürliches Element aus F ist und $l' = I(l)$. In beiden Fällen ist, wie durch Ausrechnung folgt, $(I_{\langle h^{-1}\rangle})^p = 1$, sodass der Index 1 wird, ebenso wie für I selbst. Die erste dieser beiden Fixpunktklassen ist aber die schon bestimmte, indem $h = 1$ in der Form $l\,l'^{-1}$ enthalten ist, nämlich für $l = 1$. Es kommt also nur eine Fixpunktklasse entsprechend der zweiten Lösungsform hinzu. Alle aus der p-tel Drehung durch Deformation hervorgehenden Abbildungen haben also zwei topologisch notwendige Fixpunktklassen, jede vom Index 1.

Beispiel 3: Ich bespreche noch ein etwas allgemeineres Beispiel, bei dem die Konstruktion eines Fundamentalbereichs für U etwas eingehendere Untersuchungen erforderlich macht. $g(x)$ sei eine Abbildungsfunktion, die den Isomorphismus $I:(a_1 \to a_1\, b_1)$ induziert. (Die übrigen $2p-1$ Erzeugenden sollen in sich selbst übergehen). Hier ist U die freie Gruppe mit den $2p-1$ Erzeugenden $b_1, a_2, b_2, \ldots, a_p, b_p$. Um einen Fundamentalbereich für U abzugrenzen, müssen wir notwendig Stücke von E zur Hülfe nehmen, aber solche Stücke müssen, um brauchbar zu sein, zu Ω gehören, d. h. fixpunktefrei sein. Wir müssen also die Struktur der Abbildung gE näher untersuchen. Dazu betrachten wir die Achsen der Elemente a_1, b_1 und $a_1\, b_1\, a_1^{-1}$. Diese liegen zu einander so wie in der Figur angegeben. (Die Figur entspricht dem Fall $p = 2$).

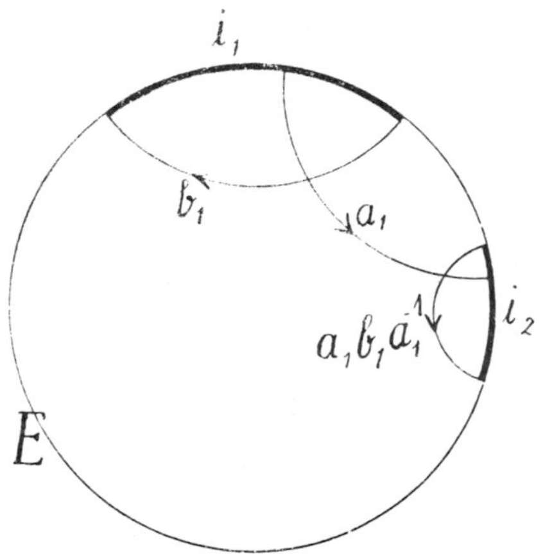

Nun ist

$$I(a_1) = a_1\, b_1$$
$$I^n(a_1) = a_1\, b_1^n.$$

Für $n \to \infty$ und $n \to -\infty$ konvergiert der eine Endpunkt der Achse von $a_1\, b_1^n$ gegen die Endpunkte der Achse von b_1, der andere gegen die Endpunkte der Achse von $a_1\, b_1\, a_1^{-1}$. Diese beiden Achsen begrenzen also fixpunktfreie Intervalle i_1 und i_2 auf E. Ihre Endpunkte sind Fixpunkte auf E, die diese Intervalle begrenzen. Die Funktionalgleichungen zeigen nun, dass bei Ausübung aller Elemente aus U die Intervalle i_1 und i_2 wieder in fixpunktfreie Intervalle übergehen, begrenzt von Fixpunkten, nämlich den Achsenendpunkten derjenigen Elemente, in die b_1 und $a_1\, b_1\, a_1^{-1}$ durch

Transformation mit eben jenen Elementen aus U übergehen. Diese Intervallmenge liegt überall dicht auf E. Zwei verschiedene Intervalle können dabei keinen Endpunkt gemeinsam haben, sodass es keine isolierten Fixpunkte auf E gibt. Die Fixpunktmenge auf E, bestehend aus 1) den abzählbar vielen Intervallendpunkten und 2) den Häufungspunkten für diese, ist also eine nirgends dichte perfekte Menge.

Wir betrachten die fixpunktfreien Intervalle etwas näher. Jedes derselben wird bei g topologisch auf sich abgebildet, so dass dabei alle Punkte in der gleichen Richtung verschoben werden. Nun sind bekanntlich topologische fixpunktfreie Abbildungen zweier offener Intervalle immer ähnlich. Hier gilt aber insbesondere, dass die Abbildung in einem beliebigen dieser Intervalle aus der Abbildung $g i_1$ oder $g i_2$ durch Transformation mit einem F-Element hervorgeht. Und für die Abbildungen $g i_1$ und $g i_2$ selbst gilt, dass sie aus einander durch Transformation mit einer (einem bestimmten Isomorphismus entsprechenden) Abbildungsfunktion hervorgehen. Hierdurch ergibt sich: Bildet man die Gesamtheit der Transformierten von h_1 mit den Elementen von F, so lässt sich die Abbildung $g E$ in jedem Punkt von E durch Superposition einer endlichen oder unendlichen Folge aus dieser Gesamtheit darstellen. Die zu benutzende Folge variiert dabei von Punkt zu Punkt.

Nachdem man so einen Einblick in die Struktur der Abbildung $g E$ bei dem gegebenen Isomorphismus erhalten hat, gelingt es leicht, einen Fundamentalbereich für U unter Zuhülfenahme zweier fixpunktefreier Randstücke herzustellen und den zugehörigen Index durch eine reine Abzählung zu ermitteln. Das Resultat ist auch in diesem Fall $2 - 2p$, und ebenso wie im ersten Beispiel liefern auch hier die konjugierten Abbildungsfunktionen keinen neuen Beitrag zur Indexsumme, was ich an dieser Stelle nicht näher ausführe.

Zum Schluss will ich es nicht unterlassen, auf den Zusammenhang dieser Untersuchungsrichtung mit der von *I. W. Alexander*[1]) eingeschlagenen hinzuweisen. *Alexander* zielt unter Beschränkung auf topologische Abbildungen und unter Einführung anderer vereinfachender Voraussetzungen von vorneherein auf die ge-

[1]) Transact. of the American Math. Soc. vol 25 (1923).

18*

samte Indexsumme. Demgegenüber scheint mir die Aufteilung der Fixpunktmenge in endlich viele Klassen und die Bestimmung des Index jeder Klasse zu einer Vertiefung des Problems zu führen.

Den Reiz dieser Untersuchungen sehe ich neben dem eigentlich topologischen Interesse vor allem darin, dass sie auf natürliche Weise zu interessanten und fruchtbaren gruppentheoretischen Problemstellungen führen.

Reprinted from
Den sjette skandinav. Matematikerkongres København 1925, 263–276

Investigations in the Topology of Closed Orientable Surfaces, I*

JAKOB NIELSEN

Contents

---

*Translation by John Stillwell of Untersuchungen zur Topologie der geschlossenen zweiseitigen Flächen. [Acta Math., 50(1927), 189–358.]

Introduction

The basic problem of topology, the formulation of necessary and sufficient conditions for manifolds to be homeomorphic, is well known to be completely solved for two-dimensional manifolds, and in essence only for these, so they have prematurely offered themselves for deeper study. The present work is confined, for the sake of simplicity, to the study of *closed, orientable* surfaces of arbitrary genus, but it is not difficult to see how one may extend the methods to other cases of two-dimensional manifolds. The subject of the investigation is principally the *continuous mapping* of the surface onto itself and the resulting *fixed point problem*, a rational formulation of which is obtained through the concept of the fixed point *class* and its *index*. The results of G. D. Birkhoff and J. W. Alexander which are relevant here will be fully discussed. The explicit, combinatorial solution of the fixed point problem, in the generality in which it can be conceived, is not yet reached. Nevertheless, many rather far-reaching special cases may be handled by a method which moreover also applies to each specially posed problem, as a few typical examples will show. It may therefore be justified to publish these investigations in detail. I have had the occasion to present the first results in two lectures in Hamburg 1924[1] and Copenhagen 1925[2].

The tools of the investigation are the basic notions of fundamental group and universal covering surface. This shifts all point set theoretic considerations to the plane, and the combinatorial part of the investigation then assumes a clear group theoretic form. At the same time it becomes clear that the difficulties which obstruct the general solution of the topological problems considered here are the same as those which arise with the simplest questions in the theory of infinite discontinuous groups. Whether the solution of the topological questions is to be expected from a further development of general group theory, or whether a direct geometric solution of these questions will suggest new methods for group theory, remains uncertain[3]; in the present work the latter tendency predominates.

H. Poincaré[4] has provided surface topology with powerful tools by giving the universal covering surface the non-Euclidean metric and creating the concept of the fundamental group. M. Dehn[5] has systematically applied these tools to the solution of important problems, extending them by the introduction of the "group diagram", and established their close relationship with general group theory. The works of both these authors have been my starting point, however the first section of the present work contains an exposition of what is needed for our purposes, so that the reader can essentially avoid the earlier works[6]. It now proves useful to *close* the universal covering surface by addition of its "points at infinity"; in

[1] "Über topologische Abbildungen geschlossener Flächen". Abhandlungen aus dem mathematischen Seminar der Hamburgischen Universität, 3, 1924.

[2] "Zur Topologie der geschlossenen Flächen", address to the 6th Scandinavian Congress of Mathematicians, Copenhagen 1925, published in the Congress report, Jul. Gjellerups Forlag, Köbenhavn 1926, [N15].

[3] Cf. a remark of M. Dehn, Math. Annalen **71**, p. 119.

[4] See in particular: Cinquième Complément à l'Analysis Situs, Rend. Pal. XVIII, 1904.

[5] Math. Annalen 69, 71 and 72.

[6] To complement our presentation of the fundamentals we refer to the following: M. Dehn: Über diskontinuierliche Gruppen, Math. Ann. 71, 1911, in particular p. 119–122, as well as the introduction and first three paragraphs of the dissertation of Dehn's student: H. Gieseking: Analytische Untersuchungen über topologische Gruppen, Münster 1912.

order to make these conveniently accessible, a conformal mapping of the hyperbolic plane onto the interior of a disc is used. Just as proper elements of the fundamental group are given by finite products of the generators, the attempt to characterize individual boundary points leads us to consider improper group elements, given by *infinite sequences of generators.* In many cases the investigation also leads to *weaker* covering surfaces, which depend on the surface mapping under consideration. As mentioned above, the investigation brings with it certain simply expressed group theoretic problems, which in many cases are treated successfully by geometric considerations. It seems to me that the method is capable of further development, and may perhaps be of service in general group theory.

CHAPTER I

The Universal Covering Surface as a Coordinate Surface

1. Canonical cut systems. We begin with an intuitive consideration, as preparation for the precise definition of the object of our investigation in §3. Consider a closed surface of genus p. Figure 1 shows the case $p = 3$. The p pairs of directed canonical curves $A_1, B_1; A_2, B_2; \ldots; A_p, B_p$ are shown issuing from a point Q of the surface. If one cuts the surface along these curves then the dissected surface may be spread out on the plane, so that it proves to be homeomorphic to a $4p$-gon. Figure 2 shows one such, whose upper side corresponds to the outer side of the surface in Figure 1. If one traverses the boundary of the dissected surface with the

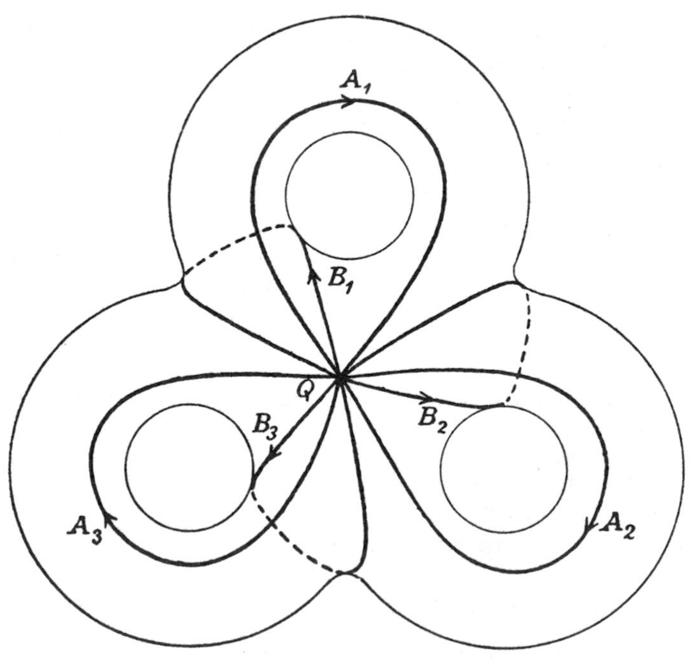

Figure 1

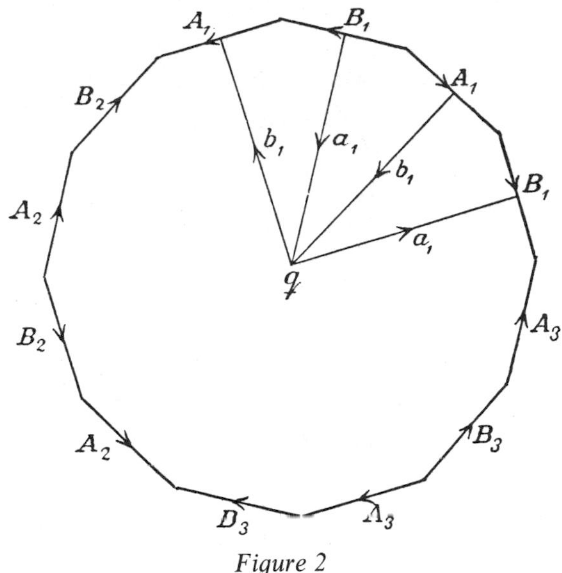

Figure 2

surface itself on one's left, seen from the outside, and if one denotes the traversal of e.g. A_1 against the direction of the arrow by A_1^{-1}, then Figure 2 shows that the canonical curves occur in the order

$$B_1^{-1}A_1^{-1}B_1A_1B_2^{-1}A_2^{-1}B_2A_2\dots B_p^{-1}A_p^{-1}B_pA_p.$$

One now chooses a point q in the interior of the $4p$-gon and connects it with two boundary points of the $4p$-gon which correspond to the same point (other than Q) of the curve B_1. The result on the surface is a closed curve cutting B_1 once, which will be denoted a_1 and oriented in such a way that it crosses B_1 from right to left. A curve b_1 is constructed analogously, crossing A_1 from left to right and meeting a_1 only at the point q. The two curves are shown in Figure 2. Using the same point q one constructs further curves $a_2, b_2, \dots, a_p, b_p$ by cyclic repetition of the process. If one imagines these curves shown on Figure 1, when the point q is taken on the rear side of the surface, and the surface is then turned round so that the rear side faces the viewer, then the system a_i, b_i appears as in Figure 3. This system is a new canonical cut system; if one again dissects the surface along it to produce a $4p$-gon, then the boundary of the latter reads

$$a_1b_1a_1^{-1}b_1^{-1}a_2b_2a_2^{-1}b_2^{-1}\dots a_pb_pa_p^{-1}b_p^{-1}$$

when one traverses it with the polygon interior on one's *right*. In this polygon the curves A_i, B_i represent rays from an interior point Q (analogous to the a_i, b_i in Figure 2); the ordering of directed curve segments around Q is therefore the same as that determined by the neighbourhood of Q in Figure 1.

We call the systems A_i, B_i and a_i, b_i "dual canonical cut systems".

2. Linear fractional transformations and the hyperbolic metric. Let x be a complex variable. The unit circle $|x| = 1$ in the x-plane is denoted by E, the interior of E by Φ, and the exterior

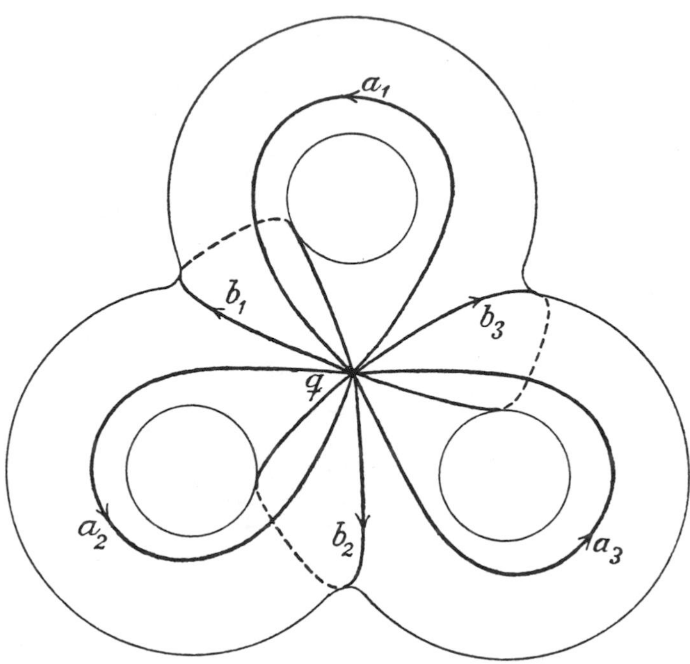

Figure 3

of E by Ψ. In order to account for the connectivity properties of the x-plane closed by a point $x = \infty$ we shall sometimes speak of $\Phi + E + \Psi$ as the x-sphere K.

We shall deal with transformations of K which map Φ, E and Ψ into themselves. In particular, we consider linear fractional transformations of x with this property. Each such transformation can be written in the form

$$S(x) = \frac{\alpha x + \bar{\beta}}{\beta x + \bar{\alpha}}, \qquad \alpha\bar{\alpha} - \beta\bar{\beta} > 0, \tag{1}$$

where α and β are arbitrary complex numbers which are determined up to a common real factor, and $\bar{\alpha}$ denotes the complex number conjugate to α. Moreover, each transformation of this form has the desired property. The above transformation is denoted by the symbol $\binom{\alpha}{\beta}$. The inverse transformation is $S^{-1} = \binom{\bar{\alpha}}{-\beta}$.

The invariant points or, as we shall say, the "fundamental points" of S, i.e. solutions to $S(x) = x$, are

$$\xi_{1,2} = \frac{1}{2\beta}(\alpha - \bar{\alpha} \pm \sqrt{\Delta}), \qquad \Delta = (\alpha - \bar{\alpha})^2 + 4\beta\bar{\beta}.$$

The pencil of circles through ξ_1 and ξ_2 is mapped into itself by the circle-preserving transformation S; the same is therefore true of the orthogonal family of circles. Now if $\Delta > 0$ both fundamental points lie on E, and since angles are preserved, each circle in the pencil through the fundamental points is mapped into itself by S. The orthogonal circles are permuted among themselves, each of their points being moved along a circle of the pencil away from the "negative" fundamental point and towards the "positive" fundamental point. If $\Delta < 0$ the fundamental points lie inversely with respect to E, and in this case E belongs to the orthogonal family. Finally, if $\Delta = 0$ the fundamental points coincide on E. As is well known, these three cases are called hyperbolic, elliptic and parabolic transformations.

In the case of the hyperbolic transformation we call the circle perpendicular to E through the fundamental points the "axis" of S.

The transformations of the form (1) constitute a group, which will be denoted by Γ.

Let S and T be two transformations in Γ. Then the fundamental points of TST^{-1} are images of the fundamental points of S under T. Transformations which are conjugate in Γ are therefore of the same kind.

We now view the interior Φ of E as conformal image of the "hyperbolic plane", i.e. the plane of N. Lobatschevsky, in the usual way: two figures in Φ are called congruent when one is mapped onto the other by a transformation in Γ. Occasionally we also view Ψ as a conformal image of the same plane (with reversal of angle), by regarding the points which are inverses with respect to E as images of the same point of the plane.

For the sake of completeness we now recall the usual non-Euclidean distance; let x_1 and x_2 be two points in Φ, and let S be a transformation in Γ which maps x_1 to 0, so that $S = \binom{\alpha}{-\bar{\alpha}\bar{x}_1}$ and hence

$$S(x_2) = \frac{\alpha(x_2 - x_1)}{\alpha(1 - \bar{x}_1 x_2)}.$$

The distance $d(x_1, x_2)$ is now defined as the logarithm of the cross ratio formed on the diameter:

$$d(x_1, x_2) = \log\frac{1 + |S(x_2)|}{1 - |S(x_2)|},$$

thus

$$\sinh^2\frac{d}{2} = \left(\frac{e^{d/2} - e^{-d/2}}{2}\right)^2 = \frac{|S(x_2)|^2}{1 - |S(x_2)|^2} = \frac{(x_2 - x_1)(x_2 - \bar{x}_1)}{(1 - x_1\bar{x}_1)(1 - x_2\bar{x}_2)} = \delta(x_1, x_2).$$

$d(x_1, x_2)$ then satisfies the usual axioms for distance. For many purposes one can also use the simpler function $\delta(x_1, x_2)$.

The geodesics in this metric are the circles perpendicular to E (including the diameters of E); thus these correspond to the lines of the plane. A hyperbolic transformation corresponds to a translation of the plane along the line represented by the axis of the transformation. The non-Euclidean distance $d(x, S(x))$ between a point and its image is the same for all points on the axis; it is called the "displacement length" of the transformation. For all points outside the axis the distance to the image point is greater, and in fact increasingly so, the further the point is from the axis. S and TST^{-1} have equal displacement lengths since a point and its image on the axis of S are carried into a point and its image on the axis of TST^{-1} by the conjugating transformation T. The hyperbolic transformation S commutes with a non-identity transformation T:

$$ST = TS$$

if and only if S and TST^{-1} are the same. T must therefore have the same fundamental points as S; and it is consequently a hyperbolic transformation with the same axis as S. This condition is also sufficient:

Two hyperbolic transformations commute if and only if they have the same axis.

Now let S and T be two hyperbolic transformations which have the same positive fundamental point A, but different negative fundamental points. Then TST^{-1} also has the positive fundamental point A, but it has a negative fundamental point different from those of S and T. The axes of S and TST^{-1} are therefore non-Euclidean parallels with A as a common point at infinity, so they approach each other asymptotically. The transformation $S \cdot TS^{-1}T^{-1}$, the "commutator" of S and T, then also has A as fundamental point, and hence can only be hyperbolic or parabolic. If one now recalls that S and TST^{-1} have the same displacement length, and if x is a point on the axis of TST^{-1}, then one sees that $d(x, S \cdot TS^{-1}T^{-1}(x))$ becomes arbitrarily small if x is sufficiently close to the axis of S, i.e. when it is translated along the axis of TST^{-1} sufficiently far in the direction of A. Now since the distance between point and image under a hyperbolic transformation has the displacement length as a lower bound, it follows that $STS^{-1}T^{-1}$ cannot be hyperbolic:

If two hyperbolic transformations have only one fundamental point in common, then their commutator is a parabolic transformation with the common fundamental point as its single fundamental point.

The reader can, of course, also confirm these theorems by direct calculation.

In what follows, the transformations which occur are almost exclusively hyperbolic. An

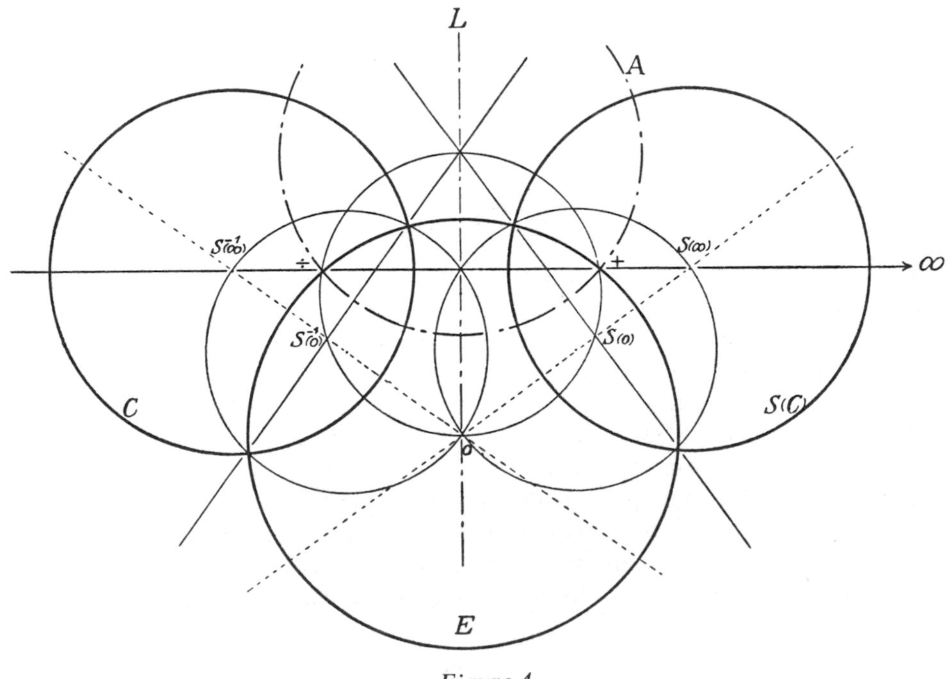

Figure 4

illustration of the simplest properties of such a transformation is given in Figure 4; a motion S (without reflection) of the hyperbolic plane into itself is given when one gives the images $S(x_1)$ and $S(x_2)$ of two points x_1 and x_2 of Φ; we must have

$$d(S(x_1), S(x_2)) = d(x_1, x_2).$$

E.g. one can determine S by the point $S(0)$ and the point $S^{-1}(0)$ which is mapped into 0. $S(0)$ and $S^{-1}(0)$ must have the same non-Euclidean, and hence in this case the same Euclidean, distance from 0. (The Euclidean rotation about the origin belongs to the group Γ.) The condition for S to be hyperbolic is that the circle through $S^{-1}(0)$, 0 and $S(0)$ intersect E. The points of intersection, denoted $\div$ and $+$ in the figure, are respectively the negative and positive fundamental points of S. Mirror images with respect to E are preserved by S. Figure 4 is mapped into itself by reflection in E as well as by reflection in the line L. The mirror image with respect to E of the circle through $S^{-1}(0)$, 0 and $S(0)$ is the line through $S^{-1}(\infty)$, ∞ and $S(\infty)$, and it therefore intersects E at the fundamental points. The circle C with centre $S^{-1}(\infty)$ and orthogonal to E is mapped onto the circle $S(C)$ with centre $S(\infty)$, orthogonal to E, since ∞ is the mirror image of the centre of C with respect to C. The image of a point of C under S is therefore its mirror image with respect to L. C is therefore the set of those points at which the Euclidean line element is congruently transformed by S. For points inside C the Euclidean line element is stretched by S, for those outside it is shortened. In particular, we remark that a subarc of E inside C is mapped by S onto a larger subarc of E. C is the non-Euclidean perpendicular bisector of the segment bounded by $S^{-1}(0)$ and 0. The axis of S must go through the fundamental points and be orthogonal to E. The centre of the axis circle A is therefore on the symmetry line L and the circle through $S^{-1}(0)$, 0 and $S(0)$. Since the line connecting it to $S^{-1}(0)$ is orthogonal to $(S^{-1}(0), 0)$, A is also orthogonal to C and $S(C)$. The arc of A between C and $S(C)$ represents the displacement length.

Obviously one can carry out the operation S by first making a reflection in L and then a reflection in $S(C)$. The result of these two operations is obviously an operation in Γ, and e.g. all points of C are correctly transformed.

The part of the x-sphere K between C and $S(C)$ is a fundamental domain for the infinite cyclic subgroup of Γ consisting of all powers of S, which we denote by $\{S\}$. If one identifies each point of C with the corresponding point of $S(C)$, the fundamental domain closes to a *torus*. If one lets all points which result from the same torus point by operations in S be mapped to it, then this defines a covering of the torus by the x-sphere punctured at the fundamental points. Conversely, we shall say that the torus results from "reduction of K by the group $\{S\}$". We denote the torus by $K \bmod \{S\}$ for short.

3. The fundamental group. This paragraph introduces a special subgroup F of Γ, consisting of infinitely many hyperbolic transformations.

Let v be an angle $<\pi/4$ with vertex O (Figure 5), and draw the circle OQR with centre P on one arm of the angle. The tangent at Q cuts OP at T. Around T one draws the circle κ' with radius TQ. The perpendicular to OR at R meets κ' at W. Suppose that the unit of length is chosen so as to make $OW = 1$. (Then, as one easily computes, $OP = \frac{1}{2}\sqrt{1 - \tan^2 v}$.) Finally, let E be the unit circle through W with centre O. E cuts TQ at V, and κ' cuts OR at X.

Since

$$(TW)^2 = (TQ)^2 = TR \cdot TO,$$

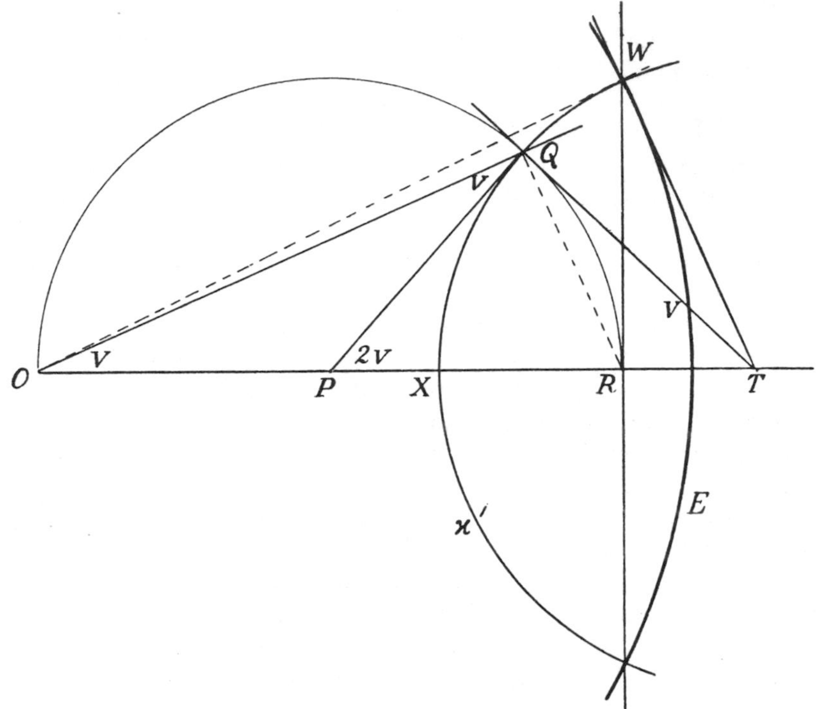

Figure 5

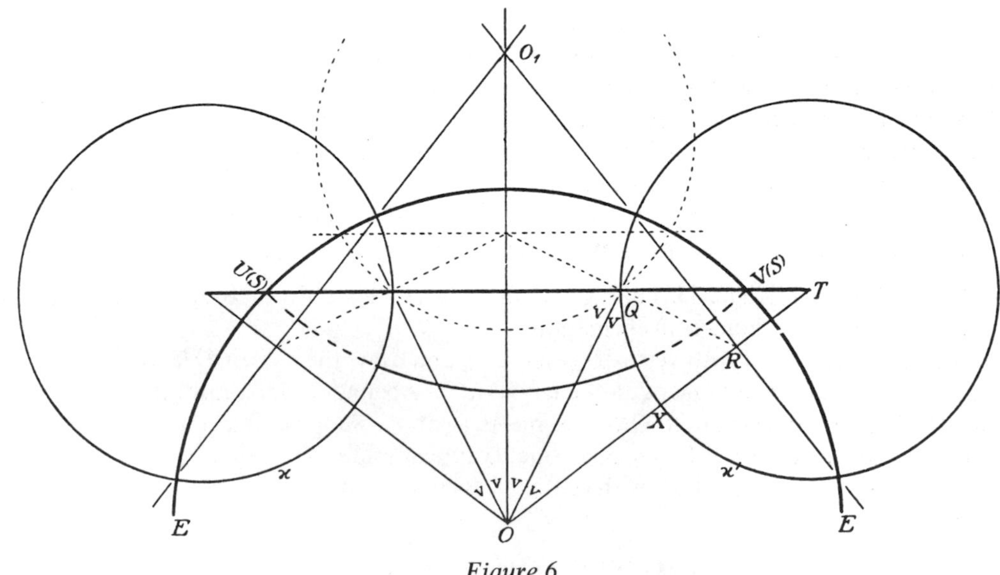

Figure 6

κ' and E meet orthogonally. Thus in the non-Euclidean metric, κ' is a line and OXQ is a right-angled isosceles triangle with base angle v.

Now we reflect Figure 5 in OQ, omitting lines which are superfluous in what follows. The mirror image of OT then becomes parallel to PQ, and hence perpendicular to TQ. We then reflect the whole figure again in this perpendicular and obtain Figure 6 as a result. If we compare Figure 6 with Figure 4, then we see that we can define a hyperbolic transformation S which carries the circle κ in Figure 6 into $\kappa' : \kappa' = S(\kappa)$. The role of C for S in Figure 4 is now played by κ. The positive fundamental point of S is $V(s)$, (denoted by V in Figure 5), the negative one is $U(S)$. The axis joining them has centre O_1, the displacement length is shown. We have $R = S(0)$, $T = S(\infty)$, and κ' is the non-Euclidean perpendicular bisector of OR.

Now we choose v so that continuation of this reflection process exactly fills the angle round O, and yields an even number of circle pairs. Namely, let p be an integer greater than 1 and

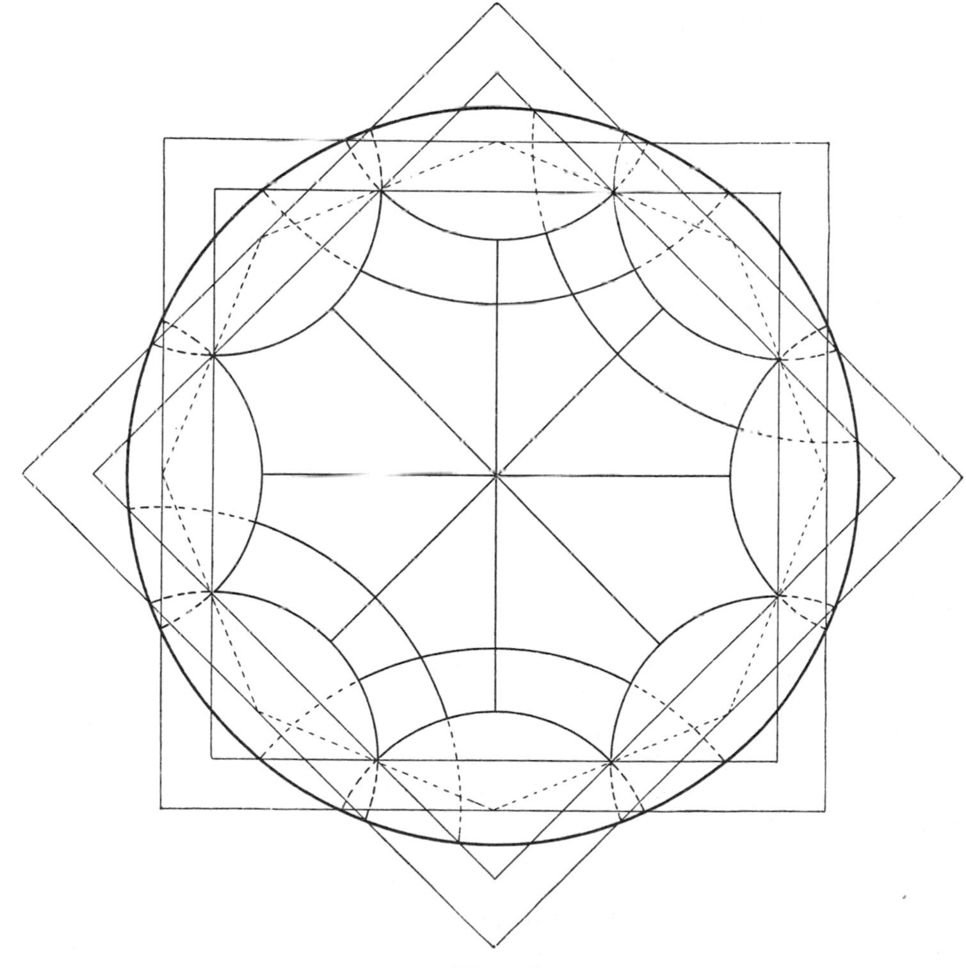

Figure 7

choose $v = \pi/4p$. Then we have $2p$ circle pairs like κ and κ', and $2p$ transformations like S, whose fundamental points pairwise separate each other. Figure 7 shows the picture for $p = 2$. The part of Φ common to the exteriors of the $4p$ circles is a non-Euclidean regular $4p$-gon with corner angle $\pi/2p$, denoted by B. The construction lines retained in Figure 7 clarify the relations explained above between the mutual positions of the images of 0 and ∞ under the $2p$ transformations, the fundamental points, the centres of axis circles, the vertices of the polygon, and the intersections of the polygon sides with E.

There is now no difficulty in obtaining the coefficients α and β in the form (1) of our $2p$ transformations; these come from the fact that the Euclidean distance of the point $x = 0$ from its image under all transformations and their inverses is $\sqrt{1 - \tan^2 \pi/4p}$. In the work cited above[1], an arithmetic expression is given. Here we can do without an explicit expression for the transformations by proceeding geometrically.

We denote the sides of the polygon B by $A_1, B_1, \ldots, A_p, B_p$, as was illustrated for $p = 3$ by Figure 2 in §1. B is placed in Φ in such a way that the positive real axis is the perpendicular bisector of B_1 and the vertex where B_1 meets A_p is the first vertex beneath the positive real axis. The $2p$ transformations are denoted by

$$a_1, b_1, \ldots, a_p, b_p$$

where a_1 maps the third side into the first, b_1 the second into the fourth, a_2 the seventh into the fifth, b_2 the sixth into the eighth, etc. The infinite group *generated* by these $2p$ transformations is called F:

$$F = \{a_1, b_1, \ldots, a_p, b_p\}.$$

It may be mentioned in passing that, as is clear from the foregoing, one can construct the image of an arbitrary point of the x-plane under an arbitrary element of F by straightedge and compass, for all p for which the division of the circle into p equal parts is so constructible.

Two point sets in the x-plane which map onto each other by operations in F will be called "equivalent with respect to F," or, usually, just "equivalent."

Any of the $4p$ operations $a_i^{\pm 1}$, $b_i^{\pm 1}$ maps B onto an equivalent region which has a polygon side in common with B. If B′ is a region equivalent to B, say B′ $= f$(B), where f is an arbitrary element of F, then $fa_i^{\pm 1}f^{-1}$, $fb_i^{\pm 1}f^{-1}$ send B′ to equivalent regions which each have a polygon side in common with B′. The images of B under all elements of F cover Φ without gaps, in a regular non-Euclidean polygon net in which $4p$ polygons meet at each vertex, and which we call N. The cells of the net accumulate towards E. If we consider the point—call it H—on the boundary of B where B_1 meets A_p, then the other cell of the net which is bounded by B_1 is a_1(B), by definition of a_1. The other side of this cell which meets H corresponds to the fourth side A_1 in B. A_1 bounds b_1(B). The next cell in the cycle around H is therefore a_1b_1(B). Now we have a free net edge meeting H which corresponds to B_1, hence the next cell to be attached is $a_1b_1a_1^{-1}$(B). If we continue the cycle in this way and recall that the $(4p-1)^{\text{th}}$ cell is b_p(B), then we come to the relation

$$a_1b_1a_1^{-1}b_1^{-1}a_2b_2a_2^{-1}b_2^{-1}\ldots a_pb_pa_p^{-1}b_p^{-1} = 1 \tag{2}$$

between the generators of F. The left side of (2) will henceforth be denoted by R. Since each relation between generators of F may obviously be derived in the well known way by successive circuits round vertices of the net, $R = 1$ is the single independent relation in F. Thus F is defined as an abstract group by $R = 1$.

If one extends the labelling A_i, B_i of the sides of B to all edges in N, then the order of labels around any vertex of the net is the same as that of the ends of the equally denoted curves round the point Q shown in Figure 1.

Figure 7 shows $4p$ segments issuing from the origin and ending at the midpoints of the polygon sides. One extends these equivalently to all the cells of N. In the neighbouring polygons B and a_1(B), two of these line segments unite into a segment connecting 0 with $a_1(0)$. We give it the orientation from 0 to $a_1(0)$ and the notation a_1. Similarly, the segment from $a_1^{-1}(0)$ to 0 is also called a_1, and analogously $b_1, a_2, \ldots, b_p$. This notation is carried over equivalently to the connecting segments between any two neighbouring points equivalent to 0. These segments yield a net n which, apart from labelling, is non-Euclideanly congruent to N. The meaning of n is the following: Let $f(0)$ be a point equivalent to $x = 0$. One can—in infinitely many ways—connect the origin to $f(0)$ by a path consisting of edges of n, and for each of these paths write a sequence of generators of F, in the order in which they are encountered as labels of edges in the path. Each one of these expressions represents the element f. If w is an edge path in the net, connecting $f(0)$ with $f_1(0)$, then

$$f_1 f^{-1} = fwf^{-1}$$

is the element of F which sends $f(0)$ to $f_1(0)$. Traversal of the path R from the origin leads back to the origin because $R = 1$. R is therefore the clockwise boundary of a cell of n, and the point H considered above is the midpoint of this cell. The ordering of labelled sides around a vertex of n is given by the neighbourhood of the point q in Figure 3.

n is the *Dehn group diagram* of F.[7]

If one now compares these relationships with the introductory considerations in §1, it is immediately clear that one can produce a closed surface of genus p from the region B by bending and identification of equivalent boundary points, and that in so doing the curve segments A_i, B_i, a_i, b_i of the "dual nets," N and n yield dual canonical cut systems on the surface. In order to have the advantage of applying metric ideas immediately to the surface, we shall define "surface" somewhat more abstractly as follows:

> If one considers each complete *system* of points of Φ which are equivalent with respect to F as *one point*, then the result of this "reduction of Φ by F" is a closed, orientable, two-dimensional manifold of genus p, endowed with a hyperbolic metric, the "surface $\phi = \Phi$ mod F." F is the "fundamental group" and Φ the "universal covering surface" of ϕ.

4. Curve types and conjugacy classes. Here we add a brief discussion of the connection between curves on ϕ and on Φ, the full clearing up of which is due to H. Poincaré (*loc. cit.*[4]): a continuous arc in Φ is represented by a continuous function k of one parameter:

$$x = k(t), \qquad 0 \le t \le 1, \qquad |k(t)| < 1.$$

[7] One can complement the above resumé with Dehn's account[5] of fundamentals.

This defines at the same time a continuous arc on ϕ. However, the same arc on ϕ is obtained from each image set

$$x = f[k(t)]$$

equivalent to the first in Φ, where f is an arbitrary element of F. If the initial and final point of the arc in Φ are equivalent via an element h of F:

$$k(1) = h[k(0)],$$

then the curve on ϕ is closed. Then in the second representation used above, fhf^{-1} is the element of F which carries the initial point of the arc in Φ to the final point. *In this way, each closed curve on ϕ is associated with a complete system of mutually conjugate elements of F, a "conjugacy class."* Two closed curves on ϕ which belong to the same conjugacy class are "homotopic", i.e. deformable into each other on ϕ, and conversely. In fact, we can choose the representations of the two curves in Φ

$$\left.\begin{aligned} x &= k(t) \\ x &= k_1(t) \end{aligned}\right\} 0 \leq t \leq 1$$

in such a way that the same element h of F maps initial point to final point for both of them:

$$k(1) = h[k(0)], \qquad k_1(1) = h[k_1(0)].$$

Now if one connects the two points $k(t)$, $k_1(t)$ with the same t-value by a non-Euclidean straight line segment, and denotes the point which divides this segment in the non-Euclidean distance ratio $\tau : 1 - \tau$ by $k_\tau(t)$, then for each fixed τ, $(0 \leq \tau \leq 1)$, $x = k_\tau(t)$ for $0 \leq t \leq 1$ is a closed curve on ϕ, since we always have

$$k_\tau(1) = h[k_\tau(0)].$$

Thus we have on ϕ a family of closed curves depending on the parameter τ, such that the "parameter distance" between two curves in the family depends continuously on τ, and for $\tau = 0$ and $\tau = 1$ respectively we have the original curves. This establishes the homotopy between the latter. Thus the conjugacy classes of F are in one-to-one correspondence with the homotopy classes of closed curves on ϕ, or what we shall call more briefly the *curve types* on ϕ. The conjugacy class consisting of the identity alone corresponds to the "null homotopic," i.e. contractible, curves on ϕ.

In the next paragraph it will be shown that F contains only hyperbolic transformations. Two of the latter which are conjugate, but different from the identity, in F have equivalent displacement segments on their axes. These segments correspond to a closed geodesic on ϕ, and we can therefore uniquely associate each curve type (with the exception of the contractible curves) with a closed geodesic of this type on ϕ.

5. The set of fundamental points. Until further notice, all terms are to be understood in the sense of the non-Euclidean metric. For each point P of Φ, $\delta(P)$ denotes the distance from P to its nearest equivalent point. The value of δ is the same for two equivalent points. Thus if we want to construct the greatest lower bound of $\delta(P)$ as P varies in Φ, it suffices to let P vary in B, which is indeed a fundamental domain for F. Now if this lower bound were zero, there would be a point of B, necessarily a boundary point in fact, left fixed by a non-identity operation of F. But when one checks the finitely many elements of F for which the image of B has a boundary point in common with B, one sees that no such fixation of a boundary point occurs. For interior points of a boundary side this is immediate from the definition of the generators, and for a vertex of B it follows from the fact that a circle round a vertex with radius smaller than half a polygon side contains no equivalent points in its interior; in fact such a circle is composed of the $4p$ corner neighbourhoods of vertices of B.

The greatest lower bound of δ in Φ is therefore positive. It follows that F cannot contain elliptic transformations, since these have fixed points in Φ. But F also cannot contain parabolic transformations: Such a transformation maps each Euclidean circle touching E at the fundamental point into itself, and therefore permutes the circles of the orthogonal family among themselves. Then, in the sense of the hyperbolic metric, each line with the fundamental point as a point at infinity is mapped into a parallel line, and the line connecting point and image makes the same angle with both lines. This makes the distance between point and image become arbitrarily small when one travels sufficiently far in the direction of the infinitely distant fundamental point along such a line.

All elements of F are therefore hyperbolic transformations and the greatest lower bound of $\delta(P)$ is the shortest length of displacement obtainable by an element of F. It may be noted that the number of conjugacy classes whose displacement length falls below a given value L is finite. Namely, for each such conjugacy class we can choose a representative (in general in different, but finitely many, ways) whose axis passes through the central fundamental domain B. Then we can choose the beginning of the displacement segment on the axis to be in B. Then the final point of the latter lies at a distance $\leq L$ from the initial point, and hence at distance $\leq L + \rho$ from the origin, where ρ is the distance of the vertices of B from the origin. The image of B under the F-element in question must therefore lie inside a circle of radius $L + \rho$ around the origin, and this is the case for only finitely many elements. We therefore have:

In each infinite sequence of F-elements, no two of which are conjugate, the displacement length tends to ∞.

Each element of F therefore corresponds to a pair of separate fundamental points on E. It follows from §2 that commuting F-elements correspond to the same fundamental point pair and conversely. Two non-commuting elements cannot have even one fundamental point in common, because by §2 their commutator would then be a parabolic transformation, and there are none such in F. The (denumerable) set G of fundamental points is therefore uniquely partitioned into pairs. Let U and V be such a pair. Those elements of F which correspond to this pair, and therefore have the same axis, constitute an abelian subgroup of F. Let f_1 and f_2 be two such elements, and let l_1 and l_2 be their displacement lengths. Then either $f_1 f_2$ or $f_1 f_2^{-1}$ has the displacement length $|l_1 - l_2|$. Since the set of displacement lengths for the axis is bounded below, there must be a smallest l which divides all the others. Let f be the element corresponding to l, and suppose V is its positive fundamental point. We call f the "primary" element for the axis, or more precisely, for the fundamental point pair UV; f^{-1} is then the primary element for VU. The subgroup in question then consists of all positive and negative powers of f. Thus F has no abelian subgroups other than the infinite cyclic ones. A displacement segment of length l on the axis obviously represents a single traversal of the corresponding closed geodesic on ϕ. However, l is not necessarily the smallest distance between two equivalent points on the axis, because the closed geodesic on ϕ can have double points. Such a double point corresponds on Φ to the intersection of the axis of f with the axis of an element which results from f by conjugation.

Now let P_1 and P_2 be two vertices of the net n, that is, two points equivalent to the origin, and let L be the non-Euclidean line connecting them. A rotation about P_2 through angle π maps not only L but also n onto itself (for the moment we disregard the labelling of n). Suppose this rotation carries P_1 to P_3. A line through P_1 carrying edges of the net is sent to a line through P_3 carrying edges of the net, and these two lines make equal angles with L. A displacement S along L which carries P_1 to P_3 therefore maps n as well as L onto itself, and the same is true of the powers of S, which send P_1 to further vertices of the net. Now since only $4p$ edges of the net issue from a given vertex, a particular edge through P_1 must be sent to a like-labelled edge, if not by S itself, then by some power S^m. This power S^m then maps the net n onto itself with preservation of labels, and therefore belongs to F. (For an even p one already has $m = 1$, cf. *loc. cit.*[1], p.248). The F-element S^m has L as axis, and we therefore have:

Each line which connects two vertices of the net is an axis in F.

Obviously none of the foregoing is altered if one takes the vertices in N instead of n, or if one connects a vertex of n to a vertex of N. However, not every axis in F is of this kind, e.g., the axes of the generators contain neither vertices of n nor vertices of N.

In particular, the lines which carry the edges of the net are axes in F.

Now let i be an interval on E and k a circle orthogonal to E through the endpoints of i. Also, let P be a point inside k which is equivalent to the origin; there is such a point, since the size of net cells approaching E decreases without bound. The diameter through P is then an axis in F which has one fundamental point in i. Moreover, if i_1 is an interval on E disjoint from i, and if P_1 is a point inside the orthogonal circle k_1 on i_1 which is equivalent to the origin, then the circle through P and P_1 orthogonal to E is an axis in F which has its fundamental points in i and i_1. In summary, we have:

The directions of F-axes through a vertex of the net lie everywhere densely around the vertex. The set of fundamental points is everywhere dense on E, and in such a way that any two intervals on E contain a pair of associated fundamental points.

6. Expansion of a point on the unit circle in terms of the fundamental group. The points of the unit circle E represent the infinitely distant points of the hyperbolic plane. A point R of E determines a direction for each point P of Φ, the direction of the non-Euclidean half line from P which has R as its point at infinity. In particular, if P is a vertex of the net n, one of the $4p$ edges issuing from P is distinguished by having the smallest angle with the half line PR, except in the special case where PR has the same direction as the main diagonal of a net cell next to P, in which case one has a choice between two equally close edges. If we first ignore this special case, we can proceed from P along the distinguished edge to the next vertex, and then continue from this vertex along its distinguished edge, etc. In this way one obtains an edge path (unique except for the special case) which tends as closely as possible to the infinitely distant point R, and this edge path corresponds to an infinite sequence of generators of F, namely, the corresponding sequence of edge labels. Apart from R, the sequence depends only on the initial point P. We choose our initial point to be the vertex which coincides with the origin, and describe the above process as follows:

The radii of the unit circle which go through the vertices of the fundamental domain B meet

E in $4p$ points which are denoted

$$R_1, R_2, \ldots, R_{4p}$$

in anticlockwise order, beginning just below the positive real axis. The intervals they determine on E,

$$R_1R_2, R_2R_3, \ldots, R_{4p-1}R_{4p}, R_{4p}R_1$$

are denoted by

$$[a_1], [b_1^{-1}], [a_1^{-1}], [b_1], [a_2], [b_2^{-1}], [a_2^{-1}], [b_2], \ldots, [a_p], [b_p^{-1}], [a_p^{-1}], [b_p].$$

The radius which contains the point $a_1(0)$ therefore meets E in the middle of the interval $[a_1]$, and so on. Now let Q be an arbitrary point of E, lying, e.g., in $[a_1]$. Then the edge from 0 to $a_1(0)$ is the distinguished one for 0. We therefore call a_1 the "first convergent" of Q and write

$$k_1(Q) = a_1.$$

Now in order to find the distinguished edge for the vertex $a_1(0)$ we shift this point to the origin, and therefore seek the first convergent of $a_1^{-1}(Q)$.

Suppose that $a_1^{-1}(Q)$ lies in $[b_2]$. Then we call a_1b_2 the "second convergent" of Q, and write

$$k_2(Q) = a_1b_2.$$

We have in this way delimited the position of Q on E more precisely, because to say that $a_1^{-1}(Q)$ lies in $[b_2]$ is to say that Q is in $a_1([b_2])$, the image of $[b_2]$ under the operation a_1; this is a subinterval of $[a_1]$ which may be denoted $[a_1b_2]$. Now one continues with $b_2^{-1}a_1^{-1}(Q)$. This point may perhaps fall in $[a_3^{-1}]$. In this case we have

$$k_3(Q) = a_1b_2a_3^{-1}$$

and Q then lies in

$$a_1(b_2([a_3^{-1}])) = a_1([b_2a_3^{-1}]) = [a_1b_2a_3^{-1}]$$

which is a subinterval of $[a_1b_2]$. By continuing this process indefinitely one obtains a sequence of convergents of Q:

$$k_1(Q), k_2(Q), k_3(Q), \ldots \text{ ad inf.},$$

which represent initial segments of the infinite sequence of generators

$$k(Q) = a_1b_2a_3^{-1} \ldots \text{ad inf.}$$

On E, these correspond to a sequence of nested intervals

$$[a_1], [a_1b_2], [a_1b_2a_3^{-1}], \ldots \text{ ad inf.},$$

which all contain the point Q in their interior. We call $k(Q)$ the "expansion of the boundary point Q in terms of the group F." It will now be shown that this expansion completely determines the boundary point, and that the role played by the boundary point in the fundamental group is reflected in its expansion.

In order to more precisely characterize those sequences of generators which can appear as expansions of boundary points, we need to look more closely at the interval subdivision of E

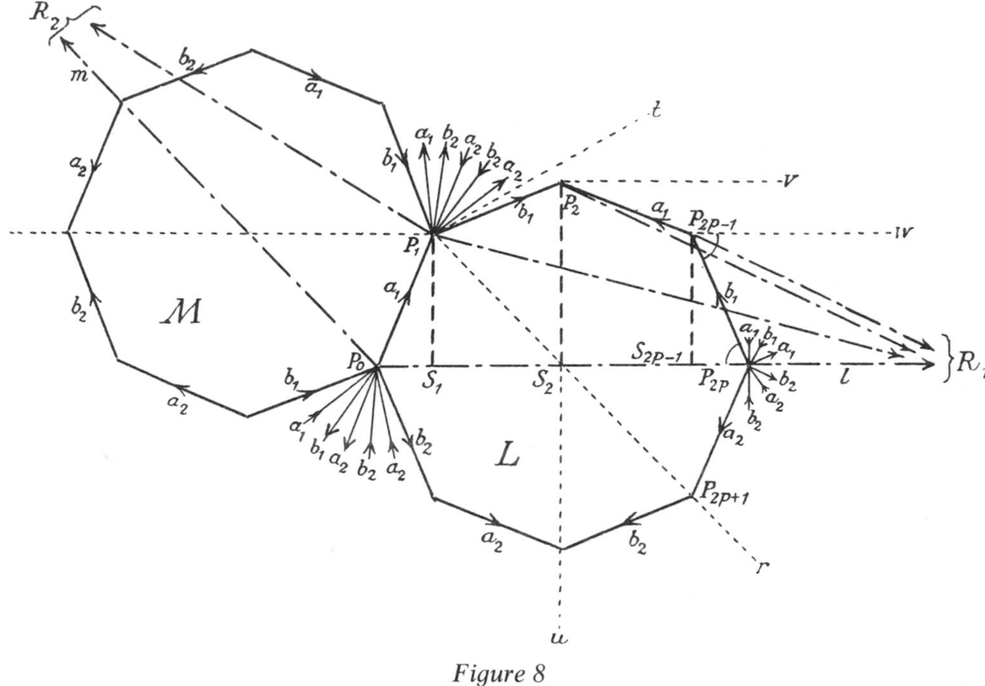

Figure 8

used above. One can derive what is necessary from the definition of the generating transformations with, say, Figure 7 on hand; but instead of this we shall use the following elementary geometric reasoning in the hyperbolic plane. Consider a vertex P_0 and two adjacent cells L and M which have, say, the edge a_1^{+1} in common. The two half lines l and m starting at P_0 and containing the main diagonals of these cells through P_0 determine a segment of the infinitely distant circle E which corresponds to the "first order" interval $[a_1]$ when P_0 corresponds to the origin. Now we go along a_1^{+1} from P_0 to the neighbouring vertex P_1. The $4p$ half lines issuing from P_1 which contain the main diagonals issuing from P_1 of neighbouring cells determine $4p$ segments on E which, *if they lie in the segment* $[a_1]$, determine "second order" intervals by subdivision of the latter (in the above example $[a_1 b_2]$). The segment $[a_1]$ is determined by parallels to l and m through P_1. These parallels travel from P_1 through the interiors of the cells considered above (for $p = 2$ see Figure 8) because the next edge b_1^{+1} of the cell L, from P_1 to P_2 has a common perpendicular with l (in the interior of L); one can see this from the fact that the normals $P_1 S_1$ and $P_2 S_2$ of P_1 and P_2 on l make *acute* angles $S_1 P_1 P_2$ and $S_2 P_2 P_1$ with b_1, namely subangles of the acute polygon angle. If one now considers the subdivision of the first order interval $[a_1]$ on E for P_0 determined by the half lines through P_1, then one finds:

(1) The segment of E corresponding to $a_1^{-1} = P_1 P_0$ is cut out by the half lines r and s, and therefore falls outside $[a_1]$. *Thus the expansion of a boundary point does not contain two mutually inverse generators in succession.*

(2) The segment of E corresponding to $b_1^{+1} = P_1 P_2$ is cut out by the half lines r and t. A *part*

of this segment, namely that between l and t, falls inside $[a_1]$. The analogous result holds at the other end of $[a_1]$, for the edge b_1^{-1} issuing from P_1.

(3) For the remaining $4p$-3 edges issuing from P_1, the corresponding segment lies entirely inside $[a_1]$.

We must therefore complete our earlier definition as follows: if h_i is the symbol for an arbitrary generator, then the second order interval $[h_1 h_2]$ is defined as the *intersection* of $[h_1]$ and $h_1([h_2])$. Then $[h_1 h_1^{-1}]$ is empty; $[h_1 h_2]$ is a proper part of $h_1([h_2])$ when h_2 is one of the two generators next to h_1 in the boundary of a cell; for the remaining $4p$-3 values of h_2, $[h_1 h_2]$ is identical with $h_1([h_2])$.

We now precisely determine the third order interval $[h_1 h_2 h_3]$, and hence the set of E-points whose third convergent is $k_3 = h_1 h_2 h_3$. This is obviously the intersection of $[h_1]$ and $h_1([h_2 h_3])$, and $[h_2 h_3]$ is now fully defined.

(1) If $[h_1 h_2] = h_1([h_2])$, so that h_2 is one of the $4p$-3 non-special values relative to h_1 (the above case 3), then $h_1([h_2])$, and hence also $h_1([h_2 h_3])$ falls wholly in $[h_1]$, so $[h_1 h_2 h_3] = h_1([h_2 h_3])$.

(2) When $h_2 = h_1^{-1}$, then $h_1([h_1^{-1} h_3])$ lies outside $[h_1]$, so $[h_1 h_1^{-1} h_3]$ is empty (above case (1)).

(3) It remains to consider the above case 2, where h_2 is one of the edges in the cell boundary next to h_1, and for brevity we confine ourselves again to the special example in Figure 8. The first order interval $[a_1^{-1}]$ relative to the point P_2 is determined by the half lines u and v (v is symmetric to u relative to the edge a_1^{-1} issuing from P_2); u cuts l at the midpoint of the cell L, v does not meet l. This is because the parallel to l through P_2 is inside the cell L, on the same grounds as before. Thus, of the $4p - 1$ possible intervals $[b_1 h_3]$ of second order relative to P_1, $4p - 2$ fall inside the first order interval $[a_1]$ relative to P_0; hence for all of these, $[a_1 b_1 h_3] = a_1([b_1 h_3])$. Thus we only have to consider $[a_1 b_1 a_1^{-1}]$. The second order interval $[b_1 a_1^{-1}]$ relative to P_1 is, as we know, only a part of $b_1[a_1^{-1}]$, namely since the segment determined by u and v is "cut" to the segment determined by r and v. In constructing the intersection of $a_1([b_1 a_1^{-1}])$ and $[a_1]$ the segment is cut further, however, inasmuch as only the segment of E determined by l and v remains as the third order interval $[a_1 b_1 a_1^{-1}]$ relative to P_0.

If we from now on omit the case where two mutually inverse generators follow each other in succession, then we have the result:

$[h_1 h_2 h_3]$ is in general identical with $h_1([h_2 h_3])$ and only in the special case a subinterval of $h_1([h_2 h_3])$, when $h_1 h_2 h_3$ occurs as a subsequence in the boundary of a cell.

It is now clear that this reasoning may be extended without change, as long as we find ourselves in the half of the cell L above l. Thus for $r \leq 2p$ we have:

The rth order interval $[h_1 h_2 \ldots h_r]$ is defined to be the intersection of $[h_1]$ and $h_1([h_2 h_3 \ldots h_r])$; it is in general identical with $h_1([h_2 h_3 \ldots h_r])$ (since the latter interval falls inside $[h_1]$), and only in the special case a proper part of $h_1([h_2 h_3 \ldots h_r])$, when $h_1 h_2 \ldots h_r$ occurs as a subsequence in the boundary of a cell.

To prove this result for the case $r = 2p$ we have only to add the remark that the parallel to l

through P_{2p-1} indeed no longer falls inside L, but—and this is all we need—in the first order segment relative to P_{2p-1} which belongs to the edge $P_{2p-1}P_{2p}$. This latter segment is bounded by the half line containing the main diagonal of L through P_{2p-1} and the symmetric half line w with respect to $P_{2p-1}P_{2p}$, and the angle between w and $P_{2p-1}P_{2p}$ equals the angle $P_{2p-1}P_{2p}P_0$, so that w and l do not meet.

We can also express this by saying that an rth order interval $[h_1 h_2 \ldots h_r]$, where $h_1 h_2 \ldots h_r$ is a subsequence in the boundary of a cell, certainly exists for $r \leq 2p$, but is, so to speak, "smaller" than a "normal" r^{th} order interval, and increasingly so as r increases. We now show that such intervals no longer exist for $r > 2p$.

This is seen immediately in Figure 8, since the first order interval relative to P_{2p} which belongs to $P_{2p}P_{2p+1}$ obviously falls outside the first order interval $[a_1^{+1}]$ relative to P_0, which is cut out by l and m. But Figure 8 shows even more; namely, l is one of the subdivision lines through P_{2p} which determine the $4p$ first order intervals, relative to this point, on E. Of these $4p$ intervals, only $2p - 1$, (wholly or partly) lie inside the $2p^{\text{th}}$ order interval relative to P_0 determined by w and l, and indeed these correspond to the $2p - 1$ edges which follow the edge $P_{2p}P_{2p-1}$ in a clockwise circuit around P_{2p}. (In the example of Figure 8 these are the edges labelled a_1^{-1}, b_1^{-1} and a_1^{+1}.) The next edge through P_{2p}, the first below l, (b_2^{+1} in the figure) belongs to an interval bounded by l; so this interval has just the infinitely distant point R_1 of l in common with the $2p^{\text{th}}$ order interval relative to P_0 being considered. Thus if we temporarily set aside the quite special case of the expansion of the subdivision points R_i and their equivalents, then the sequences of generators appearing in expansions are characterized by the following three properties:

(1) Two mutually inverse generators do not occur in succession.
(2) No subsequence of more than $2p$ successive generators from the boundary of a cell occurs.

Expressions in the generators which fulfil these conditions are called *reduced expressions* by M. Dehn (*Math. Ann.* **72** (1912), p. 414; cyclic rearrangement of the terms in an expression does not arise now.) Boundary point expansions are therefore the "straightest" paths in the net n and cannot be shortened by use of the defining relation of F; they approximate as closely as possible the straight line connecting the initial point with the infinitely distant point being expanded, in a sense we shall elucidate below. They also have beyond the Dehn reduction process, the following third property:

(3) A subsequence of $2p$ generators, forming half the boundary of a cell, is always followed by one of the $2p - 1$ generators labelling edges which continue the half boundary on the same side of the main diagonal as the half boundary itself.

It follows immediately from the above considerations that each sequence satisfying these conditions, but otherwise arbitrary, determines a sequence of nested intervals on E. Each such sequence of intervals has one—and, as we shall see, only one—point in its intersection, and the sequence is the expansion of this point.

7. Exposition of some particulars of boundary point expansions

(a) Expansion of the subdivision points R_i. We consider the expansion of the subdivision point R_1. It is the boundary between the first order intervals $[a_1]$ and $[b_p]$. Thus we have a

choice between a_1 and b_p as the first convergent of R_1, and there is no advantage in prescribing a fixed choice at this stage. However, once one has made a particular choice, e.g., $k_1(R_1) = a_1$, then the next $2p - 1$ convergents are determined (cf. Figure 8 again):

$$k_{2p}(R_1) = a_1 b_1 a_1^{-1} b_1^{-1} \ldots a_{p/2} b_{p/2} a_{p/2}^{-1} b_{p/2}^{-1}$$

when p is even, and

$$k_{2p}(R_1) = a_1 b_1 a_1^{-1} b_1^{-1} \ldots a_{(p-1)/2} b_{(p-1)/2} a_{(p-1)/2}^{-1} b_{(p-1)/2}^{-1} a_{(p+1)/2} b_{(p+1)/2}$$

when p is odd. By this stage one has arrived at the point P_{2p} in Figure 8 and there is again a choice between a_1 and b_p. One can now check that the expansion $k(R_1)$ corresponds to an infinite sequence of nested intervals only if one repeats the first choice; otherwise the $(2p + 1)^{\text{th}}$ convergent does not correspond to an interval on E. Thus we obtain an expression for R_1 which consists either of periodic repetition of the above expression, or else an expression which results from it by means of the relation $R = 1$, and we consider these expressions for $k(R_1)$ to be equivalent. The element $k_{2p}(R_1)$ has the cell diagonal l of Figure 8 as axis and $P_0 P_{2p}$ as length of displacement. R_1 is therefore the positive fundamental point of this element.

The expansion of the remaining R_i goes quite similarly. These are also the fundamental points of axes consisting purely of cell diagonals, though in general for $p > 2$ several of the latter make up a displacement segment, so that the period recommences after traversal of several half cells. These periods are easily obtained from the sequence of edges on a cell and around a vertex, and again each time in terms of two equally valid expressions. We shall omit their explicit description and give just the example of the expansion of R_2 for $p = 3$:

$$\begin{aligned} k_{18}(R_2) &= b_1^{-1} a_2 b_2 a_2^{-1} b_2^{-1} a_3 \cdot b_2^{-1} a_3 b_3 a_3^{-1} b_3^{-1} a_1 \cdot b_3^{-1} a_1 b_1 a_1^{-1} b_1^{-1} a_2 \\ &= a_1 b_1^{-1} a_1^{-1} b_3 a_3 b_3^{-1} \cdot a_2 b_2^{-1} a_2^{-1} b_1 a_1 b_1^{-1} \cdot a_3 b_3^{-1} a_3^{-1} b_2 a_2 b_2^{-1} \end{aligned}$$

and $k(R_2)$ is the periodic repetition of one or other of these two expressions for $k_{18}(R_2)$. R_2 is the positive fundamental point of $k_{18}(R_2)$, the diameter through R_2 is the axis of this element, and a displacement segment consists of three cell diagonals joined in succession.

From this cyclic correlation of cell diagonals one can easily determine the partition of the subdivision points $R_1, \ldots R_{4p}$ into equivalence classes. For $p = 2$, no two of the 8 points are equivalent; for $p = 3$, $R_2 \sim R_6 \sim R_{10}$ and hence also $R_8 \sim R_{12} \sim R_4$, and there are no further equivalences; thus for $p = 3$ there are also 8 inequivalent R_i.

Now let R^* be a point of E which does not coincide with any point R_i, but which is transferred to one of these points in the course of the expansion process. Then $k_1(R^*)$ is uniquely determined; if

$$(k_1(R^*))^{-1}(R^*) \neq R_i,$$

then $k_2(R^*)$ is also uniquely determined, etc. Suppose that $k_{r-1}(R^*)$ is still uniquely determined but that

$$(k_{r-1}(R^*))^{-1}(R^*) = R_j.$$

This means that R^* is an *interior* point of intervals of first, second, ..., $(r - 1)^{\text{th}}$ order. Then in determining the rth convergent of R^* we have to make the same choice as in determining the first convergent of R_j, and both decisions are possible without breaking off the nesting of intervals. Thus we obtain two expressions for $k(R^*)$, namely

$$k(R^*) = k_{r-1}(R^*)\cdot k(R_j),$$

where we have the two earlier expressions for $k(R_j)$. R^* is therefore a subdivision point between two s^{th} order intervals, for all $s \geq r$.

(b) Each boundary point is uniquely determined by its expansion. This theorem is the same as the assertion that an infinite sequence of nested intervals of first, second, ... order has a single point as intersection and not a segment. Let such a sequence be given by the expansion

$$k = e_1 e_2 e_3 \ldots \text{ad inf.};$$

the rth order interval is determined by the convergent

$$k_r = e_1 e_2 \ldots e_r.$$

(It is always understood to be a closed interval.) Now suppose that this sequence of intervals has the (closed) segment $X_0 Y_0$ as intersection. Then all points of this segment have the expansion k, and no other point of E has this expansion. All points of the segment $e_1^{-1}(X_0 Y_0)$ then have the expansion $e_2 e_3 \ldots$. Let $X_1 Y_1$ be the segment which contains all points with this expansion; we need not decide whether $X_1 Y_1$ equals or includes $e_1^{-1}(X_0 Y_0)$, in any case $X_1 Y_1$ is bigger than $X_0 Y_0$. This is because the first order intervals lie inside the corresponding circles forming the sides of the fundamental domain B, and these play the role that the circle C in Figure 4 plays for the transformation S there; cf. the definition of generators in §3. Thus each subinterval of $[e_1]$ is enlarged by the operation e_1^{-1}. Now we similarly construct the segment $e_2^{-1}(X_1 Y_1)$ and the segment $X_2 Y_2$ containing it, which consists of all points with the expansion $e_3 e_4 \ldots$, etc. The intervals in the sequence $X_0 Y_0$, $X_1 Y_1$, $X_2 Y_2$, ... cannot all be disjoint, since they have increasing lengths. But when two intervals in this sequence have a point in common, all points in both intervals must have the same expansion, namely that of the common point, and this contradicts the fact that the smaller of the two intervals already contains all points with this expansion.

(c) Periodic expansions determine fundamental points of F and conversely. If a boundary point P has the expansion

$$e_1 e_2 \ldots e_r e_{r+1} \ldots$$

then, by the definition of expansion, $(e_1 e_2 \ldots e_r)^{-1}(P)$ has the expansion $e_{r+1} e_{r+2} \ldots$. Now if P has an expansion consisting of periodic repetition of the subsequence $e_1 e_2 \ldots e_r$, then $(e_1 e_2 \ldots e_r)^{-1}(P)$ has the same expansion, and hence by (b)

$$(e_1 e_2 \ldots e_r)^{-1}(P) = P$$

and so P is a fundamental point of the element

$$f = e_1 e_2 \ldots e_r.$$

In fact, P must be the positive fundamental point of f, $P = V(f)$, as follows from viewing the expansion as the straightest possible edge path towards the infinitely distant point in question.

Now suppose another point P^* has the expansion

$$k(P^*) = h_1 h_2 \ldots h_m \cdot e_1 e_2 \ldots e_r \cdot e_1 e_2 \ldots e_r \ldots$$

with the previous period f. Let $h_1 h_2 \ldots h_m = h$. Then

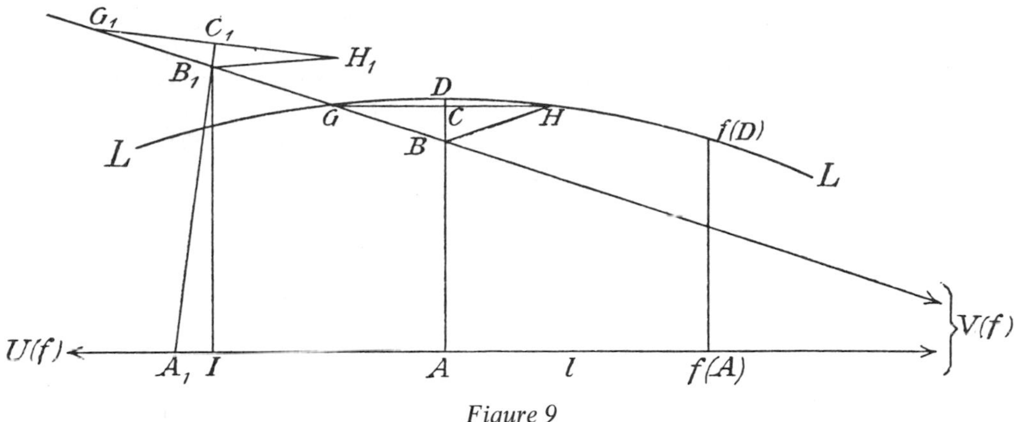

Figure 9

$$k(h^{-1}(P^*)) = k(P) - k(V(f)),$$

so

$$h^{-1}(P^*) = V(f),$$

hence

$$P^* = h(V(f)) = V(hfh^{-1}) = V(h_1 \dots h_m \cdot e_1 \dots e_r \cdot h_m^{-1} \dots h_1^{-1}).$$

Now it will be shown that, conversely, each fundamental point of F has a (purely or ultimately) periodic expansion.

Let $l = U(f)V(f)$ be the axis of an element f (Figure 9). At a point A of l one erects a perpendicular AB of such a length that the angle of parallelism $ABV(f)$ equals $\pi/4p$. One extends AB by the segment BC equal to $s/2$, the half edge of the net, and at C erects the perpendicular GCH to AC. Then GCB is a right-angled isosceles triangle with side $s/2$ and base angle $\pi/4p$, the same as the triangle formed by the midpoint, vertex and side midpoint of a cell. One can therefore think of a polygon, congruent to a cell of the net, placed in such a way that B is its midpoint and GH a side. Let L be the curve at constant distance from l which passes through G and H; it cuts the extension of AC at D; its distance from l is greater than the net side s, because $BC = s/2$ and AB is greater than $s/2$ by choice of the angle of parallelism. Let the mirror image of L in l be L'.

Now one displaces the triangle GBH in such a way that its side BG slides along the line $V(f)BG$ until the triangle reaches a position $G_1B_1H_1$ at which G_1 lies on the convex "outer" side of L, the side away from l. During this displacement H describes a curve of constant distance from $BV(f)$, thus H_1 also falls on the outer side of L. The perpendicular B_1I from B_1 to l is then greater than BA, the angle of parallelism $IB_1V(f)$ is therefore smaller than $ABV(f)$, i.e., smaller than $GBC = G_1B_1C_1$. The perpendicular bisector C_1B_1 of the triangle $G_1B_1H_1$, when extended, then cuts the line l, if at all, at a point A_1 between I and $U(f)$; in other words, the common perpendicular of G_1H_1 and l lies nearer to H_1 than to G_1, and therefore H_1 is nearer to l than G_1. (The reverse is the case when one slides the triangle in the opposite direction, so that G_1, and hence H_1, falls inside L.)

Now one imagines that G_1 is a vertex of the net and that the infinitely distant point $V(f)$ is to be expanded. The distinguished edge for G_1 then falls in the angle formed by G_1H_1 and its reflection in $G_1V(f)$ (in fact in its interior, since $V(f)$ cannot belong to two axes, so $G_1V(f)$ is not an axis, and hence in particular not a cell diagonal) and it has the same length as G_1H_1, so that its final point lies nearer to l than its initial point. Assume that the net vertex O corresponding to the origin, from which the edge path $k(V(f))$ corresponding to the expansion proceeds, lies outside the strip-like neighbourhood of l enclosed by L and L'. Then the path k, as long as it runs outside this strip, must come closer to l with each new edge. But there are only finitely many values for the distance of a vertex from l below a given value, since these distances repeat periodically with the displacement f. Thus in finitely many steps k must come inside the strip LL' if it does not already begin there, and then it can no longer leave. Now the perpendiculars AD and $f(A)f(D)$, together with l and L, bound a region which, when united with its reflection in l, forms a fundamental domain on the strip for the group of powers of f. It contains only finitely many vertices, and at the moment when k first enters a vertex obtainable by a power of f from one entered earlier, then the period is established, as the subsequence between these two events.

This proves the assertion.

(d) The basic period corresponds to the primary element. We transfer the origin O to a vertex in the periodic part of the edge path $k(V(f)$, so that we have a purely periodic expansion $k(V(f))$ to deal with. Now let h be the primary element corresponding to the axis of f with $V = V(f)$ as positive fundamental point. We construct the image $k^{(1)} = h(k(V))$ of the path $k(V)$ under h. This edge path represents the same expansion, it just begins at a different point, namely $h(O)$. $k^{(1)}$ cannot be completely disjoint from k, since k comes to coincide with a subpath of k under a power of h. But it follows from the uniqueness of the expansion that k and $k^{(1)}$ have no common points unless $k^{(1)}$ lies entirely on k. Thus $h(O)$ lies on k, and k consists of the periodic repetition of the subpath $Oh(O)$ of k representing h. This subpath cannot itself consist of periodic repetitions of a smaller subpath, since h is not a power of any other element. Thus the basic period for the axis corresponds to the primary element.

(e) All points equivalent to the subdivision points $R_1, \ldots, R_{4p}$ appear in the interval subdivision. Each point S of E equivalent to a point R_i is the final point of an axis l consisting of cell diagonals. If one erects the perpendicular to l at a vertex on l, then this is also a cell diagonal, and it first meets the net again at a distance of 2ρ from l, where ρ is the distance from the midpoint of a cell to the vertex. But 2ρ is greater than the distance from l to L in Figure 9, because this distance is smaller than $GB + BA$, and $GB = \rho$, $BA < \rho$. Namely, Figure 8 shows that the parallel to the cell diagonal through a point at distance ρ (P_2 in Figure 8) has an angle of parallelism $< \pi/4p$. Thus an edge path, with arbitrary initial point, which represents the expansion $k(S)$ must, after it has entered the strip LL' corresponding to l, necessarily pass through a vertex on l. The boundary point then falls into a position R_i when shifted by the expansion of S, so S plays the role of R^* in (a).

(f) Intuitive view of the boundary point expansion for a net of quadrilaterals v. Since the concept of boundary point expansion plays an important role in the present work, we shall give an exposition of the above in somewhat different terms, in order to make the process seem more intuitive. Visualize the $2p$ diameters through the points $R_1, \ldots, R_{4p}$ and *all lines equivalent to them* as distinguished. This divides the angle around each point equivalent to the origin into $4p$ equal sectors. The set of these non-Euclidean lines covers Φ with a new net,

the net v, whose cells are regular *quadrilaterals* with side ρ and angle $\pi/2p$, so that $4p$ meet at each vertex. Each quadrilateral cell has an edge of n as one diagonal, and an edge of N as the other. The expansion of a boundary point now proceeds by beginning at the origin, and at each subsequent vertex choosing the sector which contains the boundary point, and inside this travelling along the diagonal to the opposite vertex.

The set of infinitely distant points for the lines forming the net v is identical with the set of interval subdivision points on E. This set is a proper subset of the set of fundamental points of F, and already this subset, which contains the fundamental points for the elements of a *finite* number of conjugacy classes in F, is everywhere dense on E. The key fact is that $4p - 1$ of the sectors round a point equivalent to the origin determine m^{th} order intervals on E when the point in question can be connected to the origin by a path of $m - 1$, but no fewer, edges of the net n; however, this involves a few ekstra conditions on sequences of generators in the order of the relation by the earlier results. The fact that these intervals become arbitrarily small when the point equivalent to the origin is sufficiently close to E, hence when m is sufficiently large, is an intuitive proof of the correctness of the uniqueness theorem (b).

(g) Combinatorial solution of the homotopy problem. We first prove the following *lemma*:

> The figure consisting of the $4p$ cells which meet at a vertex does not lie on the strip LL' defined in (c).

We obviously need to prove this lemma only in the case where the centre line l of the strip goes through the vertex P in question. In (e) we have already proved that the distance between l and L is less than the length 2ρ of the cell diagonal. We shall now prove that the final point of a segment of length 2ρ drawn from a point of l at an angle $\pi/4p$ to the normal also falls outside L. This will prove the lemma, because one of the cell diagonals issuing from P makes an angle with the normal to l which is $\leq \pi/4p$. In the following proof we refer to Figure 10.

PP_3 is the cell diagonal which makes an angle $\pi/4p$ with the perpendicular PP_1 on l. $PM =$

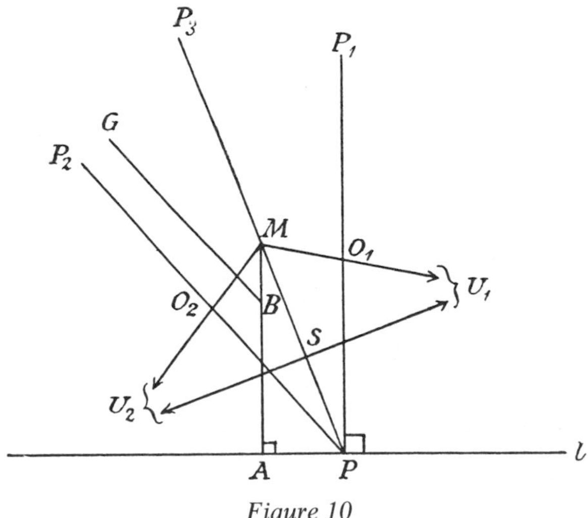

Figure 10

$MP_3 = \rho$. PP_1 and PP_2 are neighbouring edges of the net, O_1 and O_2 their midpoints, MO_1U_1 and MO_2U_2 their perpendicular bisectors. $PMO_1 = PMO_2 = \pi/4p$. U_2SU_1 is a line perpendicular to PM and parallel to MO_1U_1 and MO_2U_2. Thus MS equals the segment BA of Figure 9 in length, and we found in (e) that the latter is smaller than ρ. Thus S falls between M and P. Now consideration of the cell shows that the angle of parallelism corresponding to half an edge is greater than $\pi/4p$ and smaller than $\pi/2p$. PO_2 is the half edge, and $PO_2U_2 = \pi/2$. Since $p \geq 2$, O_2P makes the angle $\pi/2 - \pi/2p \geq \pi/2p$ with l. Thus O_2U_2 and l do not intersect, and hence U_1SU_2 and l also do not intersect. The distance MA of the point M from l is therefore greater than MS. Now let us carry over the broken line denoted by ABG in Figure 9 to Figure 10 with the same notation. B then falls between A and M because $AB = MS < MA$. We have $BG = MP_3 = \rho$. The curve L is defined as the curve through G at constant distance from l. P_3 must now fall on the outer side of L, as claimed, because MP_3 issues from a higher point of the perpendicular MA than BG, and therefore makes a smaller angle with this perpendicular; namely $AMP < O_2MP = MBG = \pi/4p$.

Thus the lemma is proved.

Now let l be the axis of an element f and let $k = k(V(f))$ be an edge path which represents the expansion of $V(f)$. Let the initial point be so chosen—by omitting an initial segment of the edge path if necessary—that k has a purely periodic expansion. The edge path $f^n(k)$, $n < 0$, contains k; it is, so to speak, a backwards extension of k. If one lets n tend to $-\infty$ one obtains a two-sidedly infinite, non-singular edge path accompanying the axis l inside the strip LL', which we denote K_0. *The question arises, whether K_0 is uniquely determined by l, and thus whether the period of the expansion is uniquely determined by giving the conjugacy class, up to cyclic interchange.* We know a special case where this question is answered in the negative, namely the case of axes which consist of cell diagonals. We therefore exclude this case from now on. Suppose that another choice of initial point leads us to a two-sidedly infinite edge path K_1, constructed analogously to K_0 but not identical with K_0. Because of the uniqueness of expansions, K_0 and K_1 cannot have a point in common. They jointly bound a two-sidedly infinite substrip inside LL', which repeats periodically under f. Thus by the lemma, this substrip contains no vertices in its interior. K_0 is therefore connected to K_1 only by single edges. If we consider a cell containing two of these edges, then by the first two properties of expansions (§6) these edges must *either* flank boundary sequences of $2p - 1$ edges in one of K_0 or K_1, and $2p - 1$ edges in the other, *or* flank boundary sequences of $2p$ edges in one and $2p - 2$ edges in the other. But the latter case is made impossible by the *third* property of expansions, as one sees immediately by proceeding to the next cell in the chain between K_0 and K_1. Thus each term in the infinite chain of cells has $2p - 1$ edges on K_0 and $2p - 1$ edges on K_1. But then the axis l goes through the centres of all cells, and two opposite sides of each cell, namely those which connect K_0 and K_1, are perpendicular to it. There are only finitely many inequivalent axes of this kind, which one can easily construct using perpendicular bisectors of edges, and represent by generators. They are, like the axes consisting of cell diagonals, symmetry lines of the three nets.

In summary, we have the following theorem:

> In expanding the positive fundamental points of all elements in a conjugacy class one obtains the same period (up to cyclic interchange). For only finitely many (namely, less than $8p$) conjugacy classes are there two expressions for the period; these are either

equivalent under the relation (the case of axes consisting of cell diagonals) or convertible into each other by conjugation by a generator (the case of axes consisting of perpendicular bisectors of edges).

The first exceptional case corresponds, as we know, to the indeterminacy in the expansion itself, in the case where we once have a choice between two convergents. In the second case the expansion is uniquely determined, but one obtains one or other period according as one approaches the axis of the element from one side or the other, i.e., according as one begins with the origin on the left or right as one follows the axis towards the positive fundamental point.

Here we have a new derivation of the combinatorial solution of the conjugacy problem due to M. Dehn[8]. Two closed curves on the surface are homotopic if and only if the corresponding elements of F are conjugate (§3). For these two elements one writes the corresponding expressions in the generators cyclically and converts these by cancellations and applications of the relation into new forms with the three characteristic properties of expansions. This is always possible, by the foregoing, and the result is in general unique and in fact the same for the two elements if and only if they are conjugate. For the finitely many exceptional cases one can write the two possible forms in a table, from which one can see immediately whether the given elements yield periods in this table, and hence make the decision in this case also.

As one can see, this method differs from the Dehn solution only in requiring the third property above in the reduction of words. The number of exceptional cases, which for Dehn include infinitely many conjugacy classes, even though they are already very special, is accordingly reduced to finitely many classes. The two examples of exceptional cases given by Dehn, *loc. cit.*[8], pp. 414 and 420, correspond to axes consisting of cell diagonals and edge perpendicular bisectors respectively, and hence remain as exceptional cases here. We therefore give the following example of our distinction when $p = 2$:

$$\alpha = a_1^{-1} b_1^{-1} b_2^{-1} a_1 b_1 a_1^{-1}$$

and

$$\beta = a_1^{-1} b_2 a_2 b_2^{-2} a_1^{-1}$$

are reduced words in the sense of Dehn; thus if they are to be conjugate, by Dehn's solution α must result from β by cyclic interchange and conjugation with one generator. In fact, $\alpha = b_1^{-1} \beta b_1$, as one checks by two applications of the relation. However, the expression α does not have the third property above, since a_1^{-1} must not follow $b_2^{-1} a_1 b_1 a_1^{-1}$. We must therefore convert α to

$$\alpha' = a_1^{-1} b_1^{-1} a_2 b_2^{-1} a_2^{-1} b_1;$$

this expression has all three properties, and therefore represents the period of the axis for α. On similar grounds we must convert β to

$$\beta' = b_1 a_1^{-1} b_1^{-1} a_2 b_2^{-1} a_2^{-1},$$

[8] M. Dehn, Transformation der Kurven auf zweiseitigen Flächen. Math. Ann. **72** (1912), pp. 413–421.

and now α' and β' are identical when read cyclically. The two elements therefore do not belong to an exceptional class in the sense used here: the corresponding axis is not a symmetry line of the nets.

(h) Connection between the expansions of corresponding fundamental points. Let f be an element of F whose positive fundamental point $V(f)$ has a purely periodic expansion, consisting of periodic repetition of the element $f = e_1 e_2 \dots e_r$. What does the expansion of $U(f)$, the positive fundamental point of f^{-1}, look like? The periodic repetition of $e_r^{-1} \dots e_1^{-1}$ has the first two properties of expansions, but not necessarily the third. But, as we know, one can change this expression by means of the relation so that it acquires the third property. *Example*: Suppose $p = 2$ and

$$k(V(f)) = b_1 a_1^{-1} b_1^{-2} b_2 a_1 \cdot b_1 a_1^{-1} b_1^{-2} b_2 a_1 \cdot \dots.$$

One then constructs the expression:

$$a_1^{-1} b_2^{-1} b_1^2 a_1 b_1^{-1} \cdot a_1^{-1} b_2^{-1} b_1^2 a_1 b_1^{-1} \cdot \dots,$$

which does not have the third property, since b_2^{-1} is not allowed to follow $b_1 a_1 b_1^{-1} a_1^{-1}$. Therefore, one makes the change to

$$a_1^{-1} \cdot b_2^{-1} b_1 a_2 b_2 a_2^{-1} b_2^{-1} \cdot b_2^{-1} b_1 a_2 b_2 a_2^{-1} b_2^{-1} \dots,$$

and uses this form to represent $U(f)$. One sees then that the latter need not be purely periodic, even though the expansion for $V(f)$ is.

If one constructs the two-sidedly infinite edge paths K_0 and K_1 accompanying the axis of f from the periods of $k(V(f))$ and $k(U(f))$ as in (g), then these paths lie wholly on the strip LL' covering the axis; hence by the lemma in (g) there are no vertices between them, but at most single cells, possibly in chains; they can have vertices or edges in common without coinciding completely—see the example above. When $k(V(f))$ contains no subsequence of $2p$ successive generators from the relation, then K_0 and K_1 coincide, and this *can* also be the case even when such "half cells" appear, for example when $p = 2$, for $f = a_1 b_1 a_1^{-1} b_1^{-2}$.

(i) Two boundary points are equivalent—leaving aside the exceptional cases noted already —*if and only if their expansions are ultimately equal*, i.e., if they coincide beyond certain initial segments, not necessarily of the same length. This theorem has been proved under (g) for all boundary points which are fundamental points of F, if one takes into account the ambiguity of the period, noted earlier, for the exceptional fundamental points which terminate axes consisting of cell diagonals or edge perpendicular bisectors. Thus we have only to prove the theorem for boundary points which are not fundamental points of F, whose expansions are therefore not periodic, and to check that no new exceptional cases arise.

Let P and Q be two points of E which are not fundamental points, and let f be an element of F for which $Q = f(P)$. There can be at most one such element, because if also $Q = f_1(P)$ then $f^{-1}f_1(P) = f^{-1}(Q) = P$, so that $f^{-1}f_1 = 1$ and $f = f_1$, since P is not a fundamental point. (On the other hand, for two equivalent fundamental points the element of F which makes them so is not uniquely determined; there are infinitely many such.) Now if one compares the expansions of Q and P, one can transfer Q to P by simultaneously transferring the origin O to $O_1 = f^{-1}(O)$. The edge paths k and $k^{(1)}$ which represent the expansions $k(P)$ and $k(Q)$ then proceed from the different vertices O and O_1 to the same infinitely distant point P. One connects O and O_1 to P by half lines h and h_1. Let the line ending in P, beside which h and h_1 lie symmetrically, be called s. Let L be a curve at constant distance from h on the side opposite to h_1, let L_1 be a curve at constant distance from h_1 on the side opposite to h, and let the distances hL and h_1L_1 equal the distance lL in Figure 9. Let t be a perpendicular to s at a point A, which meets L and L_1 at B and B_1 respectively. The region bounded by t, L and L_1, with vertices B, B_1 and P, will be called w. The edge paths k and $k^{(1)}$ from O and O_1 respectively run always within the distance lL of Figure 9 from h and h_1, respectively. Thus, beyond a certain point, both run entirely in w. Once k and $k^{(1)}$ have a point in common they coincide completely thereafter, since P is not the endpoint of an axis consisting of cell diagonals. Thus we have only to prove that k and $k^{(1)}$ cannot be disjoint. In this case they would jointly bound a subregion of w. Now since h, s and h_1 approach each other asymptotically, the length BB_1 converges to twice the distance lL of Figure 9 as A approaches P on s. One can therefore choose A in such a way that the figure considered for the lemma in (g) has no place in the region w. Then k and $k^{(1)}$ cannot contain a vertex on w between them. From this it follows exactly as in (g) that the abovementioned subregion, which k and $k^{(1)}$ bound in w, consists of a chain of cells, with a symmetry line of edge perpendicular bisectors. This is impossible, since P is not the endpoint of such a line.

(j) Fundamental sequences

Definition. An infinite sequence

$$f_1, f_2, \ldots$$

of elements of F is called a *fundamental sequence* when the sequence of points

$$f_1(0), f_2(0), \ldots$$

converges to a point on E.

If P is the limit of the latter sequence we say that the fundamental sequence is "for P." Then if x is an arbitrary point of Φ (or of Ψ) the sequence of points

$$f_1(x), f_2(x), \ldots$$

also converges to P. Because if a circle κ of radius ε around P intersects the unit circle in Q and R, then one can draw a circle λ through Q and R which separates κ from P and whose points have a greater non-Euclidean distance from κ than x has from the origin; this is possible because κ and λ are curves of constant distance from the same non-Euclidean line. Thus when N is chosen so large that $f_n(0)$ belongs to the 2-gon bounded by λ and the arc QPR of E for all $n > N$, $f(x)$ has a Euclidean distance $<\varepsilon$ from P for all $n > N$.

Now, on a circle through the origin O and the fundamental points of f_n the points $U(f_n)$, O, $f_n(0)$ and $V(f_n)$ lie in that order. Thus $V(f_n)$ has a Euclidean distance $<\sqrt{1-|f_n(0)|^2}$ from $f_n(0)$ and hence a distance $<\sqrt{1-|f_n(0)|^2}+r_n$ from P, where r_n denotes the distance between $f_n(0)$ and P. *Thus the sequence*

$$V(f_1), V(f_2), \ldots$$

converges to P.

Now the following *converse* holds: Each sequence

$$f_1, f_2, \ldots$$

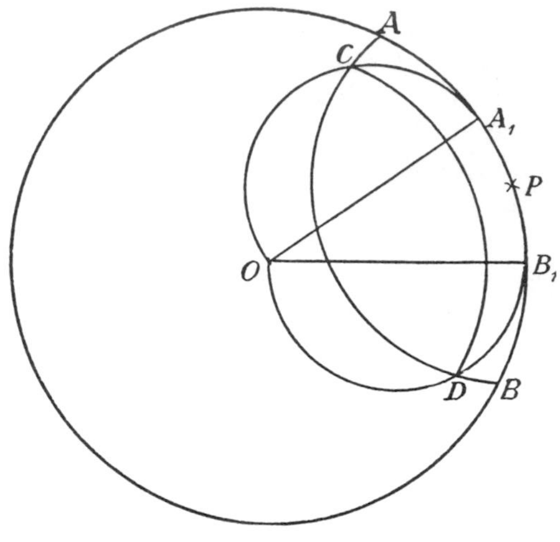

Figure 11

of elements of F, *in which no element appears infinitely often* and for which the sequence of boundary points

$$V(f_1),\ V(f_2), \ldots$$

converges to a point P of E, is a fundamental sequence for P.

Proof. The sequence $|f_1(0)|$, $|f_2(0)|$, ... converges to 1, since no element appears infinitely often in the sequence, and a circle round O with radius <1 contains only finitely many points equivalent to O. Let any ε be given. A circle AB is drawn around P with $PA = PB = \varepsilon$ (Figure 11). Let $A_1 B_1$ be a subsegment of AB with midpoint P. Semicircles on OA_1 and OB_1 meet the circle AB at C and D. Now let N be chosen so large that, for all $n > N$, not only does $V(f)$ belong to $A_1 B_1$, but also $|f_n(0)| > OC$. Then the point $f_n(0)$ is inside the quadrilateral of circular arcs $PA_1 CDB_1 P$, and hence has distance $<\varepsilon$ from P. Q.E.D.

A special fundamental sequence for the point P consists of the convergents $k_1(P)$, $k_2(P)$, ..., Namely, these elements are all different, and the points $k_1(0)$, $k_2(0)$ all belong to a strip bounded by two curves whose distance from the radius OP is the constant lL of Figure 9. The sequence $k_1(0)$, $k_2(0)$, ... therefore converges to P.

Thus the sequence $V(k_n)$ of positive fundamental points of the convergents also converges to P. When the expansion of $V(k_n)$ agrees with that of P for all convergents up to order $m(n)$, then $m(n)$ increases with n beyond all bounds.

CHAPTER II

Automorphisms of the Fundamental Group and the Corresponding Transformations of the Unit Circle

8. Definition of automorphisms. A sufficient condition for their determination. By an automorphism of the group F one means an isomorphic mapping of F onto itself, i.e., a correspondence between elements of F, with the following three properties:

(a) to each element f of F there corresponds one and only one element f' of F,
(b) each element f' of F corresponds to one and only one element f of F,
(c) the equation $f_1' \cdot f_2' = f_3'$ holds if and only if $f_1 \cdot f_2 = f_3$.

It follows that when one sets f_1 and f_2, and hence f_3, equal to the identity, the elements f_1', f_2' and f_3' are equal to each other and hence to the identity; thus the identity corresponds to itself; it also follows that the element f^{-1} corresponds to the element f'^{-1}. A primary element corresponds to a primary element. Two conjugate elements correspond to two conjugate elements, hence an automorphism induces a map of the set of conjugacy classes into itself.

We denote an automorphism of F by the letter I or by $f \mapsto f'$, and often we also denote the element f' which corresponds to f under I by f_I. The correspondence $f' \mapsto f$ is also an automorphism, denoted by I^{-1}. If $I_1(f \mapsto f')$ and $I_2(f' \mapsto f'')$ are automorphisms, then so too is $f \mapsto f''$, and we denote it by $I_1 I_2$. Under this composition rule, the automorphisms of F constitute a group, denoted by $\mathfrak{J}$.

An automorphism $I(f \mapsto f')$ is completely determined by giving the elements

$$a_1', b_1', \ldots, a_p', b_p'.$$

We think of these as given by words in the generators a_i, b_i. Let I^{-1} be given in an analogous way by a_i'', b_i''. Then if one substitutes the words $a_1', \ldots, b_p'$ for $a_1, \ldots, b_p$ in a_i'', one obtains a_i, after one has simplified the word, if necessary, by application of the defining relation

$$R(a_1, \ldots, b_p) = a_1 b_1 a_1^{-1} b_1^{\ 1} \ldots a_p b_p a_p^{-1} b_p^{-1} = 1.$$

Thus the original generators may be expressed in terms of the primed generators. Since $R(a_1, \ldots, b_p) = 1$, one also has $R(a_1', \ldots, b_p') = 1$; if one replaces the a_i', b_i' here by their expressions in the a_i, b_i, then the latter equation can be derived from the former, though perhaps only in several steps.

The converse problem arises, to decide whether $2p$ given elements $a_1', \ldots, b_p'$ correspond to the generators $a_1, \ldots, b_p$ under an automorphism of F. We shall first give only a *sufficient* condition for this to be the case. This criterion will be applicable to all the examples which occur later.

In what follows one has to distinguish between the terms "word" and "group element." Two words $s(a_1, \ldots, b_p)$ and $t(a_1, \ldots, b_p)$ represent the same element if and only if the word st^{-1} represents the identity; the latter is the case if and only if st^{-1} is representable as a product of conjugates of $R(a_1, \ldots, b_p)$ and R^{-1} identically in $a_1, \ldots, b_p$:

$$st^{-1} \equiv \prod_{i=1}^{m} h_i(a_1, \ldots, b_p) R^{\varepsilon_i}(a_1, \ldots, b_p) h_i^{-1}(a_1, \ldots, b_p),$$

where the h_i are arbitrary words.

The criterion in question is now the following: Let $s_1, \ldots, s_{2p}$ be $2p$ words which express the $2p$ elements $a_1', \ldots, b_p'$:

$$a_1' = s_1(a_1, \ldots, b_p)$$
$$b_p' = s_{2p}(a_1, \ldots, b_p)$$

and suppose the following two conditions are satisfied:

(α) $$R(s_1, \ldots, s_{2p}) \equiv h(a_1, \ldots, a_p)R^{\pm 1}(a_1, \ldots, b_p)h^{-1} \quad (1)$$

identically in the generators.

(β) There are $2p$ words $t_1(a_1, \ldots, b_p), \ldots, t_{2p}(a_1, \ldots, b_p)$ such that

$$t_1(s_1, \ldots, s_{2p}) \equiv a_1$$

$$\ldots$$

$$t_{2p}(s_1, \ldots, s_{2p}) \equiv b_p$$

identically in the generators.

Then the correspondence

$$a_1 \mapsto a_1', \ldots, b_p \mapsto b_p'$$

generates an automorphism of F.

Proof. We call the operation which replaces a_1 by the word $s_1(a_1, \ldots, b_p)$, etc. down to b_p by s_{2p} the "substitution S," and similarly call replacement of a_1 by $t_1(a_1, \ldots, b_p)$, etc. down to b_p by t_{2p} the "substitution T." Then TS is the identity substitution, by (β). Thus S represents an automorphism of the *free* group on the $2p$ generators $a_1, \ldots, b_p$. Then we also have $ST = 1$, because

$$S = S \cdot TS = ST \cdot S$$

so ST is the identity substitution. Applying T to the identity (1) therefore yields:

$$R(a_1, \ldots, b_p) \equiv h(t_1, \ldots, t_{2p})R^{\pm 1}(t_1, \ldots, t_{2p})h^{-1},$$

hence

$$R(t_1, \ldots, t_{2p}) \equiv l(a_1, \ldots, b_p)R^{\pm 1}(a_1, \ldots, b_p)l^{-1}, \quad (2)$$

where

$$h^{-1}(t_1, \ldots, t_{2p}) \equiv l(a_1, \ldots, b_p).$$

Now we have to show that the correspondence

$$a_1 \mapsto a_1', \ldots, b_p \mapsto b_p' \quad (3)$$

satisfies the conditions (a), (b) and (c) above. Let f be an arbitrary element of F and let $w(a_1, \ldots, b_p)$ be one of the words which represent f:

$$f = w(a_1, \ldots, b_p).$$

For condition (c), f must correspond to the element which one obtains when one replaces the generators $a_1, \ldots, b_p$ in w by $a_1', \ldots, b_p'$. We therefore write

$$f' = S(w) = w(s_1, \ldots, s_{2p}) \quad (4)$$

and we have to show that we always arrive at the same f', regardless of the word chosen to express f. But each word for f has the form

$$w_1(a_1, \ldots, b_p) = w(a_1, \ldots, b_p)\prod_i r_i(a_1, \ldots, b_p)R^{n_i}r_i^{-1},$$

where the η_i can be confined to the values $+1$ and -1. Thus

$$S(w_1) = w_1(s_1, \ldots, s_{2p}) = S(w)\prod_i r_i(s_1, \ldots, s_{2p})R^{\eta_i}(s_1, \ldots, s_{2p})r_i^{-1}$$

and hence by (1)

$$\equiv S(w)\prod_i r_i(s_1, \ldots, s_{2p})\cdot h(a_1, \ldots, b_p)R^{\pm\eta_i}(a_1, \ldots, b_p)h^{-1}\cdot r_i^{-1}$$

$$\equiv S(w)\prod_i h_i R^{\varepsilon_i} h_i^{-1}$$

and the latter word represents the same element f' as $S(w)$. Thus condition (α) is satisfied.

Now let f' be an arbitrary element of F and let

$$f' = v(a_1, \ldots, b_p)$$

be one expression for it. We have to show that the correspondence (3) given by (4) is one-to-one. We therefore write

$$f = T(v) = v(t_1, \ldots, t_{2p}).$$

Then

$$S(v(t_1, \ldots, t_{2p})) = S(T(v(a_1, \ldots, b_p))) = v(a_1, \ldots, b_p).$$

Another representation

$$v_1 = v\cdot\prod_i r_i R^{\eta_i} r_i^{-1}$$

of the same element f' results from

$$T(v_1) \equiv T(v)\cdot\prod_i r_i(t_1, \ldots, t_{2p})R^{\eta_i}(t_1, \ldots, t_{2p})r_i^{-1}$$

by S, and by (2) this is

$$= T(v)\prod_i h_i R^{\varepsilon_i} h_i^{-1},$$

and hence represents the same element f as $T(v)$. Thus condition (b) is satisfied. Condition (c) now follows from the character of S as a substitution.

The correspondence (3) determined by the substitution S therefore generates an automorphism. Q.E.D.

The inverse automorphism is given by $S^{-1} = T$; this sends the generators to the elements

$$a_1'' = t_1(a_1, \ldots, b_p)\,;\ldots;\, b_p'' = t_{2p}(a_1, \ldots, b_p).$$

9. Preservation of the cyclic order of fundamental points under automorphisms. Let $I(f \mapsto f')$ be an automorphism. To a primary element f_0 there corresponds a primary element f_0', and to the positive powers of f_0, the positive powers of f_0'. Two elements whose directed axes coincide therefore correspond to two elements whose directed axes also coincide. Thus we can say that *I induces a one-to-one mapping of the set G of fundamental points onto itself*, when we associate the negative and positive fundamental points, $U(f)$ and $V(f)$, with $U(f')$ and $V(f')$ respectively. We now have the following:

Theorem 1. *The cyclic ordering of the set G on E is preserved by this mapping* $G \to G_I$.

Proof. To prove this result we go back to the net n used in Chapter I, the vertices of which are all the points of Φ equivalent to the origin. The given automorphism I induces a one-to-one mapping of the set of net vertices onto itself, in which the point $f(0)$ corresponds to the point $f'(0)$. The origin O corresponds to itself. Now one connects O to $a_1'(0)$ by a line segment, which is labelled a_1' and provided with an arrow from O to $a_1'(0)$. The analogous construction is made for the remaining primed generators and their inverses. Let the longest of these $4p$ segments have length l. (Everything is understood to be in the sense of the non-Euclidean metric.) The images of the star consisting of these $4p$ segments under all the elements of F form a "net" n' whose vertices are the vertices of n. Of course, n' is not a net in the strict sense, since the edges may intersect outside the vertices, but it obviously has the following properties, which are all we need: if one travels from O to P in n along an edge path w, then one arrives at the point P' corresponding to P under I when one travels from O in n' along the path w' consisting of the primed edges corresponding to the unprimed edges in w; if one arrives at P_1 from P by the path w_1 in n, then one arrives at P_1' from P by the path w_1' in n'. If the names of the paths are also viewed as words, then

$$f = ww_1w^{-1}$$

is the element which carries P to P_1, and

$$f' = w'w_1'w'^{-1}$$

is the element which carries P' to P_1'.

The edges of the nets n and n' are in one-to-one correspondence with each other.

Now let M' be the set of vertices at distance $\leq l$ from O, let M be the image of M' under I^{-1}, and let L be the radius of a circle about the origin which contains M in its interior. Any two vertices whose distance is $\geq L$ are therefore carried by I into two vertices whose distance is $> l$. Now when two edges of n' have an endpoint or interior point in common, there is an endpoint of one of them at distance $< l$ from an endpoint of the other; thus they are I-images of two edges of n whose distance apart is $< L$.

Now let f_1 and f_2 be two elements of F, let V_1 and V_2 be their positive fundamental points, and let $k^{(1)}$ and $k^{(2)}$ be the expansions of V_1 and V_2, which we can view as edge paths issuing from O and tending to V_1 and V_2 respectively (Figure 12). (We do not exclude the possibility that $f_2 = f_1^{-1}$, in which case V_1 and V_2 form an associated pair of fundamental points.) Let P_1 and P_2 be points on $k^{(1)}$, $k^{(2)}$ respectively, beyond which these expansions are periodic. Let λ_1 and μ_1 be two curves at constant distance from the axis of f_1 such that each point outside the strip $\lambda_1\mu_1$ has distance $> L$ from the subpath P_1V_1. Define λ_2 and μ_2 analogously. C is a circle about the origin whose radius is so large that it intersects these four distance curves, and so that each point outside C has a distance $> L$ from the subpath P_1OP_2 of the edge path W formed from $k^{(1)}$ and $k^{(2)}$. The only piece of each distance curve we need is the one from E to its first intersection with C. The part of Φ which lies outside the strips $\lambda_1\mu_1$ and $\lambda_2\mu_2$, and outside C, consists of two regions ω_1 and ω_2, each of which has a boundary segment which is a subarc of E from V_1 to V_2.

Now on n' we construct the edge path W' which corresponds to W. This consists of the subpaths OP_1' and OP_2', where $P_1' = I(P_1)$ and $P_2' = I(P_2)$, moreover, it has a section periodic

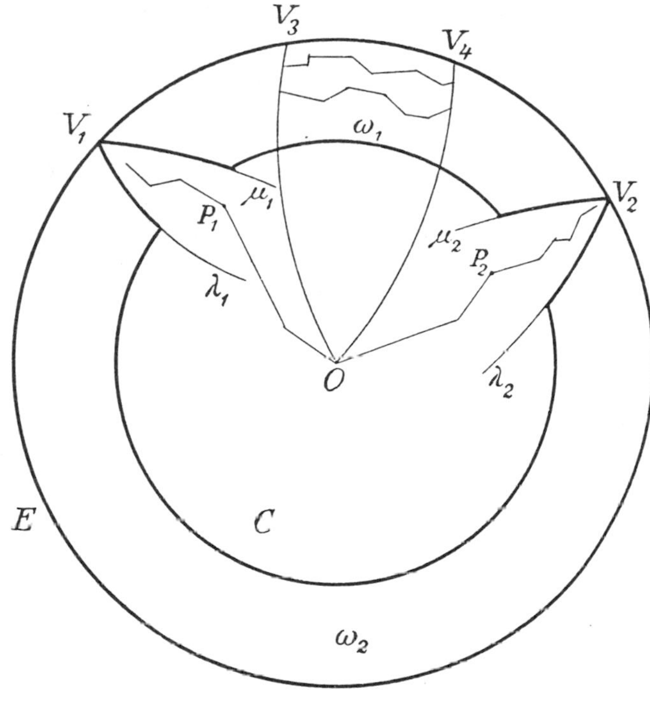

Figure 12

under f_1' from P_1' to $V_1' = V(f_1')$, which may be enclosed between curves ρ_1 and σ_1 at constant distance from the axis of f_1'; analogously, a section periodic under f_2', which runs from P_2' to $V_2' = V(f_2')$ and may be enclosed between curves ρ_2 and σ_2 at constant distance from the axis of f_2'. Also, let Γ be a circle which intersects these four distance curves and completely encloses the subpath $P_1'OP_2'$ of W'. The part of Φ outside the strips $\rho_1\sigma_1$ and $\rho_2\sigma_2$, and outside Γ, consists of two regions τ_1 and τ_2, each of which has a boundary segment which is a subarc of E from V_1' to V_2'. *Each continuous path which connects a point of τ_1 to a point of τ_2 meets W'.*

Now let f_3 and f_4 be two elements whose positive fundamental points V_3 and V_4 belong to the same subarc V_1V_2 of E, say, the one in the boundary of ω_1. We construct the sequence of points $f_3^r(0)$ and $f_4^r(0)$, where r runs through all positive integers. These points lie on circular arcs which go through V_3, O, $U(f_3)$ and V_4, O, $U(f_4)$ respectively. One determines a number r_0 so large that all points of both sequences belong to the region ω_1 for $r > r_0$. Then, for all $r > r_0$, edge paths q_r may be chosen so as to connect $f_3^r(0)$ with $f_4^r(0)$ on the net n, while running entirely inside ω_1. By construction, q_r has a distance $>L$ from W. *Thus the image q_r' of q_r under I does not meet W'.* The points $f_3'^r(0)$ and $f_4'^r(0)$ are connected on n' by q_r'. Now $f_3'^r(0)$ converges to $V_3' = V(f_3')$ as $r \to \infty$, and $f_4'^r(0)$ to V_4'. Thus, for sufficiently large r, these point sequences belong to one of the regions τ_1 and τ_2, and they must belong to *the same one* of these two regions, since q_r' does not meet W'. Thus V_3' and V_4' lie on the same subarc of E determined by V_1' and V_2', and the cyclic order of four fundamental points on E is preserved under automorphisms. Q.E.D.

Theorem 1 contains a theorem of M. Dehn (never published, to my knowledge) which Dehn told me of some years ago:

> *The axes of two elements of F intersect if and only if the axes of their images under an automorphism of F intersect.*

Essential steps of Dehn's proof are repeated in the proof above. From this theorem, Dehn drew the conclusion that the topological intersection of curve types is preserved under automorphisms. Indeed: let f be a primary element and s a displacement segment on its axis. Let P_1 and P_2 be two equivalent points on s, $P_2 = h(P_1)$. Then the axis of hfh^{-1} meets that of f at P_2, and this is preserved under automorphisms. *Thus the number of inequivalent intersection points of axes of a conjugacy class is the same for two classes which correspond under an automorphism.* When there are no such points P_1 and P_2 on s, i.e., when the conjugacy class corresponds to a curve type *without essential double points*, then the same is true for any class which corresponds to it under an automorphism. In exactly the same way one sees that: *the number of inequivalent intersection points of axes from two conjugacy classes*—the topological intersection number of the two curve types—*is preserved under automorphisms.*

Suppose three points 1, 2, 3 on E lie in that order in an anticlockwise circuit of E. The corresponding points under I may then appear clockwise or anticlockwise in the order 1′2′3′. The preservation of the cyclic order of any four points easily implies that the situation for any one triple 123 determines that for all other triples. According as the orientation of the fundamental point set G on E is preserved or not by $G \to G_I$, we speak of an *automorphism of the first kind* or *second kind.* The subgroup of automorphisms of the first kind is a normal subgroup of index 2 in the group $\mathfrak{J}$, and is denoted by $\mathfrak{J}_0$.

10. The topological boundary mapping induced by an automorphism. In §5 it was shown that the fundamental point set G is everywhere dense on E. Thus each point P of E is the limit of a sequence of fundamental points $G_1, G_2, \ldots$, converging monotonically to P, and the corresponding sequence $G_1', G_2', \ldots$ under the automorphism I is likewise monotonic by §9. Call its limit point P'. By §9, any other sequence of fundamental points converging to P yields a corresponding sequence converging to the same point P'. When P is itself a fundamental point, then P' is the point corresponding to P under the automorphism, as follows immediately from the preservation of the cyclic order. When P is not a fundamental point, then neither is P', as follows by considering I^{-1}. Now if, in all cases, we let the image of an arbitrary point P of E be P', this gives a *topological mapping of E onto itself, induced by I.* Orientation is preserved by this mapping $E \to E_I$ if and only if I is an automorphism of the first kind.

The situation can be summarized as follows: The primary elements of F constitute a cyclically ordered set; under an automorphism, this set experiences a one-to-one order-preserving mapping into itself, which can be viewed as embedded in a topological mapping of the circle onto itself. Then one can draw conclusions about properties of automorphisms from easily accessible properties of topological transformations of the circle.

11. Some theorems about automorphisms of *F*. Let I be an automorphism of F. An element f which corresponds to itself under I, $f_I = f$, is called an *element fixed by I*, an element f which goes to its inverse under I, $f_I = f^{-1}$, is called an *element inverted by I.* The elements fixed by I constitute a *subgroup* of F, denoted by $H(I)$ or simply H. When I is the identity automorphism I_0, then $H(I_0) = F$, so each fundamental point is a fixed point of the mapping $E \to E_{I_0}$, and the latter is therefore the identity mapping.

The mapping $E \to E_I$ will be denoted by $g(E)$. Its fixed points—when there are any—constitute a closed set of points on E. The complementary set consists of finitely or denumerably many intervals free of fixed points.

(a) Let I be a non-identity automorphism of the *first kind*. Then $g(E)$ preserves orientation, and each interval free of fixed points is therefore mapped onto itself. All points of such an interval are therefore pushed in the same direction, and the interval remains fixed point free under g^n. The fixed point set of g^n is therefore identical with that of g. This holds in particular for fundamental points in the fixed point set, so that I^n has exactly the same fixed elements as I. However, it is assumed that $g(E)$ really has fixed points, so that intervals free of fixed points can arise in the complement. The fixed points of $g(E)$ need not necessarily include fundamental points, as we shall see in examples later. Anyway, it follows that $H(I^n) = H(I)$, as long as $H(I) \neq 1$.

Conversely, when f is an element for which $f_I \neq f$ but $f_{I^n} = f$, it follows that $g(E)$ has no fixed points, and hence that $H(I) = 1$. This is the case, in particular, when I inverts elements, since such elements are fixed by I^2. It is also the case when $I^n = 1$: for all non-identity automorphisms of finite order, $g(E)$ has no fixed points.

Now we consider more closely the case of an automorphism I of the first kind for which there is an element f such that

$$f_I \neq f, \qquad f_{I^n} = f.$$

Let n be the smallest positive exponent with this property. Let V be the positive fundamental point of f, and g the boundary mapping induced by I. We construct the point set W consisting of the points

$$V, g(V), g^2(V), \ldots, g^{n-1}(V);$$

these n points are all different, since f is not fixed for any power of I less than n. Let V_1 and V_2 be two of these points which occur in succession in an anticlockwise circuit of E and suppose that

$$V_2 = g^r(V_1), \qquad 1 \leq r \leq n-1.$$

Then g^r induces a cyclic permutation of W of one step, and g is a power, say the λth, of g^r, where

$$\lambda r \equiv 1 \pmod{n}.$$

We can, if necessary replace I by I^r and therefore take $r = 1$. Thus g cyclically permutes the n segments determined by W on E, and no power of g below the nth can have fixed points, so that

$$H(I^m) = H(I) = 1 \qquad \text{for} \quad 1 \leq m \leq n-1.$$

The fixed point set of g^n in one of the segments is carried by powers of g into the fixed point sets of g^n in the other segments. We therefore have

$$H(I^m) = 1 \qquad \text{for} \quad m \not\equiv 0 \pmod{n},$$
$$H(I^{m \cdot n}) = H(I^n) \neq 1.$$

We call n the *characteristic exponent* of I; n is independent of the above choice of the element

f. For automorphisms of finite order the characteristic exponent is equal to the order; in case $H(I^m) = H(I)$ for all m, the characteristic exponent of I is defined to be 1:

Theorem 2. *An automorphism of the first kind with characteristic exponent* >1 *induces a boundary mapping without fixed points.*

(b) Let I be an automorphism of the *second kind*. The induced boundary mapping reverses orientation, and therefore has exactly two fixed points. Thus I *has at most one primary fixed element.* I^2 is an automorphism of the first kind with characteristic exponent 1, since $g^2(E)$ in any case has two fixed points. Fixed elements of I^2 correspond either to inverted elements of I, or else they appear in pairs whose elements are exchanged by I. If I is of finite order, then it is necessarily of order 2.

Example 1.

$$p = 2; \qquad I \begin{cases} a_1 \mapsto b_2 \\ b_1 \mapsto a_2 \\ a_2 \mapsto b_1 \\ b_2 \mapsto a_1 \end{cases}$$

$I^2 = 1$. Primary fixed element $a_1 b_1 a_1^{-1} b_1^{-1}$. All elements are involutarily paired, e.g., $a_1^{-1} b_1^{-1} a_2 b_2$ with its inverse.

We now prove the following:

Theorem 3. *Each subgroup of finite order in* $\mathfrak{J}_0$ *is cyclic.*[9]

Proof. Let $\mathfrak{H}$ be a group of order n whose elements are automorphisms of the first kind, let P_1 be an arbitrary point of E, and let $P_1, P_2, \ldots, P_n$ be the points to which P_1 is sent by the boundary mappings induced by elements of $\mathfrak{H}$. These points are all different; otherwise some automorphism in $\mathfrak{H}$ would leave one of these points fixed, whereas all elements of $\mathfrak{H}$ are of finite order, and therefore induce mappings without fixed points by Theorem 2. Each element of $\mathfrak{H}$ induces a permutation of these n points which preserves cyclic order and orientation on E. Thus an element of $\mathfrak{H}$ which sends one point to its neighbour generates all of $\mathfrak{H}$. Q.E.D.

12. Inner automorphisms. The map

$$l \mapsto f^{-1} l f, \tag{5}$$

where f is a fixed element of F and l runs through the group F, is well known to be an automorphism of F, for which the name "*inner automorphism*" or "*cogredient automorphism*" is usual. If one determines it by substitutions

$$\begin{gathered} a_1 \mapsto w^{-1}(a_1, \ldots, b_p) a_1 w(a_1, \ldots, b_p) \\ \cdots \\ b_p \mapsto w^{-1}(a_1, \ldots, b_p) b_p w(a_1, \ldots, b_p), \end{gathered}$$

where w is a word representing the element f, then this prescription satisfies the conditions (α)

[9] This theorem follows immediately from L. E. J. Brouwer, Über topologische Involutionen, §3, Amsterd. Proceedings Vol. XXI, no. 9 (1919).

and (β) of §8. Namely, one has

$$R \mapsto w^{-1} R w$$

and

$$w \mapsto w^{-1} w w \equiv w$$

so that

$$w a_1 w^{-1} \mapsto a_1$$

and analogously.

We denote the automorphism (5) by I_{0f}. If I is the automorphism $l \mapsto l'$, then $I \cdot I_{0f}$ is the automorphism

$$l \mapsto f^{-1} l' f$$

and it is denoted by I_f.

We have

$$I_{0f_1} \cdot I_{0f_2} = I_{0f_1 f_2}$$

and

$$(I_{0f})^{-1} = I_{0f^{-1}}.$$

Also

$$(I_f)^{-1} = I^{-1}_{f^{-1}_{I^{-1}}}$$

and

$$I^{-1} : l \mapsto l_{I^{-1}},$$
$$I^{-1} I_{0f} : l \mapsto f^{-1} l_{I^{-1}} f,$$
$$I^{-1} I_{0f} I : l \mapsto f_I^{-1} l f_I.$$

The automorphism

$$I^{-1} I_{0f} I = I_{0f_I} \tag{6}$$

is therefore again an inner automorphism:

> The group of inner automorphisms is a normal subgroup of $\mathfrak{J}_0$ and $\mathfrak{J}$. It is denoted by $\mathfrak{F}$. $\mathfrak{F}$ is isomorphic to F.

The automorphisms I and I_f will be called *related*. This is a symmetric and transitive property. The collection of automorphisms I_f, where I is a fixed automorphism and f runs through the group F is therefore a *coset* of the normal subgroup $\mathfrak{F}$ in $\mathfrak{J}$, which will be called an *automorphism family*. In the publications *loc. cit.*[1] and *loc. cit.*[2] I have used the term "isomorphism class" for this concept; since it seems more usual to understand a *class* of elements of a group to be a collection of mutually conjugate elements, I have now reserved the word "class" for the latter sense (as in §1) and I use the word "family" for the collection above. The class of the identity consists only of the identity automorphism. The family of the identity consists of the group $\mathfrak{F}$ of all inner automorphisms.

13. The functional equation of the boundary mapping. The automorphism (5) sends an arbitrary fundamental point G_1 to $f^{-1}(G_1)$. The induced topological boundary mapping, $g(E)$ in the notation of §11, is therefore $f^{-1}(E)$ in this case. Thus if H denotes the group, isomorphic to $\mathfrak{J}$, of all the boundary mappings $g(E)$ induced by automorphisms, and H_0 the normal subgroup, corresponding to $\mathfrak{J}_0$, of mappings which preserve the orientation of E, then F, viewed as a group of mappings of E, is a normal subgroup of H_0 corresponding to the normal subgroup $\mathfrak{F}$ of $\mathfrak{J}_0$,

One notices that when $g_1(E)$ is induced by the automorphism I_1 and $g_2(E)$ by I_2, then $I_1 I_2$ induces the boundary mapping $g_2(g_1(E))$. In future we shall omit brackets around the function arguments when there is no danger of misunderstanding.

Now if g corresponds to the automorphism I, then equation (6) of §12, interpreted as a statement about boundary mappings, and with f^{-1} in place of f, reads:

$$gfg^{-1} = f_I. \tag{7}$$

Applied to an arbitrary point P of E, (7) says:

$$gfP = f_I gP \tag{8}$$

One can convince onself of the correctness of this *fundamental functional equation* (8) somewhat more directly by viewing P as the limit of a sequence of fundamental points and first checking (8) for fundamental points. If I is given by the correspondence

$$a_1 \mapsto a_1', \ldots, b_p \mapsto b_p',$$

then (8) is equivalent to the system of $2p$ functional equations

$$\begin{aligned} ga_1 P &= a_1' gP \\ &\cdots \\ gb_p P &= b_p' gP. \end{aligned} \tag{9}$$

14. Structure of fixed point free intervals. Let $I \neq I_0$ be an automorphism of the first kind, g the induced boundary mapping, $H(I)$ the group of elements fixed by I, and suppose $H(I) \neq 1$. Let h be an element fixed by I. Then by (8)

$$ghP = hgP. \tag{10}$$

Thus g commutes with all elements of H. The operation h therefore carries fixed points of gE to fixed points, and non-fixed points to non-fixed points, and hence maps a fixed point free interval onto one of the same kind. When an operation h_1 of H sends a point of the fixed point free interval i_1 to another point of the same interval, then i_1 must be mapped entirely into itself by h_1; the fixed points bounding i_1 must therefore be the fundamental points of h_1. Such an interval will be called a *periodic interval.* Because, if P is a point of i_1 and if one knows gi_1 on the segment from P to $h_1 P$, then by (10) one knows g on the whole of i_1.

On the other hand, if i is a fixed point free interval and if one of the end points of the interval is a fundamental point, then it must be the fundamental point of a fixed element h, and then i and hi have points in common, so that $hi = i$. Thus the other end point of the interval is the other fundamental point of h, and i is periodic:

Theorem 4. *There are two kinds of fixed point free intervals, "aperiodic" and "periodic." An* *aperiodic interval is bounded by non-fundamental points and contains no pair of points which are equivalent with respect to the fixed element group H. A periodic interval i is bounded by an associated pair of fundamental points. If h is the primary (fixed) element associated with this pair, then there is no other pair of points in i, equivalent with respect to H, than those which correspond under a power of h.*

Thus when P is an arbitrary point of E and HP denotes the set of points which result from P by operations in H, then HP has no accumulation point in the interior of a fixed point free interval. However, each fundamental point of a fixed element h is an accumulation point of HP, since the set HP is mapped into itself by h. One can always choose P to be a non-fixed point, and then HP consists purely of non-fixed points by (10). Thus each fundamental point of a fixed element is an accumulation point of non-fixed points, and hence *there can be no interval consisting entirely of fixed points*, since each interval contains fundamental points (§5). If there are fixed points on E other than the fundamental points of h, then the fundamental points of h are also accumulation points of fixed points:

The set of fixed points of $g(E)$ is closed and nowhere dense. In the case where $H(I)$ is an infinite cyclic group, and only then, are there isolated fixed fundamental points.

15. Examples

Example 2. We take as an example the automorphism of F which sends a_1 to $a_1 b_1$ and leaves the remaining generators fixed; it may be denoted Λ:

$$\Lambda : (a_1 \mapsto a_1 b_1).$$

The criterion of §8 shows that Λ is an automorphism. The induced boundary mapping is denoted by $\lambda(E)$.

$H(\Lambda)$ certainly contains the group generated by $b_1, a_2, \ldots, b_p$. Thus Λ is of the first kind. Let M be the fixed point set of $\lambda(E)$. All operations in H map M into itself and the set $E - M$ of intervals into itself (§14). We shall now determine the structure of M and $E - M$ more precisely.

We have

$$\Lambda^n : (a_1 \to a_1 b_1^n)$$

for positive and negative n and therefore

$$a_{1_{\Lambda^n}}^{-1} = b_1^{-n} a_1^{-1}.$$

The positive fundamental point of this element has an expansion which consists of periodic repetition of $b_1^{-n} a_1^{-1}$, because this satisfies the characteristic conditions for expansions (§6). As $n \to \infty$, this expansion tends towards the expansion of $U(b_1)$ (i.e., for increasing n it has longer and longer convergents in common with the latter), for $n \to -\infty$ it tends toward that of $V(b_1)$. Thus the image of $U(a_1)$ under positive and negative powers of λ converges to the endpoints of the axis of b_1, so that the latter bound a periodic interval i, whose points are pushed by λ away from $V(b_1)$ and towards $U(b_1)$.

Similarly, one has

$$a_{1_{\Lambda^n}} = a_1 b_1^n,$$

hence as $n \to \infty$ the expansion of $V(a_{1_\wedge n})$ converges to that of $V(a_1 b_1 a_1^{-1})$, and as $n \to -\infty$ to that of $U(a_1 b_1 a_1^{-1})$. Thus $V(a_1)$ lies in a periodic interval j bounded by the axis of $a_1 b_1 a_1^{-1}$.

The images of i and j under all operations in H are again fixed point free periodic intervals, bounded by the axes of those elements to which b_1 and $a_1 b_1 a_1^{-1}$, respectively, are sent by conjugation with the elements of H. *It will be shown that there are no other fixed point free intervals.* Let the set of points which belong to none of these (open) intervals be M_1. Thus we have to show that M_1 is identical with M.

Let P be an arbitrary point of M_1 and let

$$k = e_1 e_2 e_3 \ldots$$

be its expansion. The e_i denote single generators or their inverses. Denote the point $(e_1 e_2 \ldots e_r)^{-1} P$ by P_r. Now the intervals i and j have the following positions relative to the interval subdivision for the group expansion (§6): $U(b_1)$ lies in the first order interval $[b_1^{-1}]$, $V(b_1)$ in $[b_1]$, thus $[a_1^{-1}]$ lies in the interior of i. Since P is outside i, we therefore have $e_1 \neq a_1^{-1}$. Also, $V(a_1 b_1 a_1^{-1})$ and $U(a_1 b_1 a_1^{-1}) = V(a_1 b_1^{-1} a_1^{-1})$ both lie in the interior of $[a_1]$, hence j also lies entirely in $[a_1]$. Now when $e_1 \neq a_1$, then e_1 is a fixed element, and hence $P_1 = e_1^{-1} P$ belongs to M_1, since M_1 is mapped into itself by all operations in H. Let e_n be the first of the e_i, if any, which is not a fixed element; then $e_n = a_1$ and not $= a_1^{-1}$. Then P_{n-1} lies in $[a_1]$ but outside j. Hence P_n belongs to i or is an endpoint of i, because

$$a_1^{-1} j = E - i - V(b_1) - U(b_1).$$

Thus $e_{n+1} = b_1^{\pm 1}$ since e_{n+1} is not inverse to e_n. Suppose $e_{n+1} = b_1$. Then there are two possibilities: either all $e_{n+m} = b_1$; in which case

$$P_n = V(b_1), \qquad P_{n-1} = V(a_1 b_1 a_1^{-1})$$

and hence P is also a fixed point of λ. Or else P_n belongs to the intersection of $[b_1]$ and i. Then the image of P_n under a sufficiently high power of b_1^{-1} falls outside $[b_1]$. Hence there is an $m \geq 1$ such that

$$e_{n+1} = e_{n+2} = \cdots = e_{n+m} = b_1,$$
$$e_{n+m+1} = a_1^{-1}.$$

Then

$$P_{n+m+1} = a_1 b_1^{-m} a_1^{-1} P_{n-1}$$

and this is again a point of M_1, since $a_1 b_1^{-m} a_1^{-1}$ belongs to H. The expansion of P_{n+m+1} therefore has the same property that we observed above for the beginning of the expansion of P, and we have: In the expansion of an arbitrary element P of M_1 the only terms which can appear, apart from the generators $b_1, a_2, \ldots, b_p$ which are all fixed elements, are subsequences $a_1 b_1^m a_1^{-1}$ for arbitrary positive or negative values of m. Such subsequences also represent fixed elements. Thus $k(P)$ has arbitrarily high convergents which are fixed elements, and since (by §7, j) the points $V(k_n)$ converge to P, P is an accumulation point of fixed points, and hence itself fixed. Q.E.D.

The set M of fixed points (which we now know is identical with M_1) is nowhere dense by §14. All intervals are periodic. No fixed point is the common endpoint of two intervals, since there are more than two intervals, and hence any two intervals are bounded by different axes. Each fixed point is therefore an accumulation point of interval endpoints. *The set of fixed points is therefore perfect* (*and nowhere dense*). It consists of the fundamental points of all those elements into which b_1 and $a_1 b_1 a_1^{-1}$ are carried by conjugation with fixed elements, and all the accumulation points of such points.

Now let h be an arbitrary fixed element. The expansion of $V(h)$ is given by an expression of the form

$$k(V(h)) = \alpha\beta\beta\beta\ldots$$

where α and β, by the foregoing, are words in $b_1, a_2, \ldots, b_p$ as well as complexes $a_1 b_1^m a_1^{-1}$. Now since $R = 1$ we have

$$a_1 b_1 a_1^{-1} = b_p a_p b_p^{-1} a_p^{-1} \ldots b_2 a_2 b_2^{-1} a_2^{-1} b_1,$$

hence $h = \alpha\beta\alpha^{-1}$ is expressible in terms of $b_1, a_2, \ldots, b_p$ and we have:

The fixed element group $H(\Lambda)$ is generated by $b_1, a_2, \ldots, b_p$.

It is easy to construct a *fundamental domain* for $H(\Lambda)$ in Φ; we shall come back to this later (§48). Here we only remark that one obviously obtains a fundamental domain for $H(\Lambda)$ *on the set $E - M$* by constructing a segment in i whose endpoints correspond under b_1, and a segment in j whose endpoints correspond under $a_1 b_1 a_1^{-1}$.

The group $H(\Lambda)$ is a free group, i.e., there are no relations between $b_1, a_2, \ldots, b_p$ other than identities. The later construction of a fundamental domain for $H(\Lambda)$ in Φ will enable us to see this without effort. One can also convince oneself directly by omitting all edges labelled a_1 in the net n which is the Dehn group diagram for F. The subcomplex connected to the origin in the resulting diagram is a "tree," i.e., it contains no closed paths, from which the assertion follows. The fixed point set M on E is identical with the set of accumulation points of vertices of this tree.

When we are given two intervals, and a topological mapping of each onto itself without interior fixed points, then it is known[10] that these mappings are always *similar*, i.e., there is a topological mapping of one interval onto the other under which the topological self-mapping of one interval is transformed into that of the other. In our case we have in particular: If h is an element fixed by Λ and i_1 is the fixed point free interval bounded by the axis of hb_1h^{-1}, then λi_1 results from λi by conjugation with h, since, by (10), the equation $\lambda = h\lambda h^{-1}$ holds on all of E. The corresponding fact holds for all intervals equivalent to j under H. If we compare λi and λj, these are *not* transformable into each other by an element of F, because i is not equivalent to j via F; namely, j is equivalent to $E - i - V(b_1) - U(b_1)$ via a_1, as we have already used above. However, there is a boundary mapping induced by an automorphism which transforms λi into λj: namely, let I be the following automorphism of the first kind:

$$I: \left| \begin{array}{l} a_1 \to a_1^{-1} \\ b_1 \to a_1 b_1^{-1} a_1^{-1} \\ \left.\begin{array}{l} a_i \to b_1^{-1} a_i b_1 \\ b_i \to b_1^{-1} b_i b_1 \end{array}\right\} i = 2, \ldots, p \end{array} \right|$$

[10] H. Kneser: Reguläre Kurvenscharen auf den Ringflächen, §3, Math. Ann. **91** (1924).

and let $g(E)$ be the induced boundary mapping. Since b_1 goes to $a_1 b_1^{-1} a_1^{-1}$ and $U(a_1)$ to $V(a_i)$, we have $gi = j$. Also, one checks immediately by substitution that

$$\Lambda I = I \Lambda.$$

We write this commutativity of two elements of a group in the form

$$\Lambda \rightleftarrows I.$$

In consequence, the boundary mappings also satisfy:

$$\lambda \rightleftarrows g,$$

and hence $g\lambda g^{-1} = \lambda$. Thus λi is transformed into λj by conjugation with g. In summary, we have:

> The self-mappings of any two fixed point free intervals for Λ are transformed into each other by the boundary mapping induced by an automorphism of the first kind (an operation in H_0). In particular, if the two intervals are equivalent the transformer belongs to F, and indeed to $H(\Lambda)$.

Example 3. As a further example, we consider an automorphism *related to* Λ,

$$\Lambda' = \Lambda_{a_1} = \Lambda \cdot I_{0a_1};$$

thus

$$\Lambda' : \left| \begin{array}{l} a_1 \to b_1 a_1 \\ b_1 \to a_1^{-1} b_1 a_1 \\ a_2 \to a_1^{-1} a_2 a_1 \\ \quad \cdots \\ b_p \to a_1^{-1} b_p a_1 \end{array} \right| .$$

Since I_{0a_1} induces the boundary mapping $a_1^{-1}(E)$, Λ' induces the boundary mapping

$$\lambda' = a_1^{-1} \lambda.$$

Since $U(a_1)$ lies in i and $V(a_1)$ in j, $\lambda' i$ is a subinterval of i and $\lambda' j$ is an interval which contains j as a subinterval. Since a topological mapping of a closed interval onto a proper part of itself always has at least one fixed point, λ' must have at least one fixed point in each of i and j. *It will now be shown that λ' has only two fixed points, and that these are not fundamental points, so that two aperiodic intervals result.* The powers of Λ' are:

$$\begin{aligned} \Lambda' &= \Lambda_{a_1} \\ \Lambda'^2 &= \Lambda^2_{a_{1_\Lambda} a_1} \\ \Lambda'^3 &= \Lambda^3_{a_{1_{\Lambda^2}} a_{1_\Lambda} a_1} \\ &\cdots \\ \Lambda'^n &= \Lambda^n_{a_{1_{\Lambda^{n-1}}} a_{1_{\Lambda^{n-2}}} \cdots a_{1_\Lambda} a_1}, \end{aligned}$$

thus

$$\Lambda'^n = \Lambda^n_{a_1 b_1^{n-1} a_1 b_1^{n-2} \cdots a_1 b_1 a_1},$$

Also (§12)

$$\Lambda'^{-1} = \Lambda^{-1}_{a_{1_{\Lambda^{-1}}}^{-1}},$$

and hence, in analogy with the above, for $n > 0$:

$$\begin{aligned}\Lambda'^{-n} &= \Lambda^{-n_1}_{a_{1_{\Lambda^{-n}}}^{-1} a_{1_{\Lambda^{-(n-1)}}}^{-1} \ldots a_{1_{\Lambda^{-1}}}^{-1}}\\ &= \Lambda^{-n}_{b_1^n a_1^{-1} b_1^{n-1} a_1^{-1} \ldots b_1^2 a_1^{-1} b_1 a_1^{-1}}.\end{aligned}$$

Thus one has

$$b_{1_{\Lambda'^n}} = a_1^{-1} b_1^{-1} a_1^{-1} b_1^{-2} \ldots a_1^{-1} b_1^{-(n-1)} a_1^{-1} \cdot b_1 \cdot a_1 b_1^{n-1} a_1 \ldots b_1^2 a_1 b_1 a_1,$$

$$b_{1_{\Lambda'^{-n}}} = a_1 b_1^{-1} a_1 b_1^{-2} \ldots a_1 b_1^{-n} \cdot b_1 \cdot b_1^n a_1^{-1} b_1^{n-1} a_1^{-1} \ldots b_1 a_1^{-1},$$

and hence the fundamental point expansions:

$$k(V(b^{\pm 1}_{1_{\Lambda'^n}})) = a_1^{-1} b_1^{-1} a_1^{-1} b_1^{-2} \ldots a_1^{-1} b_1^{-(n-1)} a_1^{-1} \cdot b_1^{\pm 1} b_1^{\pm 1} b_1^{\pm 1} \ldots,$$

$$k(V(b^{\pm 1}_{1_{\Lambda'^{-n}}})) = a_1 b_1^{-1} a_1 b_1^{-2} \ldots a_1 b_1^{-(n-1)} a_1 \cdot b_1^{\pm 1} b_1^{\pm 1} b_1^{\pm 1} \ldots,$$

Therefore, as $n \to \infty$, $V(b^{\pm 1}_{1_{\Lambda'^n}})$ converges to the point P^+ with the expansion:

$$k(P^+) = a_1^{-1} b_1^{-1} a_1^{-1} b_1^{-2} a_1^{-1} b_1^{-3} \ldots \text{ad inf.}$$

and as $n \to -\infty$ it converges to the point P^- with the expansion:

$$k(P^-) = a_1 b_1^{-1} a_1 b_1^{-2} a_1 b_1^{-3} \ldots \text{ad inf.}$$

Since these expansions are not periodic, P^+ and P^- are not fundamental points. As the expansions show, P^+ lies in $[a_1^{-1}]$ and P^- in $[a_1]$; they are therefore separated from each other by $U(b_1)$ and $V(b_1)$. The element b_1 is not fixed by Λ', and the endpoints of the axis of b_1 therefore spread over fixed point free intervals if powers of λ' are applied. Now since the latter two intervals are bounded by the same point pair $P^+ P^-$, these are the only two intervals on E. Thus $\lambda(E)$ generates two aperiodic intervals, separated by two fixed points which are not fundamental points, Q.E.D.

In particular, no fundamental point is a fixed point, *thus no element apart from the identity is fixed by* Λ', $H(\Lambda') = 1$. However, when one views generator sequences which appear as expansions of boundary points as *limit elements of* F, then there are *exactly two fixed limit elements under* Λ', namely $k(P^+)$ and $k(P^-)$: application of the substitution Λ' to these shows that they remain unchanged.

16. Correspondence between fundamental sequences under automorphisms. Let I be an automorphism, g the induced boundary mapping and

$$f_1, f_2, \ldots$$

a fundamental sequence for the boundary point P. Then by §7(j),

$$V(f_1), V(f_2), \ldots$$

converges to P. Hence by §10,

$$V(f_{1I}), V(f_{2I}), \ldots$$

converges to gP. Also, no element appears infinitely often in the sequence

$$f_{1I}, f_{2I}, \ldots$$

since this is true of the initial sequence. Thus the new sequence is also a fundamental sequence, namely, for gP:

The set of fundamental sequences is mapped one-to-one onto itself by an automorphism, and this correspondence between fundamental sequences determines the boundary mapping.

Thus even though the convergents $k_1, k_2, \ldots$ of the limit element $k(P)$ do not in general go to convergents of a limit element, the elements

$$k_{1I}, k_{2I}, \ldots$$

still constitute a fundamental sequence for gP.

Now suppose that gE has fixed points. Let f be an element outside $H(I)$. Then the elements of the sequence

$$f, f_I, f_{I^2}, f_{I^3}, \ldots$$

are all different, since I^n has no fixed elements other than those of I. Also, the sequence of points

$$V(f), V(f_I), V(f_{I^2}), \ldots$$

converges to an endpoint of the fixed point free interval to which $V(f)$ belongs. Thus one has:

If f is not a fixed element of I and if gE has fixed points, then $f, f_I, f_{I^2}, \ldots$ and $f, f_{I^{-1}}, f_{I^{-2}}, \ldots$ are fundamental sequences for the endpoints of the fixed point free interval containing the positive fundamental point of f.

The fixture of these interval endpoints is shown by the fact that these sequences are mapped to themselves, apart from initial segments, by arbitrary powers of I.

This gives a *method for determining fixed points of the boundary mapping*, and possibly for finding *fixed elements* of the automorphism: namely, since the point sequence $V(f_{I^n})$ tends to an interval endpoint as $n \to \infty$, which we can take to be "positive," for increasing n, $k(V(f_{I^n}))$ must have increasingly high convergents in common with the expansion of the positive endpoint of the interval. Now when the interval is periodic, this expansion of the endpoint is periodic, and *this fact must be expressed as a pattern in the expansion $k(V(f_{I^n}))$ for increasing n*, which is recognizable in a finite number of steps.

As an example, we go back to the automorphism Λ (§15, Example 2). Let f be an element outside $H(\Lambda)$. In $k(V(f))$ there are no fixed convergents of arbitrarily high order, since $V(f)$ is not a fixed point of λE. Let r be the largest number such that the $(r-1)^{\text{th}}$ order convergent k_{r-1} is an element fixed by Λ. Then

$$k_r = k_{r-1} a_1^{\pm 1}$$

and hence

$$k_{r_I n} = k_{r-1} a_1 b_1^n \qquad \text{or} \qquad k_{r-1} b_1^{-n} a_1^{-1}.$$

Thus the point sequence $k_{r,n}(0)$ converges to the positive fundamental point of $k_{r-1}a_1b_1a_1^{-1}k_{r-1}^{-1}$ or of $k_{r-1}b_1^{-1}k_{r-1}^{-1}$ as $n \to \infty$, and hence to the fundamental points of elements which result from b_1^{-1} or $a_1b_1a_1^{-1}$ by conjugation with fixed elements. One can easily see that this interval is in fact the one which contains $V(f)$, e.g., from the fact that the edge path representing $k(V(f))$ definitely leaves the tree consisting of edges labelled $b_1, a_2, \ldots, b_p$ in passing from k_{r-1} to k_r. We can also say, more concisely, that the interval to which the point belongs is determined by its last fixed convergent.

17. The fixed element groups of related automorphisms. Let I be an automorphism of the first kind, g the induced boundary mapping, and $H(I) \neq 1$. Suppose there is a fixed element h whose fundamental points are not interval endpoints. Only the fundamental points of h remain fixed under the operation hE. Since this operation maps the fixed point set of g into itself, each fixed point free interval is mapped onto another. Thus hg has only the fundamental points of h as fixed points, and it therefore has only two periodic intervals, and the automorphism

$$I \cdot I_{0h^{-1}} = I_{h^{-1}}$$

has only the powers of the primary element for h as fixed elements; the group $H(I_{h^{-1}})$ is therefore cyclic.

Example 4. The automorphism related to Λ,

$$\Lambda \cdot I_{0a_2^{-1}} = \Lambda_{a_2^{-1}}$$

is given by

$$\begin{cases} a_1 \mapsto a_2a_1b_1a_2^{-1}, & \\ a_2 \mapsto a_2, & \\ a_i \mapsto a_2a_ia_2^{-1}, & i = 3, \ldots, p, \\ b_i \mapsto a_2b_ia_2^{-1}, & i = 1, \ldots, p. \end{cases}$$

The fundamental points of a_2 bound intervals which are not fixed point free under Λ, since a_2 is not conjugate to b_1. The element a_2 is primary. Hence

$$H(\Lambda_{a_2^{-1}}) = \{a_2\},$$

and the induced boundary mapping yields two periodic intervals.

Now suppose that h is a primary element fixed by I, whose axis bounds a fixed point free interval i. The positive fundamental point of h is, say, the negative endpoint of i. Apart from the fundamental points of h there may be other fixed points of gE, which belong to

$$E - i - U(h) - V(h) = j.$$

The fixed point free intervals of gE contained in j are then permuted among themselves by any power h^n, and hence each automorphism I_{h^n}, $n \neq 0$, has j as a fixed point free interval. A negative power of h pushes the points of i in the same direction as does g. Hence I_{h^n}, $n > 0$, has no fixed point in i, and

$$H(I_{h^n}) = \{h\}.$$

On the other hand, there can be one—and then obviously *only one*—negative value of n for which I_{h^n} has a fixed point in i, and hence for which $H(I_{h^n})$ possibly includes more than the cyclic group $\{h\}$.

Example 5. Under Λ, the negative endpoint of the interval denoted by i in §15 coincides with the positive fundamental point of b_1. The automorphism $\Lambda_{b_1^{-1}}$ has, e.g., $a_1^{-1}b_1a_1$ or $a_1^{-1}a_2a_1$ as fixed elements. Thus we have: $H(\Lambda_{b_1^n})$ is identical with the cyclic group $\{b_1\}$ for $n \neq 0, -1$, but for these two special values it is larger.

We now consider the problem of finding fixed elements of related automorphisms somewhat more generally:

Let I be an arbitrary automorphism. Then for *arbitrary* elements f and h of F, and $n > 0$, we have:

$$f_{(I_h^{-1})^n} = hh_I h_{I^2} \ldots h_{I^{n-1}} \cdot f_{I^n} \cdot h_{I^{n-1}}^{-1} \ldots h_{I^2}^{-1} h_I^{-1} h^{-1}$$

and since

$$(I_{h^{-1}})^{-1} = I_{h_{I^{-1}}}^{-1},$$

$$f_{(I_{h^{-1}})^{-n}} = h_{I^{-1}}^{-1} h_{I^{-2}}^{-1} \ldots h_{I^{-n}}^{-1} \cdot f_{I^{-n}} \cdot h_{I^{-n}} \ldots h_{I^{-2}} h_{I^{-1}}.$$

If one sets

$$hh_I \ldots h_{I^{n-1}} = r_n$$

and

$$h_{I^{-1}}^{-1} h_{I^{-2}}^{-1} \ldots h_{I^{-n}}^{-1} = s_n$$

then it follows that

$$r_{n_{I_h^{-1}}} = r_{n+1} \cdot h^{-1}$$

$$s_{n_{I_h^{-1}}} = s_{n-1} \cdot h^{-1}.$$

Now we consider the case where the expansions of the positive fundamental points of r_n and s_n converge toward limit elements:

$$\left.\begin{array}{l} k(V(r_n)) \to r \\ k(V(s_n)) \to s \end{array}\right\} n \to \infty.$$

Then the two equations above show that r and s must be the expansions of the fixed points of $I_{h^{-1}}$, in conformity with the fact that the representations

$$r = hh_I h_{I^2} \ldots \text{ad inf.},$$

$$s = h_{I^{-1}}^{-1} h_{I^{-2}}^{-1} h_{I^{-3}}^{-1} \ldots \text{ad inf.}$$

are formally unchanged by application of $I_{h^{-1}}$. (Cf. the origin of the expansions of the two fixed points P^+ and P^- in Example 3, §15.) Here it remains possible for the limit elements r and s to be aperiodic or periodic. In the latter case we are led to proper fixed elements of $I_{h^{-1}}$. In both cases one can argue further as follows: when f is an element fixed by I or a power of I, then the positive fundamental point of $f_{(I_h^{-1})^n}$ converges to r as $n \to \infty$, and to s as $n \to -\infty$. Thus r

and s determine a fixed point free interval for $I_{h^{-1}}$ on E. Hence, if there are two fundamental points fixed by I, which are separated by r and s, it follows that $I_{h^{-1}}$ has only the fixed points r and s on E.

One sees that the cases considered above are comprehended by this general consideration: if h is an element fixed by I, then r and s are the periodic expansions of the fundamental points of h. If the two intervals they determine on E contain fundamental points fixed by I, then it follows that $H(I_{h^{-1}})$ is cyclic. However, if the axis of h bounds a fixed point free interval for I, then the complementary interval, even if it contains fundamental points fixed by I, is certainly fixed point free for $I_{h^{-1}}$.

Finally, we mention a theorem which is often useful—one can immediately check its correctness by substitution:

From $l_I = l$ and $h_{I^n} = lhl^{-1}$ one finds the element $hh_I \ldots h_{I^{n-1}} \cdot l$ fixed by $I_{h^{-1}}$.

E.g., with $p = 2$ and the automorphism

$$I: \begin{cases} a_1 b_1 \\ b_1 \\ b_2^{-1} \\ b_2 a_2 \end{cases}$$

one has the fixed element

$$l = b_2 a_2 b_2^{-1} a_2^{-1} = a_1 b_1 a_1^{-1} b_1^{-1}.$$

One has, incidentally,

$$H(I) = \{b_1, l\}.$$

One finds

$$a_{2_{I^6}} = l a_2 l^{-1},$$

$$b_{2_{I^6}} = l b_2 l^{-1},$$

and an element fixed by $I_{h^{-1}}$, where h is an arbitrary element of the group $\{a_2, b_2\}$, can then be found. E.g., for $h = b_2$ one finds a cyclic fixed element group, generated by the element fixed by $I_{b_2^{-1}}$:

$$b_2 b_{2_I} \ldots b_{2_{I^5}} \cdot l = b_2^2 a_2 b_2 a_2^2 b_2^{-2} a_2^{-1} b_2^{-1} a_2^{-2}.$$

18. Partition of automorphisms into isogredience classes. In §12 the cosets of $\mathfrak{J}$ modulo the normal subgroup $\mathfrak{F}$ of all inner automorphisms were called "automorphism families," and the systems of mutually conjugate automorphisms in $\mathfrak{J}$ were called "automorphism classes." We now turn our attention to *collections of automorphisms which are mutually conjugate via inner automorphisms.* Because of the special significance of this concept in the present investigation, the name *isogredience class* is introduced for such a collection. The boundary mappings induced by two "isogredient automorphisms" are conjugate to each other via an element of F; they will likewise be called "isogredient."

Two isogredient automorphisms belong by definition to the same automorphism class. They belong however also to the same family, because one has

$$I_{0f} I I_{0f}^{-1} = I_{f_I f^{-1}}. \tag{11}$$

Thus the partition of all automorphisms into isogredience classes is a refinement of the partition into families. Equation (11) corresponds to the equation

$$fgf^{-1} = ff_I^{-1}g \tag{12}$$

for boundary mappings when g belongs to I; one can also derive this from the functional equation (8).

For convenience of application we formulate this as

Theorem 5. *A necessary and sufficient condition for the isogredience of two automorphisms I_1 and I_2, where $I_2 = I_{1h^{-1}}$, is that there be a solution f to the equation $h = ff_{I_1}^{-1}$,*

and then put next to it:

Theorem 6. *A necessary and sufficient condition for the isogredience of two automorphisms I_1 and I_2 in the same class, $I_2 = I_3^{-1} I_1 I_3$, is that I_3 be related to an automorphism which commutes with I_1.*

The condition is sufficient. Namely, if there is a representation

$$I_3 = I_4 \cdot I_{0f}, \qquad I_4 \rightleftarrows I_1,$$

then the assertion follows by substitution in the above form for I_2. However, the condition is also necessary. Namely, it follows from

$$I_3^{-1} I_1 I_3 = I_{0f} I_1 I_{0f^{-1}}$$

that

$$(I_{0f^{-1}} I_3^{-1}) \cdot I_1 \cdot (I_3 I_{0f}) \cdot I_1^{-1} = 1.$$

Hence the automorphism $I_4 = I_3 I_{0f}$, related to I_3, commutes with I_1.

Thus an automorphism I is certainly transformed isogrediently by conjugation with automorphisms in the families of I_0, and the powers of I, but in general, also by other families.

For the family $\mathfrak{F}$ itself, the partition into isogredience classes is identical with the partition of the elements of F into conjugacy classes: I_{0f_1} is isogredient with all $I_{0ff_1f^{-1}}$, for an arbitrary f in F, and only with these.

An automorphism I induces a *partition of the elements of F* into systems which may be called *isogredience classes relative to I*:

Definition. Two elements f_1 and f_2 of F are called *isogredient relative to I*, when $I_{f_1^{-1}}$ and $I_{f_2^{-1}}$ are isogredient automorphisms.

Then when g belongs to I, $f_1 g$ and $f_2 g$ are isogredient mappings, hence there is an f such that

$$f_2 g = ff_1 gf^{-1} = ff_1 f_I^{-1} g$$

because of (8). We therefore have:

Theorem 7. *A necessary and sufficient condition for f_1 and f_2 to be isogredient relative to I is that there be a solution f to the equation*

$$f_2 = ff_1 f_I^{-1}.$$

Relative to the automorphism Λ used as example earlier, b_1 is isogredient to the identity, because

$$b_1 = a_1^{-1} \cdot a_{1\Lambda}.$$

The automorphism $\Lambda_{b_1^{-1}}$ of Example 5, §17, therefore belongs to the same isogredience class as Λ. Hence the induced boundary mapping $b_1\lambda$ results from λ by conjugation with a_1^{-1}:

$$b_1\lambda = a_1^{-1}\lambda a_1.$$

The fixed point set for $b_1\lambda$ therefore results from that for λ by the operation a_1^{-1}, and we can complete Example 5 by deducing from

$$H(\Lambda) = \{b_1, a_2, \ldots, b_p\}$$

that

$$H(\Lambda_{b_1^{-1}}) = \{a_1^{-1}b_1a_1, a_1^{-1}a_2a_1, \ldots, a_1^{-1}b_pa_1\}.$$

In general, one concludes the following from the isogredience of induced boundary mappings:

Theorem 8. *If I_1 and I_2 are two isogredient automorphisms, $I_2 = I_{1f_{I_1}f^{-1}}$, then the fixed element group $H(I_2)$ is a subgroup of F conjugate to $H(I_1)$:*

$$H(I_2) = fH(I_1)f^{-1},$$

or in other words,

$$H(I_2) = H(I_1)_{I_{0f^{-1}}}.$$

Theorem 9. *The elements f_{I^n}, for arbitrary n, all belong to the same isogredience class relative to I.*

This is shown by the identities ($n > 0$):

$$f_{I^n} \equiv \quad f_{I^{n-1}}^{-1}f_{I^{n-2}}^{-1}\ldots f_I^{-1}f^{-1}\cdot f\cdot f_I f_{I^2}\ldots f_{I^n},$$

$$f_{I^{-n}} \equiv f_{I^{-n}}f_{I^{-(n-1)}}\ldots f_{I^{-1}}\cdot f\cdot f^{-1}f_{I^{-1}}^{-1}\ldots f_{I^{-(n-1)}}^{-1}$$

when one takes Theorem 7 into consideration.

Example 6. The automorphism $\Lambda_{a_1b_1^n}$ is isogredient to the automorphism $\Lambda' = \Lambda_{a_1}$ of example 3 for arbitrary n, hence the induced boundary mapping contains two aperiodic intervals separated by two fixed points P'^+ and P'^- which are not fundamental points:

$$H(\Lambda_{a_1b_1^n}) = 1.$$

For positive n, the boundary mapping results from that for Λ' by conjugation with

$$a_1b_1^{n-1}a_1b_1^{n-2}\ldots a_1b_1a_1.$$

Then one obtains the following expansion for P'^+:

$$k(P'^+) = b_1^{-n}a_1^{-1}b_1^{-(n+1)}a_1^{-1}\ldots.$$

It is easy to set up the expansion of P'^-, as well as the expansions of both points for negative n, and to show that they are fixed limit elements of the automorphism in question.

19. Automorphisms of finite order. Let $I \neq I_0$ be an automorphism of the first kind and suppose $I^n = I_0$. Let n be the smallest positive number satisfying this equation, in other words, the order of I. Then by §11 we have

$$H(I^m) = 1, \qquad m = 1, 2, \ldots, n-1,$$

and the induced boundary mappings g^m are fixed point free, whereas $g^n = 1$.

We now consider the automorphisms related to I. Let $I_{f^{-1}}$ be one such. Then

$$k = f \cdot f_I \cdot f_{I^2} \ldots f_{I^{n-1}} \tag{13}$$

is a fixed element for $I_{f^{-1}}$, as application of $I_{f^{-1}}$ with use of $f_{I^n} = f$ shows. This also follows from the general results of §17.

Now for $r > 0$ the general formulae of §17 give:

$$(I_{f^{-1}})^r = I^r_{f^{-1}_{I^{r-1}} \ldots f^{-1}_I f^{-1}}. \tag{14}$$

For our I above we conclude from (14) that, when $r = n$:

$$(I_{f^{-1}})^n = I_{0k^{-1}}. \tag{15}$$

Now we have to distinguish between the cases $k = 1$ and $k \neq 1$.

(a) $k \neq 1$. Then $H(I_{f^{-1}}) \neq 1$, hence $H(I_{f^{-1}}) = H((I_{f^{-1}})^n)$. But, by (15), the boundary mapping induced by $(I_{f^{-1}})^n$ is the operation k from F, hence

$$H((I_{f^{-1}})^n) = \{k^*\}$$

where k^* is the primary element corresponding to k. Then also

$$H(I_{f^{-1}}) = \{k^*\}.$$

The induced boundary mapping fg determines two periodic intervals, which correspond to the primary fixed element k^*, and

$$(fg)^n(E) = k(E).$$

(b) $k = 1$. Then by (15)

$$(I_{f^{-1}})^n = I_0,$$

and n is also the smallest such exponent, because it follows from

$$(I_{f^{-1}})^r = I_0,$$

by (14), that

$$I^r = I_{0ff_I \ldots f_{I^{r-1}}}.$$

Hence I^r has a fixed element, which is only the case for $r = m \cdot n$. Thus $fg, (fg)^2, \ldots, (fg)^{n-1}$ are fixed point free and $(fg)^n = 1$. Consequently $H(I_{f^{-1}}) = 1$:

Theorem 10. *All automorphisms related to an automorphism of finite order have either a cyclic group or the identity as fixed element group. By Theorem 8, the same applies to isogredient automorphisms.*

The latter statement also comes to light formally in (13) when one replaces the f there by the element hfh_I^{-1}, which is isogredient to f relative to I by Theorem 7.

In dealing with the problems in section 4 we shall determine the automorphisms related to I which induce fixed point free boundary mappings. Thus we shall be concerned with the isogredience classes relative to I of elements f of F which satisfy the equation

$$ff_I f_{I^2} \dots f_{I^{n-1}} = 1. \tag{16}$$

It will turn out that this equation has only a finite number of non-isogredient solutions, and that the number of isogredience classes relative to I which appear as solutions may be read off from the presentation of I.

20. Mapping of complete systems of equivalent points on E induced by automorphisms. Let P be a point of E. We denote the set of points fP, where f runs through F, by FP. FP is everywhere dense on E; because if i is an arbitrary interval on E and if f is an element whose positive fundamental point lies in i and whose negative fundamental point is different from P (such an element exists by §5), then for sufficiently large n, $f^n P$ lies in i. If I is an arbitrary automorphism, g the induced boundary mapping and if $gP = Q$, then $g(FP) = FQ$ by the functional equation (8). One complete system of equivalent points is therefore mapped onto another by g.

Now suppose that

$$g(FP) = FP.$$

Then we have, say, $gP = fP$. This does not completely determine the element f in the case where P is a fundamental point. We define the "characteristic element," $f^{(P)}$, for the point P of E, when P is a fundamental point, to be the primary element of F which has P as positive fundamental point, and otherwise to be the element 1. Then the "equivalence factor" f of the point P under the mapping g is determined only up to an arbitrary power of the characteristic element:

$$gP = fP = ff^{(P)\alpha}P = f^{(fP)\alpha}fP.$$

Now when h is an arbitrary element of F, it follows from (8) that:

$$ghP = h_I gP = h_I fP = h_I f h^{-1} \cdot hP.$$

The equivalence factors of all points in FP under I therefore constitute an isogredience class relative to I^{-1}. This is completely determined when P is not a fundamental point. Now when FP contains a fixed point of g, say the point hP, then by the last equation,

$$h_I f h^{-1} = f^{(hP)\alpha},$$

hence

$$f = h_I^{-1} h \cdot h^{-1} f^{(hP)\alpha} h = h_I^{-1} h \cdot f^{(P)\alpha}$$

or

$$ff^{(P)-\alpha} = h_I^{-1} h.$$

A necessary and sufficient condition for FP to contain fixed points under I is that the equivalence factor of an arbitrary point of FP, when multiplied by a suitable power of its characteristic element, falls into the isogredience class of the identity relative to I^{-1}.

CHAPTER III

Continuous Mappings

21. Topological mappings of the surface ϕ onto itself. *T*-functions. Suppose the surface $\phi(=\Phi \bmod F$, §3) undergoes a topological, i.e., one-to-one and continuous, mapping τ onto itself. Let q be a point of ϕ, $q' = \tau q$ its image, x_0 one of the coordinate values of q, x'_0 one of q'. Now let x be an arbitrary point of Φ and W a continuous arc in Φ which goes from x_0 to x. W lies over a continuous are w on ϕ. Suppose the continuous arc which lies over τw and begins at x'_0 ends at x'. Then one concludes easily, from the fact that Φ is the universal covering surface of ϕ, that x' depends only on x, not on W. Thus each point x is uniquely associated with an image point $x' = t(x)$. A neighbourhood of x (which one can choose to be so small that it contains no pair of points equivalent under F) corresponds to a neighbourhood of x'. To two different values of x correspond different values of x'. It follows, by consideration of the inverse mapping $\tau^{-1}\phi$, that all these relations are reversible. Thus the function $x' = t(x)$ is a *topological mapping $t\Phi$ of Φ onto itself, which carries any two equivalent points into equivalent points.* Conversely, each mapping $t\Phi$ with this property determines a topological mapping $\tau\phi$. All functions $t(x)$ of this kind constitute a *group* which we denote by T; from now on each such function $t(x)$ will be called a *T-function.* We say that $t\Phi$ "lies over" $\tau\phi$. In determining t, the choice of q, x_0 and x'_0 was to some extent arbitrary. Namely, suppose that t_1 is a second T-function lying over $\tau\phi$. Then $t_1 t^{-1}$ lies over the identity mapping of ϕ, and hence is an element f of F, a "covering transformation of the universal covering surface." Thus $t_1 = ft$ and we have:

> Each topological mapping of ϕ onto itself determines *a family of related T-functions*, the members of which result from each other by composition with operations of F.

22. The functional equation of *T*-functions. Induced automorphisms. Mapping classes. Let t be a T-function and f an element of F. Then tft^{-1} is a T-function which lies over the identity mapping of ϕ, and hence it is an element f' of F:

$$tft^{-1} = f'. \tag{17}$$

This shows that F is a *normal subgroup* of T and the correspondence $f \mapsto f'$ is an *automorphism* I of F, which is said to be *induced by t*. Thus t satisfies the *functional equation*

$$tfx = f_I tx \tag{18}$$

throughout Φ, which coincides with the functional equation (8) of the boundary mapping function $g(E)$ considered earlier (§13) and, like the latter, is equivalent to a corresponding system (9) of $2p$ functional equations. Thus the value of t in a fundamental domain for F determines t throughout Φ.

By (18), t commutes with all elements of the fixed element group $H(I)$, and only with these elements of F.

Now let h be an element of F. By (17), the T-function ht related to t then induces the automorphism $I_{h^{-1}}$ related to I, because

$$htft^{-1}h^{-1} = hf'h^{-1}.$$

Thus the whole family of T-functions lying over $\tau\phi$ induces an automorphism family whence we obtain a partition principle for the τ:

> Each topological mapping of ϕ onto itself is associated with a unique family of automorphisms of F. Two mappings of ϕ are reckoned to belong to the same *mapping class*, when they are associated with the same *automorphism family*.

In particular, a mapping $\tau\phi$ is said to *belong to the class of the identity* when it is associated with the family $\mathfrak{F}$ of inner automorphisms. Then one of the associated T-functions induces the identity automorphism, and hence commutes with all elements of F.

If I_1 is induced by t_1 and I_2 by t_2, then $t_2 t_1$ induces the automorphism $I_1 I_2$, as (18) shows. Thus the group T has homomorphic image $\mathfrak{J}$,[11] and the identity I_0 of $\mathfrak{J}$ corresponds to the normal subgroup T_0 of T whose elements induce the identity automorphism I_0. Thus the collection $T_{\mathfrak{F}}$ of T-functions which induce automorphisms in the family $\mathfrak{F}$ of the identity, likewise constitute a normal subgroup of T, since $\mathfrak{F}$ is a normal subgroup of $\mathfrak{J}$. Finally, if (τ) denotes the group of all topological mappings of ϕ onto itself and $(\tau)_0$ those τ which belong to the class of the identity, then $(\tau)_0$ is a normal subgroup of (τ), and the *mapping class group* is *given by the four isomorphic quotient groups*

$$\frac{(\tau)}{(\tau)_0} = \frac{T}{T_{\mathfrak{F}}} = \frac{\mathfrak{J}}{\mathfrak{F}} = \frac{H}{F}, \tag{19}$$

where H, as in §13, denotes the group of boundary mappings induced by automorphisms.

23. Each automorphism is induced by T-functions.

Theorem 11. *The automorphism families induced by topological mappings of ϕ contain all automorphisms of F.*

This theorem is due to M. Dehn, who earlier communicated to me a complete proof for $p = 2$; his proof operates on the surface ϕ itself, using the Dehn theorems given at the end of §9. Here the proof will be carried out from a given automorphism by constructing an associated T-function in Φ, in fact we map two suitably chosen fundamental domains for F onto each other topologically so that the relation required by the functional equation (18) holds between pairs of equivalent boundary points. We begin by replacing the fundamental domain B defined in §3 by one more suitable for the present problem, which results from an easily seen modification of B.

Let i be one of the numbers 1, 2, ..., p. The axes of a_i and b_i meet at a point D_i. Now we continue with the following: the image of the axis of b_i under a_i, i.e., the axis of $a_i b_i a_i^{-1}$, the image of the axis of a_i under the latter operation, i.e., the axis of $a_i b_i a_i b_i^{-1} a_i^{-1}$, the image of the axis of $a_i b_i a_i^{-1}$ under the inverse of the latter operation, i.e., the axis of $a_i b_i a_i^{-1} \cdot b_i \cdot a_i b_i^{-1} a_i^{-1}$, the image of the axis of $a_i b_i a_i b_i^{-1} a_i^{-1}$ under the inverse of the latter operation, i.e., the axis of $k_i a_i k_i^{-1}$, where k_i denotes the commutator $a_i b_i a_i^{-1} b_i^{-1}$. The last two elements obtained are $k_i b_i k_i^{-1}$ and $k_i a_i k_i^{-1}$. If one now continues this process indefinitely, twice conjugating directly with the last element obtained, then twice conjugating with the inverse of the last element obtained, and so on, then after each four steps one obtains elements of the form $k_i^n b_i k_i^{-n}$, $k_i^n a_i k_i^{-n}$ for increasing n. Thus as the process continues the fundamental points of all

[11] Here we assume the fact, to be proved in the next paragraph, that for each automorphism there is a T-function which induces it.

these elements converge to $V(k_i)$. The intersections of the axes of successive elements of the series are denoted by

$$D_i, D_i^{(1)}, D_i^{(2)}, \ldots.$$

If one repeats the same process with the rôles of a_i and b_i interchanged, then one obtains a new sequence of elements, whose fundamental points converge to $U(k_i)$ and which give rise to a series of intersections

$$D_i, D_i^{(-1)}, D_i^{(-2)}, \ldots.$$

The subdomain Z_i of Φ which lies outside these axis circles and inside the axis circle of k_i is convex in the sense of the non-Euclidean metric, and mapped into itself by the group $\{k_i\}$. This is because k_i pushes the axis of k_i along itself, and the other boundary piece ρ_i of Z_i, consisting of infinitely many sides, is pushed along itself in such a way that each side leaps the next three and goes into the fourth. In particular,

$$k_i D_i^{(-2)} = D_i^{(+2)}.$$

The figure Z_i is mapped into itself by reflection in OD_i, so $D_i^{(2)}$ and $D_i^{(-2)}$ have the same distance from E and therefore lie on the circles $k_i C_i$ and C_i orthogonal to E and the axis of k_i, which play the rôle of $S(C)$ and C in Figure 4.

Now let the vertices of the fundamental domain B used in §3, beginning at the intersection of the sides B_1 and A_p (which was called H in §3), be denoted in anticlockwise order by:

$$Q_1 Q_{11} Q_{12} Q_{13} Q_2 Q_{21} Q_{22} Q_{23} \cdots Q_p Q_{p1} Q_{p2} Q_{p3}.$$

Then b_i^{-1} sends the point Q_{i+1} to Q_{i1} and $a_i b_i a_i^{-1}$ sends the point Q_{i1} to Q_i, hence k_i sends the point Q_{i+1} to Q_i (i mod p):

$$k_i Q_{i+1} = a_i b_i a_i^{-1} \cdot b_i^{-1} Q_{i+1} = a_i b_i a_i^{-1} Q_{i1} = Q_i.$$

Q_i and Q_{i+1} have the same distance from E, hence Q_{i+1} lies on C_i and Q_i on $k_i C_i$. The circles C_i and $k_{i+1} C_{i+1}$ intersect at Q_{i+1}. Let the subarc of C_i running from Q_{i+1} to $D_i^{(-2)}$ be l_i. *Then the new fundamental domain Ω of F is bounded by l_i, $k_i l_i$ and the subpolygon $D_i^{(-2)} D_i^{(-1)} D_i D_i^{(1)} D_i^{(2)}$ of the boundary piece ρ_i of Z_i* for $i = 1, 2, \ldots, p$.

The boundary pieces of Ω are pairwise equivalent under a_i, b_i and the operation k_i composed from them. As a result, the domains equivalent to Ω arrange themselves in fives around the points equivalent to D_i; namely, one of the four angles which result from intersection of the axes of a_i and b_i is divided by the segment l_i. Around each of the points equivalent to Q_i, p domains equivalent to Ω arrange themselves, as will become clear in what follows.

Figure 13 illustrates the relation between the fundamental domains B and Ω for the parts of these domains cut off by the non-Euclidean line $Q_1 Q_2$. The figure corresponds to the case $p = 2$, which is special to the extent that the axis of k_1 is a diameter of E and it coincides with the line $Q_1 Q_2$; otherwise, the relative positions and relations between individual parts of the figure are the same for $p > 2$. The part of B considered consists of V + I + II + III + IV, the corresponding part of Ω is V + I′ + II′ + III′ + IV′. Here, as one easily reads from the figure,

$$\text{I}' = a_1 \text{I}, \qquad \text{II}' = b_1 \text{II},$$

$$\text{III}' = a_1 b_1 \text{III}, \qquad \text{IV}' = b_1 a_1 \text{IV}.$$

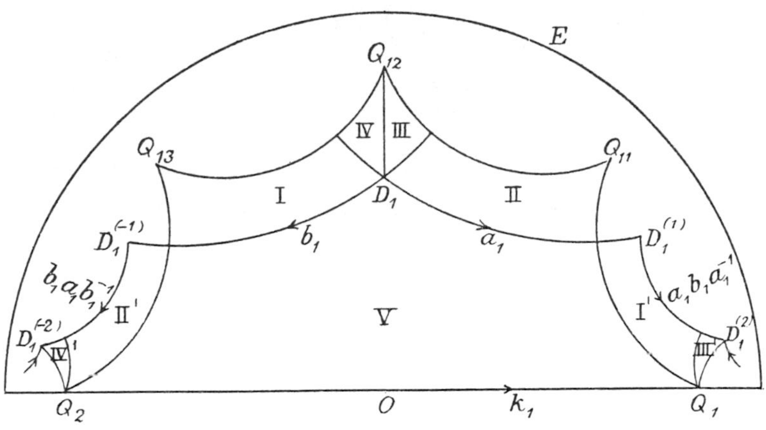

Figure 13

Since the boundary of this part of B consists of lines of the net N, in the notation of §3 the sides A_1 and B_1, Ω is formed by line segments which separate IV′ from II′, II′ from V, V from I′ and I′ from III′. The lines carrying arrows are the axes of the elements shown next to them.

The axes of $k_1, k_2, \ldots, k_p$ are disjoint from each other, as, e.g., the expansions of their fundamental points show; only for $p = 2$ the two axes coincide, because

$$R \equiv k_1 k_2 = 1.$$

The segment $k_1 l_1$ is common to $k_1\Omega$ and Ω; $k_1\Omega$ is traversed by the axes of

$$k_1, k_1 k_2 k_1^{-1}, \ldots, k_1 k_p k_1^{-1},$$

which all end between $V(k_1)$ and $U(k_p)$ on E. Ω is bordered by the "axes paths" $\rho_1, \ldots, \rho_p$ (boundary pieces of $Z_1, \ldots, Z_p$). Thus $k_1\Omega$ is bordered by the axis paths

$$k_1\rho_1 = \rho_1, k_1\rho_2, \ldots, k_1\rho_p.$$

The free side of $k_1\Omega$ (for $p > 2$) issuing from Q_1 corresponds to the side $k_2 l_2$ of Ω where $k_2\Omega$ meets it; this free side therefore borders $k_1 k_2 \Omega$ and the latter domain is therefore bordered by the axis paths

$$k_1 k_2 \rho_1, k_1 k_2 \rho_2 = k_1 \rho_2, k_1 k_2 \rho_3, \ldots, k_1 k_2 \rho_p.$$

The free side $k_1 k_2 k_3 l_3$ of $k_1 k_2 \Omega$, corresponding to the side $k_3 l_3$ of Ω, issues as a free side from Q_1 (for $p > 3$), and now borders $k_1 k_2 k_3 \Omega$. This process is continued in a cycle around Q_1 until one comes to $k_1 k_2 \ldots k_{p-1}\Omega$. But since

$$R \equiv k_1 k_2 \ldots k_p = 1, \tag{20}$$

the latter is identical with $k_p^{-1}\Omega$, and it therefore borders Ω along l_p. Thus Q_1 is surrounded in clockwise order by a cycle of p domains

$$\Omega, k_1\Omega, k_1 k_2 \Omega, \ldots, k_1 k_2 \ldots k_{p-1}\Omega = k_p^{-1}\Omega.$$

It now follows by symmetry that the angle between the segments l_{i-1} and $k_i l_i$ issuing from Q_i has magnitude $2\pi/p$. In particular for $p = 2$ these segments lie on the same line.

The p domains above meet along the $p(p-1)$ axis paths which result from $\rho_1, \rho_2, \ldots, \rho_p$ by the operations

$$1, k_1, k_1 k_2, \ldots, k_1 k_2 \ldots k_{p-1}.$$

The cyclic order of the endpoints of these axis paths on E is easily derived from the order of the p domains around Q_1 and the positions of the ρ_i relative to Ω.

Finally, for $p > 2$, we need a system of auxiliary lines: Let m_i be a rectilinear, directed segment, which runs from $D_{i-1}^{(-2)}$ to $D_i^{(2)}$ (i mod p). Thus m_i cuts an equilateral triangle Δ_i with base m_i and apex Q_i from Ω. Let Ω_0 be the part of Ω which remains after the omission of these p congruent triangles. The triangles

$$\Delta_1, k_1\Delta_2, k_1 k_2 \Delta_3, \ldots, k_1 k_2 \ldots k_{p-1}\Delta_p$$

form a regular p-gon Λ with midpoint Q_1. The sides of the latter are

$$m_1, k_1 m_2, k_1 k_2 m_3, \ldots, k_1 k_2 \ldots k_{p-1} m_p.$$

Its vertices are the points

$$D_1^{(2)}, k_1 D_2^{(2)}, k_1 k_2 D_3^{(2)}, \ldots, k_1 k_2 \ldots k_{p-1} D_p^{(2)} = k_p^{-1} D_p^{(2)} = D_p^{(-2)}.$$

The fundamental domain Ω is replaceable by $\Omega_0 + \Lambda$.

Now let K be the subgroup of F which is generated by $k_1, k_2, \ldots, k_p$. Because of (20), K is a *free* group with $p - 1$ generators. We let $K\Omega$ denote the union of domains equivalent to Ω under K. If one now constructs cycles around $Q_2, \ldots, Q_p$ in analogy with the above cycle of p domains equivalent to Ω around Q_1, and repeats the process at the vertices equivalent to the Q_i in the domains already obtained, etc., then one exhausts all and only $K\Omega$. If k is an arbitrary element of K, then $k\Omega$ is bordered by the axis paths $k\rho_1, \ldots, k\rho_p$. Each axis path $k\rho_i$ cuts across E, and one of the two parts into which it divides Φ is a (non-Euclidean) convex region. The intersection of all these convex regions, as $i = 1, 2, \ldots, p$ and k runs through K, is a *convex region* Π whose boundary is formed by all the $k\rho_i$, because each crosscut $k\rho_i$ of E has all the others on its concave side. Now it is obvious that

$$K\Omega = K(\Omega_0 + \Lambda) = \Pi.$$

Thus if f is an element in F but outside K, the convex region $f\Pi$ has at most boundary points of Π in common with Π. In this way Φ is divided by K and its cosets into convex regions.

In the case $p = 2$, where the "appendage" Λ vanishes and Ω is identical with Ω_0, K is cyclic and Π is bounded by ρ_1 and ρ_2, which have their common endpoints on E and translate into themselves via k_1 and its inverse, k_2.

The boundary of B results from the canonical cut system of ϕ in Figure 1. The boundary of Ω results from another frequently used dissection of the surface, namely p disjoint pairs of loop cuts whose crossings are tied to an arbitrary point of the surface by p "reins." If one thinks of closed geodesics on the surface ϕ which represent the axes of the a_i and b_i in Figure 1, then one obtains p disjoint pairs of loop cuts. Their crossing points are connected to Q by geodesic segments l_i and to each other by geodesic segments m_i. The result is Figure 14 ($p = 3$), in which the A_i and B_i are left out and the axes are denoted by the names of the corre-

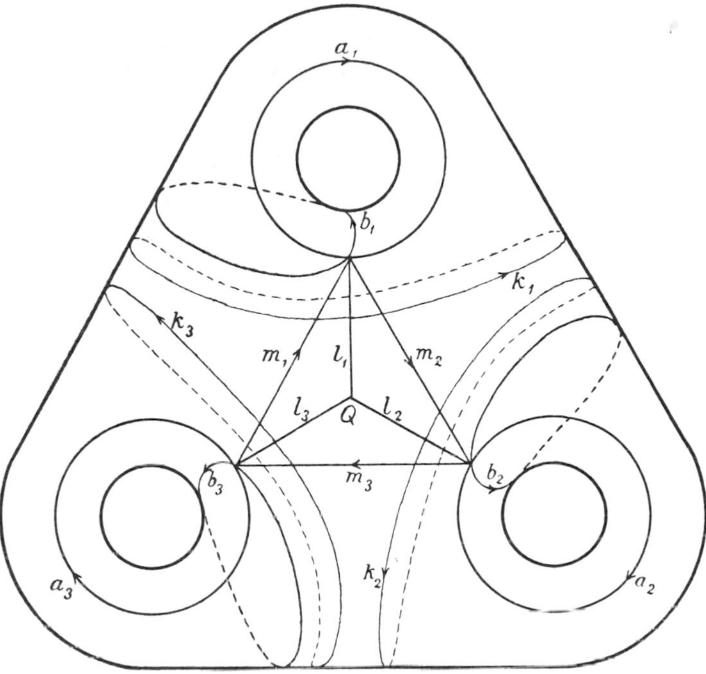

Figure 14

sponding transformations. The positions of the closed geodesics representing the axes of the k_i are also indicated, each cutting a handle off the surface. In the case $p = 2$ there is only one such cut, which goes through Q; the m_i then coincide with the l_i.

After these preparatory considerations we now carry out the proof of the above assertion. Let I $(f \mapsto f')$ be an arbitrary automorphism of F. We shall construct a domain Ω' corresponding to Ω; corresponding elements will be denoted by adding primes. The result of §9, that the cyclic order of fundamental points is preserved by I, will be used constantly.

The axes of a_i' and b_i' intersect at D_i'. In analogy with the foregoing we introduce a sequence of elements

$$b_i',\ a_i',\ a_i'b_i'a_i'^{-1},\ a_i'b_i'a_i'b_i'^{-1}a_i'^{-1}, \ldots;$$

and let the corresponding intersections of the axes be D_i', $D_i^{(1)\prime}$, $D_i^{(2)\prime}$, ... The fundamental points of this and the analogous sequence converge to the fundamental points of $k_i' = a_i'b_i'a_i'^{-1}b_i'^{-1}$. To the convex region Z_i there corresponds a convex region Z_i', as the region between all these axes, bounded by the axis of k_i' and an "axis path" ρ_i'.

Now let $K' = K_I$ be the subgroup of F generated by $k_1', \ldots, k_p'$ and let k' be the symbol for an arbitrary element of K'. Each axis path $k\rho_i$ now corresponds to an axis path $k'\rho_i'$. The set of endpoints of the axis paths $k'\rho_i'$ results from the set of endpoints of the $k\rho_i$ by the boundary mapping gE induced by I. It therefore has the same cyclic order. Thus each $k'\rho_i'$ has all the others on its concave side. Together, the $k'\rho_i'$ determine a *convex region* Π', just as the $k\rho_i$ determine the convex region Π.

Now the points $D_i^{(2)\prime}$ and $D_i^{(2)\prime}$ lie on the boundary of Π' in the same cyclic order as do the corresponding unprimed points on the boundary of Π. Thus if one draws the rectilinear directed segments m_i' from $D_{i-1}^{(-2)\prime}$ to $D_i^{(2)\prime}$ they cut a convex polygon Ω_0' from Π'. Likewise, the segments

$$m_1', k_1'm_2', k_1'k_2'm_3', \ldots, k_1'k_2'\ldots k_{p-1}'m_p'$$

are the sides of a *convex polygon* Λ' which meets Ω_0' along m_1'; namely, the vertices

$$D_1^{(2)\prime}, k_1'D_2^{(2)\prime}, k_1'k_2'D_3^{(2)\prime}, \ldots, k_1'k_2'\ldots k_{p-1}'D_p^{(2)\prime} = k_p'^{-1}D_p^{(2)\prime} = D_p^{(-2)\prime} \qquad (21)$$

lie on the boundary of Π' in the same cyclic order as do the corresponding unprimed points on the boundary of Π. By using all elements of K' and comparing with the corresponding elements of K, it follows that

$$K'(\Omega_0' + \Lambda') = \Pi'.$$

If f is an element in F but outside K, then the operation f shifts Π to the other side of one of the axis paths $k\rho_i$ bordering Π, since $f\Pi$ has at most boundary points of Π in common with Π. It follows from the preservation of the cyclic order of fundamental points under gE that then $f'\Pi'$ is also separated from Π' by the corresponding $k'\rho_i'$, and hence that these two regions have no interior point in common.

Now we can conclude the proof by establishing a suitable topological relation between $\Omega_0 + \Lambda$ and $\Omega_0' + \Lambda'$. For the sake of symmetry, and in order to conveniently obtain a related result which will be necessary later, we first replace Ω by an analogous domain: let Q_1' be an *arbitrary* point in the interior of Λ'. Connect it with the vertices (21) of Λ' by rectilinear segments, which will be denoted in order by

$$k_1'l_1', k_1'k_2'l_2', k_1'k_2'k_3'l_3', \ldots, k_1'k_2'\ldots k_p'l_p' = l_p'.$$

This divides Λ' into p triangles

$$\Delta_1', k_1'\Delta_2', \ldots, k_1'\ldots k_{p-1}'\Delta_p'.$$

Then let Ω' be the region which results from Ω_0' by attachment of the triangles $\Delta_1' + \Delta_2' + \cdots + \Delta_p'$, Δ_i' has m_i' as base and

$$Q_i' = k_{i-1}'^{-1}k_{i-2}'^{-1}\ldots k_1'^{-1}Q_1'$$

as apex. *Ω' is bounded by l_i', k_i', l_i' and the subpolygon $D_i^{(-2)\prime}D_i^{(-1)\prime}D_i'D_i^{(1)\prime}D_i^{(2)\prime}$ of ρ_i' ($i = 1, 2, \ldots, p$). The boundary pieces of Ω' are pairwise equivalent under a_i', b_i' and the operations k_i' composed from them.*

Ω' is a convex region. No side of Ω' contains a pair of points equivalent under F, apart from the endpoints of the sides on the ρ_i'; this follows from the fact that Φ is covered simply and without gaps by the regions $F\Omega'$. (Attention has already been drawn to the fact that a_i' and b_i' correspond to curve types without double points, in the result of Dehn mentioned at the end of §9.)

It is clear that a circuit round Ω and the corresponding circuit round Ω' have the same orientation in Φ, or the opposite orientation, according as I is of the first or second kind.

Now let s be any directed side of Ω and let fs be its equivalent (thus f is one of the elements $a_i^{\pm 1}$, $b_i^{\pm 1}$ or $k_i^{\pm 1}$). Let s' be the corresponding side of Ω', bearing correctness of orientation in mind. Then its equivalent is $f's'$. Let t be any topological mapping of s onto s'. Then $f'tf^{-1}$ is a topological mapping of fs onto $f's'$. If one lets s run through single representatives of the pairs of equivalent sides of Ω, then one obtains a topological mapping of the boundary of Ω onto the boundary of Ω'. Now let t be a topological mapping of Ω onto Ω' which agrees with the one already defined on the boundary of Ω. Then, on the boundary of Ω, t satisfies the functional equation

$$tfx = f'tx \tag{22}$$

for the elements f in question which establishes the equivalence between boundary sides. If one now transports this mapping $t\Omega$ to all domains $f\Omega$ equivalent to Ω, now for arbitrary f in F, with the help of equation (22), then one obtains a topological mapping t of Φ onto itself, thus, a T-function which induces I. Thus our goal is reached.

For later purposes we shall choose this T-function to be of a special kind. The edges of the net N, the images of the canonical curves A_i and B_i, form $4p$ rectilinear crosscuts in Ω, in sets of four which lead from each point Q_i to neighbouring sides of Ω on ρ_i and ρ_{i-1}; cf. Figure 13. No two of these crosscuts meet each other outside the points Q_i. They divide Ω into $4p + 1$ convex polygons. Their images under t are Jordan arc crosscuts in Ω'. Now one can choose the mapping t with the following requirements in mind:

1. With the earlier choice of the point Q'_1 in Λ' one takes care that Q'_1 does not lie on an edge of the net N.
2. With the choice of the topological mapping of the individual boundary sides of Ω onto the corresponding sides of Ω' one takes care that the images of the endpoints of crosscuts of Ω, other than the Q_i, do not fall on edges of the net. This is possible, because the axes of the a'_i and b'_i do not lie on edges of the net N; the latter lines are axes of elements whose curve types have essential double points.
3. The extension of the topological mapping to the interior of Ω is made as follows: each of the crosscuts of Ω is first mapped in any way topologically onto the rectilinear segment which connects the images of its endpoints, already determined. This is possible because Ω' is convex. Now the boundary of each convex subpolygon of Ω is mapped topologically onto the boundary of the corresponding, likewise convex, subpolygon of Ω'. One extends this mapping in any topological manner to the interiors of the subpolygons.

Under the topological mapping of ϕ, corresponding to I, established by this T-function, the images A'_i and B'_i of the canonical curves A_i and B_i consist of a finite number of geodesic segments whose endpoints lie outside the A_i and B_i. The $2p$ curves A'_i, B'_i therefore have only *finitely* many points of intersection with the $2p$ curves A_i, B_i.

24. *T*-functions with prescribed functional equation and prescribed values at finitely many points of Φ. In this paragraph, Theorem 11 of the preceding paragraph will be extended as follows:

Theorem 12. *For any automorphism of F and a finite number of inequivalent points of Φ, there is a T-function which induces the given automorphism and takes arbitrarily prescribed (inequivalent) values at the given points.*

Lemma 1. *There is a topological mapping τ of ϕ onto itself, in the class of the identity, which sends the n given (different) points $d_1, d_2, \ldots, d_n$ to n given (different) points $d'_1, d'_2, \ldots, d'_n$.*

This can be seen immediately: One connects d_i to d_i' by a geodesic polygonal path c_i so that no two of the c_i have a common point, and encloses c_i in a geodesic polygon, with interior δ_i, so that no two of the polygons meet. Then one maps each δ_i onto itself by a topological mapping τ which sends d_i to d_i' and which is the identity on the boundary of δ_i. Outside the δ_i, τ is also the identity. Then there is a T-function lying over τ which induces the identity automorphism I_0, namely the one which is the identity on points of Φ lying over points of ϕ outside the δ_i.

Lemma 2. *There is a T-function which induces I_0 and which sends an arbitrary point D of the fundamental domain* B *to $a_1 D$.*

We can assume that D is the origin, since otherwise we need only conjugate with an I_0-inducing T-function which sends D to O (Lemma 1). Let the sides of B again be labelled anticlockwise, beginning at the intersection of B_1 and A_p. Let P, Q, R be three interior points of the third side of B, in this order in an anticlockwise circuit. Connect O to Q and to the point $a_1 Q$ on the first side of B by a straight segment. P is connected to $a_1 P$ and R to $a_1 R$ by simple polygonal crosscuts in B which contain the path $QO(a_1 Q)$ between them. These two crosscuts are topologically mapped onto the segments $v = 1$, $0 \le u \le 1$ and $v = -1$, $0 \le u \le 1$ in a u-v-plane, and also PQ onto $u = 0$, $1 \ge v \ge 0$, QR onto $u = 0$, $0 \ge v \ge -1$; by requiring that whenever x is mapped to (u, v) then $a_1 x$ is mapped to $(u + 1, v)$, the mapping of the piece $(a_1 P)(a_1 Q)(a_1 R)$ of the first side of B onto $u = 1$, $1 \ge v \ge -1$ is determined. Finally, $QO(a_1 Q)$ is mapped onto $v = 0$, $0 \le u \le 1$ and the mapping is extended in any way to a topological mapping γ of the strip $PR(a_1 R)(a_1 P)$ of B onto the rectangle $0 \le u \le 1$, $-1 \le v \le 1$. The above-mentioned requirement extends γ to a mapping of the whole part ω of Φ generated from the strip by the group $\{a_1\}$ onto the whole parallel strip $-1 \le v \le 1$. Now this parallel strip is mapped onto itself topologically by

$$u' = u + 1 - |v|, \qquad v' = v$$

and the latter is pulled back by means of γ to a mapping t of ω onto itself. In the regions equivalent to ω, no two of which have a point in common, t is defined by

$$tfx = ftx$$

and outside these regions it is defined to be the identity. Then t is a T-function which induces I_0 and which sends O to $a_1 O$.

Lemma 3. *If $D_1, D_2, \ldots, D_n$ are n inequivalent points in Φ and $D_1', D_2', \ldots, D_n'$ are n further points such that D_1 is equivalent to $D_1', \ldots, D_n$ to D_n', then there is an I_0-inducing T-function which sends D_1 to $D_1', \ldots, D_n$ to D_n'.*

One first applies Lemma 2 to obtain $2pn$ T-functions $t_{i,h}$ which all induce I_0 and have the following properties: Let h be an arbitrary generator of F, let D_i^0 be the point in B equivalent to D_i, then $t_{i,h}$ shall send the point D_i^0 to hD_i^0 and all D_j, $j \ne i$, to themselves. To do this it is only necessary to arrange the construction of $t_{i,h}$ so that the strip of B in which $t_{i,h}$ is not the identity (the strip that was called $PR(a_1 R)(a_1 P)$ in the proof of Lemma 2) contains none of the points D_j^0, $j \ne i$, which is obviously always possible. Then $t_{i,h_1} \cdot t_{i,h_2}$ sends the point D_i^0 to $h_1 h_2 D_i^0$. One can therefore construct a combination t_i from $t_{i,a_1}^{\pm 1}, \ldots, t_{i,b_p}^{\pm 1}$ which sends D_i^0 to D_i, and a combination t_i' which sends D_i^0 to D_i'. Finally, set $t^{(i)} = t_i' t_i^{-1}$. Then:

(1) $t^{(i)}$ is a "product" of T-functions and hence again a T-function.
(2) $t^{(i)}$ induces I_0, since all its factors do.
(3) $t^{(i)}$ sends D_i to D_i'.
(4) $t^{(i)}$ sends all $D_j, j \neq i$, to themselves, and all points equivalent to the D_j likewise.

Then

$$t^{(1)}t^{(2)}t^{(3)}\dots t^{(n)}$$

is the desired function.

Lemma 4. *If $D_1, \dots, D_n$ are n inequivalent points in Φ and if $D_1', \dots, D_n'$ are any other inequivalent points, then there is an I_0-inducing T-function which sends D_1 to $D_1', \dots, D_n$ to D_n'.*

Suppose the given points lie over $d_1, \dots, d_n, d_1', \dots, d_n'$ of ϕ. Let τ be a topological mapping of ϕ in the class of the identity which sends d_1 to $d_1', \dots, d_n$ to d_n', and let t be an I_0-inducing T-function which lies over τ (Lemma 1). Then $tD_1 = D_1'', \dots, tD_n = D_n''$ where D_1'' is equivalent to $D_1', \dots, D_n''$ equivalent to D_n'. Also, let t_1 be an I_0-inducing T-function which sends D_1'' to D_1', $\dots, D_n''$ to D_n' (Lemma 3). Then $t_1 t$ is the desired function.

Theorem 12 stated at the beginning of this paragraph now follows immediately:

Let I be an arbitrary automorphism of F, also let $D_1, \dots, D_n$ and $D_1', \dots, D_n'$ be any two sets of n inequivalent points of Φ. Let t be a T-function which induces I (§23). Let t_1 be a T-function which induces I_0 and which sends the n inequivalent points $tD_1, tD_2, \dots, tD_n$ to $D_1', D_2', \dots, D_n'$ (Lemma 4). Then the T-function $t_1 t$ sends D_1 to $D_1', \dots, D_n$ to D_n', and $t_1 t$ induces the automorphism $I \cdot I_0 = I$.

25. *T*-functions which leave the origin fixed. Let I be an arbitrary automorphism of F. Then Theorem 12 gives a T-function t which induces I and satisfies the condition $t(O) = O$. From the net n, t creates a topological image $tn = n'$ in Φ. If, as before, we label the edge of n which leads from O to $a_1 O$ with a_1 and an arrow, and analogously for the remaining generators, then ta_1 is a Jordan arc which leads from O to $a_1' O$, and it may be labelled a_1', and analogously. If one carries this labelling to all equivalent edges of the net, then the two topologically regular nets n and n', inscribed within each other, represent the automorphism I in an intuitive way, as was done in §9 with the nets denoted n and n' there. The difference is that the edges of n' are no longer rectilinear in general, however, they no longer meet each other outside the vertices.

The point of ϕ over which the origin O lies was denoted by q in §1. Let w be a path on ϕ which issues from q and returns to it. Suppose the curve W in Φ which lies over w and issues from O ends at $f(0)$. This associates each closed curve w on ϕ *with initial point* q with a unique element f of F, and two curves which are homotopic *with q fixed* are associated with the same element of F.

The T-functions with fixed point O constitute a group, which is in one-to-one correspondence with the group of topological mappings of ϕ *with fixed point* q. The image of the surface curve a_1 issuing from q under the mapping $\tau\phi$, over which $t\Phi$ lies, is the surface curve $\tau a_1 = a_1'$ issuing from q, over which the edge a_1' of the net n' lies, and with which the element a_1' is therefore associated. Thus the new generators can be read immediately from the images of the canonical curves.

As long as one does not distinguish a point of ϕ, the closed curves on ϕ are associated with *conjugacy classes* of F, and the topological mappings of ϕ with automorphism *families* of F.

With the choice of an arbitrary point q of ϕ the closed curves on ϕ with initial point q become associated with the individual elements of F, and the topological mappings of ϕ with fixed point q with individual automorphism of F. *And all automorphisms arise in this way.*

26. A normal form for the representation of automorphisms of *F*. The nets n and n' used in the preceding paragraph can be applied to find a representation of the automorphism I. Any edge path on n which leads from O to $a_1'O$ gives a word in $a_1, \ldots, b_p$ which represents the element a_1', and similarly for the remaining generators. It will now be shown that each automorphism may be represented in an especially simple form which was used for the criterion of §8, namely, that there be.a representation

$$a_i' = \alpha_i(a_1, \ldots, b_p), \qquad b_i' = \beta_i(a_1, \ldots, b_p) \tag{23}$$

which represents an automorphism of the *free* group $\{a_1, \ldots, b_p\}$ and for which $R(\alpha_1, \ldots, \beta_p)$ is a conjugate of $R(a_1, \ldots, b_p)^{\pm 1}$.

Let u_1 be a point of Φ which lies on neither n nor n', and let $u_1, u_2, \ldots$ be the points equivalent to u_1. Each continuous directed arc in Φ, which avoids all points u_i, determines an angle variation at each point u_i. Two edge paths on n (or n') with the same initial and final point, which subtend the same angle at *all* u_i can only differ by subarcs which run out and back over the same path, since each cell of the net contains a u_i. Edge paths without such subarcs may be called "simple." Then each edge of n' corresponds uniquely to a simple edge path on n, which connects the same vertices and subtends the same angles at all u_i, and conversely. (If one replaces each subarc of an edge of n' which is a crosscut of a cell M of n by that part of the boundary of M which does not include the u_i in M in combination with the crosscut, then one obtains a connected path on n which reduces uniquely to the desired simple edge path on n.) In this way one obtains the expressions (23) for the $a_1', \ldots, b_p'$. Interchanging n and n' gives

$$a_i = \bar{\alpha}_i(a_1', \ldots, b_p'), \qquad b_i = \bar{\beta}_i(a_1', \ldots, b_p').$$

Then it is obvious that

$$\bar{\alpha}_i(\alpha_1, \ldots, \beta_p) \equiv a_i$$

holds *identically* in the a_i, b_i, and analogously for the other equations. Now if one makes a circuit from O around the cell boundary

$$a_1' b_1' a_1'^{-1} b_1'^{-1} \ldots a_p' b_p' a_p'^{-1} b_p'^{-1}$$

in n', then one subtends an angle $+2\pi$ or -2π at a point u_μ in the interior of this cell, according as t preserves orientation or not (taking clockwise rotation to subtend the positive angle) and an angle 0 at all the remaining u_i. Thus in a circuit round the edge path

$$\alpha_1 \beta_1 \alpha_1^{-1} \beta_1^{-1} \ldots \alpha_p \beta_p \alpha_p^{-1} \beta_p^{-1}$$

of n the angle subtended is $\pm 2\pi$ at u_μ and 0 at all the other u_i. This edge path on n therefore consists of the boundary of the cell of n which contains u_μ, and possible open edge paths of n which start at the cell boundary and are traversed forth and back. But then

$$R(\alpha_1, \ldots, \beta_p) \equiv h(a_1, \ldots, b_p) R(a_1, \ldots, b_p)^{\pm 1} h^{-1}$$

identically in the a_i, b_i, whence the earlier normal form is attained.

One can visualize this situation on the surface ϕ itself as follows: Let u be the point of ϕ over which the u_i lie. Let the surface ϕ *punctured* at u be called ϕ^*. Then, on ϕ^*, $R(a_1, \ldots, b_p)$ is a curve which contracts to the puncture, and hence it is no longer homotopic to 0. The fundamental group of ϕ^* is the free group with generators $a_1, \ldots, b_p$. The curves $a'_1, \ldots, b'_p$ then correspond to the primitive system $\alpha_1, \ldots, \beta_p$ of this group, and the image of R is again a simple curve, contractible to the puncture. If in particular one chooses t so that $\tau\phi$ has a fixed point at u, as one can by Theorem 12, then t lies over a topological mapping of ϕ^*.

27. *S*-functions. Continuous transition between two continuous mappings of the surface belonging to the same automorphism family. An "S-function," or more precisely, "S-function for the automorphism I," $s(x)$, is a *continuous* function, defined for $|x| < 1$ and with absolute values < 1, which satisfies the functional equation

$$sfx = f_I sx. \tag{24}$$

Thus the S-functions include the T-functions, but unlike the latter they are not in general invertible, hence the S-functions do not constitute a group.

An S-function $s\Phi$ lies over a unique continuous mapping $\sigma\phi$ of ϕ onto itself. The continuous mappings of ϕ which arise in this way will be called σ-mappings. Over the same mapping $\sigma\phi$ lie all "S-functions related to s," $hs\Phi$, where h is an arbitrary element of F. The latter induces the automorphism $I_{h^{-1}}$. The subdivision of topological mappings $\tau\phi$ into classes, made in §22, can therefore be extended to σ-mappings:

> Two σ-mappings of ϕ are reckoned to belong to the same *mapping class* when they are associated with the same *automorphism family*.

It will now be shown *that the set of σ-mappings of ϕ making up a mapping class is connected*,[12] so that this division into classes is subsumed under Brouwer's concept of a mapping class.[13]

Thus, let σ_0 and σ_1 be two σ-mappings of ϕ, and let s_0 and s_1 be two S-functions, lying over σ_0 and σ_1 respectively, which induce the *same* automorphism I. Let x be an arbitrary point of Φ. Finally, let $s_t(x)$ be the point which divides the non-Euclidean segment connecting $s_0(x)$ to $s_1(x)$ in the non-Euclidean distance ratio $t:(1-t)$, where $0 \leq t \leq 1$. The function $s_t(x)$ defined for $|x| < 1$ in this way is continuous and has absolute values <1; it also satisfies the functional equation

$$s_t fx = f_I s_t x.$$

Because, in order to obtain $s_t fx$, one must draw the segment from $s_0 fx$ to $s_1 fx$ and divide it in the ratio $t:(1-t)$. But, since $s_0 fx = f_I s_0 x$ and $s_1 fx = f_I s_1 x$, this segment is the image under f_I of the segment from $s_0 x$ to $s_1 x$, whence the assertion follows. Thus s_t is an S-function associated with I for each $0 \leq t \leq 1$, and for $t = 0$ and $t = 1$ it equals the given functions s_0

[12] In other words, any two of the mappings are homotopic. (Translator's note).

[13] L. E. J. Brouwer: Sur la notion de "Classe" de transformations d'une multiplicité, Proceedings Vth Intern. Congr. of Math. Cambridge 1913, and: Énumeration des classes de représentations d'une surface sur une autre surface, Comptes rendus, t. 171, p. 89 (1920). The latter work includes the above theorem—obviously two σ-mappings of ϕ associated with different automorphism families are not continuously deformable into each other.

and s_1 respectively. The one-parameter family σ_t, $0 \leq t \leq 1$, of σ-mappings of ϕ over which the s_t lie, then represents the required continuous transition between σ_0 and σ_1.

In particular, one can always choose the σ-mapping σ_1 for a given σ_0 to be a *topological* mapping, so that s_1 is a T-function for I. Thus each σ-mapping can be continuously deformed into a topological mapping. The fact that the *Brouwer mapping degree*[14] is invariant under continuous deformation then implies that the mapping degree of a σ-mapping of ϕ equals $+1$ or -1, according as the associated automorphism family consists of automorphisms of the first or second kind.

28. Continuous incorporation of the boundary mapping. Let I be an arbitrary automorphism of F, $s\Phi$ an S-function for I, and gE the boundary mapping induced by I (§10). *Then $s\Phi$ is completed by gE*, i.e., the two functions s and g together determine a continuous mapping of the closed disc $\Phi + E$ onto itself.

The proof of this assertion depends on the following: Let P be an arbitrary point of E and let $P_1, P_2, P_3, \ldots$ be a sequence of points of Φ which converges to P but is otherwise arbitrary; *then the sequence $sP_1, sP_2, sP_3, \ldots$ converges to gP.* Let $\bar{P}_n$ be the point equivalent to P_n in the central fundamental domain B of F, and let

$$P_n = f_n \bar{P}_n, \qquad n = 1, 2, 3, \ldots.$$

Since the non-Euclidean distance between O and $\bar{P}_n$ remains bounded, so too does the distance between $f_n O$ and $f_n \bar{P}_n = P_n$, hence $f_n O \rightarrow P$ as $n \rightarrow \infty$. Thus the elements f_n constitute a fundamental sequence for P (§7, j). Then the elements f_{nI} constitute a fundamental sequence for gP (§16), hence $f_{nI} O \rightarrow gP$ as $n \rightarrow \infty$. Now, all points $s\bar{P}_n$ belong to the region s**B**, and they therefore have bounded distance from O, so it follows that $f_{nI} s\bar{P}_n \rightarrow gP$ as $n \rightarrow \infty$. But then, by the functional equation for s, we have

$$f_{nI} s\bar{P}_n = sf_n \bar{P}_n = sP_n,$$

so that $sP_n \rightarrow gP$ as $n \rightarrow \infty$, Q.E.D.

In the special case where I is the identity automorphism I_0 and gE is therefore the identity boundary mapping, it is easy to see the continuous completion of $s\Phi$ by gE directly, without using the concept of fundamental sequence. The function $\delta(x_1, x_2)$ was defined in §2; since it equals $\sinh^2$ of half the non-Euclidean distance between x_1 and x_2, one has

$$\delta(f(x_1), f(x_2)) = \delta(x_1, x_2)$$

for each f in F. Now, in the case being considered, the function s commutes with every f. Hence

$$\Delta(x) = \delta(x, s(x)) = \delta(f(x), f(s(x))) = \delta(f(x), s(f(x))) = \Delta(f(x))$$

for every f in F. $\Delta(x)$ therefore has the same value at equivalent points and hence it is bounded in Φ, since it is bounded in B. It then follows from

$$\Delta(x) = \frac{|s(x) - x|^2}{(1 - |x|^2)(1 - |s(x)|^2)} < \text{const.}$$

that $s(x) \rightarrow x$ as $|x| \rightarrow 1$, and hence the continuous completion is the identity mapping on E.

[14] L. E. J. Brouwer: Über Abbildung von Mannigfaltigkeiten, §1; Math. Annalen, Bd. **71**, p. 97 ff.

For the sake of uniformity, we now often denote gE by sE, so that we can speak of $s(\Phi + E)$. However, one should bear in mind that sE does not depend on the particular function $s\Phi$, but only on the automorphism induced by s. Thus each class of σ-mappings of Φ corresponds to a family of related boundary mappings, and the method of the present work is essentially to derive those properties of mappings which belong to the *class* from the associated family of boundary mappings. Thus, in place of the investigation of *continuous* mappings in *two* dimensions, we have the investigation of *topological* mappings in *one* dimension.

One can extend $s(x)$ to a function defined on the whole complex x-sphere K, by also viewing the exterior Ψ of E as universal covering surface of ϕ. For this purpose we agree to let x and $1/\bar{x}$ (where $\bar{x}$ is the complex conjugate of x) lie over the same point of ϕ. Then $s(x)$ is defined in Ψ by reflection in E, so that in general

$$s(x) = \frac{1}{\overline{s(1/\bar{x})}}.$$

The continuous mapping sK of K onto itself is continuously deformable into a topological mapping of K onto itself, and therefore has degree $+1$ or -1. It follows that the mapping sK has each point of K in its image.

29. Regular and singular points with respect to a subgroup of F.

Definition. Let F_1 be an arbitrary subgroup of F. A point P of the complex sphere K is called *regular with respect to* F_1 when there is a neighbourhood of P which contains no pair of points equivalent under F_1, and otherwise *singular with respect to* F_1.

All points of Φ and Ψ are regular with respect to F_1, since they are regular with respect to F. The singular points with respect to F_1 are therefore to be found on E. Each regular point has a neighbourhood consisting of regular points. Each fundamental point of an element of F_1 is obviously singular with respect to F_1, and so is each accumulation point of such points. Let m be the set of fundamental points of elements of F_1 and M the "closed hull of m" which results from adding its accumulation points. *It will be shown that the set of singular points with respect to* F_1 *is identical with the set* M.

Let P be a singular point with respect to F_1 and let i be a segment of E containing P. The assertion to be proved says that i contains a fundamental point of F_1.

Lemma. *Let O, P, Q, R be four points in that cyclic order on E. Let α and β be two hyperbolic transformations with the same displacement length and negative fundamental point at O. Let Q be the positive fundamental point of α and let R be the positive fundamental point of β. Then the displacement arc $P - \alpha P$ on E is smaller than the displacement arc $P - \beta P$.*

Because if γ is a hyperbolic transformation with O as positive, and P as negative, fundamental point and a displacement length such that $\gamma Q = R$, then $\beta = \gamma\alpha\gamma^{-1}$ and $\beta P = \gamma\alpha\gamma^{-1} P = \gamma\alpha P = \gamma(\alpha P)$; thus the arc $P - \alpha P$ is extended by γ to $P - \beta P$.

Two applications extend this lemma to the case where the negative fundamental point of β is not at O but at a point O_1 belonging to the arc OR of E not containing P and Q.

Now let A and B be the endpoints of the above segment i (Figure 15). Let α be a hyperbolic transformation with A as positive, and B as negative, fundamental point, whose displacement length is smaller than any occurring in the group F. Let CPD be an interval containing

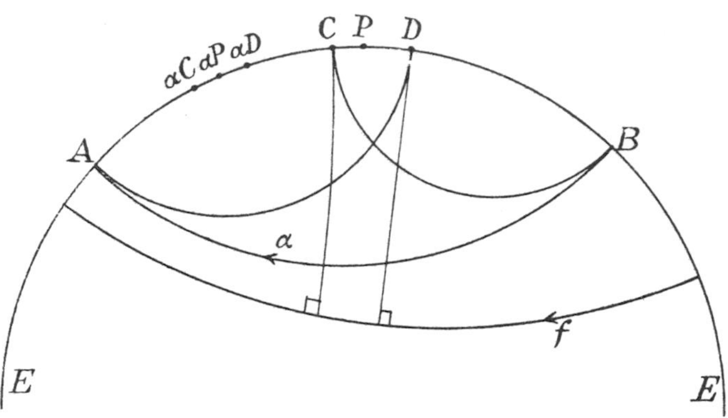

Figure 15

P which is so small that it does not meet its own image under α. (Thus C separates P from αD.) Connect C to B and D to A by circles orthogonal to E. This determines a neighbourhood u of P, namely that consisting of the curvilinear triangle in the figure, together with its reflection in E. Now let f be an element of F_1 whose fundamental points both do not belong to the interval $AB = i$. Then, by the lemma, the points of i are pushed farther by f than by α. If one drops the non-Euclidean perpendiculars from C and D to the axis of f, then u is contained between them, and fu therefore has no point in common with u. Now since P is assumed to be singular with respect to F_1, there is an element f_1 in F_1 such that $f_1 u$ has points in common with u, and at least one of the fundamental points of f_1 must therefore lie in i. Q.E.D.

When the fundamental points of F_1 lie everywhere densely on E, then M is identical with E. Thus for F itself the whole of E is singular. Now suppose that the fundamental points of F_1 are not everywhere dense on E. Then the set $E - M$ of regular points of E consists of finitely or denumerably many intervals, whose union is reproduced by application of any element of F_1, since the set m of fundamental points of F_1, and hence M, is also reproduced. It now follows, as in §14, that when one of the intervals forming $E - M$ contains a pair of points equivalent under F_1, it is bounded by an axis of an element of F_1, and it then contains only such pairs of equivalent points as correspond under powers of this element. The interval may then be called a "periodic interval with respect to F_1." "Aperiodic intervals with respect to F_1" contain no pairs of points equivalent under F_1, and their endpoints are not fundamental points of F_1.

30. Mappings of covering surfaces of ϕ corresponding to subgroups of F. Let F_1 be a subgroup of F. If one regards each complete system of points of Φ equivalent under F_1 as a single point, then one obtains a covering surface of ϕ, which will be called the "covering surface of ϕ corresponding to F_1" or "Φ mod F_1." Φ is the universal covering surface of Φ mod F_1, and F_1 is the covering transformation group of Φ over Φ mod F_1.[15]

[15] Φ mod F_1 is in general not a "regular" covering surface of ϕ (see, e.g., Weyl, Die Idee der Riemannschen Fläche, p. 50), so that one cannot speak of a group of covering transformations for Φ mod F_1 over ϕ.

Now let s be an S-function and let I be the automorphism of F induced by s. A necessary and sufficient condition for s to lie over a continuous mapping of Φ mod F_1 onto itself is that s send any two points equivalent under F_1 to two points equivalent under F_1, and hence, by (24), that I map F_1 into itself.

In particular, this condition is always satisfied when one takes F_1 to be a subgroup consisting of elements fixed by I. Let H_1 be such a group, and set Φ mod $H_1 = \Phi_1$. Suppose that h_1, $h_2, h_3, \ldots$ is a (finite or infinite) system of fixed elements of I, which generates H_1. Now let s_1 be an I-inducing S-function which leaves the origin O fixed (§24). A system of arcs in Φ which begin at O and end at $h_1(O)$, $h_2(O)$, ... respectively lies over a fundamental system of closed curves on Φ_1. Since $s_1 \rightleftarrows h_i$, $i = 1, 2, \ldots$, these have images under $s_1\Phi_1$ which are homotopic to the originals, and the same is therefore true for all closed curves on Φ_1. Now since, by §27, there is a continuous deformation of $s\Phi$ into $s_1\Phi$, and hence also of $s\Phi_1$ into $s_1\Phi_1$, throughout which the functional equation remains valid, the same is also true for the mapping $s\Phi_1$ originally considered.

In particular, one can take H_1 to be the largest group of its kind, namely the group of all fixed elements of I, which was previously denoted by $H(I)$. The boundary of Φ mod H consists of the point set E mod H. However, those points of E which are regular with respect to H (§29), lose their boundary point character when one, as at the end of §28, views the whole complex sphere K as a doubly counted universal covering surface of ϕ, and then reduces this mod H. A point of E which is regular with respect to H then becomes an ordinary point of $K_1 = K$ mod H. An interval on E which is periodic with respect to H (§29) does not give rise to any boundary on K_1, since it lies over a non-singular closed curve on K_1. However, the endpoints of intervals on E which are aperiodic with respect to H—if there are any—do give boundary points of K_1.

Now suppose that there are no aperiodic intervals with respect to H. Then K_1 has no boundary points. K_1 is covered by $K - M$, where the boundary set M is the set of singular points with respect to H. H is the associated covering transformation group. Now when H is generated by a finite number of operations, say v, and when one can enclose the fundamental point pairs of these operations by disjoint circles, which correspond under the generators, then H is a *free* group with v generators and $K_1 = K$ mod H is a closed surface of genus v (cf. the conclusion of §2 for $v = 1$).

The S-function $s(x)$ considered above, which is defined on the whole of K by §28, corresponds to a continuous mapping sK_1 of K_1. Again we take an S-function s_1, continuously deformable into s, which leaves the origin O fixed. It was shown above that each arc in Φ which connects O to a point equivalent to O under H lies over a closed curve of K_1 which is homotopic to its own image under s_1K_1. Here we can obviously weaken the condition that the arc run in Φ to the condition that it run in $K - M$, with the result that all closed curves of K_1 issuing from the point corresponding to O are homotopic to their images under s_1K_1, i.e., *that the mapping s_1 of the closed surface K_1 belongs to the class of the identity.* The same then holds for the mapping sK_1 originally considered, since it is continuously deformable into s_1K_1, and hence belongs to the same class.

One sees that K_1 is the *weakest* covering surface of ϕ on which the mapping by the given S-function s belongs to the class of the identity, because K_1 results from reducing K modulo *the largest group of elements of F which commute with s*, namely $H(I)$.

CHAPTER IV

The Fixed Point Problem

31. Classification of the fixed points of a continuous mapping. Let σ be a continuous mapping (σ-mapping, §27) of ϕ and let q be a fixed point of $\sigma\phi$:

$$\sigma q = q.$$

Let x_0 be a point of Φ lying over q, and let s be the S-function lying over σ, for which

$$sx_0 = x_0.$$

Now if f is an element of F, the S-function fsf^{-1}, which also lies over σ, has a fixed point (lying over q) at fx_0. If s induces the automorphism I, then fsf^{-1} induces the automorphism

$$I_{0f} \cdot I \cdot I_{0f^{-1}} = I_{f_I f^{-1}}$$

isogredient with I, cf. (11), §18. Accordingly, we now speak of S-functions which are conjugate via an element of F, s and fsf^{-1}, as *isogredient S-functions.* Thus two such functions have equivalent fixed point sets.

Each fixed point of σ is therefore associated with an *isogredience class* of automorphisms of F (§18), which is a subset of the automorphism family associated with σ.

Definition. The collection of fixed points of $\sigma\phi$ associated with the same isogredience class is called a *fixed point class.*

Let q_1 and q_2 be two fixed points of $\sigma\phi$, let s_1 be an S-function lying over σ which has a fixed point lying over q_1, and let s_2 be another which has a fixed point lying over q_2. Now, when q_1 and q_2 belong to the same fixed point class, s_1 and s_2 are isogredient S-functions by definition, and hence they have equivalent fixed point sets. Thus s_1 also has a fixed point lying over q_2. If, on the other hand, q_1 and q_2 do not belong to the same fixed point class, then s_1 and s_2 are not isogredient, thus in particular not identical, and s_1 has no fixed point lying over q_2. When the automorphism induced by s_1 has fixed elements, then these reproduce the fixed point set of $s_1\Phi$ by (24). We therefore have:

Theorem 13. *When an S-function $s\Phi$ lying over $\sigma\phi$ has any fixed points at all, its fixed point set covers the points of one and only one fixed point class of $\sigma\phi$, and completely, indeed in general infinitely often.*

The question of *one-to-one* representation of all fixed points of $\sigma\phi$ will be answered in §33.

A geometric definition of the division of fixed points of $\sigma\phi$ into classes, in terms of ϕ itself, can be expressed as follows: Let q_1 and q_2 be two fixed points of $\sigma\phi$, and let k be a directed arc on ϕ from q_1 to q_2. Then σk also runs from q_1 to q_2, and hence $k(\sigma k)^{-1}$ is a closed curve on ϕ. *If k can be chosen so that $k(\sigma k)^{-1}$ is null homotopic, i.e., contractible on ϕ, then, and only then, do q_1 and q_2 belong to the same fixed point class of $\sigma\phi$.* Because then and only then does a mapping $s\Phi$ over $\sigma\phi$ have fixed points lying over q_1 and q_2 at the same time, and one chooses k to be an arc in Φ joining them.

In particular, all points of a continuum on ϕ which is pointwise fixed by $\sigma\phi$ belong to the same class.

The diffusion of the set of fixed points in a class depends to a certain extent on the particular form of the mapping σ in the small; one might conjecture that any fixed point class might be made *either* to vanish *or* reduced to a single point by continuous deformation. We first seek a sufficient condition for non-vanishing of a fixed point class under continuous deformation of σ, in which case it is characteristic of the mapping *class* (§35–37), and then grasp the problem of determining the number of, and mutual relations between, the characteristic and topologically necessary fixed point classes for a given mapping class.

32. Finiteness of the class number. Let $\sigma\phi$ be an arbitrary σ-mapping of ϕ. Let $s\Phi$ be any S-function lying over $\sigma\phi$ and let I be the automorphism induced by s. Let B be a fundamental domain for F, say the central cell of the net N defined in §3. Let q_1 be a fixed point of $\sigma\phi$, x_1 the point of B which lies over it. Then sx_1 is equivalent to x_1, say $sx_1 = f_1 x_1$. Hence sB meets the cell f_1B of the net. But sB is a subregion of a circle concentric with, but smaller than, E, which therefore meets only finitely many cells of the net, say

$$f_1\mathrm{B}, f_2\mathrm{B}, \ldots, f_m\mathrm{B}.$$

The above process therefore associates each fixed point of $\sigma\phi$ with one of these m elements. Two fixed points associated with the same element, say f_i, appear simultaneously as fixed points of $f_i^{-1}s\Phi$, and therefore belong to the same class. Thus $\sigma\phi$ has *at most* m classes of fixed points;

Theorem 14. *The number of fixed point classes of a σ-mapping is finite.*

We now determine the class number more precisely. Let x_1 be a point of Φ which lies over a fixed point of $\sigma\phi$, and suppose:

$$sx_1 = f_1 x_1.$$

Let h be an arbitrary element of F. Because of (24), we then have

$$shx_1 = h_I f_1 h^{-1} \cdot hx_1.$$

The equivalence factors (§20) of the complete system of equivalent points Fx_1 therefore constitute an isogredience class relative to I^{-1}, just as for the case of the boundary points in §20 (indeed, these satisfy the same functional equation). Now if $sx_2 = f_2 x_2$, the two fixed points over which x_1 and x_2 lie obviously belong to the same class if and only if there is a solution h of the equation

$$f_2 = h_I f_1 h^{-1}$$

and thus exactly when f_1 and f_2 belong to the same isogredience class relative to I^{-1} (cf. §18). We therefore have: *The number of fixed point classes of $\sigma\phi$ equals the number of isogredience classes relative to I^{-1} which are represented by the elements $f_1, f_2, \ldots, f_m$ obtained in the above process.*

33. One-to-one representation of all fixed points of $\sigma\phi$. Let $\sigma\phi$ be any σ-mapping. If one constructs the whole family of S-functions lying over σ:

$$s_1{}^*\Phi, s_2{}^*\Phi, s_3{}^*\Phi, \ldots,$$

then each fixed point of $\sigma\phi$ also appears as a fixed point of an $s_i{}^*\Phi$. However, each appears infinitely often. When $s_i{}^*$ and $s_k{}^*$ are isogredient, i.e., conjugate via an operation in F, then they have equivalent fixed point sets (§31). Thus we must first choose a representative from each class of isogredient S-functions: Suppose

$$s_1\Phi, s_2\Phi, s_3\Phi, \ldots$$

is a complete system of mutually non-isogredient S-functions lying over $\sigma\phi$, and let

$$I_1, I_2, I_3, \ldots$$

be the isomorphisms of F they induce. But here too—as was already brought out in Theorem 13—a fixed point of $\sigma\phi$ in general appears infinitely often. Namely, when there are elements fixed by I_n, so $H(I_n) > 1$, each fixed point set of $s_n\Phi$ reproduces itself under $H(I_n)$, because of the functional equation. Thus one must count only those fixed points of $s_n\Phi$ which lie in a *fundamental domain of* $H(I_n)$. These now really do lie over different fixed points of $\sigma\phi$. Namely, any two equivalent points of this fundamental domain correspond under an element of F which is not fixed by I_n; the two points therefore cannot both be fixed points of $s_n\Phi$, because of the functional equation (24):

Theorem 15. *When $s_n\Phi$, $n = 1, 2, \ldots$ is a complete system of non-isogredient S-functions over $\sigma\phi$, I_n is the automorphism of F induced by s_n, and β_n is a fundamental domain for the fixed element group $H(I_n)$, then the collection of all fixed points of all $s_n\beta_n$ represents the set of fixed points of $\sigma\phi$ in a one-to-one manner. By Theorem 14, $s_n\beta_n$ contains fixed points for only finitely many values of n, and for each of these values of n, $s_n\beta_n$ completely represents a fixed point class of $\sigma\phi$.*

It is useful to express this fact in another form, more convenient in practice, which starts with the choice of a particular S-function. Let $s\Phi$ be an S-function lying over $\sigma\phi$, and let I be the automorphism induced by s. Then $fs\Phi$ represents the whole family of S-functions lying over $\sigma\phi$ when f runs through F; $f_1 s$ and $f_2 s$ are isogredient if and only if f_1 and f_2 are isogredient *relative* to I, hence (by Theorem 7, §18) when there is an f such that

$$f_2 = ff_1 f_I^{-1}.$$

One therefore has to choose a system of elements of F:

$$f_0 = 1, f_1, f_2, \ldots$$

which includes exactly one member from each isogredience class relative to I, and then investigate each $f_n s\Phi$ for fixed points. Only finitely many of them possess fixed points, and in fact a fixed point class is represented completely and without repetition in a fundamental domain of $H(I_{f_n^{-1}})$.

34. Finite extent of the individual fixed point classes. In the notation of the preceding paragraph, the following is true:

Theorem 16. *There is a circle of radius $\rho < 1$, concentric with E, which includes the fixed point sets of all $s_n\beta_n$ (Theorem 15).*

Proof. We first consider a particular value of n and set $H(I_n) = H$. Then let

$$k_0 = 1, k_1, k_2, \ldots$$

be a sequence of elements of F such that

$$F = H + k_1 H + k_2 H + \dots$$

is the coset decomposition of F modulo H. The fundamental domain B of F then yields the fundamental domain

$$\beta_n = \mathrm{B} + k_1 \mathrm{B} + k_2 \mathrm{B} + \cdots$$

for H. We have to show that fixed points of $s_n \Phi$ can occur in $k_r \mathrm{B}$ for only finitely many values of r.

In fact, let $s_n P = P$ and suppose P is in $k_r \mathrm{B}$. Then

$$s_n(k_r^{-1} P) = k_{r_{I_n}}^{-1} k_r (k_r^{-1} P)$$

and hence $k_{r_{I_n}}^{-1} k_r \mathrm{B}$ meets $s_n \mathrm{B}$. Now if $s_n \mathrm{B}$ meets the fundamental domains

$$f_1 \mathrm{B}, f_2 \mathrm{B}, \dots, f_m \mathrm{B},$$

and only these, we must have, say

$$k_{r_{I_n}}^{-1} k_r = f_\mu .$$

But for each of the m values of μ there is *at most* one value of r, because if it were also true that

$$k_{s_{I_n}}^{-1} k_s = f_\mu$$

for $r \neq s$, then it would be true that

$$k_{r_{I_n}}^{-1} k_r = k_{s_{I_n}}^{-1} k_s$$

and hence

$$k_r k_s^{-1} = (k_r k_s^{-1})_{I_n}$$

so that $k_r k_s^{-1}$ would belong to H, contradicting the fact that $k_r H$ and $k_s H$ are different cosets of H. Thus there is a circle in the interior of E which includes the fixed point set of $s_n \beta_n$ for the value of n under consideration. But since $s_n \Phi$ has fixed points for only finitely many values of n there is also—by the choice of β_n—a circle which satisfies the condition simultaneously for all n, as claimed.

Now we consider the mapping extended to the boundary, $s_n(\Phi + E)$. The set m of fundamental points of $H(I_n)$, and hence also its closed hull M, i.e., the set of singular points with respect to $H(I_n)$ (§29), is contained in the set of fixed points of $s_n E$. If $s_n \Phi$ has any fixed points at all, then obviously each point of m, and hence also each point of M, is an accumulation point of fixed points of $s_n \Phi$. Theorem 16 now says that: *the set M of singular points with respect to $H(I_n)$ is identical with the set of all accumulating points on E of fixed points of $s_n \Phi$.* Because when the fixed point set of $s_n \beta_n$, for arbitrary choice of β_n, has no accumulation point on E, no regular point of E with respect to $H(I_n)$ can be an accumulation point of fixed points of $s_n \Phi$. It should be emphasized, however, that the possibility of regular points with respect to $H(I_n)$ on E being fixed points of $s_n E$ nevertheless remains open.

35. The direction function ξ. Index. In the determination of fixed points of a mapping of the Euclidean plane one often benefits from the concept of angle variation. In the non-Euclidean metric used here the following is a natural replacement of this notion:

Let x and y be two different points of Φ. (The concept can also be transported to Ψ by reflection in E.) One draws the circle orthogonal to E through x and y. As one traverses this circle in the direction from x to y, but past y, let z be the point at which it meets E. Thus z is the infinitely distant point of the half line from x through y in the non-Euclidean metric. We can express z as a function $z(x, y)$ of x and y explicitly by

$$z(x, y) = \frac{x\bar{y} - \bar{x}y + \sqrt{(x-y)(\bar{x}-\bar{y})(1-x\bar{y})(1-\bar{x}y)}}{\bar{y}(1+x\bar{x}) - \bar{x}(1+y\bar{y})}.$$

Here the above determination of direction corresponds to the rule of always taking the positive square root. When x and y coincide the function $z(x, y)$ loses its meaning. When x and y are different points on E, the definition remains valid, and we simply have $z(x, y) = y$. (We only need the case where x and y are both in Φ or both on E.) If f is a linear fractional transformation which maps E into itself, then obviously

$$z(f(x), f(y)) = f(z(x, y)). \tag{26}$$

Now let $s(\Phi + E)$ be an S-function. Denote the fixed point set of $s(\Phi + E)$ by Π and set

$$\Phi + E - \Pi = \Omega.$$

Then

$$z(x, s(x))$$

is a continuous function on Ω with absolute value 1. From it we construct the continuous real function

$$\xi(x; s) = \frac{1}{2\pi i}\log z(x, s(x)),$$

whose value at the point x is determined only mod 1. With each continuous directed path in Ω we now associate $\int d\xi$ along the path, and hence with each closed directed curve in Ω an integer, the *index* of the directed curve. This quantity is constant under continuous deformation of the curve in Ω, and hence in particular it is zero when the curve is contractible to a point of Ω without meeting Π.

The positive direction around E is now defined to be that in which $(1/2\pi i)\log z$ increases, thus with the usual orientation of the complex plane it is that for which one has Φ on one's left. This definition of positive direction around a curve may be extended to all Jordan curves in $\Phi + E$. Let C be a Jordan curve in Ω with non-zero index, say the value j for its positive direction. Then C contains a subset Π_1 of Π in its interior. Each Jordan curve C_1 in Ω, which contains Π_1 and no other points of Π in its interior, then also has index j, as one can see by introducing a Jordan curve C_2 which lies inside both C_1 and C while still containing Π_1 in its interior. One can therefore take j to be the *index of the isolated subset* Π_1 of Π. In this connection the sign of j is also meaningful, inasmuch as it is obviously independent of the orientation chosen for the complex plane.

One can think of the ξ-values as associated with the points of an infinite sheeted, unbranched covering surface of Ω.

36. Invariance of the index under topological transformations. Let C be a closed curve in

Ω (with the notation of the previous paragraph), say

$$C: x = x(u), \quad 0 \leq u \leq 1, \quad x(0) = x(1), \quad s(x(u)) \neq x(u).$$

Then $\xi(x(u); s)$ is a continuous function of u. One fixes a branch of the function by choice of the initial value for $u = 0$, and writes the function afterwards as $\xi(u)$. Then $\xi(1) - \xi(0)$ is the index of C.

Now let λ be an arbitrary orientation-preserving topological mapping of $\Phi + E$ onto itself. *It will be shown that the index of λC with respect to the mapping $\lambda s \lambda^{-1}$ is equal to the index of C with respect to s.* By the Tietze deformation theorem[16] there is a continuous sequence λ_v of topological mappings, with parameter $0 \leq v \leq 1$, of $\Phi + E$ onto itself, such that λ_0 is the identity and $\lambda_1 = \lambda$. Now let

$$s_v = \lambda_v s \lambda_v^{-1}$$

and

$$\lambda_v(x(u)) = \tilde{x}(u, v),$$

where $\tilde{x}(u, v)$ is continuous in v for each fixed u, and in fact uniformly for all u, so that $\tilde{x}(u, v)$ is continuous on the square $0 \leq u \leq 1, 0 \leq v \leq 1$. The same then holds for

$$\lambda_v(s(x(u))) = \lambda_v s \lambda_v^{-1} \cdot \tilde{x}(u, v) = s_v(\tilde{x}(u, v))$$

and hence also for

$$\xi(u, v) = \frac{1}{2\pi i} \log z(\tilde{x}(u, v), s_v(\tilde{x}(u, v)));$$

namely, $s_v(\tilde{x}(u, v)) \neq \tilde{x}(u, v)$ for $0 \leq u \leq 1, 0 \leq v \leq 1$, since these two points are the images of different points $x(u)$ and $s(x(u))$ under λ_v. Thus if one determines the function $\tilde{\xi}(u, v)$ by $\tilde{\xi}(0, 0) = \xi(0)$, then $\tilde{\xi}(u, 0) = \xi(u)$. The difference

$$\tilde{\xi}(1, v) - \tilde{\xi}(0, v),$$

which depends continuously on v, is therefore a constant, since it has to be an integer, hence

$$\tilde{\xi}(1, 1) - \tilde{\xi}(0, 1) = \tilde{\xi}(1, 0) - \tilde{\xi}(0, 0) = \xi(1) - \xi(0). \quad \text{Q.E.D.}$$

The above condition that λ be orientation-preserving can now easily be relaxed, since the index of C is obviously preserved when s is conjugated with the reflection in the real axis, $\lambda(x) = \bar{x}$.

If one views the surface ϕ as lying in ordinary space, then Φ is obtained as its universal covering surface by mapping the dissected surface ϕ onto a fundamental domain of F. This involves considerable arbitrariness. Any two mappings of Φ onto ϕ obtained in this way are related to each other by a T-function. If the two mappings are used to construct S-functions corresponding to a σ-mapping $\sigma\phi$, then these S-functions are conjugate to each other via the T-function. *The above invariance theorem therefore gives us the right to speak of the index of an isolated fixed point set of $\sigma\phi$ as belonging to the fixed point class itself.*

---

[16] Pal. Rend. **38** (1914).

In particular, equivalent fixed point subsets of isogredient S-functions have the same index. We shall later make further use of the right to conjugate an S-function by an arbitrary T-function without changing the indices of its fixed point sets.

37. Index of a fixed point class. Index sum Ξ. The general fixed point problem. Again suppose a mapping $\sigma\phi$ is given. Let $s\Phi$ be an S-function lying over it, and I the automorphism induced by s. Let β be a fundamental domain of $H(I)$. Then a fixed point class of $\sigma\phi$ is represented one-to-one by the subset Π_1 of fixed points of $s\Phi$ lying in β (§33). In this connection, Π_1 meets only finitely many domains equivalent to the fundamental domain B of F (§34). Now when Π_1 is isolated, i.e., when there is a Jordan curve C in $\Phi + E$ which does not meet the fixed point set Π of $s(\Phi + E)$, and which contains Π_1 and no other points of Π in its interior, then the index of Π_1, i.e., of C (§35), *is called the index of the fixed point class under consideration.* This isolation property is not always attained, namely it is not when there are elements fixed by I. Because then it can happen that two equivalent points of Φ simultaneously belong to a continuum in Φ consisting of fixed points of $s\Phi$; it is often of interest to consider precisely this behaviour. Cf. Example 12, §48. However, in order to be able to associate an index with each fixed point class, we shall in general make the following "Assumption α" about $\sigma\phi$, under which, as one easily sees, the general fixed point problem suffers no improper restriction.

Assumption α. *Among the finitely or denumerably many regions on ϕ which comprise the complement of the fixed point set of $\sigma\phi$, there is one with the same genus as ϕ.*

In this region one can set up a canonical cut system for ϕ and use it as the basis for the mapping of ϕ onto the fundamental domain B of F; then none of the S-functions lying over $\sigma\phi$ have fixed points on the net N. (In other words: One can reach this convenient state of affairs, when the relation between ϕ and Φ is already established, by Assumption α and, if necessary, conjugation of the S-function by a T-function, without affecting the index.) Then one can construct the index of the boundary of each cell of the net under s. If one presents the fundamental domain of $H(I)$ in the form

$$\beta = \mathrm{B} + k_1\mathrm{B} + k_2\mathrm{B} + \cdots,$$

as in §34, then all but finitely many of these cells have index zero, and the index of the fixed point class is the sum of the indices over all cells. In particular, if one chooses $k_1, k_2, \ldots$ so that β is a simply connected domain, then one can construct a curve C in β, from edges of N, which encloses all fixed points of $s\Phi$ in β. Then the index of C obviously coincides with the sum of the indices of the cells enclosed by C.

For a mapping $\sigma\phi$ satisfying Assumption α, the index of each individual fixed point class is therefore defined, and since there are only finitely many such, so also is *the sum of the indices of all fixed point classes.* The latter is denoted by $\Xi = \Xi(\sigma)$.

Now it is to be expected (and this will be shown in the sequel, though not in general, but in a series of special cases) that the number of fixed point classes with non-zero index, and their individual indices, hence also the index sum Ξ, will depend only on the mapping *class* of σ. The general fixed point problem then acquires the following form:

General Fixed Point Problem. *To determine the number of fixed point classes with non-zero index, and the index of each class, for each mapping class of ϕ, from the associated automorphism family of F.*

Now we consider once again the conjugation of a given mapping by a topological mapping. Let $\tau\phi$ be a topological mapping of ϕ onto itself, let t be a T-function lying over τ, and let J be the automorphism induced by t. Then the S-function $s_1 = tst^{-1}$ induces the automorphism $I_1 = J^{-1}IJ$. Hence I_1 belongs to the same *automorphism class* as I, but not in general to the same isogredience class. The indices of the fixed point classes determined by s and s_1 are the same by §36. The same also holds for the other fixed point classes of σ and $\tau\sigma\tau^{-1}$. Namely, t transforms the whole family of S-functions related to s into the whole family related to s_1:

$$t\cdot fs\cdot t^{-1} = f_J\cdot tst^{-1} = f_J s_1, \tag{27}$$

and it thereby reproduces the partition into isogredience classes with respect to the automorphisms under consideration: The equations

$$f_2 = f\cdot f_1\cdot f_I^{-1}$$

and

$$f_{2J} = f_J\cdot f_{1J}\cdot f_{JI_1}^{-1}$$

follow from each other by application of J and J^{-1} respectively. Thus when the solution of the fixed point problem depends only on the given automorphism we have: In solving the general fixed point problem for a given automorphism it is permissible to conjugate with an arbitrary automorphism. Under this transformation the family of the given automorphism remains together, as (27) shows, thus we have the corollary: *The solutions of the general fixed point problem correspond to the conjugacy classes of automorphisms.* Here one has an explicit formulation of the obvious fact that the solution of the fixed point problem for a mapping class does not depend on the choice of a canonical cut system on the surface, nor on the choice of a generator system for the group F.

38. Continuous mappings of the torus belonging to the class of the identity. The present work has the topology of surfaces of genus $p > 1$ as its object. However, for the sake of introduction and comparison, it is advisable to present a theorem about the torus. Since one can easily carry over our earlier concepts to the case $p = 1$, we shall retain our present style of notation.

The Euclidean x-y-plane Φ is the universal covering surface of the torus ϕ, so points whose coordinates differ by integers lie over the same point of ϕ. Let $s\Phi$ be a continuous mapping of Φ onto itself, given by continuous mapping functions

$$x' = f(x, y),$$
$$y' = g(x, y),$$

which satisfy the functional equations

$$f(x+1, y) = f(x, y) + 1, \qquad f(x, y+1) = f(x, y),$$
$$g(x+1, y) = g(x, y), \qquad g(x, y+1) = g(x, y) + 1.$$

Then $s\Phi$ lies over a continuous mapping $\sigma\phi$, and σ is continuously deformable into the identity mapping. Namely, if one lets each image point sP of Φ move toward the original P on a straight line at constant speed equal to the distance between P and sP, then one obtains a family of mappings $s_t\Phi$ depending continuously on the time t, $0 \le t \le 1$, and given by mapping functions f_t and g_t which satisfy the same functional equations as f and g. Hence $s_t\Phi$ lies over a continuous mapping $\sigma_t\phi$; σ_0 is the given mapping σ, and σ_1 is the identity.

Now suppose that $\sigma\phi$ satisfies Assumption α (§37). In the part of the complement of the fixed point set of $\sigma\phi$, which has genus 1, we construct a canonical pair of loop cuts, and choose the correspondence between ϕ and Φ so that the lines with integer x coordinate lie over one curve of the pair, and the lines with integer y coordinate lie over the other. The direction function ξ now corresponds to $(1/2\pi)$ arc tan $[(y' - y)/(x' - x)]$, so that we are dealing with the usual angle variation. Because of the functional equations, the quantity $(y' - y)/(x' - x)$ has the same value at corresponding points on opposite sides of the unit square. The total angle variation for a circuit of the unit square is therefore zero. One fixed point class of $\sigma\phi$ is represented in a one-to-one manner by the fixed points of $s\Phi$ in a square. This fixed point class therefore has index zero. Other fixed points of $\sigma\phi$ are covered by points P for which P and sP have integer coordinate differences, not both zero. In order to grasp these, we investigate the mappings related to s, i.e., mappings $s_1\Phi$ defined by the mapping functions

$$x' = f_1(x, y) = f(x, y) + a,$$

$$y' = g_1(x, y) = g(x, y) + b,$$

where a and b are any two integers. But then f_1 and g_1 satisfy the same functional equations as f and g, and the same argument as above gives the index zero for the boundary of a square cell. The fact that $\sigma\phi$ has only finitely many fixed point classes follows, as for $p > 1$ (§32), from the fact that the image of the square $0 \leq x \leq 1$, $0 \leq y \leq 1$ under s meets only finitely many squares equivalent to the original. We can summarize these results as follows:

Theorem 17. *If σ is a continuous mapping of the torus onto itself which belongs to the class of the identity and satisfies Assumption α,* i.e., *if there are two simple curves on the torus which meet each other at only a single point but do not meet any fixed points of σ, and which are homotopic to their respective images under σ, then $\Xi(\sigma) = 0$, because each of the (finitely many) fixed point classes which possibly appear has index zero.*

The example $x' = x + \frac{1}{2}$, $y' = y$ shows that in this case there need not necessarily be fixed points at all.

The general fixed point problem on the torus is completely settled and has been presented in sufficient detail[17] for us to be able to confine ourselves here to the solution given in Theorem 17, for the general fixed point problem for the mapping class of the identity.

39. The direction group (F). If one represents the ξ-values $-\infty < \xi < \infty$ on a line Z, then Z is being viewed as the universal covering manifold of E. The group F, regarded as a group of topological mappings of E onto itself, then corresponds to a group of topological mappings of Z onto itself, which we call the "direction group" and denote by (F).

[17] First in two works in Math. Annalen, vol. **82** (1921): J. Nielsen, *Über die Minimalzahl der Fixpunkte bei den Abbildungstypen der Ringflächen* [N7], and L. E. J. Brouwer, *Über die Minimalzahl der Fixpunkte bei den Klassen von eindeutigen stetigen Transformationen der Ringflächen*, in which the concepts "fixed point class" and "index" are not yet prominent. The class concept was then introduced and the proof greatly simplified, at the same time extending the result to arbitrary continuous mappings of the (n-dimensional) torus in the Danish work of the author: *Ringfladen og Planen*, Matematisk Tidsskrift B (1924), [N10]. Recently, H. Hopf has arrived at a similar formulation, independently of the author: Heinz Hopf, Über Mindestzahlen von Fixpunkten, Math. Zeitschr. vol. 26 (1927).

We now give an abstract presentation of (F) by means of generators and defining relations.

Let f be an arbitrary element of F, z_0 an arbitrary point of E, ξ_0 one of the ξ-values lying over z_0, and ξ_1 one of the ξ-values lying over fz_0. Then a topological mapping of Z onto itself is determined by the requirements that it shall lie over fE and send ξ_0 to ξ_1. Since the intervals from ξ_0 to $\xi_0 + 1$ and ξ_1 to $\xi_1 + 1$ each cover E once, this mapping is periodic with period 1. A point ξ_v of Z which lies over the fundamental point $V(f)$ is therefore sent to $\xi_v + v$, where v is an integer, and this displacement number is the same for all ξ values lying over $V(f)$ and $U(f)$. The topological mapping of Z obtained may then be denoted by

$$(f)_v.$$

It satisfies the functional equation

$$(f)_v(\xi + 1) = (f)_v\xi + 1. \tag{28}$$

The operations of (F) lying over the identity operation I of F constitute a subgroup of (F) consisting of the powers of

$$(\mathrm{I})_1,$$

i.e., the displacement of Z by unity in the positive direction. These commute with all elements of (F), because (28) says

$$(f)_v(\mathrm{I})_1\xi = (\mathrm{I})_1(f)_v\xi. \tag{29}$$

Then for all values of n one has

$$(\mathrm{I})_1^n = (\mathrm{I})_n$$

and

$$(f)_v(\mathrm{I})_n = (\mathrm{I})_n(f)_v = (f)_{v+n}, \tag{29a}$$

so that we have found a system of generators for (F) in the $2p + 1$ operations

$$(a_1)_0, (b_1)_0, \ldots, (a_p)_0, (b_p)_0, (\mathrm{I})_1.$$

Here

$$(a_1)_0^{-1} = (a_1^{-1})_0$$

and so on.

Now we seek a complete relation system. First, (29) gives the commutation relations

$$\left.\begin{array}{l}(a_i)_0 \rightleftarrows (\mathrm{I})_1\\ (b_i)_0 \rightleftarrows (\mathrm{I})_1\end{array}\right\} i = 1, \ldots, p. \tag{30}$$

Because of (30), each element γ of (F) can be written in the form

$$\gamma = \Pi((a_i)_0, (b_i)_0)\cdot(\mathrm{I})_1^n.$$

Then, if γ is to be the identity, i.e.,

$$\gamma = (\mathrm{I})_0,$$

$\Pi((a_i)_0, (b_i)_0)$ must lie over the identity mapping of E, hence we must have

$$\Pi(a_i, b_i) = 1$$

in F. Then (cf. §8) we must have, identically in the a_i, b_i,

$$\Pi(a_i, b_i) \equiv \prod_{j=1}^{m}{}^{*} h_j(a_i, b_i) R(a_i, b_i)^{\varepsilon_j} h_j(a_i, b_i)^{-1}, \tag{31}$$

where the star indicates that we are dealing with a new product. The identity (31) can be written in the $(a_i)_0$, $(b_i)_0$ just as well as in the a_i, b_i. Thus it all comes down to determining the element

$$R((a_i)_0, (b_i)_0) = (a_1)_0(b_1)_0(a_1^{-1})_0(b_1^{-1})_0 \ldots (a_p)_0(b_p)_0(a_p^{-1})_0(b_p^{-1})_0.$$

But this element lies over the identity of F, hence

$$R((a_i)_0, (b_i)_0) = (\mathrm{I})_1^x, \tag{32}$$

where x is to be determined. Since $(\mathrm{I})_1$ commutes with all elements of F, all conjugating factors $h_j((a_i)_0, (b_i)_0)$ on the right-hand side of (31) drop out, so that (32) and (30) enable γ to be brought into the form

$$\gamma = (\mathrm{I})_1^{n + x \cdot \Sigma \varepsilon_j}$$

in which the exponent must then be zero.

Now, in order to determine the exponent x in (32), one has to clarify the formation of the product $(f)_0(f_1)_0$: fE gives a negative displacement to points on the left of the directed axis $U(f)V(f)$, a positive displacement to points on the right, and these displacements are smaller than the circumference of E. Under $(f)_0 Z$, the points over $U(f)$ and $V(f)$ do not move, and the remaining points experience negative and positive displacements < 1. Thus

$$-1 < (f)_0\xi - \xi < 1$$

and also

$$\nu - 1 < (f)_\nu\xi - \xi < \nu + 1. \tag{33}$$

Now one puts arrows on the axes of f and f_1, pointing from the negative to the positive fundamental point, and considers the product $(f)_0(f_1)_0$ case by case as follows:

(1) f and f_1 have the same axis. Then the points of Z lying over the common fundamental points are fixed by $(f_1)_0$ as well as $(f)_0$, and hence

$$(f)_0(f_1)_0 = (ff_1)_0. \tag{34}$$

(2) The axes of f and f_1 intersect. Then the points of Z lying over $U(f_1)$ and $V(f_1)$ respectively are fixed by $(f_1)_0$ and experience displacements of opposite signs under $(f)_0$. Hence, by (33), (34) again holds.

(3) The axes of f and f_1 do not meet. Then they determine between them a subregion Γ of $\Phi + E$. When the arrows on the axes have opposite senses in the boundary of Γ, then points of Z which lie over $f_1^{-1}V(f)$ and $V(f_1)$ respectively experience displacements of opposite signs under $(f)_0(f_1)_0$, and hence (34) again holds.

(4) The axes of f and f_1 have the same sense in the boundary of Γ, which is (a) opposite to the positive sense of E, or (b) identical with it. Here the answer depends on the position

of the fundamental points of ff_1. We need the answer only in the case where one knows that at least one of these fundamental points lies on an arc of E in the boundary of Γ. In case (a), each such point first receives a positive displacement from f_1 and then from f either a positive displacement (in case it still lies in the boundary of Γ in its new position) or a negative displacement (in case it has come to the left of the axis of f) which is smaller than the positive displacement, hence the total displacement is always positive. But when a fundamental point of ff_1 has received a positive displacement, we have

$$(f)_0(f_1)_0 = (ff_1)_v$$

for a positive value of v. Now, since the displacement of $V(f_1)$ is less than the circumference of E, it follows from (33) that $v = 1$, hence

$$(f)_0(f_1)_0 = (ff_1)_1 \tag{35a}$$

in case (a), and analogously

$$(f)_0(f_1)_0 = (ff_1)_{-1} \tag{35b}$$

in case (b).

In all cases the expansions of the fundamental points give full information as to which of the above cases occurs. Since any two of the axes of a_1, b_1 and $a_1 b_1$ intersect, it follows from (34) that

$$(a_1)_0(b_1)_0(a_1^{-1})_0 = (a_1 b_1 a_1^{-1})_0. \tag{36}$$

The axes of $a_1 b_1 a_1^{-1}$ and b_1^{-1} correspond to the case (4a), and the fundamental points of $a_1 b_1 a_1^{-1} b_1^{-1} = k_1$ lie in the region Γ between them, hence by (35a) one has

$$(a_1 b_1 a_1^{-1})_0(b_1^{-1})_0 = (a_1 b_1 a_1^{-1} b_1^{-1})_1,$$

and therefore, since the equation holds for the remaining indices,

$$\begin{aligned} R((a_i)_0, (b_i)_0) &= (k_1)_1(k_2)_1 \ldots (k_p)_1 \\ &= (k_1)_0(k_2)_0 \ldots (k_p)_0(\mathrm{I})_1^p, \end{aligned}$$

where $k_i = a_i b_i a_i^{-1} b_i^{-1}$.

Now (for $p > 2$) the axes of k_1 and k_2 again correspond to case (4a), and the fundamental points of $k_1 k_2$ lie in the region between them, hence by (35a) one has

$$(k_1)_0(k_2)_0 = (k_1 k_2)_1 = (k_1 k_2)_0(\mathrm{I})_1,$$

and the same conclusion holds (when $p > 3$) for $k_1 k_2$ and k_3, etc., up to the last step. Thus one has

$$R((a_i)_0, (b_i)_0) = (k_1 k_2 \ldots k_{p-1})_0(k_p)_0(\mathrm{I})_1^{p+(p-2)}.$$

But the elements $k_1 k_2 \ldots k_{p-1}$ and k_p are inverse to each other because $R(a_i, b_i) = 1$, hence by case (1)

$$(k_1 k_2 \ldots k_{p-1})_0(k_p)_0 = (\mathrm{I})_0$$

and hence

$$x = 2p - 2.$$

Consequently, we have the following abstract presentation for the direction group (F):

$$(F)\begin{cases} \left.\begin{array}{ll} \textit{Generators:} & (a_i)_0, (b_i)_0, (\mathrm{I})_1 \\ \textit{Relations:} & (a_i)_0 \rightleftarrows (\mathrm{I})_1, \\ & (b_i)_0 \rightleftarrows (\mathrm{I})_1, \end{array}\right\} i = 1, 2, \ldots, p \\ (a_1)_0(b_1)_0(a_1)_0^{-1}(b_1)_0^{-1} \ldots (a_p)_0(b_p)_0(a_p)_0^{-1}(b_p)_0^{-1} = (\mathrm{I})_1^{2p-2}. \end{cases}$$

This group was first given by H. Hotelling[18], in a somewhat different setting and by different methods. However one sees without difficulty that the two approaches lead to the same group. Hotelling views (F) as the fundamental group of the closed orientable three-dimensional manifold, each of whose "points" is a *point of the surface* ϕ *paired with the direction of one of its tangents*. The above exponent $2p - 2$ is then the (single) torsion coefficient of this manifold.

40. Increment in the direction function along equivalent paths and on E. Let s be an S-function, I the automorphism induced by s, Π the fixed point set of $s(\Phi + E)$ and $\Phi + E - \Pi = \Omega$. We abbreviate $\xi(x; s)$ to $\xi(x)$. Let w be a directed arc in Ω from x_1 to x_2. Choose a ξ_1 from among the values of $\xi(x_1)$ (recall that $\xi(x_1)$ is determined only mod 1). Then let

$$\xi_2 = \xi_1 + \int_w d\xi. \tag{37}$$

Let f be an element of F which is fixed by I, and hence commutes with s. Then it follows from (26) (§35) that, in the whole of Ω,

$$z[f(x), s(f(x))] = z[f(x), f(s(x))] = f(z[x, s(x)]),$$

and hence

$$\xi(f(x)) = (f)_v \xi(x) \tag{38}$$

where the undetermined integer v takes the indeterminacy of ξ mod 1 into account. Then

$$\int_{fw} d\xi = \int_w d((f)_v \xi) = (f)_v \xi_2 - (f)_v \xi_1 \tag{39}$$

is valid for each v; the independence of the difference from v is evident because

$$(f)_v \xi = (f)_0 \xi + v.$$

Equations (38) and (39) then give

$$\int_w d[\xi(f(x)) - \xi(x)] = \int_w d[(f)_v \xi - \xi] = (f)_v \xi_2 - \xi_2 - [(f)_v \xi_1 - \xi_1]$$

for each value of v. If one now replaces the path of integration w by any other path in Ω from x_1 to x_2, then ξ_2 changes at most by an integer, and by (28) this has no influence on the difference $(f)_v \xi_2 - \xi_2$.

[18] Harold Hotelling, Three-dimensional Manifolds of States of Motion, Transact. Amer. Math. Soc. vol. **27** (1925).

Thus

$$\int d[\xi(f(x)) - \xi(x)]$$

is independent of the path of integration in Ω.

Now let x_1 and x_2 be two points of E in Ω which are not separated by the fixed point set of sE, so that they belong to the same fixed point free interval on E, and let w be the arc of E which connects them in Ω. Then one can again form (37). Since sw is contained in the same fixed point free interval as w, and here $z(x, s(x)) = s(x)$, no point of the interval (ξ_1, ξ_2) on the Z axis lies over a fixed point of sE. Thus if sE has any fixed points at all, $\xi_2 - \xi_1$ depends only on the position of the points $s(x_1)$ and $s(x_2)$. If in particular w lies in a fixed point free periodic interval, bounded by the fundamental points of the fixed element f, and if

$$x_2 = f(x_1)$$

then by (38)

$$\xi_2 = (f)_v \xi_1,$$

and here v must be 0; otherwise there would be a point of the Z axis between ξ_1 and ξ_2 lying over a fundamental point of f, and hence over a fixed point of sE. Thus integration along E gives

$$\int_{x_1}^{f(x_1)} d\xi = (f)_0 \xi_1 - \xi_1. \tag{40}$$

41. Classes of index zero in case of a cyclic fixed element group. Let s be an S-function, let I be the induced automorphism, and suppose $H(I)$ is cyclic, with primary element f, so that $H(I) = \{f\}$. Then sE can have two fixed point free intervals, periodic with respect to f (§14). Again let Ω be the fixed point free part of $\Phi + E$. Then the net N belongs entirely to Ω on the basis of Assumption α (§37). By §37, the index of the fixed point class represented by $s\Phi$ is defined by $\int d\xi$ over the boundary of a fundamental domain for $\{f\}$. It is therefore necessary to begin by finding such a fundamental domain whose boundary is fixed point free.

Let $U(f)$ and $V(f)$ be the single fixed points on E. The fixed point set of $s\Phi$ in a fundamental domain of $\{f\}$ may be enclosed in a circle lying in the interior of E, by §34. Let λ and μ (Figure 16) be two curves of constant distance from the axis of f which contain this circle between them. Then the 2-gon formed by the circular arcs λ and μ contains the whole fixed point set of $s\Phi$, since the latter is reproduced by $\{f\}$. Let $C = z_1 z_2$ be a non-Euclidean perpendicular to the axis of f which does not lie on any of the edges of N nor go through any of its vertices. The cells of the net which C passes through are then arranged along C in such a way that two simple edge paths bound their union. Let AB be a piece of one of these paths which leads from a point A on λ, across the 2-gon $\lambda\mu$ to a point B on μ.

One connects A rectilinearly with z_1 and B with z_2 (the endpoints of C on E). This gives a path w from z_1 to z_2 free of fixed points. Obviously w does not meet its own image under f. The region between w and fw is a fundamental domain of $\{f\}$ with the four "vertices" z_1, z_2, fz_2, fz_1, in counterclockwise order. Its boundary belongs to Ω since the two subarcs of E are also fixed point free.

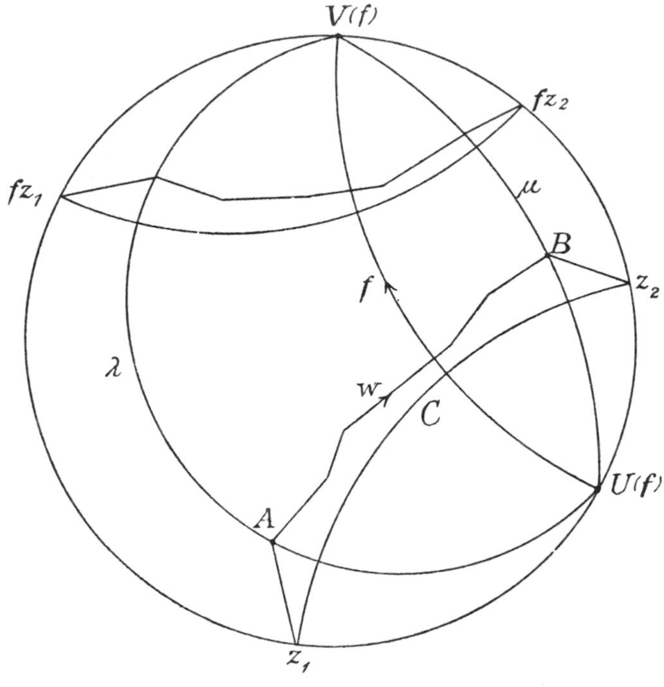

Figure 16

Now let ξ_1 be a value of $\xi(z_1)$. Also let

$$\xi_2 = \xi_1 + \int_w d\xi.$$

Then (39), with $v = 0$, gives

$$\int_{fw} d\xi = (f)_0\xi_2 - (f)_0\xi_1.$$

The contributions to $\int d\xi$ from the two arcs on E follow from (40), since the assumptions made there are satisfied here. The four contributions to $\int d\xi$ in counterclockwise order are therefore

$$\int_{z_1}^{z_2} = \xi_2 - \xi_1,$$

$$\int_{z_2}^{fz_2} = (f)_0\xi_2 - \xi_2,$$

$$\int_{fz_2}^{fz_1} = (f)_0\xi_1 - (f_0)\xi_2,$$

$$\int_{fz_1}^{z_1} = \xi_1 - (f)_0\xi_1,$$

and their sum is zero:

Theorem 18. *When s induces an automorphism with cyclic fixed element group and sE has two fixed point free periodic intervals, then the fixed point class of s*Φ *has index zero.*

It may be mentioned that nothing needs to be assumed concerning the direction of displacement of the points of the fixed point free intervals under sE.

42. The Birkhoff formula. In this paragraph the fixed point problem is settled (for $p > 1$) for continuous mappings in the class of the identity.

Let $\sigma\phi$ be a σ-mapping, belonging to the class of the identity, which satisfies Assumption α (§37). Then we can assume that no fixed points of the S-functions over σ lie on the net N. Let s be the S-function which lies over σ and induces the identity automorphism I_0; thus s commutes with all elements of F.

Now we shall determine the index of the fixed point class of $\sigma\phi$ represented by $s\Phi$. Since the fixed element group $H(I_0)$ is identical with F, we have to evaluate $\int d\xi$ over the counterclockwise boundary of a fundamental domain of F, since this curve plays the rôle of the curve C (§37). We choose the central cell B, whose boundary was assumed to be free of fixed points. Denote the vertices of B, beginning with the first below the positive real axis and in counterclockwise order, by $Q_0, Q_1, \ldots, Q_{4p-1}$. Then

$$Q_1 = a_1 b_1^{-1} a_1^{-1} Q_0,$$
$$Q_2 = a_1^{-1} Q_1,$$
$$Q_3 = b_1 Q_2 = a_1^{-1} Q_0,$$
$$Q_4 = b_1 a_1 b_1^{-1} Q_3 = b_1 Q_1,$$

and after this the same terms repeat, but with indices 2, 3, etc. We fix a value, ξ_0, for $\xi(Q_0)$. Then let

$$\xi_1 = \xi_0 + \int_{Q_0}^{Q_1} d\xi$$

$$\xi_2 = \xi_1 + \int_{Q_1}^{Q_2} d\xi$$

$$\ldots$$

$$\xi_{4p} = \xi_{4p-1} + \int_{Q_{4p-1}}^{Q_0} d\xi,$$

where the integrals are taken over the boundary edges of B. The desired index is then $\xi_{4p} - \xi_0$. Now since

$$Q_1 = a_1 b_1^{-1} a_1^{-1} Q_0,$$

ξ_1 results from ξ_0 by one of the topological mappings of the ξ-axis onto itself which correspond to the element $a_1 b_1^{-1} a_1^{-1}$, in the previous notation:

$$\xi_1 = (a_1 b_1^{-1} a_1^{-1})_{v_1} \xi_0 \tag{41}$$

and similarly there is an integer v_2 such that

$$\xi_2 = (a_1^{-1})_{v_2} \xi_1. \tag{42}$$

Now

$$\xi_3 = \xi_2 + \int_{Q_2}^{Q_3} d\xi = \xi_2 - \int_{Q_0}^{Q_1} d((a_1^{-1})_{v_2} \xi) = \xi_2 - [(a_1^{-1})_{v_2} \xi_1 - (a_1^{-1})_{v_2} \xi_0]$$

on the basis of (39), and hence by (42), (41), (36), (29a) and (30):

$$\begin{aligned}
\xi_3 &= (a_1^{-1})_{v_2} \xi_0 \\
&= (a_1^{-1})_{v_2} (a_1 b_1 a_1^{-1})_{-v_1} \xi_1 \\
&= (a_1^{-1})_{v_2} (a_1 b_1 a_1^{-1})_{-v_1} (a_1)_{-v_2} \xi_2 \\
&= (a_1^{-1})_0 (a_1 b_1 a_1^{-1})_0 (a_1)_0 (\mathrm{I})_{-v_1} \xi_2 \\
&= (b_1)_{-v_1} \xi_2.
\end{aligned}$$

We then also have

$$\begin{aligned}
\xi_4 &= \xi_3 + \int_{Q_3}^{Q_4} d\xi \\
&= \xi_3 - \int_{Q_1}^{Q_2} d((b_1)_{-v_1} \xi) \\
&= \xi_3 - (b_1)_{-v_1} \xi_2 + (b_1)_{-v_1} \xi_1 \\
&= (b_1)_{-v_1} (a_1 b_1^{-1} a_1^{-1})_{v_1} \xi_0 \\
&= (b_1)_0 (a_1 b_1^{-1} a_1^{-1})_0 \xi_0 \\
&= (b_1)_0 (a_1)_0 (b_1)_0^{-1} (a_1)_0^{-1} \xi_0.
\end{aligned}$$

Now since the last formula is repeated with increasing subscripts, we get the desired index

$$\begin{aligned}
\xi_{4p} - \xi_0 &= R((a_i)_0 (b_i)_0)^{-1} \xi_0 - \xi_0 \\
&= (\mathrm{I})_1^{2-2p} \xi_0 - \xi_0 \\
&= 2 - 2p
\end{aligned}$$

on the basis of the last relation in the group (F).

The index of the fixed point class represented by sΦ is therefore $2 - 2p$.

It remains to determine the indices of other possible fixed point classes. Let f be an arbitrary element of F. The S-function $fs\Phi$ related to $s\Phi$ induces the inner automorphism $I_{of^{-1}}$, cf. §12 and the paragraphs following it. The induced boundary mapping is fE with the fixed points $U(f)$ and $V(f)$, and it has two periodic, fixed point free, intervals (§14) determined by them.

The fixed element group $H(I_{of^{-1}})$ is the cyclic group $\{f^*\}$ where f^* is the primary element corresponding to the axis of f (§17). Hence by Theorem 18 the associated fixed point class—if $fs\Phi$ has fixed points at all—has index zero.

In summary we therefore have:

Theorem 19. *The finitely many fixed point classes of a mapping $\sigma\phi$ belonging to the class of the identity all have index zero, except the single topologically necessary class, which has index $2-2p$.*

In particular, the index sum also satisfies

$$\Xi(I_0) = 2 - 2p, \tag{43}$$

where the use of the representative I_0 of the automorphism family of the identity as argument for Ξ indicates that the formula is really valid for the mapping *class*.

By Theorem 17, (43) is also valid for $p = 1$.

Formula (43) is due to G. D. Birkhoff,[19] who proved it for one-to-one analytic mappings of an analytic surface of genus p with the help of a lemma of Poincaré, and from it also derived the existence of at least one fixed point in the general topological case. The above investigation extends the validity of the formula to arbitrary continuous mappings in the class of the identity, and also shows that, even though arbitrarily many fixed point classes can appear (§24), the index sum $2-2p$ always derives its value from a single fixed point class. Finally, it replaces the assumption of finitely many fixed points by the much weaker Assumption α. When one wants not only to prove the existence of fixed points, but also to evaluate the role of individual fixed point classes by indices, then an assumption like Assumption α is unavoidable; one need only realize that, in the limiting case where all points of ϕ are fixed points of $\sigma\phi$, the concept of index has long ago lost its meaning.

B. v. Kerekjártó[20] has shown by means of an example that there are topological mappings in the class of the identity with only *one* fixed point. One can construct one easily using the following lemma, told to me by H. Kneser.

> A topological mapping of a circle onto itself, with at least one fixed point, can be extended to a topological mapping of the disc which is fixed point free in the interior.

In fact, when O is a fixed point and P is an arbitrary point of the circle, P' its image (which can coincide with P) and p is an arbitrary point of the chord OP, then one can require the image p' of p to lie on the chord OP' with, say

$$\frac{Op'}{OP'} = \left(\frac{Op}{OP}\right)^2.$$

In place of the circular disc one naturally can have any Jordan disc. Now one maps the individual boundary edges of the fundamental domain B of F onto themselves in such a way that only the endpoints are fixed, and also in such a way that the mappings of equivalent edges are transformed into each other by the appropriate F-element. This topological mapping of

[19] Dynamical systems with two degress of freedom, Trans. Amer. Math. Soc. vol. **18**, especially p. 286 ff.

[20] Vorlesungen über Topologie I, Springer-Verlag, 1923, p. 214.

the boundary of B onto itself, which has just the $4p$ vertices as fixed points, is then extended by the lemma to a mapping of B onto itself which is fixed point free in the interior. The topological mapping of ϕ which this determines belongs to the class of the identity and has only one fixed point, over which the vertices of B lie. Thus here we have the topologically necessary fixed point class, consisting of only *one* point. The index of this point is $2p - 2$.

43. The Alexander formula. In this paragraph we review an extraordinarily valuable theorem of J. W. Alexander[21] which (in the terminology used here) permits the sum Ξ of the indices of all fixed point classes to be given explicitly. A rather special assumption about the mapping is made; but it is clear that the result is independent of this assumption when, as is the case in later applications, one knows from elsewhere that Ξ depends only on the automorphism family under consideration. Hence it suffices to determine Ξ in a special case.

The derivation which follows is closely connected to Alexander's own.

Suppose an arbitrary automorphism I is given. By §23, one can construct a T-function $t\Phi$ with the following properties:

1. t induces I.
2. If $\tau\phi$ denotes the topological mapping of ϕ over which $t\Phi$ lies, and if Σ is the canonical cut system

$$\Sigma = A_1, B_1, \ldots, A_p, B_p$$

   on ϕ with common point Q, then no fixed point of $\tau\phi$ lies on Σ.
3. τQ does not lie on Σ.
4. $\tau\Sigma$ meets Σ at only finitely many points, and each of these points is a transverse intersection of a curve of Σ by one on $\tau\Sigma$.

At the end of §23 this was achieved by arranging that the image under t of the net N lying over Σ consist of geodesic segments with endpoints outside N.

Properties 2, 3 and 4 then hold for $\tau^{-1}\phi$ also.

The basic cell B of the net N is cut by $t^{-1}N$ into finitely many simply connected subregions

$$O_1, O_2, \ldots, O_v.$$

Then B is likewise cut by tN into v subregions, each of which is equivalent to one of the regions tO_i. Let

$$t_1O_1, t_2O_2, \ldots, t_vO_v$$

be these new subregions of B, where t_i is a T-function related to t, hence also lying over $\tau\phi$. It can be that t_i is the same T-function for different values of i. The number of these T-functions which determine different fixed point classes is at most equal to the number of non-isogredient T-functions over $\tau\phi$ for which B has points in common with its image. Without having to determine this number, we can now argue as follows: Fixed points of t_iO_i belong to the same fixed point class of $\tau\phi$, as part of the fixed point set of $t_i\Phi$, and the index of this fixed point subset equals $\int d\xi(x, t_i)$ taken over the positive, fixed point free, boundary of O_i. Even when it takes several of the O_i to completely include a fixed point class of $\tau\phi$, it is still true that the sum

[21] J. W. Alexander, Invariant points of a surface transformation of given class, Transact. Amer. Math. Soc., vol. **25**.

Ξ of the indices of all fixed point classes of $\tau\phi$ is measured by the sum of the indices of positive boundaries of the regions O_i under $O_i \to t_i O_i$, as we shall now show.

As in §42, one again denotes the vertices of B by

$$Q_0, Q_1, \ldots, Q_{4p-1}.$$

Let Q^- and Q^+ be the points of B lying over $\tau^{-1}Q$ and τQ, respectively. Let h be a directed curve of the system Σ, H the directed edge of B lying over h which has B on its left, and let fH be its other equivalent on B. Let H^- and H^+ be the sets of Jordan arcs in B lying over $\tau^{-1}h$ and τh, respectively. Now suppose a point p on ϕ traverses the directed curve $\tau^{-1}h$. The point P of B lying over P then traverses H^-, jumping finitely often from boundary points of B to their equivalents. During this process, P is always reckoned to belong to the region O_i which has H^- in its positive boundary, so i takes different values. Meanwhile, $t_i P$ continuously and monotonically traverses the boundary side H or fH, according as τ is orientation preserving or not, and hence according as the *degree* ε of τ is $+1$ or -1. We take the former case first. Let Q_ν be the initial point, and hence $Q_{\nu+1}$ the final point of H. The direction ray $P \to t_i P$ begins at the position $Q^- Q_\nu$, ends at the position $Q^- Q_{\nu+1}$, and always cuts the edge line containing H. The corresponding direction point z is therefore confined to a subarc (H) of E determined by this edge on the side away from B. Each jump of P corresponds to a jump of z on the subarc (H) of E which is completely determined by magnitude and sign. If one first includes these jump values, then the direction point $z[P, t_i P]$ traverses the positive subarc $z(Q^-, Q_\nu) \to z(Q^-, Q_{\nu+1})$ on E as P traverses H^-. If one lets h run over the curves of Σ and their inverses, then z completes a positive circuit of E, and similarly a negative circuit when $\varepsilon = -1$. One can think of the jumps as being generated by continuous variation of the direction ray if one lets the point P, each time it reaches the boundary of B, wander rectilinearly through the interior of B to the equivalent boundary point, keeping the image point of P fixed throughout this motion, so that the direction ray turns round it. With this arrangement we can write

$$\int_{\Sigma^-} d\xi = \varepsilon,$$

where Σ^- denotes the set of all H^-, in other words the image of $\tau^{-1}\Sigma$ in B.

Now let P describe a positive circuit around B. The image point $t_i P$ (with changing i) then lies on Σ^+, i.e., over $\tau\Sigma$, and therefore jumps each time P crosses Σ^-. If one again includes the jump values, by now fixing the original point P while the image point traverses the "jump chord" of the convex region B, then $\int d\xi$ along H is measured by arcs $z(Q_\nu, Q^+) \to z(Q_{\nu+1}, Q^+)$ on $E - (H)$, so that it is independent of ε and always positive; altogether, the $4p$ sides H yield a positive circuit of E. Therefore

$$\int_{\Sigma + \Sigma^-} d\xi = 1 + \varepsilon, \tag{44}$$

where the integral contains not only the contributions from positive circuits round the boundaries of all O_i, but also the jump values. It therefore remains to find the sum of the jump values and to subtract it.

We consider a point x on the boundary edge H of B where Σ^- meets the boundary of B. It lies over an intersection of the curve h of Σ with, say, the curve $\tau^{-1}k$ of $\tau^{-1}\Sigma$ (Figure 17). Let

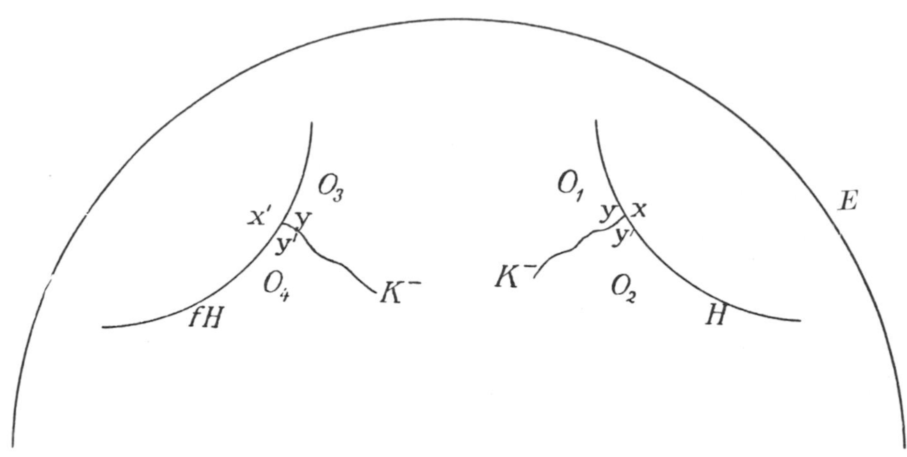

Figure 17

the point on the boundary of B equivalent to x be x'; the set K^- in B of curve segments lying over $\tau^{-1}k$ issues from the latter. We denote the four regions O_i meeting x and x' by O_1, O_2, O_3, O_4 as in the figure. The point $y = t_1 x$, which is the image of x for a positive circuit of O_1, lies on the boundary of B as the intersection of a boundary side over k with H^+. Let the equivalent boundary point be y'. Then one easily sees that

$$\begin{aligned} t_1 x &= y, \\ t_2 x &= y', \\ t_3 x' &= y, \\ t_4 x' &= y'. \end{aligned} \tag{45}$$

In order not to assume anything about the positions of y and y', since one only knows that they are equivalent and different from x and x', these two points are not shown in Figure 17; the letters denoting them are shown in the corners at x and x' only to recall (45). If one now visualizes the generation of jump values described above, in which one keeps one of the original point P or the image $\bar{P}$ fixed while the other wanders across a chord of the convex region B, then one finds the following for the four jump values belonging to the cut in question.

	P	$\bar{P}$	$P\bar{P}$ **rotates around**
In traversing K towards x	traverses xx'	is fixed at y	y
" " fH ...	is fixed at x'	traverses yy'	x'
" " K towards x'	traverses $x'x$	is fixed at y'	y'
" " H ...	is fixed at x	traverses $y'y$	x

Thus the ray $P\bar{P}$ begins at the position xy and finally returns to it, so that the direction point z of the ray describes an integral number of circuits of E. Now x, x', y and y' are the vertices of a convex polygon; none of them coincides with a vertex of B and at most two are on the same

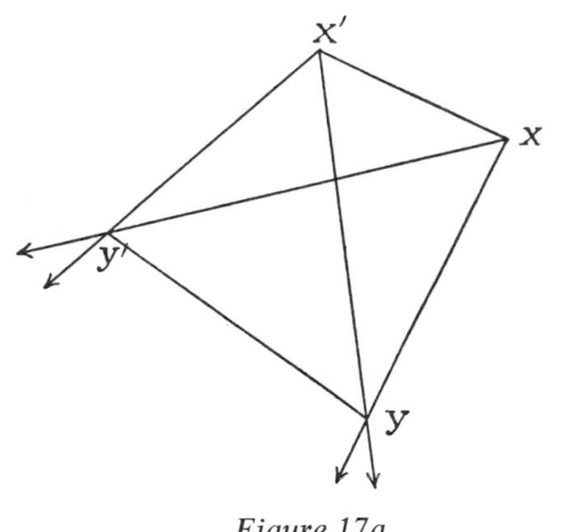

Figure 17a

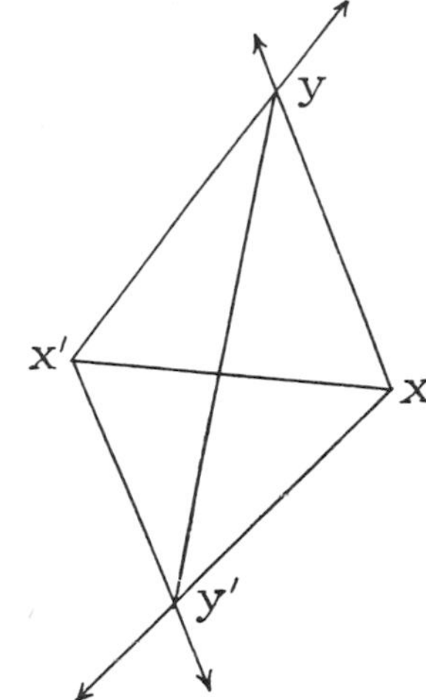

Figure 17b

edge. Thus when the chords xx' and yy' do not intersect, the jump sum is zero, as one easily concludes from the above table and Figure 17(a); when the chords intersect, however, Figure 17(b) shows that the jump sum is a full rotation in the sense given by the sequence

$$xy'x'y. \tag{46}$$

(Incidentally, this determination of the jump sum for the intersection of Σ with $\tau\Sigma$ is independent of the metric used, as is the above derivation of the number $1 + \varepsilon$.) But the point pairs xx' and yy' separate each other on the boundary of B if and only if the curves h and k being used are a conjugate pair of the canonical cut system Σ on ϕ. One can assign the intersection number $+1$ or -1 to a crossing of one curve by another on an orientable surface by declaring that A_1 crosses B_1 with intersection number $+1$. Now in order that the jump sum for the intersection considered above make a *positive* complete rotation, y' must lie on the edge of B meeting H and fH, by (46). If a point P travels from x on K^- in the interior of B, then the image point $t_2 P$ traverses a piece of this boundary edge in the positive or negative direction according as $\varepsilon = +1$ or $\varepsilon = -1$. *Correspondingly, h is cut by $\tau^{-1}k$ in the same or opposite sense as by k.* The reverse holds for a negative full rotation. Following Poincaré, one lets $N(k, h)$ denote the algebraic sum of intersection numbers for the crossings of the directed curve h by the directed curve k. Then the jump sum over all intersections between $\tau^{-1}k$ and h, as the number of complete circuits of the direction point z around E, and hence as an integer summand of $\int d\xi$, is given by

$$N(\tau^{-1}k, h) \cdot N(k, h) \cdot \varepsilon = N(k, \tau h) \cdot N(k, h),$$

since application of τ multiplies the intersection number by ε. This also holds when k and h are not a conjugate pair, since $N(k, h) = 0$ in this case. Then since

$$N(A_i, B_i) = -N(B_i, A_i) = 1$$

the jump sum is the number

$$\sum_{i=1}^{p} [N(A_i, B_i) \cdot N(A_i, \tau B_i) + N(B_i, A_i) \cdot N(B_i, \tau A_i)]$$
$$= -\sum_{i=1}^{p} [N(\tau B_i, A_i) + N(B_i, \tau A_i)]$$

and by subtracting this from the above value (44) we obtain the following formula for the index sum Ξ:

$$\Xi_A = 1 + \varepsilon + \sum_{i=1}^{p} [N(\tau B_i, A_i) + N(B_i, \tau A_i)]. \tag{47}$$

This is the formula of J. W. Alexander. The subscript A on Ξ is to indicate that the formula holds in general only under the above assumptions concerning the mapping $\tau\phi$. For its extension to the cases $p = 0$ and $p = 1$, see the work of Alexander cited above[21]. For $p = 1$ it also follows from the work of the author cited above as[17].

The intersection number $N(h, k)$ of two curves h and k on ϕ which meet only finitely often is determined by Poincaré[22] as follows. One represents h and k in F and lets the generators of F commute:

$$h \sim \prod_{i=1}^{p} a_i^{\alpha_i} b_i^{\beta_i},$$
$$k \sim \prod_{i=1}^{p} a_i^{\alpha_i'} b_i^{\beta_i'}.$$

Then

$$N(h, k) = \sum_{i=1}^{p} \begin{vmatrix} \alpha_i & \beta_i \\ \alpha_i' & \beta_i' \end{vmatrix}. \tag{48}$$

The minimal intersection number and the intersection number of each intersection point ("intersection class") is represented by the intersection of the *axes* of h and k, and it is easily derived from the expansions of the fundamental points $V(h)$ and $V(k)$. In this way one can discern how $N(h, k)$ is made up from the individual intersection classes, i.e., intersection numbers $+1$ and -1 corresponding to intersections of axes.

If the intersection number of h and k is not finite, then one can *define* $N(h, k)$ by (48). Thus the number depends only on the conjugacy classes of F representing the curves.

Now if the automorphism family associated with $\tau\phi$ is determined by the automorphism I:

$$[a_i \mapsto a_i', b_i \mapsto b_i']$$

and if (letting generators commute)

$$\begin{aligned} a_e' &\sim \prod_{i=1}^{p} a_i^{\alpha_{ei}} b_i^{\beta_{ei}}, \\ b_e' &\sim \prod_{i=1}^{p} a_i^{\gamma_{ei}} b_i^{\delta_{ei}}, \end{aligned} \tag{49}$$

[22] Cinquième Complément à l'Analysis Situs, Rend. d. Circ. mat. d. Palermo XVIII (1904).

then, by (48)

$$N(\tau B_i, A_i) = -\delta_{ii},$$

$$N(B_i, \tau A_i) = -\alpha_{ii}$$

and hence (47) becomes:

$$\Xi_A = 1 + \varepsilon - \sum_{i=1}^{p} (\alpha_{ii} + \delta_{ii}). \tag{50}$$

Thus Ξ_A depends only on the automorphism family, and indeed, apart from ε, only on the *trace* of the exponent sum matrix determined from (49). If the conjectured theorem that the index sum Ξ depends only on the mapping class is proved, we shall have $\Xi_A = \Xi$.

If I is the identity automorphism, then

$$\varepsilon = \alpha_{ii} = \delta_{ii} = 1, \qquad i = 1, \ldots, p,$$

and (50) gives

$$\Xi_A(I_0) = 2 - 2p.$$

This is (under the present special assumptions) the Birkhoff formula (43), which was derived in the previous paragraph under more general assumptions (namely, Assumption α only).

If one conjugates the mapping $\tau\phi$ in question with an arbitrary topological mapping $\tau_1\phi$, then $\tau_1\tau\tau_1^{-1}$ has the same ε value as τ and satisfies the same assumptions which lead to the formula (50). Also, by §36, $\tau_1\tau\tau_1^{-1}$ has the same Ξ value, hence the same Ξ_A value, as τ. It therefore follows that the trace $\Sigma(\alpha_{ii} + \delta_{ii})$ of the exponent sum matrix does not change under conjugation of the automorphism with an arbitrary automorphism.

44. Complete solution of the fixed point problem for orientation-preserving mapping classes whose automorphism families contain automorphisms of finite order. Let $s\Phi$ be an S-function lying over $\sigma\phi$ which satisfies Assumption α and for which the automorphism I (of the first kind) induced by s is of finite order:

$$I^n = I_0.$$

Then $H(I) = 1$ and sE is fixed point free (§11). The fundamental domain of $H(I)$ is Φ itself. $\int d\xi$ over the boundary E of Φ gives the index $+1$ for the fixed point class of $\sigma\phi$ represented by $s\Phi$. Now we consider the S-function fs related to s; it induces $I_{f^{-1}}$. From f and I one constructs the element k of equation (13), §19.

Now if $k \neq 1$, then $H(I_{f^{-1}})$ is cyclic, and fsE has two periodic fixed point free intervals (§19). Thus, by Theorem 18 (§41), the fixed point class represented by $fs\Phi$ has index zero. In particular, $fs\Phi$ can be fixed point free.

However, if $k = 1$ then $(I_{f^{-1}})^n = I_0$ (§19), and the fixed point class represented by $fs\Phi$ has index $+1$, like $s\Phi$. The condition for this really to be a new fixed point class for $\sigma\phi$ is, by §33, that f not be isogredient to 1 relative to I. Consequently, the number of fixed point classes of $\sigma\phi$ with non-zero index is equal to the number of isogredience classes relative to I which are solutions of equation (16) (§19). This number depends only on I, and the index sum Ξ is therefore the same for all mappings in the mapping class concerned. Hence it can be found under the special assumptions which lead to the Alexander formula (47), or to (50). We therefore have:

Theorem 20. *If an S-function induces an automorphism of the first kind with order n, then the continuous surface mapping which it represents has exactly*

$$2 - \sum_{i=1}^{p} (\alpha_{ii} + \delta_{ii})$$

fixed point classes with index 1, *and any other fixed point classes, if any, have index zero. The numbers* α_{ii} *and* δ_{ii} *are defined by* (49).

The following purely group theoretic corollary is immediate:

Theorem 21. *If I is an automorphism of F of the first kind, with order n and exponent sums* (49), *then the equation*

$$xx_I x_{I^2} \ldots x_{I^{n-1}} = 1$$

has exactly

$$2 - \sum_{i=1}^{p} (\alpha_{ii} + \delta_{ii})$$

non-isogredient solutions relative to I. Hence

$$\Sigma(\alpha_{ii} + \delta_{ii}) \leq 1$$

since there is always the solution $x = 1$.

Example 7. The automorphism

$$I\colon [a_i \mapsto a_{i+1}, b_i \mapsto b_{i+1}] \qquad (i \bmod p)$$

has order p. Here

$$\alpha_{ii} = \delta_{ii} = 0$$

hence

$$\Xi_A = 2.$$

The equation

$$xx_I \ldots x_{I^{p-1}} = 1 \tag{51}$$

has, apart from the identity, the solution

$$x = a_1 b_1 a_1^{-1} b_1^{-1}.$$

(One can also see directly by group theoretic methods that (51) only has solutions of the forms

$$ff_{I^{-1}}$$

and

$$fa_1 b_1 a_1^{-1} b_1^{-1} f_{I^{-1}}$$

for an arbitrary f in F.) These two isogredience classes relative to I correspond to the two fixed point classes with index $+1$ whose existence follows from $\Xi_A = 2$. The easiest way to see that these are really different isogredience classes is to see that they lead to fixed points of different classes: as S-function associated with I one takes the $1/p$ turn:

$$s(x) = e^{2\pi i/p} \cdot x,$$

so $s\Phi$ has the single fixed point $x = 0$. The vertices of the fundamental domain B are permuted by $s\Phi$, and therefore lie over a fixed point of $\sigma\phi : a_1 b_1 a_1^{-1} b_1^{-1} s\Phi$ has only Q_0 as fixed point. These special S-functions therefore represent the two topologically necessary fixed point classes by single points, and there are no other fixed point classes.

One notes that no use is made of Assumption α in determining the index of the Ξ_A topologically necessary fixed point classes, since the isolation of these fixed point classes is secured independently. It seems possible that this isolation could also be established for the other fixed point classes which possibly appear (with index zero, corresponding to an $I_{f^{-1}}$ whose associated $k \neq 1$), so that in this paragraph Assumption α would be satisfied. I shall not investigate this question here, except to mention the following.

Let γ be a closed, directed, non null homotopic curve on ϕ and let c be a corresponding element of F. The condition for γ to be homotopic to its image $\sigma\phi$ is that there be an f in F for which

$$c_I = f^{-1} \cdot c \cdot f.$$

Thus $I_{f^{-1}}$ has the fixed element $c \neq 1$, and by §19 the latter is a power of the primary element k^* which belongs to the axis of the element $k(f)$ defined by (13), §19:

Theorem 22. *All (non-contractible) curves homotopic to their images under $\sigma\phi$ are represented by the powers of a primary element of F, of which some power can be brought into the form*

$$ff_I \ldots f_{I^{n-1}}.$$

Apart from this precise relationship to primary elements, one can say more briefly that (13) represents all curve types invariant under $\sigma\phi$ when f runs through F.

45. On the solution of the automorphism equation. When one wants, in the case of the previous paragraph, to distinguish the Ξ fixed point classes of $\sigma\phi$ with non-zero index individually, using the fixed point sets of covering S-functions, then under the assumption that

$$I^n = I_0$$

(where n is the smallest positive number of this kind) one has to set up a complete system of solutions of the automorphism equation

$$v(x) = 1 \tag{52}$$

which are non-isogredient relative to I, where x denotes an element of F and we use the abbreviation

$$xx_I x_{I^2} \ldots x_{I^{n-1}} = v(x).$$

Now when

$$n = q \cdot m, \qquad q > 1, \qquad m > 1,$$

one sets

$$\mu(x) = xx_I \dots x_{I^{m-1}}$$

and

$$\Pi(y) = yy_{I^m} \dots y_{I^{m(q-1)}}.$$

Then if

$$v(x_1) = 1$$

and if one sets

$$\mu(x_1) = y_1,$$

it follows that

$$\Pi(y_1) = 1.$$

Conversely, one can try to solve (52) by first finding a solution y_1 of the equation

$$\Pi(y) = 1 \tag{53}$$

and then solving the equation

$$\mu(x) = y_1. \tag{54}$$

Here it is important that solutions of (53) which are isogredient relative to I^m lead to solutions of (52) which are isogredient relative to I (the converse is evident):

Theorem 23. *It follows from*

$$v(x_1) = 1,$$
$$\mu(x_1) = y_1, \qquad \mu(x_2) = y_2,$$
$$y_2 = fy_1 f_{I^m}^{-1}$$

that

$$x_2 = fx_1 f_I^{-1}.$$

Proof. Let $fx_1 f_I^{-1} = z$. It follows from the definition of $\mu(x)$ that

$$\mu(z) = \mu(fx_1 f_I^{-1}) = f\mu(x_1)f_{I^m}^{-1} = fy_1 f_{I^m}^{-1} = y_2 = \mu(x_2) \tag{55}$$

which, written in expanded form, is

$$zz_I \dots z_{I^{m-1}} = x_2 x_{2I} \dots x_{2I^{m-1}}.$$

If one applies I to both sides of this equation the result is

$$z^{-1}\mu(z)z_{I^m} = x_2^{-1}\mu(x_2)x_{2I^m},$$

hence by (55)

$$z^{-1}y_2 z_{I^m} x_{2I^m}^{-1} y_2^{-1} x_2 = 1,$$

or in other words:

$$(zx_2^{-1})_{I^m} = y_2^{-1}(zx_2^{-1})y_2.$$

Thus the automorphism $(I^m)_{y_2^{-1}}$ has the fixed element zx_2^{-1}. Now

$$((I^m)_{y_2^{-1}})^q = (I^{mq})_{\Pi(y_2)^{-1}} = I^{mq} = I^n = 1,$$

since

$$\Pi(y_2) = f\Pi(y_1)f^{-1} = fv(x_1)f^{-1} = 1.$$

Therefore, since $m < n$, $(I^m)_{y_2^{-1}}$ is a non-identity automorphism of finite order, and as such it has only the identity as fixed element. Thus

$$zx_2^{-1} = fx_1 f_I^{-1} x_2^{-1} = 1,$$

and the assertion is proved.

For $f = 1$, Theorem 23 gives the corollary:

It follows from $v(x_1) = 1$ *and* $\mu(x_1) = \mu(x_2)$ *that* $x_1 = x_2$.

and from this that:

It follows from $\mu(x_1) = 1$ *that* $x_1 = 1$.

If $v(x_1) = 1$, then x_1^{-1} is a solution of the automorphism equation which results from replacing I by I^{-1} in $v(x) = 1$. Since $I^n = I_0$ one finds

$$x_1^{-1} x_{1I^{-1}}^{-1} \ldots x_{1I^{-(n-1)}}^{-1} = x_1^{-1}(v(x_1))^{-1} x_1 = 1.$$

Thus isogredient solutions relative to I lead to isogredient solutions relative to I^{-1}.

It follows from Theorem 23 that distinct fixed point classes of $\sigma\phi$ remain distinct for $\sigma^m\phi$ when m is a divisor of n. Thus

$$\Xi(I^m) \geq \Xi(I) \tag{56}$$

when m divides the order of I. The following example shows that both inequality and equality can occur. This results from the fact that a solution y_1 of (53) need not necessarily give a solution for x in (54).

Example 8. This example of the solution process also clarifies the way in which the factor decomposition of exponents proceeds. Let $p = 2$, and for simplicity write

$$a_1 = a, \qquad b_1 = b, \qquad a_2 = c, \qquad b_2 = d;$$

thus

$$R = aba^{-1}b^{-1}cdc^{-1}d^{-1} = 1.$$

Now consider the automorphism of the first kind

$$I: \begin{cases} c^{-1} \\ b^{-1}a^{-1} \\ a^{-1}b^{-1}cd \\ c^{-1} \end{cases}.$$

By forming powers and using $R = 1$ one finds:

$$I^2: \begin{cases} c^{-1}d^{-1} \\ ac \\ b^{-1}cd \\ d^{-1}c^{-1}ba \end{cases} \qquad I^4: \begin{cases} d^{-1}c^{-1} \cdot a^{-1} \cdot dc \\ c^{-1}d^{-1} \cdot b^{-1} \cdot cd \\ c^{-1} \\ d^{-1} \end{cases}$$

and from this, that $I^8 = I_0$. Using exponent sums one finds

$$\Xi(I) = \Xi(I^2) = 2, \qquad \Xi(I^4) = 6.$$

In analogy with (53), one first seeks a system of representatives for the 6 isogredience classes relative to I^4 which are solutions of the equation

$$yy_{I^4} = 1,$$

and without difficulty one finds the 6 solutions

$$1, c, d, acd, bdc, a^{-1}b^{-1}cd.$$

These six elements really belong to distinct isogredience classes relative to I^4, as follows from the fact that f and $f_{I^4}^{-1}$, for each f, have the same exponent sums in all generators, and the exponent sums for the same generator in two elements f_1 and $ff_1f_{I^4}^{-1}$ must therefore have an even difference. Now since $\Xi(I^2) = 2$ one knows that, apart from the element 1, exactly one of the six elements above must be representable in the form xx_{I^2}. One finds that the equation

$$xx_{I^2} = acd,$$

corresponding to (54), has the solution

$$x = dc.$$

Since $\Xi(I) = 2$ one also knows that there must be exactly one solution of the equation

$$xx_I = dc,$$

and for this one finds the solution

$$x = ac.$$

By the above remark, we can confine ourselves to inverse automorphisms up to the fourth power of I, hence it remains to investigate

$$I^3: \begin{cases} d^{-1}c^{-1}bac \\ c^{-1}a^{-1}dc \\ d \\ d^{-1}c^{-1}b \end{cases}.$$

One finds $\Xi(I^3) = 2$. Now since dc is a solution not isogredient to 1 for I^2, $c^{-1}d^{-1}$ is one for $I^6 = I^{-2}$, and the class of solutions for I^3 not isogredient to 1 comes from solution of

$$xx_{I^3} = c^{-1}d^{-1}.$$

One finds

$$x = c^{-1}.$$

This completes the treatment of the example. The results obtained are: If $s\Phi$ is an I-inducing S-function which lies over $\sigma\phi$, then $\sigma^8\phi$ belongs to the class of the identity and hence by Theorem 19 it has a topologically necessary fixed point class of index $2 - 2p = -2$. For the powers of σ one has:

$\sigma\phi$ has two classes of index 1, represented by s and acs,
$\sigma^2\phi$ has two classes of index 1, represented by s^2 and dcs^2,
$\sigma^3\phi$ has two classes of index 1, represented by s^3 and $c^{-1}s^3$,
$\sigma^4\phi$ has six classes of index 1, represented by

$$s^4, cs^4, ds^4, acds^4, bdcs^4, a^{-1}b^{-1}cds^4,$$

$\sigma^5\phi$ has two classes of index 1, represented by s^5 and cs^5,
$\sigma^6\phi$ has two classes of index 1, represented by s^6 and $c^{-1}d^{-1}s^6$,
$\sigma^7\phi$ has two classes of index 1, represented by s^7 and $c^{-1}a^{-1}s^7$,
$\sigma^8\phi$ has one class of index -2, represented by s^8.

With the higher powers the factors repeat, as well as the class numbers and indices; the exponents of s continue to increase. In fact, nothing is known about s^8 except that it induces the identity automorphism.

46. Automorphisms of the first kind which invert elements. Let I be an automorphism of the first kind, induced by $s\Phi$, and let $k \neq 1$ be an element of F such that

$$k_I = k^{-1}.$$

Then I has the characteristic exponent 2 (§11) and the boundary mapping sE is fixed point free (Theorem 2, §11). Thus the fixed point set of $s\Phi$ has index 1. Now we consider $k^m s\Phi$. This S-function induces $I' = I_{k^{-m}}$. One has

$$k_{I'} = k^m k^{-1} k^{-m} = k^{-1}.$$

Thus the fixed point set of $k^m s\Phi$ also has index 1. It remains to investigate when $k^m s$ is isogredient to s. It follows by applying I to

$$k^m = l l_I^{-1}$$

that

$$k^{-m} = l_I l_{I^2}^{-1}$$

and multiplication of these two equations yields

$$1 = l l_{I^2}^{-1}.$$

Thus l must be a fixed element of I^2. Now I^2 also has all powers of k as fixed elements. For $l = k^r$ we get

$$k^m = k^r k_I^{-r} = k^{2r}$$

and similarly

$$k^{2r+1} = k^r k k_I^{-r}.$$

Thus the even powers of k are isogredient to 1 relative to I, and the odd powers are isogredient to k. If k is itself to be isogredient to 1, then this must occur because of a fixed element l of I^2 which is not a power of k. Hence k is definitely not isogredient to 1 when I^2 has no fixed elements other than powers of k:

Theorem 24. *If s induces an isomorphism I of the first kind for which there is an element $k \neq 1$ such that*

$$k_I = k^{-1},$$
$$H(I^2) = \{k\},$$

then the surface mapping represented by s has two fixed point classes of index 1, *represented by s and ks.*

Example 9. We consider $p = 2$, with the same notation for the generators as in the previous example, the automorphism of the first kind:

$$I: \begin{cases} c \\ b^{-1}cd \\ b^{-1}ca \\ ca \end{cases}.$$

In order to trace possible fixed elements of I, we apply the method sketched in §16, which consists in iterating the application of I to a non-fixed element. Six repetitions give

$$a \mapsto c \mapsto b^{-1}ca \mapsto d^{-1}ac \mapsto a^{-1}b^{-1}ca \mapsto c^{-1}d^{-1}ac \mapsto a^{-1}k,$$

where

$$k = c^{-1}ba^{-1}b^{-1}ca = c^{-1}a^{-1}dcd^{-1}a.$$

By continuing the process one finds:

$$k_I = k^{-1}.$$

Thus the iteration process leads automatically to this inverted element, and in fact it is independent of the element chosen initially. This follows easily from the structure of the boundary mapping induced by I. One has immediately:

$$a_{I^2} = k^{-1}ak$$

and since $a_I = c$,

$$c_{I^2} = kck^{-1},$$

hence for an arbitrary integer n,

$$a_{I^{2n}} = k^{-n}ak^n, \qquad c_{I^{2n}} = k^n ck^{-n}.$$

The expansion of $U(k)$ and $V(k)$ shows that the axis of k is a diameter of E. The latter two equations show that its endpoints are the only boundary points fixed by I^2, and that they bound two fixed point free intervals; the cyclic displacement direction on E under I^2 is the same for both intervals. As iteration is applied to any element different from k, increasing powers of k or k^{-1} gradually separate out, so that attention is necessarily drawn to k.

Now since k is primary and hence $H(I^2) = \{k\}$, one has two fixed point classes with index $+1$, by Theorem 24. Because

$$\alpha_{11} = 0, \qquad \delta_{11} = -1, \qquad \alpha_{22} = 1, \qquad \delta_{22} = 0,$$

(50) gives

$$\Xi_A(I) = 2.$$

Although this exhibits two fixed point classes, the index sum of which equals the one computed from the Alexander formula, it is of course not a proof that there are no other topologically necessary fixed point classes whose contributions to the index sum cancel out. (However, one can easily carry out such a proof by constructing an example of an I-inducing S-function with only two fixed point classes. We omit this.)

When one wants to carry the investigation into the domain of pure group theory, by studying the automorphisms related to I, one can first attempt to simplify the form of I by conjugation with a suitable automorphism. The reason why a simpler form of I has not been used from the outset is to show by this example how the iteration process itself draws the attention to the conjugating factor by which a formal simplification can be reached.

In the above calculation the power I^{12} distinguishes itself. One has

$$I^{12}: \begin{cases} k^{-1}ak \\ kck^{-2}c^{-1}bk \\ kck^{-1} \\ k^{-1}ak^2a^{-1}dk^{-1} \end{cases}.$$

One therefore considers the automorphism of the first kind

$$T: \begin{cases} a^{-1} \\ c^{-1}b \\ c^{-1} \\ a^{-1}d \end{cases}$$

which satisfies the relation

$$T^2 = I_0,$$

and one finds

$$J = TIT^{-1}: \begin{cases} c \\ d \\ ab \\ a^{-1} \end{cases}.$$

Thus one can replace the investigation of I by that of J. Here

$$k = cdc^{-1}d^{-1} = bab^{-1}a^{-1}$$

appears immediately as an inverted element, corresponding to a diameter axis.

$$J^2: \begin{cases} ab \\ a^{-1} \\ cd \\ c^{-1} \end{cases}$$

shows plainly how the boundary mapping induced by J^2 commutes with a half rotation of E.

Here one has

$$J^{12n}: \begin{cases} k^{-n}ak^n \\ k^{-n}bk^n \\ k^nck^{-n} \\ k^ndk^{-n} \end{cases}.$$

Thus the images of the surface curve corresponding to the element a, under all powers of the mapping $\sigma\phi$, belong to a cycle of 12 different topological types which are permuted cyclically by σ. The same holds for all curves whose corresponding closed geodesics do not intersect the closed geodesic κ represented by the axis of k, which are represented in F by a and b alone or c and d alone. However, if one takes another curve, say a curve λ corresponding to $c^{-1}b$, then one has:

$$(c^{-1}b)_{J^{12n}} = k^nc^{-1}k^{-2n}bk^n.$$

Thus the type of the curve $\sigma^r\lambda$ is not periodic. The closed geodesic on ϕ homotopic to $\sigma^r\lambda$ approaches more closely to κ, and along a greater part of its length, as r increases. One has

$$\left.\begin{aligned} V[(c^{-1}b)_{J^{2n}}] \to V(k) \\ U[(c^{-1}b)_{J^{2n}}] \to U(k) \end{aligned}\right\} \quad \text{as } n \to \infty.$$

Example 10. After the foregoing example of a non-periodic automorphism with characteristic exponent 2, we conclude with a brief discussion of one with characteristic exponent >2. Again $p = 2$.

$$I: \begin{cases} b^{-1}a^{-1}dc^{-1} \\ c^{-1} \\ c^{-1}b \\ b^{-1}a^{-1} \end{cases}.$$

Here $\Xi_A(I) = 4$. One finds $c_{I^3} = c$, thus I has characteristic exponent 3. Then I induces a fixed point free boundary mapping, and hence the index is $+1$. One also finds that

$$I_{c^{-1}}, I_{a^{-1}} \quad \text{and} \quad I_d$$

have characteristic exponent 3, because the third, but not the first, powers of these automorphisms have the respective fixed elements

$$b, b^{-1}c, c.$$

Now the four elements

$$1, c, a, d^{-1}$$

are pairwise non-isogredient relative to I; one proves this easily by considering exponent sums, whence one comes to linear equations with integer unknowns which clearly have no solutions. Thus the four related automorphisms above give four different fixed point classes with index 1.

47. Index determination for a non-cyclic fixed element group. Let $s\Phi$ be an S-function lying over $\sigma\phi$, I the automorphism induced by s, and $H = H(I) \neq 1$ the fixed element group. Assume H is not cyclic, so that it contains more than one primary element. Suppose also that there are non-fixed elements, so that $H \neq F$. Then by §11 and §14 we have that I is of the first kind and the fixed point set Π_E of sE is closed and nowhere dense. Thus when Π_E contains isolated points, they cannot be fundamental points of F.

Suppose $\sigma\phi$ satisfies Assumption α (§37), so that the net N is fixed point free. Let Π_Φ be the fixed point set of $s\Phi$. The set of all those points of the central cell B equivalent to points of Π_Φ can be covered by a Jordan polygon c in the interior of B. Let β be a fundamental domain of H. By §34, the subset of Π_Φ in β is covered by finitely many polygons equivalent to c. Call their union C, and let HC be the subset of Φ which results from C under all the operations of H. HC covers all of Π_Φ, and $\Phi - HC$ is a fixed point free, connected region.

The set of accumulation points of HC on E is identical with the set M of singular points with respect to H (§29). M is a perfect, nowhere dense set, identical with a subset of Π_E or Π_E itself.

A point of E not belonging to Π_E has, since it does not belong to M, a fixed point free neighbourhood, and it may therefore be connected by a fixed point free segment to a point of $\Phi - HC$. Two such points, P and Q, may then be connected by a fixed point free polygonal crosscut in $\Phi - HC$, and two other points of the same kind which do not separate P and Q may be connected by a similar crosscut which does not meet the first.

We shall now assume that it is possible to cut out a fundamental domain β for H by 2ν fixed point free crosscuts of this kind in Φ. The boundary of β then consists of these 2ν crosscuts and the 2ν segments of E between them, likewise assumed to be fixed point free. It is also permissible for such a segment to shrink to a point, as happens when two consecutive crosscuts have a common endpoint on E. Since β is to be a fundamental domain for H, the crosscuts must necessarily correspond in pairs under elements of H. One concludes immediately that H is a free group with ν generators

The index of the fixed point class represented by $s\Phi$ is now measured by $\int d\xi$ over the positive boundary of β, since it is fixed point free by hypothesis (§37). This integral is evaluated as follows.

Let q be one of the 2ν crosscuts, with endpoints M and N. Suppose q is traversed from M to N in a positive circuit round β. Let h be the element of H which sends q to the equivalent boundary crosscut of β. Since $h \rightleftarrows s$ one has that the associated contributions to the integral satisfy

$$\int_M^N d\xi + \int_{hN}^{hM} d\xi = \int_M^N d(\xi - (h)_\mu \xi),$$

where the independence of this value from the integer μ follows from (39), §40. We therefore set $\mu = 0$ and, denoting the ξ-value of M by ξ_M, let

$$\xi_N = \xi_M + \int_M^N d\xi.$$

Then §40 gives

$$\int_M^N + \int_{hN}^{hM} = \xi_N - \xi_M - [(h)_0\xi_N - (h)_0\xi_M]$$

$$= \xi_N + \tau - \xi_M - [(h)_0(\xi_N + \tau) - (h)_0\xi_M]$$

for an arbitrary integer τ. One chooses τ here so that

$$0 \leq \xi_N + \tau - \xi_M < 1.$$

The segment of the ξ-axis from ξ_M to $\xi_N + \tau$ lies over the smallest positive are on E from sM to sN, which may be denoted by

$$\text{pos}(sM \to sN).$$

Then

$$(h)_0(\xi_N + \tau) - (h)_0\xi_M$$

lies over

$$\text{pos}(shM \to shN) = U - \text{pos}(shN \to shM),$$

where U denotes a complete positive circuit of E. The desired value of the integral is therefore measured on E by

$$\text{pos}(sM \to sN) + \text{pos}(shN \to shM) - U.$$

Now if O is the next "vertex" after N in the boundary of β, namely the initial point of the crosscut which follows q, and if one reflects that sE is a topological, orientation preserving mapping, then it follows that

$$\int_N^O d\xi$$

is measured by

$$\text{pos}(sN \to sO).$$

If one now sums these contributions over a circuit around β, the sum of all these smallest positive arcs gives exactly one positive circuit of E, since the cyclic order of the $2v$ "vertices" of β is preserved by s. Since v complete circuits have to be subtracted from this:

The desired index is $1 - v$.

§41 shows that this result is also correct for $v = 1$.

If one applies the idea of §30 to this case, then (with the notation used there)

$$K_1 = K \bmod H$$

is a closed surface of genus v, on which s induces a mapping sK_1 in the class of the identity. By Theorem 19, the latter has a fixed point class with index $2 - 2v$, and possibly other classes of index zero. The above considerations now show that the essential class is the one represented by sK. None of the E-points lying over K_1 is a fixed point (the points of M do not lie over $K \bmod H$) and the index $2 - 2v$ is distributed equally between Φ and Ψ.

Notice, however, that points of K which are sent to H-equivalent points by s also appear as fixed points of sK_1. This does not disturb the above result, because they constitute fixed point classes with index zero. This also holds in the case ($\nu = 1$) of §41, because of Theorem 17, §38.

48. Examples. We shall illustrate the process of the preceding paragraph by a few examples, which at the same time afford some remarks which are important for the general method.

Example 11. We consider the automorphism Λ of example 2, §15. Let s be an S-function which induces Λ and satisfies Assumption α. The boundary mapping sE has been completely dealt with in §15. We shall now, as we announced there, construct a fundamental domain for

$$H(\Lambda) = \{b_1, a_2, b_2, \ldots, a_p, b_p\},$$

of the type described in §47.

Let Q be an arbitrary point of the fixed point free interval i and let P be an arbitrary point of j (§15). Let q_1 be a fixed point free crosscut in Φ which connects them. Then $b_1 q_1$ connects the points $b_1 Q$ and $b_1 P = P_1$. We now introduce the notation:

$$\begin{aligned}
P_2 &= a_2^{-1} P_1,\\
P_3 &= b_2^{-1} P_2,\\
P_4 &= a_2 P_3,\\
P_5 &= b_2 P_4,\\
P_6 &= a_3^{-1} P_5,\\
P_7 &= b_3^{-1} P_6,\\
P_8 &= a_3 P_7,\\
P_9 &= b_3 P_8,\\
&\ldots\\
P_{4(p-2)+2} &= a_p^{-1} P_{4(p-2)+1},\\
P_{4(p-2)+3} &= b_p^{-1} P_{4(p-2)+2},\\
P_{4(p-1)} &= a_p P_{4(p-2)+3},\\
P' &= b_p P_{4(p-1)}.
\end{aligned}$$

Then

$$P' = b_p a_p b_p^{-1} a_p^{-1} \ldots b_2 a_2 b_2^{-1} a_2^{-1} b_1 P = a_1 b_1 a_1^{-1} P,$$

and therefore P' lies in j.

The points mentioned lie on E in the cyclic order

$$P, Q, b_1 Q, P_1, P_4, P_3, P_2, P_5, P_8, P_7, P_6, \ldots, P_{4(p-2)+1}, P_{4(p-1)}, P_{4(p-2)+3}, P_{4(p-2)+2}, P',$$

and none of them is a fixed point of sE, since P and Q are not fixed points and only operations of H have been applied. Now one connects

$$P_{4n+1} \text{ to } P_{4(n+1)} \text{ by } q_{n+2},$$
$$P_{4n+2} \text{ to } P_{4n+3} \text{ by } a_{n+2}^{-1}q_{n+2},$$
$$P_{4(n+1)} \text{ to } P_{4n+3} \text{ by } r_{n+2},$$
$$P_{4n+2} \text{ to } P_{4(n+1)+1} \text{ by } b_{n+2}r_{n+2}$$

for $n = 0, 1, \ldots, p - 2$, where the connecting paths are fixed point free crosscuts in Φ, no two of which meet. The latter condition can be satisfied by first taking crosscuts which are circles orthogonal to E. These do not meet in Φ and they are paired correctly by the above elements of H. Unfortunately they may contain fixed points of $s\Phi$. However, since their endpoints are not in the boundary fixed point set M, they pass through only finitely many of the polygons whose union was denoted by HC in the previous paragraph. One then replaces each segment cutting such a polygon by a polygonal path in the neighbourhood of the polygon boundary, using equivalent paths for equivalent crosscuts. In this way one can avoid the crosscuts meeting.

These $v = 2p - 1$ crosscut pairs

$$q_1, q_2, \ldots, q_p, r_2, \ldots, r_p$$

border a fundamental domain β of H. H is the free group with $2p - 1$ generators, as already remarked in §15.

The index associated with Λ now comes to

$$1 - v = 2 - 2p$$

by §47, and this is also the value of Ξ_A in formula (50).

The surface $K_1 = K \bmod H(\Lambda)$ has genus $v = 2p - 1$. It contains two closed curves which are covered by $E - M$; they are represented by the segments of E belonging to the boundary of β, namely $P'P$ in j and $Q(b_1 Q)$ in i.

Each mapping $\sigma\phi$ in the mapping class determined by Λ therefore has a fixed point class of index $2 - 2p$. One sees easily that there need not be any more fixed point classes by considering a special mapping in the class. Enclose a curve of type b_1 on ϕ by a region homeomorphic to an annulus; outside this region the mapping is the identity, but the region is mapped onto itself in the same way as the annulus

$$r \leq |\rho e^{i\theta}| \leq R$$

in the complex plane is mapped onto itself by the transformation

$$\rho' = \rho,$$

$$\Theta' = \Theta + \frac{\rho - r}{R - r} 2\pi.$$

The interior of the region is then fixed point free, while the rest of the surface constitutes a connected fixed point set, and hence a single fixed point class.

Although the fixed point problem is now completely solved for the mapping class considered in this example, it is still of interest to further explore a purely group theoretic solution (which does not depend on the construction of a special mapping). Thus we shall show that any other fixed point classes have index zero. This means considering S-functions fs as f runs through representatives of all isogredience classes relative to Λ. This is easy as long as f is chosen from H. Then, by §17, $\Lambda_{f^{-1}}$ has in general a cyclic fixed element group, and hence gives index zero by §41. Only when f is conjugate to b_1^{+1}, so that the axis of f determines a fixed point free interval on E under Λ, has $\Lambda_{f^{-1}}$ a larger fixed element group. But then f is isogredient to 1; see the remark after Theorem 7, §18. Now what if f does not lie in H? Suppose, e.g., that $f=a_1^{-1}$. By example 3, §15, $\Lambda' = \Lambda_{a_1}$ leads to two fixed points on E which are not fundamental points and were previously denoted by P^+ and P^-. Apart from that, $a_1^{-1}sE$ is fixed point free, so that $H(\Lambda') = 1$.

By §34, P^+ and P^- are not accumulation points of fixed points. Thus a circle concentric with E and of sufficiently large radius <1 encloses the (possibly existing) fixed point set of $a_1^{-1}s\Phi$; the index which belongs to this S-function is therefore well-defined. If it were non-zero, then a new fixed point class would be found for *each* mapping in the class of Λ, in contradiction to the above example. But how can one, *without* going back to this example, conclude solely from the functional equation

$$a_1^{-1}sfx = f_{\Lambda'}a_1^{-1}sx$$

for $a_1^{-1}s$ that the index must be zero?

The following answer to this question extends beyond the present special example and brings, as the following paragraph will show, an essentially new factor into the determination of index.

P^+ has the expansion (§15):

$$a_1^{-1}b_1^{-1}a_1^{-1}b_1^{-2}a_1^{-1}b_1^{-3}\dots\text{ad inf.}$$

and P^- has the expansion

$$a_1b_1^{-1}a_1b_1^{-2}a_1b_1^{-3}\dots\text{ad inf.}$$

We consider the fixed element of Λ

$$\gamma = a_2b_1$$

and call its axis A. The expansions of the fundamental points of γ are periodic repetitions of a_2b_1 and $b_1^{-1}a_2^{-1}$, respectively. These fundamental points therefore separate P^+ and P^- from each other. Now (cf. §15):

$$\gamma_n = \gamma_{\Lambda'^n} = a_1^{-1}b_1^{-1}a_1^{-1}b_1^{-2}\dots a_1^{-1}b_1^{-(n-1)}a_1^{-1}\cdot\gamma\cdot a_1b_1^{n-1}\dots a_1b_1a_1.$$

We call the axis of this element A_n. These axes A_n approach P^+ as n increases.

The image $a_1^{-1}sA$ is a Jordan arc which connects the fundamental points of

$$\gamma_1 = a_1^{-1}\gamma a_1$$

and which is mapped into itself by the displacement γ_1, so that it runs between two curves at constant distance from the axis A_1 of γ_1. Let the maximum non-Euclidean distance between $a_1^{-1}sA$ and A_1 be l. Now γ_n is a conjugate of γ, and hence A_n is equivalent to A, so it follows from the functional equation that the figure formed by $a_1^{-1}sA_n$ and A_{n+1} is equivalent to the one formed by $a_1^{-1}sA$ and A_1, and therefore $a_1^{-1}sA_n$ and A_{n+1} also have the maximal distance l.

Now the above form of γ_n shows that the two elements γ_n and γ_{n+1} result from the elements γ and

$$\delta_n = b_1^{-n} a_1^{-1} \cdot \gamma \cdot a_1 b_1^n$$

by conjugation with the same element. The figure formed by the axes A_n and A_{n+1} is therefore congruent to the one formed by A and the axis D_n of δ_n. D_n results from A by the displacement $b_1^{-n} a_1^{-1}$, and hence the endpoints of D_n converge to $U(b_1)$ as n increases. The shortest non-Euclidean distance between D_n and A therefore grows beyond all bounds as n increases. The same must then hold for the shortest distance between A_n and A_{n+1}. And since $a_1^{-1} s A_n$ runs within a distance l from A_{n+1}, A_n and $a_1^{-1} s A_n$ can meet for at most finitely many values of n.

We now choose a value of n such that A_n runs inside a fixed point free neighbourhood of P^+ and does not meet its image $a_1^{-1} s A_n$. Then $a_1^{-1} s A_n$ runs wholly inside the neighbourhood of P^+ bounded by A_n, because $a_1^{-1} s$ sends boundary points away from P^- and towards P^+, so that the endpoints of $a_1^{-1} s A_n$ lie closer to P^+ than their preimages. Thus for each point x of A_n, the non-Euclidean half-line from x to $a_1^{-1} sx$ meets the boundary E at a point $z(x)$ on the small boundary arc which is determined by the endpoints of A_n and contains P^+. The converse relationships hold with P^-. By an analogous argument, one can bound P^- by an arc B orthogonal to E such that $a_1^{-1} sB$ does not meet B, in fact it runs entirely on the side of B away from P^-. Then, for each x on B, $z(x)$ lies on the large boundary arc determined by B which does *not* contain P^-. If one now allows a point x to run round E, but in such a way that it detours round the fixed points P^+ and P^- on the small arcs A_n and B respectively, then $z(x)$ is everywhere defined, and the total variation of z is zero, since $z(x)$ never falls into P^-. Any fixed point class represented by $a_1^{-1} s\Phi$ therefore has index zero. Q.E.D.

One can thus extend the function $z(x)$, defined for all non-fixed points, continuously to P^+, by defining this point to be its own z-point. However, P^- remains "essentially singular"; one can obtain any value of $z(x)$ with absolute value 1 by allowing x to converge suitably to P^-. We shall return to this situation in the next paragraph.

Example 12. We again consider an example of a mapping class with negative index sum Ξ, where the negative value results from contributions by more than one fixed point class with negative index.

Let s be an S-function which satisfies Assumption α and which induces the automorphism of the first kind (with $p = 2$):

$$I: \begin{cases} c^{-1}bab \\ b \\ c \\ c^{-1}bdc \end{cases}.$$

For arbitrary n, one has immediately

$$I^n: \begin{cases} (c^{-1}b)^n ab^n \\ b \\ c \\ (c^{-1}b)^n dc^n \end{cases}$$

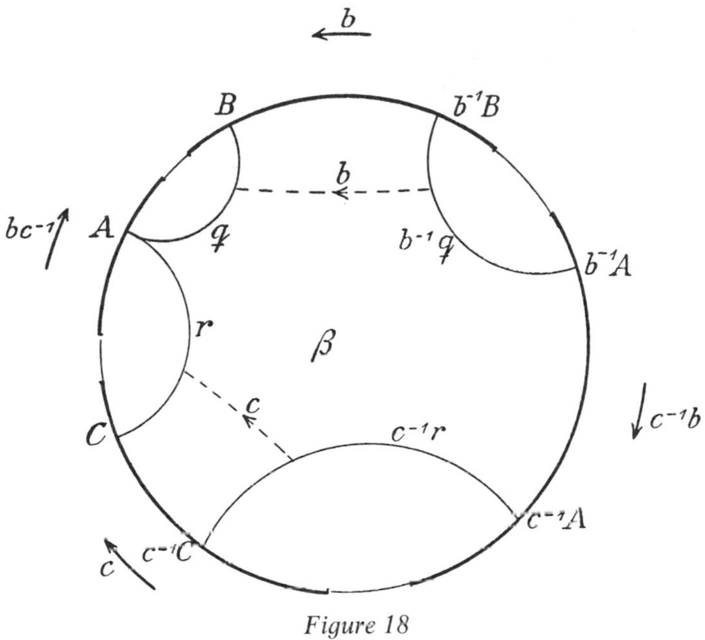

Figure 18

and

$$\Xi_A(I^n) = -2.$$

One sees that the expansion of $V(a_{I^n})$ converges to the expansion of $V(c^{-1}b)$ as $n \to \infty$, and to the expansion of $U(c^{-1}b)$ as $n \to -\infty$, from which one concludes that the axis of $c^{-1}b$ bounds a fixed point free periodic interval under sE. By consideration of $U(a_{I^n})$ and $U(d_{I^n})$ one gets the same conclusion for the axes of b and c. Since bc^{-1} results from $c^{-1}b$ by conjugation with c (or b^{-1}), the same is true for the axis of bc^{-1}. The relative positions of these intervals are shown in Figure 18, where they are drawn heavily on E. The displacement direction in each interval under application of the element bounding the interval is shown by an arrow outside E. Now let A, B, C be arbitrary points in the intervals for bc^{-1}, b, c respectively. A is connected to B by a fixed point free crosscut q, and to C by r. The figure also shows $b^{-1}q$ and $c^{-1}r$, and the correspondence between crosscut pairs is shown by dotted lines.

$H(I)$ is the free group $\{b, c\}$, as the fundamental domain β just constructed shows, and as can also be seen easily by pure group theory. If one applies all operations in $\{b, c\}$ to β, then Φ will be filled up completely. If one observes the images of q and r under all operations in $\{b, c\}$, one sees that each fixed point on the boundary is separated from the midpoint of Φ by infinitely many of these images, and conversely, each such point is a fixed point on E. Each such point is therefore uniquely associated with a sequence in the symbols b and c and their inverses. (Such an "expansion of the fixed point on E with respect to the fixed element group" occurs generally in the case of §47.) One notes that $c^{-1}A = c^{-1}b(b^{-1}A)$, so that the closed surface $K_1 = K \bmod H$ of genus 2 contains three different closed curves covered by $E - M$.

Thus we have found the index $1 - v = -1$ for $s\Phi$ by the method of §47, and since $\Xi_A = -2$ there is still one fixed point class to be found.

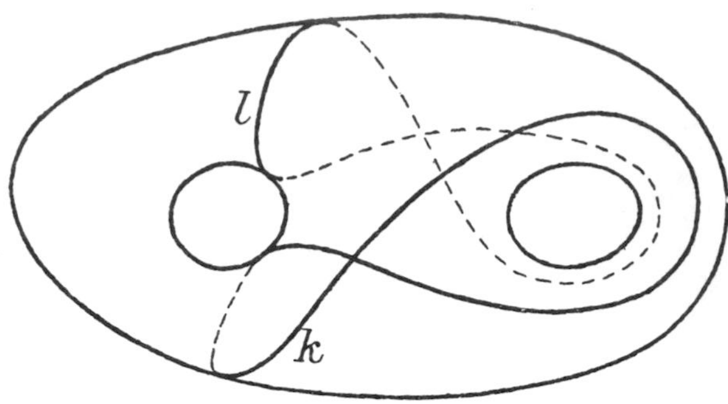

Figure 19

We consider the S-function cs, which induces $I' = I_{c^{-1}}$! By the same process as above, one finds easily that the four fixed elements of I'

$$c,\ cd^{-1} \cdot b^{-1}c \cdot dc^{-1},\ d^{-1}a \cdot b \cdot a^{-1}d,\ d^{-1} \cdot b^{-1}c \cdot d$$

have axes which bound four fixed point free periodic intervals (the interval for c is now the one complementary to that used above), and one obtains a fundamental domain for $H(I')$ from $\nu = 2$ pairs of crosscuts based on these four intervals. One therefore finds that $cs\Phi$ has index -1.

Is this now a new fixed point class? In other words: Is it certain that there is no solution to the equation

$$xx_I^{-1} = c?$$

This question may be answered affirmatively, in such a way as to complete the discussion of this example, by construction of a particular mapping in the class under consideration.

One draws on ϕ a curve k of type bc and a curve l of type $d^{-1}aba^{-1}dc$. It is arranged that k and l do not meet, and that each of them has only the single double point required by type. Figure 19 shows such a pair of curves. They are assumed to be based on a canonical cut system a, b, c, d analogous to that in Figure 3. Then the complement set

$$\phi - k - l$$

consists of three regions, each of which is homeomorphic to the interior of an annulus. A topological mapping $\tau\phi$ is now defined as follows: All points of k and l shall be fixed points, and each of the three annular regions shall undergo a topological transformation of the kind used in the previous example. If one takes the double point of k as origin of the canonical cut system, and if one chooses the "direction of twist" in the three annular regions suitably, then one easily sees that the curve a becomes a curve of type $c^{-1}bab$, and the curve d becomes a curve of type $c^{-1}bdc$. We are therefore dealing with a mapping in the class determined by I. Since one has only the two connected fixed point sets k and l, there are *at most* two fixed point classes. On the other hand, we concluded from the Alexander formula that there is more than one class. Hence k and l really belong to *different* classes. One can consider, e.g., the annular region whose boundary curves are both of type c, and take a map in Φ for which one boundary curve consists of points fixed by $s\Phi$. Then the other boundary curve cannot have the same property, because a path in the annulus from one boundary to the other is not homotopic to its image under $\tau\phi$. One also sees that the other boundary curve must undergo exactly the displacement given by the element c^{-1} (or an equivalent element). This shows again the relationship noticed above between the fixed point classes represented by $s\Phi$ and $cs\Phi$.

One sees that $\tau\varphi$ does not satisfy Assumption α. But this is not necessary to make our conclusions possible. One infers from this that, in constructing examples of mappings, it is not always appropriate to represent the necessary fixed point classes by single points. On the contrary, one often benefits from representing the fixed point classes by irreducible continua which divide the surface into complementary regions, as simply connected as possible, in which the mapping is easy to describe.

This leads immediately to the question of what is the greatest possible extent of the fixed point set, or what is the *least degree of multiple connectivity* in the complement of the fixed point set for a given mapping class.

49. Example of a fixed point class with negative index for an automorphism without fixed elements. All previous cases of fixed point classes with negative index came about because the automorphism in question had more than one fixed primary element. It will now be shown that a negative class index can occur in an essentially different way.

Example 13. With $p = 2$, we consider the following automorphism I and its powers up to the fourth:

$$I:\begin{cases} c^{-1}a^{-1} \\ b^{-1}a^{-1} \\ b^{-1}a^{-1}d \\ c^{-1} \end{cases},\qquad I^2:\begin{cases} d^{-1}abac \\ abac \\ aba \\ d^{-1}ab \end{cases}$$

$$I^3:\begin{cases} a^{-1}b^{-1}a^{-1}c^{-1}a^{-1}b^{-1}a^{-1}d \\ c^{-1}a^{-1}b^{-1}a^{-1}c^{-1}a^{-1}b^{-1}a^{-1}d \\ c^{-1}a^{-1}b^{-1}a^{-1}c^{-1}a^{-1} \\ a^{-1}b^{-1}a^{-1} \end{cases}$$

$$J = I^4:\begin{cases} acabacd^{-1}abacaba \\ d^{-1}abacabacd^{-1}abacaba \\ d^{-1}abacabacd^{-1}abac \\ acabac \end{cases}$$

We write $I^4 = J$ for brevity. It can be seen immediately that the expansions of the positive fundamental points of the generators are periodic repetitions of the above expressions. The sum of the smallest negative boundary arcs from $V(a)$ to $V(a_I)$, $V(a_I)$ to $V(a_{I^2})$, $V(a_{I^2})$ to $V(a_{I^3})$ and from $V(a_{I^3})$ to $V(a_J)$ is greater than 2π; for the remaining generators the corresponding sum is smaller than 2π. It follows that the boundary mapping g induced by I is fixed point free, and that the boundary mapping $g^4 = \gamma$ for J has fixed points. To I therefore belongs the index $+1$.

We now investigate the fixed point set of γ. To do this we must construct the powers of J, in order to decide what are the limiting positions of boundary points under powers of γ. Now it is identically true that:

$$a_{J^n} = a \cdot (a^{-1} a_J) \cdot (a^{-1} a_J)_J \cdot (a^{-1} a_J)_{J^2} \dots (a^{-1} a_J)_{J^{n-1}}.$$

Before applying powers of J to the word

$$a^{-1} a_J = cabacd^{-1} abacaba,$$

one observes that the subword $d^{-1} a$ satisfies

$$(d^{-1} a)_J = d^{-1} a \cdot bacaba.$$

Then one realizes, from the absence of negative exponents, that no cancellation occurs in the above expression for a_{J^n}, and that the expansion of $V(a_{J^n})$ consists of periodic repetition of this word. Thus, as $n \to \infty$, $V(a_{J^n})$ converges to the point P_1 with the expansion

$$k(P_1) = a \cdot (a^{-1} a_J) \cdot (a^{-1} a_J) \dots \text{ad inf.}$$

If one constructs

$$J^{-1}: \begin{cases} ab^{-1} d^{-1} ab^{-1} cd^{-1} ab^{-1} d^{-1} a \\ a^{-1} dba^{-1} dc^{-1} ba^{-1} dba^{-1} c^{-1} ba^{-1} dba^{-1} dc^{-1} ba^{-1} dba^{-1} \\ a^{-1} dba^{-1} d \\ c^{-1} ba^{-1} dba^{-1} dc^{-1} ba^{-1} d \end{cases},$$

then one sees similarly that, as $n \to -\infty$, $V(a_{J^n})$ converges to the point P_2 with the expansion

$$k(P_2) = a \cdot (a^{-1} a_{J^{-1}}) \cdot (a^{-1} a_{J^{-1}})_{J^{-1}} \dots \text{ad inf.}$$

$V(b_{J^n}^{-1})$ converges to the same point as $n \to -\infty$. Namely, $b_{J^{-1}}^{-1}$ begins with a, and one easily sees that the same expansion then arises. Analogously, one now confirms the following behaviour for other fundamental points under the powers J^n:

as $n \to \infty$ $\quad \left.\begin{matrix} V(b^{-1}) \\ V(a^{-1}) \end{matrix}\right\} \to K(P_3) = a^{-1} \cdot (aa_J^{-1}) \cdot (aa_J^{-1})_J \dots \text{ad inf.},$

as $n \to -\infty$ $\quad \left.\begin{matrix} V(a^{-1}) \\ V(b) \end{matrix}\right\} \to k(P_4) = a^{-1} \cdot (aa_{J^{-1}}^{-1}) \cdot (aa_{J^{-1}}^{-1})_{J^{-1}} \dots \text{ad inf.},$

as $n \to \infty$ $\quad \left.\begin{matrix} V(b) \\ V(d^{-1}a) \end{matrix}\right\} \to k(P_5) = d^{-1} a \cdot (a^{-1} d(d^{-1} a)_J) \cdot (a^{-1} d(d^{-1} a)_J)_J \dots \text{ad inf.},$

as $n \to -\infty$ $\quad \left.\begin{matrix} V(d^{-1}a) \\ V(d^{-1}) \end{matrix}\right\} \to k(P_6) = d^{-1} \cdot (dd_{J^{-1}}^{-1}) \cdot (dd_{J^{-1}}^{-1})_{J^{-1}} \dots \text{ad inf.},$

as $n \to \infty$ $\quad \left.\begin{matrix} V(d^{-1}) \\ V(c^{-1}ba^{-1}) \end{matrix}\right\} \to k(P_7) = c^{-1} \cdot (cc_J^{-1}) \cdot (cc_J^{-1})_J \dots \text{ad inf.},$

as $n \to -\infty$ $\quad \left.\begin{matrix} V(c^{-1}ba^{-1}) \\ V(d) \end{matrix}\right\} \to k(P_8) = c^{-1} b \cdot (b^{-1} c(c^{-1} b)_{J^{-1}}) \cdot (b^{-1} c(c^{-1} b)_{J^{-1}})_{J^{-1}} \dots \text{ad inf.}$

as $n \to \infty$ $\quad V(d) \to k(P_1).$

In order to make clear the positions of these points relative to each other and relative to the eight fundamental points used, we now give the initial terms of their expansions:

$$k(P_1) = acaba\ldots,$$
$$k(P_2) = ab^{-1}d^{-1}ab^{-1}\ldots,$$
$$k(P_3) = ab^{-1}a^{-1}c^{-1}\ldots,$$
$$k(P_4) = a^{-1}dba^{-1}d\ldots,$$
$$k(P_5) = d^{-1}abac\ldots,$$
$$k(P_6) = d^{-1}ab^{-1}cd^{-1}\ldots,$$
$$k(P_7) = c^{-1}a^{-1}b^{-1}a^{-1}\ldots,$$
$$k(P_8) = c^{-1}ba^{-1}db\ldots.$$

Thus the eight points lie on E in the same order as their subscripts, and each of the eight intervals they determine contains one of the eight fundamental point used above. These intervals are therefore fixed point free, and only the eight points $P_1, \ldots, P_8$ are fixed by γ. Then there can be no elements fixed by J, because such elements reproduce the fixed point set, and if more then two fixed points were present the fundamental points of a fixed element would be accumulation points of fixed points. We therefore have the result:

The automorphism $J = I^4$ has no proper fixed elements, and exactly 8 fixed limit elements. The induced boundary mapping $\gamma = g^4$ has 8 aperiodic fixed point free intervals, separated by the fixed points $P_1, \ldots, P_8$.

One sees that the boundary mapping γE sends points of E towards the points P_1, P_3, P_5, P_7 and away from the points P_2, P_4, P_6, P_8. The former may be called the *positive*, and the latter the *negative*, fixed points of γE.

The set M consisting of these 8 points must be mapped into itself by gE, because

$$g^4(gM) = g(g^4 M) = gM$$

and since there is no *finite* point set, other than M, mapped into itself by g^4, it follows that $gM = M$. Since I commutes with $J = I^4$, one finds that

$$(aa_J^{-1})_I = a_I a_{I \cdot J}^{-1} = c^{-1}a^{-1} \cdot a_J c_J,$$

and therefore

$$((aa_J^{-1})_{J^r})_I = c_{J^r}^{-1}(a^{-1}a_J)_{J^r} \cdot c_{J^{r+1}},$$

so that P_3 is sent by g to P_1. Thus g permutes the elements of M cyclically by two steps, g^2 by four steps, and g^3 by six steps, *hence I, I^2 and I^3 have class index* $+1$.

Now what is the index for $J = I^4$? Here the situation proceeds mutatis mutandis as in the treatment of the automorphism Λ' at the end of Example 11, §48, so that a brief indication of the proof will suffice.

Let $s(\Phi + E)$ be an arbitrary S-function for J. By §34, all the fixed points of $s\Phi$ lie inside a smaller circle concentric with E, so $P_1, \ldots, P_8$ are isolated fixed points. One considers the positive fixed point P_1 and determines a sequence of equivalent axes which enclose P_1 and contract towards it. Then one sees easily from the form of J that the least non-Euclidean distance from these axes to their images under s grows beyond all bounds as the sequence proceeds, so that the point $z(x)$ is very near to P_1 when x traverses axes sufficiently far down the sequence.

Here one makes little use of the special form of the automorphism. Namely, suppose A and B are the axes of two conjugate elements f and kfk^{-1}. Now when the position of x relative to sx is the same for two points x, one of which traverses A and the other B, then it follows from the functional equation that k is a fixed element of the automorphism in question. In all infinite sequences of axis pairs in which the first axes of all pairs are equivalent, and likewise the second axes are all equivalent, and for which the relative position of the first axis to the second occurs only finitely often, the least non-Euclidean distance between the axes of pairs grows beyond all bounds. Thus when the automorphism in question has no fixed elements, one needs its special form only to exclude the possibility that the position of A_i relative to sA_i is the same for two axes A_i ($i = 1, 2$) without this being true for the position of x relative to sx as x traverses the axes A_i, in which case there is a question of conjugation of the image point by a power of the element represented by the axis. *In this way, the method developed in* §48 *for the example* Λ' *attains a quite general significance.*

We can therefore continuously extend the function $z(x)$ for the S-function belonging to the automorphism J to the four positive fixed points on the boundary, and proceed as we did with Λ' at the four negative fixed points. By taking the index integral $\int d\xi$ over E with detours of small arcs round the eight fixed points, one then obtains the index $1 - v$, where v is the number of negative fixed points, hence in our case 1–4:

$J = I^4$ *has index* -3.

The fixed point classes with negative index which occur in the absence of fixed elements require no assumption like Assumption α on the structure of the fixed point set. They have, in common with the fixed point classes of negative index which were derived earlier on the basis of a fixed element group with more than one generator, the property of "stability under iteration." Each power of the mapping has the same class with the same index, whereas all known classes with positive index lose it under certain powers of the mapping.

50. Boundary mappings of related automorphisms. Let I be an automorphism of the first kind and let $h \neq 1$ be any element of F. As in §17 we construct the sequence of elements

$$r_n = hh_I h_{I^2} \dots h_{I^{n-1}}, \qquad n = 1, 2, \dots.$$

Then

$$r_{n+1} = hr_{nI} = r_n h_{I^n}.$$

(1) Suppose that two elements of this sequence are equal:

$$r_m = r_\mu, \qquad \mu < m.$$

Then

$$h_{I^\mu} h_{I^{\mu+1}} \dots h_{I^{m-1}} = 1,$$

hence if we apply $I^{-\mu}$ and set $m - \mu = v$ we get

$$r_\nu = hh_I \dots h_{I^{\nu-1}} = 1,$$

and if we apply I^m,

$$r_m^{-1} r_{m+\nu} = 1,$$

so that $r_{m+\nu} = r_m$ for all m. The sequence r_n is therefore periodic. In particular, for $m = 1$ one has

$$h = r_1 = r_{\nu+1} = r_\nu h_{I^\nu} = h_{I^\nu}.$$

Consequently, h is a fixed element for I^ν, but not for I itself, since $r_\nu = 1$, hence the boundary mapping induced by I must be fixed point free. Now by (14), §19, we have generally

$$(I_{h^{-1}})^n = (I^n)_{r_n^{-1}}$$

so that, in the present case,

$$(I_{h^{-1}})^\nu = I^\nu.$$

Since h is not a fixed element of $I_{h^{-1}}$, it follows that $I_{h^{-1}}$ also induces a fixed point free boundary mapping. Thus I, as well as $I_{h^{-1}}$, has index $+1$:

Theorem 25. *A necessary condition for the equation*

$$xx_I x_{I^2} \dots x_{I^{n-1}} = 1$$

to have solutions is that I have neither proper fixed elements nor fixed limit elements, and that n be a multiple of the characteristic exponent of I.

(2) Now suppose that all the r_n are different. Then all the points $r_n(0)$ are different, and the accumulation points of this point set constitute a closed point set M on E. M experiences a one-to-one mapping into itself under the boundary mapping g induced by $I_{h^{-1}}$. Because if P is any point of M and if

$$r_{n_1}, r_{n_2}, r_{n_3}, \dots$$

is a fundamental sequence for P, then gP is the limit of the sequence

$$r_{n_i I_{h^{-1}}}(0), \qquad i = 1, 2, \dots.$$

Now, in general

$$r_{nI_{h^{-1}}} = h \cdot r_{nI} \cdot h^{-1} = r_{n+1} h^{-1},$$

and since $h^{-1}(0)$ is a fixed point in Φ, it follows that

$$r_{n_i+1}, \qquad i = 1, 2, \dots$$

is a fundamental sequence for gP. As a subsequence of r_n, the latter must have its limit in M. The fact that $g^{-1}P$ belongs to M follows from the last equation when one replaces n by $n - 1$ and applies $(I_{h^{-1}})^{-1} = I_{h_I^{-1}}^{-1}$:

$$r_{n(I_{h^{-1}})^{-1}} = r_{n-1} h_{I^{-1}},$$

Then one can apply the same argument, and the assertion is proved.

Now if M consists of just *one* point, this must be a fixed point of g, and r_n is a fundamental sequence for it.

Suppose then that M consists of *more than one, but finitely many* points. Then g cyclically permutes the points of M through a certain number of steps. Let $v \geq 1$ be the smallest number for which g^v leaves each point of M fixed. I claim that M then consists of exactly v points, so that the length of the shift in the above permutation is relatively prime to the number of points in M.

Namely, let P be a point of M and C a circle around P which contains no other points of M on itself and its interior. One determines the r_n for which $r_n(0)$ falls inside C, and orders them by increasing subscripts:

$$r_{n_1}, r_{n_2}, \ldots .$$

This sequence is then a fundamental sequence for P. Now since the sequence

$$r_{n_i+v}, \qquad i = 1, 2, \ldots$$

belongs to $g^vP = P$ by the above, all but finitely many of the points $r_{n_i+v}(0)$ lie inside C. One can therefore determine a number j such that the sequence

$$r_{n_i+v}, \qquad i = j, j+1, \ldots$$

is a subsequence of

$$r_{n_i}, \qquad i = j, j+1, \ldots .$$

Then, for the same values of i, r_{n_i+kv} is also a fundamental sequence for P for an arbitrary fixed positive k, and the same holds for the sequence

$$r_{n_i+k\cdot v}, \qquad n_i \text{ fixed}, \quad k = 0, 1, 2, \ldots$$

when we choose $i \geq j$. But then

$$r_{n_j+kv+\mu}, \qquad 0 \leq \mu \leq v-1, \quad k = 0, 1, 2, \ldots$$

is a fundamental sequence for $g^\mu P$. The elements of these v sequences include all elements of the original sequence r_n from r_{n_j} onwards, hence M is exhausted by the points $g^\mu P$. M therefore consists of v points, Q.E.D.

Thus the points of the sequence $r_n(0)$, from a certain place onwards, fall periodically into the immediate neighbourhoods of the points $P, gP, \ldots, g^{v-1}P$.

It follows that the boundary mappings $g, g^2, \ldots, g^{v-1}$ are fixed point free, and we have the following:

Theorem 26. *When the sequence of points*

$$hh_I \ldots h_{I^{n-1}}(0), \qquad n = 1, 2, \ldots$$

has exactly v accumulation points on E, then all the automorphisms $(I_{h^{-1}})^n$, $n \not\equiv 0 \pmod{v}$ have index $+1$.

The sequence denoted by s_n in §17 may be applied to similar considerations.

51. Example of a mapping class with simultaneous appearance of fixed point classes of positive and negative indices.

Example 14. We take the automorphism I of Example 13, §49, retaining the notation used there, and investigate the related automorphism

$$I' = I_a: \begin{cases} a^{-1}c^{-1} \\ a^{-1}b^{-1} \\ a^{-1}b^{-1}a^{-1}da \\ a^{-1}c^{-1}a \end{cases}.$$

In order to find fixed points of the induced boundary mapping g', we use the sequence r_n from the previous paragraph, now setting $h = a^{-1}$ and letting I be the automorphism of Example 13. Then

$$\begin{aligned} r_{2n} &= a^{-1}a_I^{-1}a_{I^2}^{-1}a_{I^3}^{-1}\ldots a_{I^{2n-1}}^{-1} \\ &= (a^{-1}a_I^{-1})(a^{-1}a_I^{-1})_{I^2}\ldots(a^{-1}a_I^{-1})_{I^{2n-2}} \\ &= cc_{I^2}c_{I^4}\ldots c_{I^{2n-2}}. \end{aligned}$$

It is clear, from the form of I^2 given previously, that this is a fundamental sequence which converges to the point P_1 with the expansion

$$\begin{aligned} k(P_1) &= cc_{I^2}c_{I^4}\ldots \text{ad inf.} \\ &= c\cdot aba\cdot d^{-1}abacabacd^{-1}abac\ldots. \end{aligned}$$

Then one also has

$$\begin{aligned} r_{2n+1} &= a^{-1}a_I^{-1}\ldots a_{I^{2n}}^{-1} \\ &= a^{-1}(a^{-1}a_I^{-1})_I(a^{-1}a_I^{-1})_{I^3}\ldots(a^{-1}a_I^{-1})_{I^{2n-1}} \\ &= a^{-1}c_Ic_{I^3}\ldots c_{I^{2n-1}}. \end{aligned}$$

These elements form a fundamental sequence which converges to the point P_3 with the expansion

$$\begin{aligned} k(P_3) &= a^{-1}c_Ic_{I^3}\ldots \text{ad inf.} \\ &= a^{-1}\cdot b^{-1}a^{-1}d\cdot c^{-1}a^{-1}b^{-1}a^{-1}c^{-1}a^{-1}\ldots. \end{aligned}$$

If one writes this

$$k(P_3) = a^{-1}(k(P_1))_I,$$

then it becomes evident that $P_3 = g'P_1$. Here we have the case of Theorem 26 where $v = 2$, *so that I' has index $+1$.*

For g'^2 we have the fixed points P_1 and P_3. Two more are found using the sequence of elements

$$s_n = a_{I^{-1}}a_{I^{-2}}\ldots a_{I^{-n}},$$

where

$$I^{-1}:\begin{cases} a^{-1}d \\ d^{-1}ab^{-1} \\ d^{-1} \\ b^{-1}c \end{cases}, \qquad I^{-2}:\begin{cases} d^{-1}ab^{-1}c \\ c^{-1}ba^{-1}dba^{-1}d \\ c^{-1}b \\ ba^{-1} \end{cases}.$$

One finds

$$\begin{aligned} s_{2n} &\to b^{-1}c\cdot(b^{-1}c)_{I^{-2}}\cdot(b^{-1}c)_{I^{-4}}\ldots\text{ad inf.} \\ &= b^{-1}c\cdot d^{-1}ab^{-1}d^{-1}a\ldots = k(P_2), \\ s_{2n+1} &\to a^{-1}d\cdot(b^{-1}c)_{I^{-1}}(b^{-1}c)_{I^{-3}}\ldots\text{ad inf.} \\ &= a^{-1}d\cdot ba^{-1}\cdot c^{-1}ba^{-1}dba^{-1}dc^{-1}ba^{-1}d\ldots \\ &= k(P_4) = a^{-1}d(k(P_2))_{I^{-1}}. \end{aligned}$$

Now one easily convinces oneself that these four points are the only fixed points of g'^2E. One only needs to apply the automorphisms I'^2, I'^4, ... and I'^{-2}, I'^{-4}, ... to any element whose positive fundamental point falls into one of the four subintervals determined by P_1 to P_4, and to see that the expansion of the fundamental point converges to one of the four expressions above. We therefore have:

Under the boundary mapping induced by $I'^2 = (I_a)^2$ there are 4 aperiodic intervals, separated by the positive boundary fixed points P_1 and P_3 and the negative boundary fixed points P_2 and P_4. The method of §49 therefore gives index -1 for I'^2.

At the same time, we have found that the fixed point classes with index $+1$ for I and I' are different: For the boundary mapping induced by I there are four points, which are cyclically permuted by one step, but for the boundary mapping induced by I' there are two points, which change places. The two mappings are therefore not similar, and I and I' are therefore not isogredient.

If one applies the same method to I_c, the sequence r_n constructed from $h = c^{-1}$ leads to 4 accumulation points on E. Thus I_c, $(I_c)^2$ and $(I_c)^3$ have index $+1$, and $(I_c)^4$ has index -3, so I_c behaves like I. If one applies the method to $I_{b^{-1}}$, one finds that this automorphism behaves like I_a, namely, $I_{b^{-1}}$ has index $+1$ and $(I_{b^{-1}})^2$ has index -1. It remains to be investigated whether these two automorphisms are isogredient to the two previous ones.

We can, however, point out the following result:

The related automorphisms I^2 and $(I_a)^2$ each determine one fixed point class of the associated mapping class, and indeed, the one for I^2 has index $+1$ while the one for $(I_a)^2$ has index -1. The Alexander index sum for this mapping class is zero. This shows that it is not possible to derive an upper bound for the number of topologically necessary fixed point classes from the Alexander formula.

Added in Proof. The simultaneous appearance of a fixed point class of index $+1$ with one of index -1, where the latter (in contrast to the example above) arises from a *free fixed element group* with two generators, may be illustrated by the following simple example ($p = 2$):

$$I: \begin{cases} a \\ b \\ cdc^{-1}, \\ c^{-1} \end{cases} \qquad I^4: \begin{cases} a \\ b \\ k^{-1}ck, \\ k^{-1}dk \end{cases}$$

where $k = aba^{-1}b^{-1} = dcd^{-1}c^{-1}$. We have $\Xi(I) = 0$ by (50), §43. One finds the fixed element group $H(I)$ to be the free group $\{a, b\}$, and the method of §47 therefore gives the index -1. One also has

$$I_c: \begin{cases} c^{-1}ac \\ c^{-1}bc \\ d \\ c^{-1} \end{cases}, \qquad (I_c)^4: \begin{cases} kak^{-1} \\ kbk^{-1} \\ c \\ d \end{cases}.$$

Thus c is a fixed element for $(I_c)^4$ but not for I_c, so that I_c has index $+1$. The two classes cancel in the index sum Ξ. One sees that there need not be more classes as follows: Construct a torus T with canonical cut system c, d and apply the mapping σT characterized by

$$c \mapsto d, \qquad d \mapsto c^{-1}.$$

It follows from[17] §38, that σT has two fixed point classes of index $+1$, and σ may be constructed with this minimal number of fixed points. Let Q be one of these fixed points of σT. Since σ preserves orientation of T, Q may be extended to a simply connected fixed point region. One now constructs a second torus T' with canonical cut system a, b and lets $\sigma T'$ be the identity. This torus is joined to T at the boundary of Q after removal of a disc from T' and Q from T. The result is a double torus $T + T'$ on which $\sigma(T + T')$ belongs to the mapping class above and has only two fixed point classes.

This attachment of T' reduces the index of the class of Q by 2, hence to -1. In order to see this, one has to observe that this fixed point class of $\sigma(T + T')$ cannot be isolated on the universal covering surface of $T + T'$. In order to determine its index, one must modify $\sigma(T + T')$ in such a way that the mapping satisfies "Assumption α" of §37.

This example leads to the following general idea: One starts with a mapping $\sigma\phi$ of a surface ϕ of genus p and a mapping $\sigma\phi'$ of a surface ϕ' of genus p'. Suppose both mappings are orientation-preserving. Both involve fixed points, which one can always achieve by invoking, if necessary, fixed point classes of index 0. Then one joins the two surfaces at two fixed point discs, obtaining a mapping $\sigma(\phi + \phi')$ of a surface $\phi + \phi'$ of genus $p + p'$. One sees that this gives great freedom in the construction of examples of surface mappings with desired combinations of fixed point classes with different indices.

One also sees that this gives a new way to attack the general fixed point problem: When it is possible to find, for a given automorphism, a *fixed element* represented by a simple curve which is null *homologous* but *not* null *homotopic*, then one can let this curve consist of fixed points and contract it in space to a point. This pinches the surface into two parts, and one can now investigate the fixed point problem for the mapping classes of the subsurfaces, which are of lower genus.

Topological Mappings of a Jordan Curve onto Itself*

JAKOB NIELSEN

The intention of the following work is to prove and illuminate a theorem of H. Kneser, which the latter author introduced and proved briefly as a lemma in the work "Reguläre Kurvenscharen auf den Ringflächen" in Math. Annalen **91**, pp. 142–144.[1] In view of the intrinsic interest of this theorem, we shall indicate a different approach to it.

A Jordan curve can be mapped topologically, i.e., in a one-to-one and continuous manner, onto a circle. A topological mapping of a Jordan curve onto itself is transformed by the above mapping into a topological mapping of the circle onto itself. We therefore consider a circle C, subject to a topological mapping t onto itself, which we assume to be orientation-preserving. The iterated mappings are denoted by t^nC for arbitrary positive or negative integers n. If P is a point on C, then its image under t^n is denoted by t^nP. The mapping t^0 is the identity mapping of C onto itself.

1. We first assume that there are "fixed points" of tC, i.e., points P for which $tP = P$. The fixed points constitute a closed set on C, and they remain as fixed points for all powers t^nC. The complementary set is composed of a finite or denumerable number of fixed point free intervals. Such an interval is mapped onto itself by t^n and remains fixed point free under t^n $(n \neq 0)$, with all points of the interval being displaced in the same direction by t^n, when $n > 1$, as by t, and in the opposite direction when $n < 0$.

2. Next suppose that tC is fixed point free but not all its power are. Let $n > 1$ be the least positive exponent for which t^nC has fixed points, and let $t^nP = P$. Thus the n points $P, tP, \ldots, t^{n-1}P$ are all different, and they are all fixed points under t^n, since

$$t^nt^mP = t^mt^nP = t^mP.$$

The mapping t^n falls into the category 1 above, and the equation shows that its fixed point set is mapped onto itself by t and all its powers. It follows that we have a mapping of periodic character, in a certain sense. The points $P, tP, \ldots, t^{n-1}P$, in that order, are cyclically exchanged by t, so that the n intervals into which they divide C are permuted by t through a number of steps which is relatively prime to n. The fixed point sets of t^n in the individual intervals are mapped by t onto each other. The fixed point set of t^n can therefore be divided into n homeomorphic sections. One can now see how the whole mapping t^n has a "periodic character": namely, if one knows t^nQ for all points Q in one of the n intervals mentioned above, then t^n is obtained in any other interval by

$$t^nt^mQ = t^mt^nQ.$$

3. We come now to the last case, in which tC and all its powers t^nC $(n \neq 0)$ are fixed point free, and it is this case which forms the substance of the theorem mentioned in the introduction.

*Translation by John Stillwell of Om topologiske Afbildninger af en Jordankurve på sig selv. [Mat. Tidsskr. B 1928, 39–46.].

[1] H. Kneser's theorem is closely connected with investigations of H. Poincaré, E. E. Levi and P. Bohl, which are specified in the work of Kneser mentioned above.

To see, first of all, that there are mapping, of this kind, one chooses tC to be a rotation of C through $2\pi\gamma$, where γ is an irrational number. Then t^nC is a rotation through $2\pi n\gamma$, and this mapping is fixed point free for $n \neq 0$, since $n\gamma$ is not an integer. We can free this example of its special metric character. Namely, if τC is a topological mapping of C onto itself which preserves orientation but is not a rotation, then $\tau t \tau^{-1}$ is a topological mapping of C onto itself which is not a rotation, but whose powers are all fixed point free, since $(\tau t \tau^{-1})^n = \tau t^n \tau^{-1}$. This almost obvious remark is of interest because the theorem we are concerned with says that this last example brings us very close to the general case in 3.

4. Now let tC be an arbitrary topological mapping of C onto itself which is fixed point free, together with all its powers, so that $t^nP \neq P$ for all P and all $n \neq 0$. We choose a particular point P and construct the points t^nP for all n. These must all be different. Let M_P be the set of their accumulation points. M_P is closed. Since $tt^nP = t^{n+1}P$, the set of points t^nP is mapped one-to-one onto itself by t, and the same is the case for the set M_P of its accumulation points. Thus $tM_P = M_P$. If M_P does not make up the whole of C, then its complement $C - M_P$ consists of open intervals. The complement must likewise be mapped onto itself by t and all its powers. Let j be one of the complement intervals. If t^nj and j were the same complement interval, then its endpoints would be fixed by t^n, contrary to hypothesis. Now let Q be an arbitrary point on C. At most one point of the set t^nQ can lie in j, because t^nQ is mapped onto t^mQ by t^{m-n} and the latter maps j onto another interval. Thus no point of the complementary set $C - M_P$ can belong to the accumulation set M_Q. Hence M_Q is a subset of M_P. Since both P and Q were arbitrarily chosen points of C, it follows that $M_Q = M_P$. The accumulation set is therefore independent of the points used to construct it. It is characteristic of the mapping t. We call it $M(t)$. In the case where $M(t)$ does not make up all of C, we can choose a point Q in the complement. Since t^nQ has M as its accumulation set, one concludes that each point of M is an accumulation point of intervals. It follows that M contains no isolated points (at which two complement intervals are adjacent), and hence the complement is everywhere dense. $M(t)$ is therefore perfect, and it either makes up all of C or it is nowhere dense.

5. Let us apply this to the example mentioned in 3, where tC is a rotation of $2\pi\gamma$, and γ is an irrational number. Let us first use a point P to construct the set of points t^nP and its accumulation set M_P. Next, let Q be the point which results from P by a rotation through an arbitrarily chosen angle α. Then t^nQ results from t^nP, and hence M_Q results from M_P, by a rotation through α. On the other hand, we know that $M_P = M_Q = M(t)$. The point set $M(t)$ therefore has the property that it is brought into coincidence with itself under rotation through an arbitrary angle, i.e., it makes up the whole circle C. We have therefore arrived in this way at the well-known theorem, which can be formulated as follows: The set of all integral multiples of an irrational number is everywhere dense modulo 1.

6. We shall now construct an example of a mapping t for which $M(t)$ does not make up the whole circle. Let $\tau\Gamma$ be a rotation of a circle Γ through $2\pi\gamma$, where γ is irrational. Let P be an arbitrary point and let τ^nP be its everywhere dense set of images. Let C be another circle, and let i_0 be an interval on it which we let "correspond to" P. Around the midpoint of the remaining part of C and within its interior we take an interval i_1, which we let correspond to τP. The two remaining parts of C correspond to the arcs on Γ bounded by P and τP, in a unique way once we have carried over an orientation. Around the midpoint of the part corresponding to the arc containing $\tau^{-1}P$, and within its interior, we take a new interval i_{-1} which corresponds to $\tau^{-1}P$, and repeat the process for $\tau^2 P$, $\tau^{-2}P$, $\tau^3 P$, ... The result is an everywhere dense set of intervals on C, the cyclic order of which is the same as that of the set of points $\tau^n P$ on Γ. Now we construct a topological mapping tC as follows: Each interval i_n is mapped topologically onto i_{n+1} in any way which preserves orientation. (If one wants to keep a specific example in mind, let, e.g., each interval be mapped proportionally onto its successor.) This defines t on all intervals and their endpoints. Every other point of C can be determined as a cut in the set of intervals, and it corresponds to the point on Γ determined by the corresponding cut in the set $\tau^n P$. The mapping of the latter point by τ now determines the mapping of the former point by t. Thus we have defined a topological mapping tC, the powers of which are all fixed point free, and it is clear that the interior points of the intervals i do not belong to $M(t)$; however, all other points of C belong to $M(t)$.

One can vary this example by choosing a point Q on Γ outside the set $\tau^n P$ and constructing $\tau^n Q$ for all n, then choosing a point R outside the sets $\tau^n P$ and $\tau^n Q$, constructing $\tau^n R$, and continuing in this way for a finite or denumerable number of steps. One numbers the points obtained on Γ in any order and then makes an analogous construction of an everywhere dense set of intervals on E with preservation of the cyclic order. One then lets intervals correspond under tC if the associated points correspond under $\tau\Gamma$, thereby constructing a tC whose $M(t)$ has complementary intervals which do not form a single cycle, but rather a finite or denumerable number of cycles.

7. Conversely, if one begins with a mapping tC whose powers are all fixed point free and whose $M(t)$ is not all of C, then one can number the everywhere densely distributed intervals comprising its complement and let them correspond to an everywhere dense set of points in the same cyclic order on a circle C'. After this is done, the given mapping tC has completely determined a topological mapping $t'C'$ of C' onto itself, whose powers are all fixed point free and for which $M(t')$ makes up all of C'.

H. Kneser's theorem can now be formulated as follows:

Given a topological mapping tC whose powers are all fixed point free and for which the accumulation set $M(t)$ makes up all of C, then there is a topological mapping τ of C onto a circle Γ such that $\tau t \tau^{-1}\Gamma$ is a rotation of Γ through $2\pi\gamma$, where γ is irrational.

Of course, one can also let C and Γ be the same circle, as in 3.

In view of what we have found above, the same theorem also holds when $M(t)$ does not make up all of C, provided one allows τ to map complement intervals of $M(t)$ on C to points on Γ.

8. We now go to the proof of the theorem, and first investigate the value of the constant γ. For a rotation, γ is the angle (apart from a factor $1/2\pi$) through which an arbitrary point on the circle is turned by the mapping, or, what comes to the same thing, $1/n$ of the angle through which a point is turned by the nth iterate of the mapping. We use the latter idea in defining γ for a given mapping, by iterating it and dividing the resulting arc lengths by the exponents. In order to keep account of whole turns we introduce a many-valued angle variable α on the circle C and set $\alpha/2\pi = x$. This makes the x-axis a "covering line" of C, and points whose x-values differ by an integer correspond to the same point of C.

Corresponding to the given topological mapping $P \mapsto tP$ on C we now have a topological mapping $x \mapsto f(x)$ of the x-axis, and in fact infinitely many; namely, if $f(x)$ is such a function, then so is $f(x)$ + an integer. One chooses a particular function and now calls it $f(x)$. It obviously satisfies the functional equation

$$f(x+1) = f(x) + 1. \tag{1}$$

The mapping t^2C now corresponds to the function $f(f(x))$, which we write $f^2(x)$, and t^nC corresponds to $f^n(x)$; the latter function obviously satisfies the functional equation which follows from (1):

$$f^n(x+k) = f^n(x) + k, \tag{2}$$

where k is an arbitrary integer. Equation (2) shows that f^n maps an interval of the x-axis of integral length k onto an interval of the same length k, and hence it maps an interval of nonintegral length onto an interval whose length has the same integral part as the original. Now we choose a particular x-value, say $x = 0$. By the hypothesis on t, $f^n(0)$ is not an integer for any $n \neq 0$. Let p be the integer for which

$$p < f^n(0) < p + 1. \tag{3}$$

Then the interval from 0 to $f^n(0)$ will be mapped by $f^n, f^{2n}, \ldots, f^{(m-1)n}$ onto intervals whose lengths lie between the same integers. Addition of all these lengths gives

$$mp < f^{mn}(0) < m(p+1). \tag{4}$$

If we now set

$$\frac{f^n(0)}{n} = \gamma_n,$$

then (3) and (4) show that

$$|\gamma_n - \gamma_{mn}| < \frac{1}{|n|},$$

whence

$$|\gamma_m - \gamma_n| = |\gamma_m - \gamma_{mn} + \gamma_{mn} - \gamma_n| < \frac{1}{|m|} + \frac{1}{|n|}$$

and we have proved that the sequence γ_n is convergent. We set

$$\gamma_n \to \gamma \qquad \text{as} \quad |n| \to \infty.$$

If we choose an arbitrary value x instead of 0, then we can restrict it to the range between 0 and 1 by (2), and we get

$$\left|\frac{f^n(x) - x}{n} - \frac{f^n(0)}{n}\right| = \frac{1}{|n|}|(f^n(x) - f^n(0)) - (x - 0)| < \frac{1}{|n|},$$

since $x - 0$ and $f^n(x) - f^n(0)$ are both positive and <1. Thus

$$\frac{f^n(x) - x}{n} \to \gamma \qquad \text{as} \quad |n| \to \infty \quad \text{uniformly in} \quad x, -\infty < x < \infty.$$

If one replaces $f(x)$ by $f(x) + k$, then γ is replaced by $\gamma + k$. If one replaces $f(x)$ by $f^n(x)$, which satisfies the same hypothesis, since t^nC satisfies the same conditions as tC, then γ is replaced by $n\gamma$. By our hypothesis, $f(x) - x$ has a fixed sign and its absolute value has a positive minimum δ. Hence $|f^n(0)| \geq |n|\delta$ and $|\gamma| \geq \delta$. Thus γ cannot be 0. Neither can it be rational. Namely, if the γ-value of $f(x)$ were the rational p/q, then the γ-value of $f^q(x)$ would be p and the γ-value of $f^q(x) - p$ would be 0. Thus γ is irrational.

9. We now construct the function $\xi = g(x)$ with which H. Kneser maps the x-axis topologically onto a ξ-axis so that the mapping f of the x-axis becomes a translation of the ξ-axis along itself through distance γ. On the x-axis we consider the points $f^n(0) + k$ for all n and k. These points are all different, since $f^m(0)$ and $f^n(0)$ never differ by an integer. One first defines $g(x)$ on the given point set by letting

$$g(f^n(0) + k) = n\gamma + k.$$

On this point set, $g(x)$ satisfies the two functional equations:

$$g(x + 1) = g(x) + 1, \tag{5}$$

$$g(f(x)) = f(x) + \gamma. \tag{6}$$

Now we have to prove that $g(x)$ is a monotonically increasing function. Let us suppose that

$$f^q(0) + h > f^n(0) + k,$$

which, by applying f^{-n} to both sides, we can write

$$f^{q-n}(0) > k - h.$$

We have to deduce from this that $q\gamma + h > n\gamma + k$, i.e.,

$$(q - n)\gamma > k - h.$$

This will be proved when we show that $f^n(0)$ and $n\gamma$ lie between the same integers, independently of the choice of n. But this follows from (4):

$$p < \frac{f^{mn}(0)}{m} < p + 1$$

and, as $m \to \infty$,

$$\frac{f^{mn}(0)}{m} \to n\gamma,$$

so that $p < n\gamma < p + 1$, and the inequality is strict because γ is irrational. This gives the desired result when combined with (3).

Now we use the hypothesis we have not used so far, namely, that the accumulation set $M(t)$ makes up all of C. This means that the set $f^n(0)$ is everywhere dense modulo 1, and that the set $f^n(0) + k$ is everywhere dense on the x-axis. On the other hand, the set of points $\xi = n\gamma + k$ is everywhere dense on the ξ-axis, since γ is irrational. Thus we have a one-to-one order-preserving map g of an everywhere dense set on the x-axis onto an everywhere dense set on the ξ-axis, and this extends continuously to a topological mapping $\xi = g(x)$ of the x-axis onto the ξ-axis. Since g then satisfies the functional equation (5) for all x, g gives a topological mapping τ of C onto a circle Γ covered by the ξ-axis. And since g also must satisfy the functional equation (6) for all x, τ transforms the given topological mapping tC, by conjugation, into a rotation of Γ through an angle $2\pi\gamma$.

Thus we have proved the theorem: A topological mapping of a Jordan curve onto itself, whose powers are all fixed point free and for which the accumulation set is the whole curve, is the topological image of the rotation of a circle through an irrational multiple of π. If the accumulation set is not the whole curve, then the theorem still holds, provided each interval in the complement of the accumulation set is mapped to a single point on the circle.

Investigations in the Topology of Closed Orientable Surfaces, II*

JAKOB NIELSEN

Contents

Introduction

In a work which has appeared under the same title[1], the author has investigated topological properties of surfaces of genus $p > 1$ in relation to the universal covering surface and the

*Translation by John Stillwell of Untersuchungen zur Topologic der geschlossenen Flächen, II. [Acta Math., 53 (1929), 1–76.]

[1] Untersuchungen zur Topologie der geschlossenen zweiseitigen Flächen, Acta Mathematica, vol. 50, [N18]. References to this work in what follows will be made by I, with the appropriate paragraph.

fundamental group, with applications to the general fixed point problem. In what follows these investigations are extended and brought to a conclusion in a certain direction. For the sake of simplicity, consideration is restricted to *topological* (i.e., one-to-one and continuous) mappings of the surface onto itself. This is no essential restriction, as follows from the theorem proved recently by H. Kneser[2], that each *continuous* self-mapping of non-zero degree of a surface of genus $p > 1$ is continuously deformable into a topological mapping.

The interior, Φ, of the unit circle E in the complex plane is used as the image of the universal covering surface of the surface ϕ of genus $p > 1$, with x as the complex variable. The covering transformations are then linear fractional transformations of x of hyperbolic type with fundamental points on E, and in their entirety they constitute the fundamental group F of the surface ϕ. A topological mapping $\tau\phi$ of the surface onto itself is now covered by a denumerable family of topological mappings of Φ onto itself, whose elements differ only by covering transformations. Such a "T-function" $t\Phi$ satisfies a system of functional equations

$$t(f(x)) = f_I(t(x)),$$

where f is an arbitrary element of F and f_I denotes the image of f under an automorphism I of F. If kt is another T-function lying over τ, so that k is an arbitrary element of F, then this function induces the automorphism $I_{k^{-1}}$ which sends f to kf_Ik^{-1}. As k varies the result is a complete automorphism family, the members of which differ only by inner automorphisms, and this family determines the class of the mapping $\tau\phi$.

In the *first two* sections of the present work a *single* T-function $t\Phi$ is under investigation. This function may be extended, by a topological mapping of the boundary circle E onto itself which depends only on I, to a topological mapping $t(\Phi + E)$ of the closed disc onto itself (I, 28). As a result, all properties of the surface mapping $\tau\phi$ which are invariant under continuous deformation of τ, in other words, properties of the mapping *class*, are expressed in the boundary mapping tE. We confine ourselves to the case where the orientation is preserved by $\tau\phi$ and hence also by $t(\Phi + E)$, i.e., where I is of the "first kind" (I, 9).

In the *first section* we obtain a *division of all T-functions into finitely many types* on the basis of the structure of the fixed point set of the boundary mapping tE. Two functions of the same type have homeomorphic boundary fixed point sets. The case in which tE is the identity boundary mapping, so that I is the identity automorphism, has already been completely investigated in I, 42, and is therefore omitted from the present account. The set of "fixed elements" of the automorphism I, i.e., those elements h of F for which $h_I = h$, constitutes a subgroup $H(I)$ of F, which always contains at least the identity. The following two cases now have to be distinguished:

(a) $H(I)$ consists of the identity alone. This type occurs, e.g., when tE is fixed point free. However, $H(I) = 1$ can also occur when there are boundary fixed points which are not fundamental points, as the example in I, 49 showed. *In this case it turns out* (§6) *that the number of these boundary fixed points is always finite and even, and in fact this number* is $\leq 4(2p - 1)$. The direction of displacement of the boundary points under tE alternates in the intervals between the fixed points; one can therefore divide the fixed boundary points into "attracting" and "repelling" types. The non-Euclidean convex polygon with zero angles determined by the

[2] In a letter from Kneser.

attracting fixed points alone (or the repelling fixed points alone) contains no two points equivalent under F, and therefore covers a certain part of ϕ simply.

(b) $H(I)$ consists of more than the identity. Thus there are proper fixed elements under I, whose fundamental points are then fixed points of tE. *In this case it turns out* (§10) *that the fixed element group H is always a free group. The number v of its free generators satisfies* $1 \leq v \leq 2p - 1$. The closed hull of the set of fixed fundamental points was determined in I, 29 as the set of singular points with respect to H. When $v > 1$ this set is perfect and nowhere dense. This set determines a non-Euclidean convex region Δ with infinitely many boundary sides when one connects the endpoints of the denumerably many complementary intervals of the set by non-Euclidean lines. Δ is called the "kernel domain" of I. *The surface ϕ now contains* (§8) *a region δ bounded by finitely many disjoint, simple, mutually non-homotopic curves, which is covered infinitely often by Δ. $H(I)$ is the group of covering transformations of Δ over δ, and hence the fundamental group of δ.* The region δ contains all the closed geodesics on ϕ (in the sense of the non-Euclidean metric inherited from Φ) covered by axes of elements of H, and δ is the smallest region with this property, *since its boundary curves are themselves such geodesics.* In special cases, δ can consist of the whole surface ϕ, cut along one or more closed geodesics.

It also turns out, (§5) *that regular points with respect to H cannot be accumulation points of boundary fixed points.* Thus when the intervals complementary to the set of singular points are not fixed point free, they can contain only isolated boundary fixed points, *whose number in a fundamental domain of H is finite and even* (§§7 and 11). Similar considerations to (a) then apply.

One notices that the distinction between cases (a) and (b) is unnecessary when one speaks of the group consisting only of the identity as the "free group with zero generators." The fixed element group $H(I)$ of a non-identity automorphism I is then always a free group with v generators, where $0 \leq v \leq 2p - 1$, and the cases with the equality sign really occur. The part of E which borders a fundamental domain of H—this fundamental domain is the whole disc for $v = 0$—then contains only a finite, and in fact even, number 2μ of boundary fixed points, where μ can be 0.

When we have a T-function with more than two fixed boundary points, we can construct a non-Euclidean convex region Ω from the non-isolated fixed points and the isolated attracting fixed points, which we call the "principle domain" of the T-function (§3). An important tool in the investigation is then the fact that Ω is mapped onto itself by each fixed element, and onto a region outside Ω by each other element of F. The kernel domain Δ is either a proper subregion of Ω or identical with Ω.

In the *second chapter* these results are applied to the *fixed point problem, so far as this concerns the individual T-function.* The set of those fixed points of $\tau\phi$ which are covered by the fixed point set of $t\Phi$ are collected into a *fixed point class* (I, 31). They occur simply in a fundamental domain of $H(I)$ (I, 33), since the fixed point set of $t\Phi$ is reproduced by all fixed elements, and by no other elements of F. The *index* of the fixed point class is defined as the total angle variation of the vector from x to $t(x)$ in a circuit around a fundamental domain of H (I, 37). Generalizing the arguments following Example 11 in (I, 48), it is now shown (§§13–14) how one can avoid possible isolated boundary fixed points on the boundary of the fundamental domain of H by small arcs when determining the index. *As a result, the index i depends only on the type of the boundary mapping, and in fact*

$$i = 1 - v - \mu,$$

where v is again the number of free generators of the fixed element group and μ is half the number of boundary fixed points in its fundamental domain. One sees that a positive index $i = 1$ can occur only in the case of a fixed point free boundary mapping ($v = \mu = 0$). If one recalls the restrictions on v and μ which follow from the above-mentioned properties of the region Ω, *then it turns out* (§15) *that the index i of a fixed point class on a surface of genus* $p > 1$ *satisfies the condition*

$$3 - 4p \leq i \leq 1.$$

The case $i = 0$ requires $v = 1, \mu = 0$ or $v = 0, \mu = 1$. *T-functions with index* 0 *therefore always have exactly two boundary fixed points*, where these two can determine two periodic, or two aperiodic, complementary intervals (I, 14).

One can make this result particularly intuitive by extending the T-function by reflection in E to a topological mapping of the whole complex x-sphere K (I, end of 28) and then constructing the surface $K_I = K \bmod H(I)$ (I, 30). K_I is a closed surface of genus v which, when S denotes the set of singular points with respect to H, is covered by $K - S$ with the boundary set S, and H is the covering transformation group of $K - S$ over K_I. A mapping tK_I of K_I onto itself is induced by $t(x)$, and belongs to the class of the identity (§12). On K_I there are finitely many disjoint simple closed curves covered by the regular points of E, and on them are 2μ isolated fixed points of tK_I, each of which has index $+1$. This fact, together with the symmetry of the two halves of the sphere and the Birkhoff formula, gives the above value for the index of the T-function. If one joins each attractor among the 2μ fixed points to a repeller by a handle on which the mapping can be continued without fixed points, then one obtains a surface K_I^* of genus $v + \mu$ which depends only on I, on which the mapping is a deformation (§16). Then, because of the reflection in E, half of the Birkhoff index $2 - 2(v + \mu)$ is the index $i = 1 - v - \mu$ of the T-function. For $i = 1$, K_I^* is a sphere, for $i = 0$ it is a torus, for $i < 0$ a surface of genus q, where $2 \leq q \leq 4p - 2$. *One can therefore say that each index determination for an (orientation preserving) mapping problem in two dimensions reduces to an application of the Birkhoff formula to a certain auxiliary surface which depends only on the class of the mapping in question.*

The *third chapter* investigates the complete family of T-functions which cover a given mapping $\tau\phi$. This requires consideration of just one representative from each isogredience class (I, 18 and 31). The decisive fact here is that the principal domains of two related automorphisms do not overlap, hence the pieces of ϕ they cover must also be disjoint (§§17–19). This completely solves the fixed point problem for a rather wide set of mapping classes, namely, all those which can be generated by "twists" along a system of disjoint simple curves; these are the ones whose kernel domains make up all of ϕ (§22).

There is also the *group theoretic consequence, that an automorphism family can contain only finitely many non-cyclic, mutually non-conjugate fixed element groups* (§19), since the numbers v_i of their free generators must satisfy

$$\sum_i (v_i - 1) \leq 2p - 2.$$

One can draw conclusions about the "frequency" of individual types of T-functions within a complete family from the disposition of the principal domains, and then relate this to the Alexander index sum Ξ (§§20–21). Since a fixed point class with non-zero index must be represented by at least one point, the *minimum number* of fixed points for a given mapping class is $\geq$ the number Z of mutually non-isogredient automorphisms with $i \neq 0$ in the given family.

Now $i \leq 1$ immediately implies $Z \geq \Xi$, which of course only says something for $\Xi > 0$. But when $\Xi < 0$ has a numerically large value (and one can easily give examples with arbitrarily large negative values of Ξ) then it emerges that "almost all" classes with negative index contribute -1 to Ξ. More precisely, the result is the following (§20):

$$Z \geq \Xi \quad \text{if} \quad \Xi \geq 0,$$
$$Z \geq 1 \quad \text{if} \quad -4(p-1) \leq \Xi < 0,$$
$$Z \geq |\Xi| - 4(p-1) \quad \text{if} \quad \Xi < -4(p-1).$$

In this way one is able *to estimate from below the class number Z with the help of the index sum* Ξ, a fact which is noteworthy because Z is an invariant of the mapping class, i.e., it is the same for any two mappings which send an arbitrary closed curve to *homotopic* curves, whereas Ξ is an invariant of a broader collection of mapping classes, in which images are required to be *homologous* rather than homotopic[3]. The question whether Ξ *may also be used to estimate Z from above* must be left open here, since it is closely connected with the unsolved problem of whether there are automorphisms for which all powers induce fixed point free boundary mappings; in other words, whether there are fixed point classes of index $+1$ which maintain this index under arbitrarily high iterations of the surface mapping.

CHAPTER I

Division of the Automorphisms into Finitely Many Types

1. Lemmas. In order not to interrupt the presentation later, we deal with some elementary results in advance.

Let h_1 and h_2 be two hyperbolic transformations which map the unit circle E in the complex x plane into itself, so that they have their fundamental points on E. Suppose they have no fundamental point in common, and that the positive fundamental points of h_1 and h_2 are not separated by the negative fundamental points. By the "arc b between the positive fundamental points" we mean the arc of E between these points which does not contain the negative fundamental points. The transformation h_2 maps b onto a subarc $h_2 b$ of itself, which adjoins the positive fundamental point of h_2. Then h_1 maps $h_2 b$ onto an arc $h_1 h_2 b$ which is an interior subarc of b. Thus some interior point of b remains fixed under $h_1 h_2$. Consideration of the inverse operation shows similarly that some interior point of the arc b_1 between the negative fundamental points of h_1 and h_2 remains fixed under $h_1 h_2$. Therefore $h_1 h_2$ is a hyperbolic transformation whose positive resp. negative fundamental point is an interior point of b resp. b_1. Applying this to the elements of the fundamental group F we formulate the

Lemma 1. If f_1 and f_2 are two elements of F with distinct fundamental points, and if the positive fundamental points of f_1 and f_2 are not separated by the negative fundamental points, then the positive fundamental point of the element $f_1 f_2$ is an interior point of the arc between the positive fundamental points, and the negative fundamental point of $f_1 f_2$ is an interior point of the arc between the negative fundamental points of f_1 and f_2.

[3] The importance of this fact has also been emphasized to me by H. Hopf.

Let h be a hyperbolic transformation with fundamental points $U(h)$ and $V(h)$ on E, let A be the axis of h and let D be the diameter of E perpendicular to A. The circles orthogonal to the pencil through $U(h)$ and $V(h)$ are permuted by h ("level lines" of the "flow" h). Among these is a circle C around $U(h)$ whose image hC around $V(h)$ is the reflection of C in D (I, end of 2). C and hC may be called the "congruence circles" of h, since the Euclidean line element on C is congruently transformed by h. Now let δ be a hyperbolic transformation with axis D and the endpoint P of D as positive fundamental point. Then $\delta h \delta^{-1}$ is a hyperbolic transformation whose displacement distance equals that of h, whose axis is δA and whose congruence circles are the circles δC and $\delta hC = \delta h \delta^{-1} \cdot \delta C$. The exterior of the circle δC is therefore mapped by $\delta h \delta^{-1}$ into the interior of the circle δhC. If one now lets the displacement length of δ grow beyond all bounds, then $\delta U(h)$, $\delta V(h)$, δC and δhC converge to P, and one sees that, for a δ with large displacement length, a point which does not belong to the immediate neighbourhood of P is moved by $\delta h \delta^{-1}$ into the immediate neighbourhood of P. More generally, the same holds for any sequence h_n of hyperbolic transformations of equal displacement length when both fundamental points of h_n converge to a point P of E as n increases. Increasing the displacement length of a hyperbolic transformation draws the congruence circles closer to the fundamental points. The above result therefore also holds when we do not assume equality for the displacement lengths of the h_n, but only a positive lower bound. In particular, this is the case for the elements of F (I, beginning 5):

Lemma 2. When f_n, $n = 1, 2, \ldots$ is a sequence of elements of F such that both fundamental points of f_n converge to a point P of E as n increases, then

$$\lim_{n\to\infty} f_n x = P \tag{1a}$$

for each point x different from P.

It follows from the definition of "fundamental sequence" (I, 7(j)) that the sequence f_n of Lemma 2 is a fundamental sequence for P. The validity of (1a) then follows immediately for all x not on E. What matters here is just the extension of (1a) to E-points, and in the case of Lemma 2, to all E-points different from P.

We now consider a fundamental sequence f_n for an E-point P more generally. The positive fundamental points of the sequence satisfy $V(f_n) \to P$ as $n \to \infty$ (I, 7(j)). Nothing has yet been said about the behaviour of the negative fundamental points $U(f_n)$. We now assume that $U(f_n) \to Q$ as $n \to \infty$. Lemma 2 deals with the case $Q = P$. We therefore assume now that Q is a point of E different from P. The sequence of axes of the f_n then converges to the non-Euclidean line QP. Let α be a circle concentric with E, but smaller, which meets QP and hence also all axes sufficiently far down the sequence. In each complete system of equivalent axes in F there are only finitely many which enter the interior of α. Each element of the sequence f_n is therefore conjugate to at most finitely many other elements of the sequence. Thus the displacement length of the sequence grows beyond all bounds (I, 5), and when C_n and $f_n C_n$ are the congruence circles of f_n, $C_n \to Q$ and $f_n C_n \to P$. Each point different from Q is therefore moved into the immediate neighbourhood of P by all f_n sufficiently far down the sequence:

Lemma 3. When f_n, $n = 1, 2, \ldots$ is a sequence of elements of F such that $V(f_n) \to P$ and $U(f_n) \to Q$ as $n \to \infty$, then

$$\lim_{n\to\infty} f_n x = P \tag{1}$$

for each point x different from Q.

This lemma includes Lemma 2 when $Q = P$. Nothing is said about the behaviour of the sequence of points $f_n Q$ in either case. If $f_n Q$ does not converge to P, then Q is called the *exceptional point* of the fundamental sequence f_n for P. If $f_n Q \to P$, so that (1) holds for all x without exception, we say that the fundamental sequence f_n for P has no exceptional point.

2. Disposition of equivalent boundary fixed point sets. Let tx be any T-function (I, 21). It maps the interior Φ of the unit circle E topologically onto itself and satisfies the functional equation

$$tfx = f_I tx. \tag{2}$$

The correspondence $f \mapsto f_I$ is an automorphism I of F (I, 22); $t\Phi$ is extended continuously to E by a topological mapping tE which depends only on I (I, 28). We confine ourselves throughout to the case where t is orientation-preserving, in other words, where I is of the first kind (I, 9); then the topological mapping $\tau\phi$ of the surface ϕ onto itself, over which $t\Phi$ lies, is also orientation-preserving. In this paragraph and the next we also assume that the boundary mapping tE has more than two fixed points. The fixed point set of the boundary mapping will be denoted by M. M is closed. The case where M is all of E, so that I is the identity automorphism, is completely treated in I, 42, and is excluded from now on. Then M is nowhere dense. Also, the group $H(I)$ of elements of F fixed by I is a proper subgroup of F, including at least the identity.

In this paragraph we investigate the position of M relative to an equivalent set fM, where f is an arbitrary element of F. First of all, (2) shows that a fixed element sends a fixed point to another fixed point. Thus if $f_I = f$, then $fM = M$. We therefore assume $f_I \neq f$.

Repeated application of (2) gives

$$t^n fx = f_{I^n} t^n x, \qquad n \gtreqless 0. \tag{3}$$

If one applies this to a point m of M the result is

$$t^n fm = f_{I^n} t^n m = f_{I^n} m. \tag{4}$$

We now investigate the sequence of elements f_{I^n}. Since $f_I \neq f$, the fundamental points $U(f)$ and $V(f)$ belong to the set $E - M$, which consists of finitely or denumerably many intervals. The endpoints of the fixed point free (under t) interval u containing $U = U(f)$ are the points

$$U^+ = \lim t^n U \quad \text{and} \qquad U^- = \lim t^{-n} U, \quad n \to \infty.$$

Similarly, the endpoints of the interval v which contains $V = V(f)$ are

$$V^+ = \lim t^n V \quad \text{and} \qquad V^- = \lim t^{-n} V, \quad n \to \infty.$$

Now since

$$\left.\begin{array}{l} t^n V(f) = V(f_{I^n}) \\ t^n U(f) = U(f_{I^n}) \end{array}\right\} n \gtreqless 0,$$

it follows from §1 that f_{I^n} is a fundamental sequence for V^+ as $n \to \infty$, with U^+ as the single possible exceptional point, and as $n \to -\infty$ it is a fundamental sequence for V^-, with U^- as the single possible exceptional point. Now if m is a point of M different from U^+ and U^-, then it follows from (4) and Lemma 3 that:

$$\lim t^n fm = \lim f_{I^n} m = \begin{cases} V^+ & \text{as } n \to \infty \\ V^- & \text{as } n \to -\infty. \end{cases}$$

This shows that fm is an interior point of v, because, since M has more than two points, v is the only fixed point free interval under t which has V^+ and V^- as endpoints. Thus we have

Theorem 1. *When the boundary fixed point set M of a T-function contains more than two points, its image fM under a non-fixed element f lies entirely in the fixed point free interval containing the positive fundamental point of f, with the possible exception of the two points of M which bound the fixed point free interval containing the negative fundamental point of f.*

3. Principal domain of an automorphism. It now remains to investigate the special position of the points U^+ and U^- more closely. This requires us to distinguish between three cases for the relative positions of the intervals u and v.

Case (a). u and v are the same interval, which therefore contains both fundamental points of f. We have $U^+ = V^+$ and $U^- = V^-$. Let the closed interval which results from v by addition of the endpoints be $\bar{v}$. If one replaces t by t^{-1}, which indeed makes no difference to the relative positions of M and fM, then V^- and V^+ change places. We can therefore assume that the four points V^-, $U(f)$, $V(f)$, V^+ lie in that order on $\bar{v}$ (Figure 1). Then fV^+ lies in v. If fV^- also lies in v, then all of fM lies in v, and the exceptional case of Theorem 1 does not occur. We therefore pursue the assumption that fV^- falls outside v, and first ask whether $fV^- = V^+$ is possible. This is settled by the following general consideration:

Let $t(\Phi + E)$ be an arbitrary T-function with a fixed point p and f a non-fixed element. In order that fp also be a fixed point of t,

$$tp = p,$$

$$tfp = fp,$$

on the other hand, by (2):

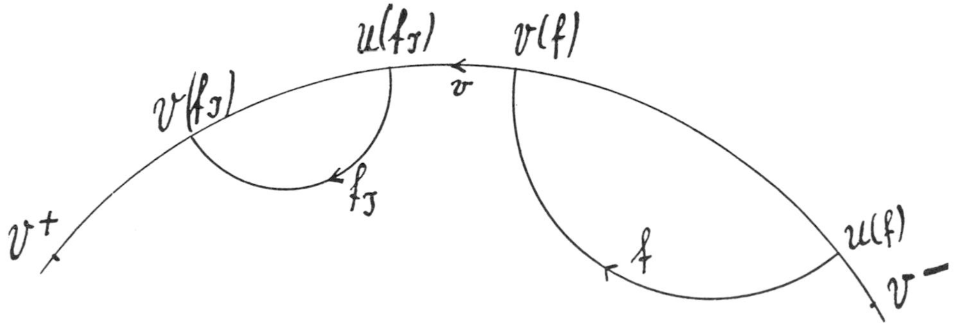

Figure 1

$$tfp = f_I tp = f_I p;$$

hence

$$f^{-1} f_I p = p.$$

Now $h = f^{-1} f_I \neq 1$, since f is not a fixed element. Thus p is a fundamental point, so it lies on E, and since it is a fixed point it belongs to the boundary fixed point set M. Consequently, h is a fixed element. Since fp is a fundamental point of fhf^{-1}, the latter must likewise be a fixed element. Thus f sends the fundamental points of the fixed element $h = f^{-1} f_I$ to the fundamental points of the fixed element $fhf^{-1} = f_I f^{-1}$, which are different.

In our case (a) it follows from this that $fV^- \neq V^+$. Because the points $U(f)$ and $V(f_I)$ enclose the points $V(f)$ and $U(f_I)$ in v, whether or not the axes of f and f_I intersect. The positive fundamental points of the elements f^{-1} and f_I are therefore not separated by the negative fundamental points, and hence by Lemma 1 the negative fundamental point of $f^{-1} f_I$ is a point of v (between $V(f)$ and $U(f_I)$). Thus $f^{-1} f_I$ cannot be a fixed element.

Finally, it remains to pursue the assumption that fV^- falls outside $\bar{v}$. The equation

$$\lim_{n \to \infty} t^n f V^- = \lim_{n \to \infty} f_{I^n} V^- = V^+$$

remains in force, since $V^- = U^-$ is not an exceptional point as $n \to +\infty$. Consequently V^+ is also an endpoint of a fixed point free interval in $E - \bar{v}$, whose points flow towards V^+ under t. We say: V^+ is an *isolated, bilaterally attracting fixed point*. Now $f^{-1} V^+$ also falls outside $\bar{v}$. Thus if one replaces t by t^{-1} for a moment, and f by f^{-1}, it follows from what has just been proved that V^- is an isolated bilateral attractor under t^{-1}, and hence *under t it is an isolated, bilaterally repelling fixed point*. In summary, we can say that in case (a) M is mapped by f into v, with at most the exception of an isolated, bilaterally repelling fixed point.

For the cases which follow one notes that the equations

$$\left.\begin{aligned} t^n f U^- &= f_{I^n} U^- \to V^+ \\ t^{-n} f U^+ &= f_{I^{-n}} U^+ \to V^- \end{aligned}\right\} n \to \infty \qquad \begin{aligned}(5)\\(6)\end{aligned}$$

hold, since U^- is not an exceptional point for f_{I^n} and U^+ is not an exceptional point for $f_{I^{-n}}$ $(n \to +\infty)$. Now when u and v are different and the directions of movement under t are different in these two intervals, it follows immediately from (5) and (6) that fU^-, as well as fU^+, lies in v, and hence so does all of fM. This settles:

Case (b). u and v have a common endpoint, and it is either $U^- = V^-$ or $U^+ = V^+$. Namely, the directions of movement under t are opposite in u and v, and hence fM lies entirely in v.

The cases $U^- = V^+$ and $U^+ = V^-$, in which the direction of movement is therefore the same in the adjacent intervals u and v, as follows from the theorem derived in the next paragraph, are incompatible with the assumption of more than two fixed boundary points.

What remains now is:

Case (c). u and v are not adjacent. U^+, U^-, V^+ and V^- are therefore four different points on E. When the direction of movement under t is different in u and v, then fM lies entirely in v, as remarked above. We therefore assume that the direction of movement is the same in u and v, and hence that U^- and V^- are separated by U^+ and V^+ (Figure 2).

It is now a possibility that $fU^- = V^+$. Then U^- is a fundamental point of a fixed element

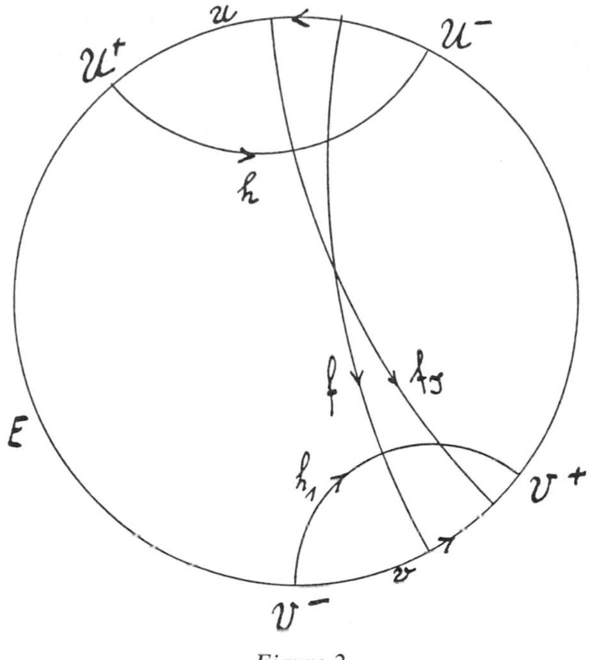

Figure 2

$h = f^{-1}f_I$, as found above, and indeed U^- must be the positive fundamental point of this element, as follows from Lemma 1 when one considers the positions of the axes involved. Thus u is a periodic interval (I, 14) and U^+ is the negative fundamental point of h. V^- and V^+ are the negative and positive fundamental point, respectively, of the fixed element

$$h_1 = fhf^{-1} = f_I f^{-1}.$$

When this situation occurs—and Example 2 in (I, 15) shows that it can—one has the case where M and fM "touch," namely at a pair of associated fundamental points, which separate the remainder of M from the remainder of fM. This is the only case in which M and fM can have points in common for a non-fixed element f. In this case the four points U^-, U^+, V^- and V^+ are singular points with respect to a non-cyclic fixed element group (I, 29) and hence non-isolated boundary fixed points.

When fU^- does not lie in $\bar{v}$, it follows in a similar way, from (5), that V^+ is an isolated, bilaterally attracting fixed point, and U^- is an isolated bilaterally repelling fixed point. It likewise follows from (6) that, when fU^+ does not lie in $\bar{v}$, U^+ is an isolated bilateral attractor and V^- is an isolated bilateral repeller.

In order to summarize these results we introduce the following concept:

Definition. If the boundary fixed point set $M = M(t) = M(I)$ of an I-inducing T-function t consists of more than two points, then we understand $M^* = M^*(t) = M^*(I)$ to be the subset of M which results by omission of all isolated, bilaterally repelling fixed points.

At most one endpoint of a fixed point free interval is omitted. Each accumulation point of omitted points is therefore an accumulation point of points not omitted; one need only replace each omitted point by one of the two others which bound intervals in conjunction with the omitted point. M^* therefore has the same accumulation points as M, and is likewise closed. It follows from (2) that a fixed element h sends one isolated, bilaterally repelling fixed point to another. Thus $hM^* = M^*$ for each fixed element h.

Now one can summarize the results of the investigation of the above three cases as follows:

Theorem 2. *When the boundary fixed point set $M(I)$ for an automorphism I consists of more than two points, $fM^*(I)$ is identical with $M^*(I)$ for each fixed element f, and for each non-fixed element f it is mapped into the closure of a subinterval of $E - M^*$. When M^* and fM^* have points in common for a non-fixed element f, these are a pair of associated fixed fundamental points which separate the remainder of M^* from the remainder of fM^*.*

The set M^* is closed, as already remarked. It contains at least two points, since M contains more than two points and at most one endpoint is omitted from each interval complementary to M. Now one connects the endpoints of each of the finitely or denumerably many intervals which make up $E - M^*$ by a circular arc orthogonal to E. The subregion Ω of Φ which lies outside all these circles is convex in the sense of the non-Euclidean metric, with boundary sides which are infinite non-Euclidean lines. It may happen that M^* consists of only two points, and in this case a domain does not arise, but only a non-Euclidean line. It will not lead to misunderstanding if, for the sake of uniform terminology, we speak of a "domain" Ω even when it consists of a line. Ω depends only on I, since the boundary mapping depends only on I.

Definition. The region $\Omega = \Omega(t) = \Omega(I)$ defined above is called the "principal domain" of the automorphism I or the T-function t. One speaks of a principal domain only when the induced boundary mapping has more than two fixed points.

Since M^* is mapped onto itself by a fixed element of I, the same is true for Ω. Since, by Theorem 2, the point sets M^* and fM^* lie in complementary arcs of E when f is a non-fixed element, it follows that Ω and $f\Omega$ have no interior point in common. However, they can meet along a common boundary side, as the conclusion of Theorem 2 shows. Thus an arbitrary element of F sends a boundary side of Ω either to another boundary side or to a line outside Ω; it therefore intersects none of the lines equivalent to itself. Each boundary side of Ω therefore lies over a simple open or closed geodesic on the surface ϕ. Two geodesics on ϕ which arise in this way are either identical or disjoint. These geodesics bound a region ω on ϕ which is covered by Ω. Two equivalent interior points of Ω must be equivalent under a fixed element. Ω therefore covers ω simply if $H(I) = 1$, and otherwise infinitely often:

Theorem 3. *The principal domain $\Omega(I)$ is the universal covering surface of a region $\omega(I)$ on the surface ϕ bounded by disjoint geodesics. The fixed element group $H(I)$ is the covering transformation group of $\Omega(I)$ over $\omega(I)$. If I has no non-trivial fixed element, then ω is simply connected; ω degenerates to a line when Ω does the same.*

In special cases, ω can make up the whole surface ϕ, cut along one or more closed geodesics. This is the case when Λ is the automorphism treated in I, 15; then $\omega(\Lambda)$ results from cutting the surface ϕ along the axis of the element b_1. Namely, M is perfect, and hence identical with M^*, and all the intervals in the complement of M are periodic and spanned by the axes of

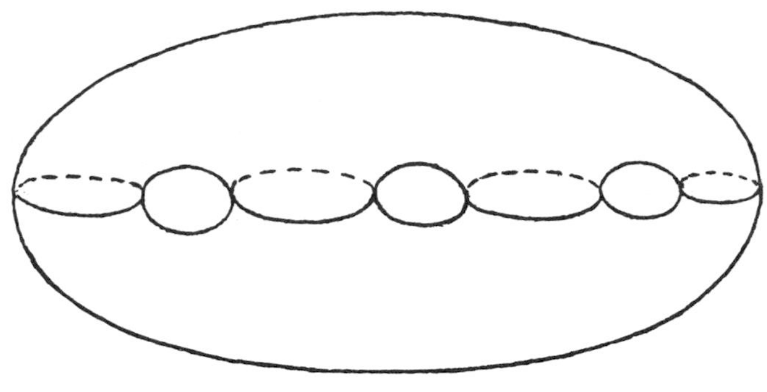

Figure 3

conjugates of b_1. Ω and $a_1^{-1}\Omega$ meet along the axis of b_1. The fundamental group of $\omega(\Lambda)$ is the free group $H(\Lambda) = \{b_1, a_2, b_2, \ldots, a_p, b_p\}$.

4. A theorem on isolated boundary fixed points. A surface ϕ of genus p may be divided into four simply connected pieces by $2p + 2$ simple closed surves, e.g., as shown in Figure 3 by the $p + 1$ horizontal curves and the $p + 1$ vertical curves. This system of $2p + 2$ curves is called Σ. Each of these curves is homotopic to exactly one closed geodesic (in the sense of the metric transported from Φ) on ϕ, namely, the one covered by the axes of the elements in the conjugacy class of the curve in question (I, 4). Each of these geodesics is simple, since it is homotopic to a simple curve in Σ. Two of these geodesics meet either once or not at all, according as this is the case for the homotopic curves in Σ or not. We can therefore assume that the curve system Σ in Figure 3 consists of closed geodesics. The axes in Φ lying over Σ divide Φ into infinitely many regions, of which any four meeting at a point make up a fundamental domain of F.

Now let P be an arbitrary point of E and let R be the radius leading to P. If a point x traverses the radius R in the direction of P, then it passes through infinitely many of the regions into which Φ is divided. The point on ϕ covered by x must therefore cross at least one of the curves in Σ infinitely often. R therefore cuts a system of infinitely many equivalent axes. Since only finitely many equivalent axes enter a circle which is concentric with E, but smaller, the sequence of equivalent axes cut by R must contract towards P. If one orients the curves of Σ, then one can decide whether R crosses the axes in question from left to right or right to left. One of the two must occur infinitely often, and one can therefore confine attention to an infinite subsequence in which the crossings are all the same:

Lemma 4. For each point P of E one can determine a sequence of conjugate, primary elements f_n of F such that $U(f_n) \to P$ and $V(f_n) \to P$, and the convergence of these two point sequences to P is from opposite sides. Moreover, the closed geodesic on ϕ covered by the axis of f_n is simple.

Now let P be an isolated boundary fixed point of a T-function t, and suppose that the boundary points flow towards P on one side, and away from P on the other, under t. This is the case when P is the single boundary fixed point of t, or when two different fixed point free intervals with the same displacement direction are adjacent to P. We now determine a sequence of equivalent axes which contract towards P, as in Lemma 4. The fundamental points

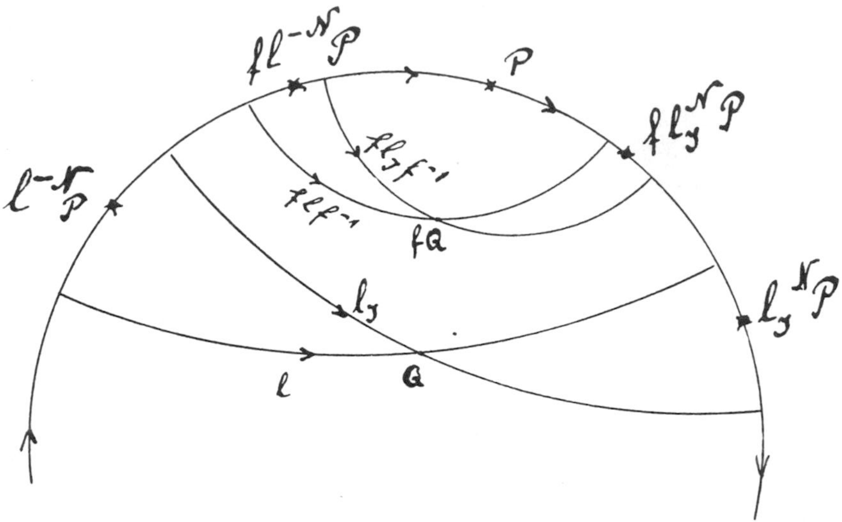

Figure 4

of the latter finally all belong to the fixed point free intervals adjacent to P. Let l be such an element of the sequence and let l_I be the element which corresponds to it under the automorphism I induced by t. By the hypothesis on the displacement direction, the axes of l and l_I intersect (Figure 4). Now let λ and λ' be the simple closed geodesics on ϕ covered by the axes of l and l_I, respectively. All axes of the sequence we are using lie over λ, the axes of the corresponding elements under I lie over λ'. The intersection of a pair of corresponding axes therefore lies over an intersection of λ and λ', and there are only finitely many of the latter. At least one of them must therefore occur infinitely often in the sequence. We can therefore refine the sequence so that all intersections of corresponding pairs of axes are equivalent. Now let Q be the intersection point of the axes of l and l_I and let fQ be the intersection of a pair which is later, i.e., closer to P. Then flf^{-1} and $fl_I f^{-1}$ are the two elements conjugate to l and l_I whose axes meet at fQ. On the other hand, the latter element corresponds to the former under I. Thus

$$f_I l_I f_I^{-1} = fl_I f^{-1}$$

and hence

$$f^{-1} f_I = l_I^n$$

or

$$f_I = fl_I^n \tag{7}$$

for some n, where we can assume l, and hence l_I, is primary (I, 5).

Since l is conjugated by f into flf^{-1}, $V(f)$ lies on the arc containing P between the fundamental points of flf^{-1}, and $U(f)$ lies on the arc not containing P between the fundamental points of l. Since l_I is conjugated by f into $fl_I f^{-1}$, it follows likewise that $V(f)$ and $U(f)$ lie between the fundamental points of $fl_I f^{-1}$ and l_I respectively. Thus $V(f)$ lies on the arc containing P between $U(fl_I f^{-1})$ and $V(flf^{-1})$, and $U(f)$ lies on the arc not containing P

between $U(l)$ and $V(l_I)$. See Figure 4, where the axis of f is not shown, in order to avoid an assumption on the relative position of $V(f)$ and P. Now since $V(f_I)$ either coincides with $V(f)$ (namely, when $P = V(f)$) or lies to the right of $V(f)$ in the figure, Lemma 1 shows that only exponents $n \geq 0$ can occur in (7).

It will now be shown that exponents n in question can also be bounded from above. First, it follows from (7) by application of I^{-1} and multiplication by l^{-n} from the right that

$$f_{I^{-1}} = fl^{-n}. \tag{7a}$$

Now let N be a positive number, so large that $l_I^N P$ lies between $V(l)$ and $V(l_I)$ and $l^{-N}P$ lies between $U(l)$ and $U(l_I)$. Then $fl_I^N P$ falls between $V(flf^{-1})$ and $V(fl_I f^{-1})$ and $fl^{-N}P$ falls between $U(flf^{-1})$ and $U(fl_I f^{-1})$. The arc between P and $V(l_I)$ is mapped by fl_I^N onto its subarc between $fl_I^N P$ and $V(fl_I f^{-1})$, and therefore the last-mentioned arc contains $V(fl^{-N})$. Now $V(f_{I^{-1}})$, $V(f)$ and $V(f_I)$ either coincide with P, or else they lie in this order on the same side of P. Thus the n in (7) and (7a) must lie below the value N. Here N depends only on the relative positions of P and the axes of l and l_I, not on f.

Now one lets f run through a sequence of elements which yields a sequence of axes for flf^{-1} contracting towards P in the above manner. Each f corresponds to one value of n in (7). Since n is bounded independently of f, we can find two different f, say f' and f'', which give the same value of n. Then it follows from

$$f'^{-1}f'_I = f''^{-1}f''_I = l''_I$$

that

$$f''f'^{-1} = (f''f'^{-1})_I.$$

Thus $f''f'^{-1}$ is a fixed element $\neq 1$. Suppose that the axis of $f''lf''^{-1}$ is closer to P than that of $f'lf'^{-1}$. The fixed element $f''f'^{-1}$ transforms the axis of $f'lf'^{-1}$ into the axis of $f''lf''^{-1}$, and hence it has its positive fundamental point on the arc containing P between the fundamental points of $f''lf''^{-1}$. But P is the only fixed point on this arc. Thus P is a fundamental point of the fixed element $h = f''f'^{-1}$. Hence tE has exactly two fixed points, namely the fundamental points of h; if there were a fixed point Q different from these, all points $h^n Q$ would also be fixed points, and these would accumulate at P as $n \to \infty$, whereas P is isolated:

Theorem 4. *There is no boundary mapping with only one fixed point. When a boundary mapping has an isolated fixed point at which the displacement direction is the same in both the adjacent fixed point free intervals, then it has exactly two fixed points, and these are the fundamental points of a cyclic fixed element group.*

Example 9 in I, 46 shows that the situation in Theorem 4 really can occur.

5. Regular boundary fixed points. When an automorphism I is given, the points of E may be divided into regular and singular points with respect to $H(I)$. A point is called regular when it has a neighbourhood u such that all elements of $H(I)$ other than 1 map u outside itself; otherwise it is called singular. All fundamental points of elements of H are singular, likewise the accumulation points of such points, and this exhausts the set $S = S(I)$ of singular points (I, 29). If $H(I) = 1$ then S is empty. If $H(I)$ is cyclic, then S consists of 2 points. If $H(I)$ contains more than a cyclic group, and hence more than one primary element, then S is perfect and nowhere dense. The complementary set $E - S$ consists of intervals which will be called "intervals of regularity." S is a subset of the boundary fixed point set M and, when M contains more than 2 points, so that M^* is defined, it is also a subset of M^*, since S in this case contains no isolated points of M, and hence no points of $M - M^*$. We now have

Theorem 5. *A regular boundary fixed point is always isolated, and either bilaterally attracting or bilaterally repelling.*

In order to prove this theorem we make use of surface area in the non-Euclidean metric, denoting the area of any polygon $\mathfrak{P}$ by $\mathrm{Ar}(\mathfrak{P})$. If $\mathfrak{P}$ is a triangle, then it is known that

$$\mathrm{Ar}(\mathfrak{P}) = k(\pi - \Sigma),$$

where Σ denotes the angle sum of $\mathfrak{P}$ and k is a constant. We set $k = 1/\pi$, thus normalising the surface area so that a triangle with zero angles (i.e., one formed by three circular arcs which are orthogonal to E and tangential to each other there) has surface area of unity. For an arbitrary triangle $\mathfrak{P}_3$ we now have

$$\mathrm{Ar}(\mathfrak{P}_3) = 1 - \frac{\Sigma}{\pi}$$

and for an arbitrary n-gon $\mathfrak{P}_n$ (using division into $n - 2$ triangles)

$$\mathrm{Ar}(\mathfrak{P}_n) = n - 2 - \frac{\Sigma}{\pi}, \tag{8}$$

where Σ denotes the angle sum of the n-gon. In I, 3 we constructed a $4p$-gon with angle sum 2π as fundamental domain for F. Thus the area of the surface ϕ of genus $p > 1$, in the sense of the metric of constant negative curvature transported from Φ, is obtained from (8) as:

$$\mathrm{Ar}(\phi) = 4(p - 1). \tag{9}$$

Now let P be a regular boundary fixed point and let u be a neighbourhood of P, bounded by a circle orthogonal to E, such that all images of u under elements $\neq 1$ in $H(I)$ are disjoint from u. Contrary to the assertion of Theorem 5, we assume that P is an accumulation point of M, and hence also of M^*. One chooses in u an arbitrary number n of points from M^* and makes them the vertices of a non-Euclidean convex n-gon $\mathfrak{P}_n$ with zero angles. $\mathfrak{P}_n$ is a subregion of the principal domain $\Omega(I)$, since its vertices are boundary points of this convex region. Now one considers $f\mathfrak{P}_n$ for all $f \neq 1$ in F. If f is not a fixed element, then $f\mathfrak{P}_n$ lies outside $\mathfrak{P}_n$, since $f\Omega$ lies outside Ω (§3). If f is a fixed element, then $f\mathfrak{P}_n$ also lies outside $\mathfrak{P}_n$, since $\mathfrak{P}_n$ is in u. Thus $\mathfrak{P}_n$ covers a subregion of ϕ simply. This contradicts (9), since by (8)

$$\mathrm{Ar}(\mathfrak{P}_n) = n - 2$$

and n was chosen arbitrarily. Thus P is an isolated boundary fixed point. Since it is regular, P is not a fundamental point. Hence it follows from Theorem 4 that P is either two-sided attracting or two-sided repelling. This proves Theorem 5.

In I, 34 it was proved that a boundary fixed point P which is regular with respect to $H(I)$ cannot be an accumulation point of fixed points of $t\Phi$, when t is an arbitrary I-inducing T-function. This theorem has now been extended to show that P cannot be an accumulation point of fixed points of $t(\Phi + E)$. If one extends t by reflection in E, as in I, 28, to a function tK defined on the whole complex x-sphere K, then P is an isolated fixed point of tK.

6. Type construction for $H(I) = 1$. Now suppose $H(I) = 1$. Thus I has no proper fixed elements. The singular set S is empty, and E consists entirely of regular points. Then there can be *only finitely many boundary fixed points*, since a regular point cannot be an accumulation point of fixed points by Theorem 5. When there are any boundary fixed points at all, then by Theorem 5, one endpoint of a fixed point free interval is bilaterally attracting and the other is bilaterally repelling. Fixed points of these two kinds must therefore alternate on E, and their total number must be even. Let it equal 2μ. In order to construct M^*, one has to leave out every second fixed point and retain only the bilateral attractors. M^* therefore consists of μ points. Observing the convention that M^* is meaningful only when M contains more than two points, this means that $\mu \geq 2$. The principal domain $\Omega(I)$ is therefore a μ-gon with zero angles. By (8) we have

$$\mathrm{Ar}(\Omega) = \mu - 2. \tag{10}$$

By Theorem 3, Ω must *simply* cover a simply connected region ω on ϕ in our case, hence

$$\mathrm{Ar}(\Omega) \leq \mathrm{Ar}(\phi).$$

It then follows from (9) and (10) that

$$\mu \leq 4p - 2.$$

The total number of boundary fixed points is therefore $\leq 8p - 4$.

Theorem 6. *The boundary mapping induced by an automorphism of the first kind without proper*[4] *fixed elements has 2μ alternately attracting and repelling fixed points, where*

$$0 \leq \mu \leq 4p - 2.$$

Thus there is a division of such automorphisms into $4p - 1$ different "types," according to the value of μ.

None of the 2μ boundary fixed points is a fundamental point, since there are no fixed elements. The region $\omega(I)$ defined on ϕ for $\mu > 1$ reduces to a line for $\mu = 2$, and it is a simple, open geodesic. In case $\mu > 2$ the boundary sides of ω are disjoint, simple, open geodesics, any one of which approaches its successor asymptotically, creating an infinitely long "cusp" of the simply connected region ω. Two such cusps do not approach each other asymptotically on ϕ. Such an approach would say that the corresponding vertices of Ω were equivalent points; because of the absence of fixed elements they would have to correspond under a nonfixed element, and this is excluded by Theorem 2, since they are not fundamental points.

The following remarks may be made on the realisation of the different types in Theorem 6: One finds an example for $\mu = 0$ in I, 44, an example for $\mu = 1$ is the automorphism Λ' of Example 3, I, 15. For $\mu = 2$, a linear ω-region, an example is the automorphism I'^2 of Example 14, I, 51. For $\mu > 2$ we refer to the automorphism $J = I^4$ of Example 13, I, 49, for which $\mu = 4$ when $p = 2$. However, I should explicitly mention that *no* example for which the equality sign

[4] I.e., different from the identity.

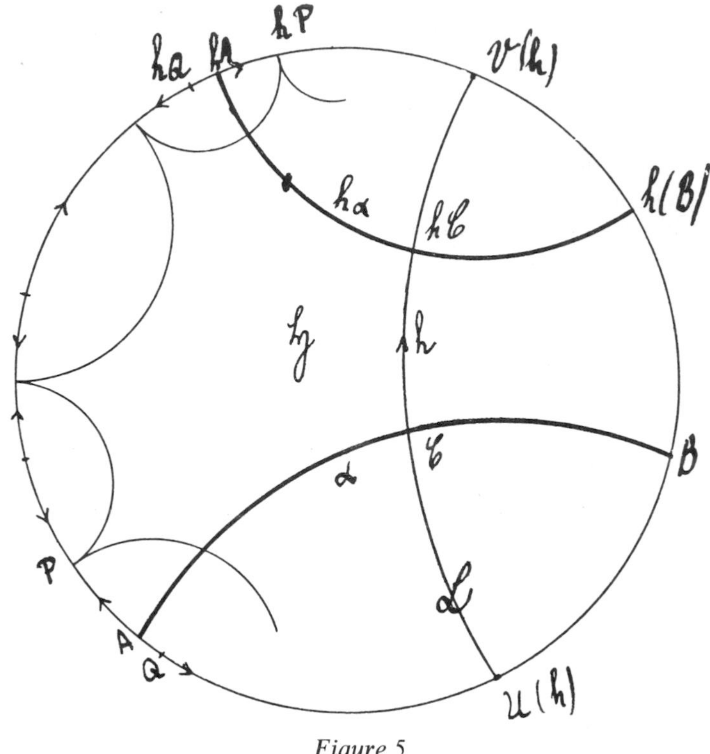

Figure 5

occurs in (11) is known to me. In this case, if it can occur at all, the region ω must fill the surface ϕ up to a null set of very complicated structure.

7. Type construction for cyclic $H(I)$. Now assume that $H(I)$ is cyclic. Let h be the primary fixed element, so that $H(I) = \{h\}$. $U(h)$ and $V(h)$ are the only singular points (Figure 5). A fundamental domain $\mathfrak{H}$ for H lies between an arc α orthogonal to E which cuts the axis of h orthogonally, but is otherwise arbitrary, and $h\alpha$. We consider one of the intervals of regularity between $U(h)$ and $V(h)$, say the one lying on the left in the figure. A and hA bound a piece of it which is fundamental with respect to H. It contains only finitely many fixed points, by Theorem 5, and they are bilaterally attracting or bilaterally repelling. The two kinds must alternate, and their number on the interval (A, hA) must be even. Because if the first point P (see Figure 5) is attracting, then so also is hP, since the T-function t commutes with h. Let the number of fixed points on (A, hA) be $2\mu_1$.

If $\mu_1 > 0$, then the boundary mapping has infinitely many fixed points, which accumulate at the singular points $U(h)$ and $V(h)$. One constructs their principal domain Ω (§3) using the attracting points alone, as is shown in Figure 5 with $\mu_1 = 3$, where some boundary sides are drawn. The part of Ω contained in $\mathfrak{H}$ and lying to the left of the axis of h is, as is clear from the figure, a $(\mu_1 + 4)$-gon with angle sum 2π, since there are μ_1 zero angles and the remaining angles form two supplementary pairs. It follows from (8) that the surface area is μ_1. Similarly,

μ_2 is the area of the polygon common to Ω and $\mathfrak{H}$ on the right of the axis, where $2\mu_2$ is the number of fixed points on the arc (B, hB). Ω and $\mathfrak{H}$ therefore have an area of size $\mu = \mu_1 + \mu_2$ in common. By Theorem 3 this region covers the region $\omega(I)$ on ϕ simply. It now follows from

$$\mathrm{Ar}(\omega) \leq \mathrm{Ar}(\phi)$$

and (9) that

$$\mu \leq 4p - 4. \tag{12}$$

The total number 2μ of boundary fixed points belonging to $\mathfrak{H}$, i.e., the number of those which are non-singular and inequivalent with respect to H, is therefore $\leq 8(p-1)$. By Theorem 2, two such distinct points cannot be sent to each other by non-fixed elements.

Theorem 7. *The number of inequivalent regular fixed points of a boundary mapping induced by an automorphism of the first kind with a cyclic fixed element group is 2μ, where*

$$0 \leq \mu \leq 4p - 4.$$

Thus the automorphisms of this kind are divided into $4p - 3$ different types, according to the value of μ.

The division into types can be refined according to the summands μ_1 and μ_2 of μ in the two intervals of regularity.

One can visualise Theorem 7 by constructing the surface $K_1 = K \bmod H$ as in I, 30, where K denotes the sphere of complex x-values. K_1 is a torus. On K_1 there are two closed curves, each of which is covered by one of the intervals of regularity infinitely often. One is covered once, e.g., by the arc (A, hA), the other once by the arc (B, hB). An I-inducing T-function t induces a mapping tK_1 on K_1 belonging to the class of the identity[5]. Along the curves mentioned, t depends only on I, and 2μ is the number of fixed points of tK_1 lying on the two curves, corresponding to the boundary fixed points in $\mathfrak{H}$.

8. Kernel domain of an automorphism. Now we finally investigate the case in which $H(I)$ contains more than a cyclic group. The set S of singular points with respect to H is perfect and nowhere dense. S is a subset of M and, since S contains no isolated points, and hence no points of $M - M^*$, it is also a subset of M^*. Now one connects the endpoints of each of the infinitely many intervals of regularity which make up $E - S$ by a non-Euclidean line. The result is a non-Euclidean convex region Δ with infinitely many sides. Since S is a subset of M^*, Δ is a subregion of the principal domain Ω. (When $M = M^* = S$ we have $\Delta = \Omega$.) If $f_I \neq f$ then Δ and $f\Delta$ do not overlap, since Ω and $f\Omega$ do not overlap. If $f_I = f$ then not only is $fM^* = M^*$, but also $fS = S$. Because S is the closed hull of the set of fixed fundamental points (I, 29), and since the set of fixed fundamental points is mapped onto itself by a fixed element, the same is the case for S.

Definition. The region $\Delta = \Delta(t) = \Delta(I)$ is called the "kernel domain" of the automorphism I or T-function t. One speaks of a kernel domain only when the fixed element group of the automorphism contains more than a cyclic group. The kernel domain is a part of the principal domain and can sometimes coincide with it.

[5] See Theorem 12.

When Δ and $f\Delta$ touch for $f_I \neq f$, then Ω and $f\Omega$ must also touch, since Δ is contained in Ω. Conversely, when Ω and $f\Omega$ touch, it follows from Theorem 2 that this happens along the axis of a fixed element, and hence Δ and $f\Delta$ also touch.

A fixed element h maps S onto itself and hence also maps the set of intervals of regularity onto itself. If i is such an interval and if i and hi have points in common then $hi = i$, so the endpoints of i are the fundamental points of h, and i can be called a "periodic interval of regularity." (Cf. I, 14, where the same consideration is applied to $E - M$, and I, 29.) Now in the next paragraph it will be shown that each interval of regularity is periodic. In this respect the set $E - S$ behaves differently from the set $E - M$, with which aperiodic intervals can also occur; indeed, in the case of §6, for example, all fixed point free intervals are aperiodic. The theorem just mentioned now yields the following fact: Each boundary side of Δ is an axis, and therefore lies over a closed geodesic on ϕ. The latter is simple, because the boundary sides of Δ meet none of their equivalent axes, as follows from the relative position of Δ and $f\Delta$ discussed above. For the same reason, any two of these curves are disjoint on ϕ. No two of them are homotopic, without being identical, since each curve type contains only one closed geodesic. *Thus they are finite in number, since ϕ has finite genus.* There are therefore only finitely many inequivalent intervals of regularity. We now have the following theorem corresponding to Theorem 3:

Theorem 8. *The kernel region $\Delta(I)$ of an automorphism I whose fixed element group contains more than a cyclic group is the universal covering surface of a part $\delta(I)$ of ϕ bounded by finitely many simple and disjoint closed geodesics. The fixed element group $H(I)$ is the covering transformation group of Δ over δ.*

Δ contains all axes of fixed elements, since their fundamental points are boundary points of the convex region Δ. Thus δ contains all closed geodesics corresponding to the fixed elements. One can easily prove the following theorem on the distribution of these axes, which we mention here without proof, since it will not be used in what follows: If i and j are two intervals on E, each of which contains singular points, then there is a fixed element with one fundamental point in i and the other in j.

9. Periodic character of the intervals of regularity. In this paragraph we prove the theorem already used above:

Theorem 9. *Each interval of regularity is periodic.*

In the case of a cyclic fixed element group the theorem is correct, since there are then exactly two singular points, which are associated fixed fundamental points, hence they determine two periodic intervals of regularity (§7).

Thus we assume $H(I)$ contains more than a cyclic group. Then there are no isolated singular points, and hence no adjacent intervals of regularity. Any two boundary sides of Δ therefore have a common perpendicular. Two separating pairs of boundary sides of Δ have common perpendiculars which intersect. The length of at least one of these must then lie above a certain constant depending only on the metric. To see this, one constructs a triangle with angles $\pi/4$, $\pi/2$ and 0 and reflects it repeatedly in the sides forming the angle $\pi/4$. In this way one obtains eight triangles around the vertex of this angle, which together form a zero angled quadrilateral. The common perpendiculars of its opposite sides bisect each other orthogonally and have

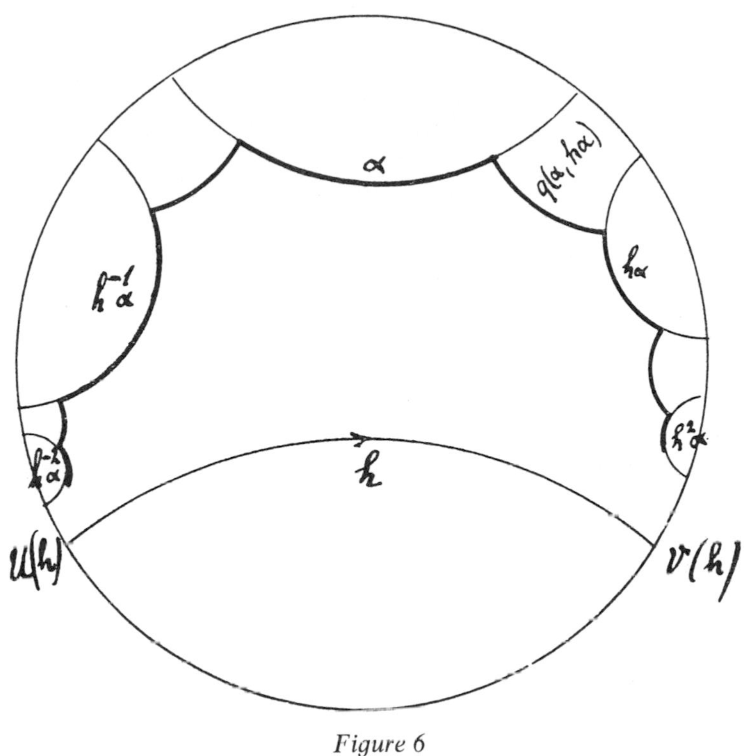

Figure 6

length κ, where $\kappa/2$ is the side of the original triangle between the angles $\pi/4$ and $\pi/2$. Thus one can use κ for the above-mentioned constant.

Now let α be any boundary side of Δ and let h be any fixed element. Then $h\alpha$ is also a boundary side of Δ. When $h\alpha$ is different from α, so that α is not the axis of h, let $q(\alpha, h\alpha)$ be the common perpendicular of α and $h\alpha$, which we call the "crosspiece of Δ associated with α and $h\alpha$." The image of $q(\alpha, h\alpha)$ under h^n is $q(h^n\alpha, h^{n+1}\alpha)$. Now one draws the boundary sides $h^n\alpha$ of Δ for all $n \gtreqless 0$ and all $q(h^n\alpha, h^{n+1}\alpha)$ (Figure 6). All these crosspieces and the intervening "side pieces," i.e., subarcs of the $h^n\alpha$, together form a simple curve in Φ whose ends tend to $U(h)$ and $V(h)$, and we call it an "h-α-chain." (It is drawn heavily in Figure 6.) Now let β be a boundary side of Δ different from all $h^n\alpha$ but on the same side of the axis of h as α. Suppose, e.g., that β lies between $h^{-1}\alpha$ and α. Then $h\beta$ lies between α and $h\alpha$, and hence $q(\beta, h\beta)$ and $q(\alpha, h\alpha)$ intersect. At most one of these two crosspieces can then be $<\kappa$. One calls a chain whose crosspiece length is $<\kappa$ a "distinguished chain," the corresponding axis a "distinguished axis," and the closed geodesic it covers on ϕ a "distinguished curve." If the h-α-chain is distinguished, then h is a primary element. Because if $h = k^n$, $n > 1$, then α and $k^n\alpha$ are separated by $k\alpha$ and $k^{n+1}\alpha$; the crosspieces $q(\alpha, k^n\alpha)$ and $q(k\alpha, k^{n+1}\alpha)$ therefore intersect; they are equally long, since k maps one onto the other, and their length must therefore be $>\kappa$. Consequently, when the h-α-chain is distinguished, there cannot be another distinguished chain on the same side of the axis of h.

For each distinguished axis there are at most two distinguished chains, since one is possible

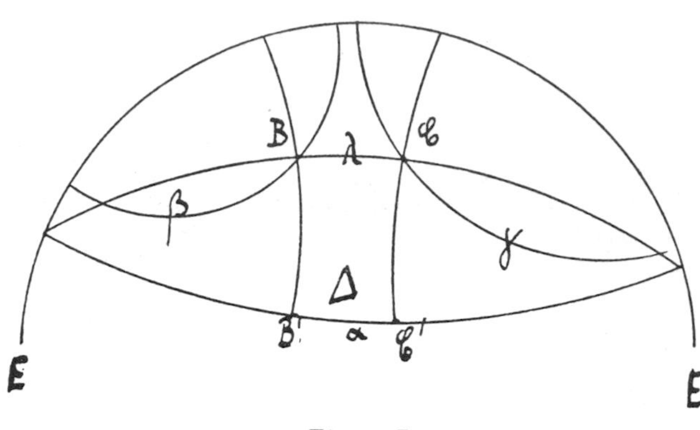

Figure 7

on each side. A fixed element maps a chain onto a chain with the same crosspiece length, and hence it sends one distinguished chain to another. When h is transformed by a non-fixed element f into a fixed element $h_1 = fhf^{-1}$, then the axes of h and h_1 are boundary sides of Δ; and when Δ lies to the right of the directed axis of h, it lies to the left of the directed axis of h_1 (Theorem 2). Thus h and h_1 each give rise to a chain only on one side. When they are both distinguished, then the two distinguished chains belong to different sides of the distinguished curve covered by the two axes. To summarise: For each distinguished curve there are one or two systems of equivalent distinguished chains, and thus one or two values for the lengths of the corresponding crosspieces.

Now one supposes, contrary to Theorem 9, that there is a boundary side α of Δ which is not an axis, so that it bounds an aperiodic interval of regularity. For each fixed element h, $h\alpha$ is a boundary side of Δ different from α. For each non-fixed element f, $f\alpha$ is a line lying entirely outside Δ, because Δ and $f\Delta$ can at most touch along an axis by Theorem 2. Thus α contains no equivalent points.

Let λ be a curve at constant distance ρ from α on the same side of α as Δ. Let Z be the strip of constant width between α and λ. We shall first show that the region common to Z and Δ has infinite area. Let β and γ (Figure 7) be two boundary sides of Δ which cut λ and therefore detach from Z two 2-gons not contained in Δ. There is a positive lower bound for the length of the segment BC on λ between β and γ. (Such a lower bound is realized, e.g., by the segment of λ between β' and γ' when β' and γ' are two lines which form a zero-angled triangle on base α.) The boundary sides of Δ cut by λ therefore do not accumulate on λ except possibly at the endpoints, which are also the endpoints of α. Now suppose that β and γ are successive boundary sides. Let B be the intersection of β and λ facing γ and let C be the intersection of γ and λ facing β. Let B' and C' be the projections of B and C on α (Figure 7). Then the figure bounded by the segment $B'C'$ on α, the segment BC on λ and the two perpendiculars BB' and CC' has an area greater than some positive number depending only on ρ. This figure is common to Z and Δ. Now even when infinitely many boundary sides of Δ enter Z, there is a figure of this kind between any two of them in succession. Thus Z and Δ have a region of infinite area in common.

Now a fundamental domain of F has finite area (see (9)). Thus the region common to Z and Δ has two equivalent interior points, P and Q. Since Δ has no interior point in common with

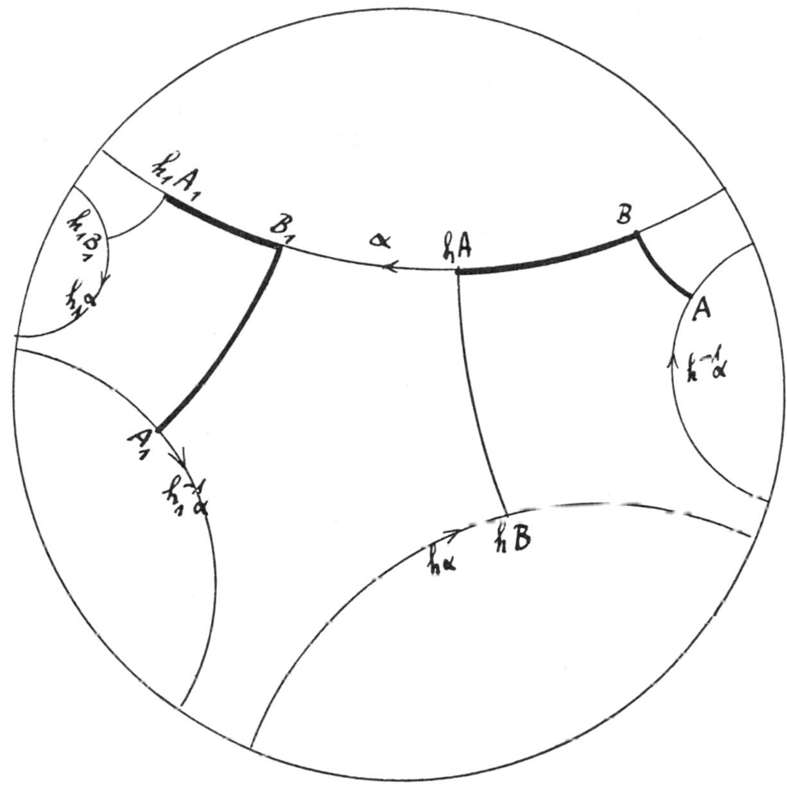

Figure 8

its image under a non-fixed element, we must have $Q = hP$, where h is a fixed element. We drop the perpendicular PP' from P to α. Its image under h is the perpendicular QQ' from Q to the boundary side $h\alpha$ of Δ, which is different from α. $PP' < \rho$, since P lies in Z, and hence $QQ' < \rho$ also. On the other hand, Q is at distance less than ρ from α, since Q is in Z. Thus α and $h\alpha$ approach each other within a distance less than 2ρ. Now one chooses $\rho < \kappa/2$ and determines a fixed element h such that (Figure 8)

$$AB = q(h^{-1}\alpha, \alpha) < 2\rho < \kappa.$$

The image of AB under h is the crosspiece $q(\alpha, h\alpha)$ from hA on α to hB on $h\alpha$. The crosspiece AB drawn heavily in the figure and the side piece BhA form, together with their images under all h^n ($n \gtreqless 0$), a distinguished chain. Thus h is primary. The heavily drawn arc $ABhA$ contains no equivalent points except the endpoints. Because the side piece contains no equivalent points, being a subarc of α; no point of the side piece can be equivalent to a point of the crosspiece, since a boundary point of Δ is not equivalent to an interior point of Δ; and finally the crosspiece cannot contain equivalent points, since it would then be cut by an equivalent crosspiece, which is excluded because its own length, and hence the length of its equivalents, is $<\kappa$. Thus h determines a simple curve type. The distinguished curve k covered by the axis of h is therefore simple.

Now one chooses $2\rho_1$ smaller than the two crosspiece lengths corresponding to k, draws the curve λ_1 at constant distance ρ_1 from α, and considers the part of the resulting strip of constant width which lies to the left of $hAhB$ in the figure. This half-strip Z_1 also has infinite area and therefore leads to the same conclusion as Z did previously. It has an infinite area in common with Δ. Thus there is an element h_1 such that $h_1\alpha$ and $h_1^{-1}\alpha$ project into Z_1, and this gives a distinguished h_1-α-chain, since $2\rho_1 < 2\rho < \kappa$. The fundamental piece $A_1B_1hA_1$ of this chain, drawn heavily in Figure 8, contains no equivalent points apart from its endpoints. The axis of the primary element h_1 therefore lies over a simple distinguished curve k_1. Moreover, k and k_1 do not meet. Because the same consideration as above shows that the fundamental pieces of the two chains, drawn heavily in the figure, contain no equivalent points apart from the pairs of endpoints. The choice of ρ_1 guarantees that k and k_1 are also not identical, since the crosspiece length of the h_1-α-chain corresponding to k_1 turns out smaller than the possible crosspiece lengths for the two sides of k.

Now one chooses $2\rho_2$ smaller than both the crosspiece lengths corresponding to k_1 and uses it to construct an h_2-α-chain, which determines a new distinguished curve k_2, and continues this process indefinitely. This gives arbitrarily many distinct, simple, disjoint, closed geodesics on ϕ. No two of them are homotopic. This result contradicts the finite genus of ϕ.

Thus Theorem 9 is proved.

10. Freeness of the fixed element group. Construction of its fundamental domain. We assume also from now on that $H(I)$ contains more than a cyclic group, and use Theorem 8 to construct the region δ on ϕ covered by the kernel region Δ. When two boundary sides of Δ correspond

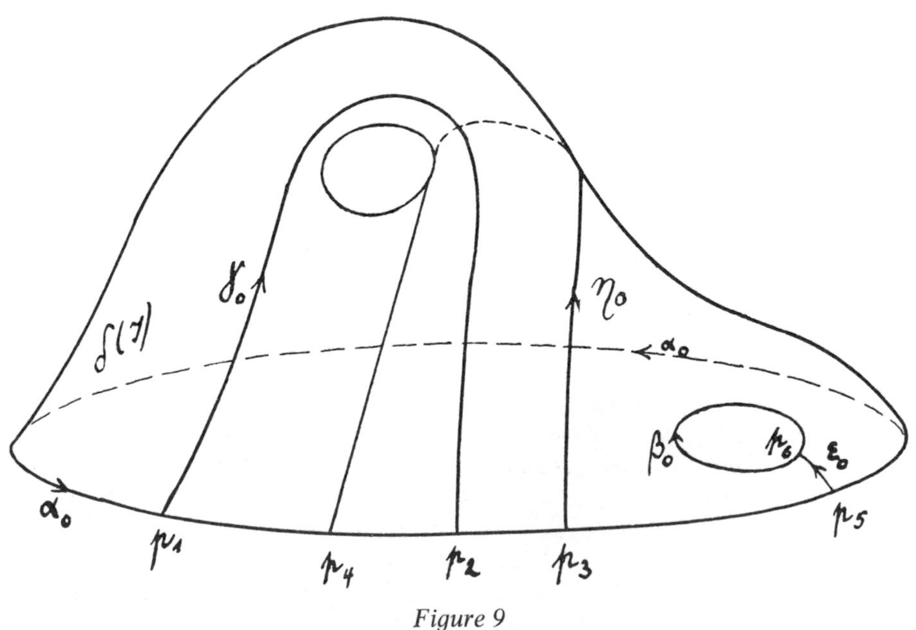

Figure 9

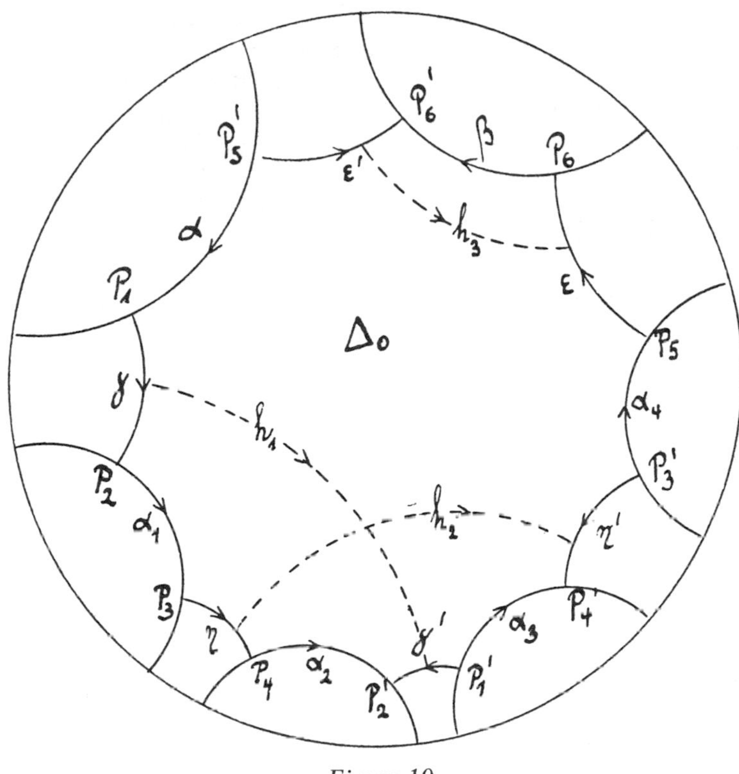

Figure 10

under a non-fixed element, the closed geodesic they cover on ϕ participates in the boundary from both sides and counts as two boundary curves. Let $r \geq 1$ be the number of boundary curves of δ and let q be the genus of δ. We have $q < p$, since there are always boundary curves. For $q = 0$, $r \geq 3$, because when δ is planar, $r \neq 1$ since a closed geodesic is not contractible, and $r \neq 2$ since two closed geodesics are not homotopic. Thus we always have

$$v = 2q + r - 1 \geq 2.$$

Now we shall construct a subdomain Δ_0 of Δ which covers δ exactly once, by dissecting δ by crosscuts as in the Enzyklopädie article[6]. See Figures 9 and 10. For the sake of simplicity, we assume in the figures and text that $q = 1$ and $r = 2$, so that $v = 3$; but the process is of course quite general.

Let α_0 and β_0 be the two boundary curves of $\delta(I)$ (Figure 9). The non-separating crosscut γ_0 from the point p_1 on α_0 to the point p_2 on α_0 reduces δ to a planar region with three boundary curves. The crosscut η_0 from p_3 on α_0 to p_4 on α_0 connects the two pieces of α_0 formed by γ_0. Finally, one makes the crosscut ε_0 from p_5 on α_0 to p_6 on β_0. These three crosscuts reduce δ

[6] M. Dehn and P. Heegaard, Analysis Situs. Math. Enzykl. III, **1**, 1, p. 200. In the formula at the bottom of p. 198, r should be read instead of $2r$. Cutting along a crosscut reduces K by 1.

to a simply connected region. Moreover, one can choose these three crosscuts to be the shortest lines of their type; then these geodesic segments are perpendicular to α_0 and β_0, since the latter are closed geodesics on ϕ. One can easily see this possibility directly; it results incidentally from the corresponding construction in Φ, which we now pursue.

Let α (Figure 10) be any one of the boundary sides of Δ which lie over α_0. If α_0 is oriented so that it has δ on its left, then α likewise has Δ on its left. Let P_1 be any one of the points of α which lie over p_1. These choices completely determine the domain Δ_0 which covers δ simply. When we traverse the boundary of the dissected δ from p_1 and initially travel along γ_0, keeping δ to our left, the path of the covering point x in Φ which begins at P_1 traverses γ while x_0 in ϕ traverses γ_0, and ends at a point P_2 over p_2 on a boundary side α_1 of Δ equivalent to α. Here γ is perpendicular to α and α_1. While x_0 travels on α_0 from p_2 to p_3, x travels on α_1 from P_2 to P_3 (over p_3), then traces a line η over η_0 etc. At the moment when x_0 passes p_5 for the second time, x meets α at P_5'; because then (but first then) x_0 has completed a path which bounds a simply connected region in conjunction with an arc of α_0, so the resulting closed path is null homotopic. The reader can easily pursue the corresponding paths of x_0 on ϕ and x on Φ with Figures 9 and 10 in hand, and see the general validity of this construction for arbitrary q and r, when one takes an arbitrary, but fixed, boundary curve α_0 of δ as the basis for a crosscut system. The boundary sides of the right-angled polygon Δ_0 which lie on boundary sides of Δ collectively form a simple covering of the boundary curves of δ. The remaining boundary sides of Δ_0 are crosspieces of Δ which correspond in pairs by fixed elements; in Figure 10 the axes of the three elements h_1, h_2 and h_3, as far as they run in Δ_0, are shown by dotted lines, and these also indicate the correspondence between γ and γ' etc. Each unit in q corresponds to a pair of elements, and each unit in $r-1$ to one element. Thus the number of these elements is v.

If one now thinks of the circles orthogonal to E which carry the crosspieces γ, γ', etc. being extended to E, then they do not meet, because of their orthogonality to the sides of Δ. The subregion $\mathfrak{H}$ of $\Phi + E$ which lies outside all these circles is a fundamental domain for the group $\{h_1, \ldots, h_v\}$ generated by $h_1, \ldots, h_v$, and this group is therefore a "free group," i.e., there is no non-identity relation between $h_1, \ldots, h_v$. Since $\{h_1, \ldots, h_v\}$ is generated by fixed elements, it is either identical with $H(I)$ or a proper subgroup of $H(I)$. In the latter case one forms a coset decomposition of H modulo $\{h_1, \ldots, h_v\}$. The number of points in $\mathfrak{H}$ equivalent to a given point in $\mathfrak{H}$ under fixed elements is then equal to the number of cosets, and hence independent of the point chosen. For a point of Δ_0 this number turns out to be 1, since Δ_0 is the intersection of Δ and $\mathfrak{H}$ and it contains no equivalent points by construction. Thus

$$H(I) = \{h_1, \ldots, h_v\}.$$

Δ_0 is a $4v$-gon with angle sum $4v \cdot \pi/2 = 2v\pi$. Hence by (8)

$$\mathrm{Ar}(\Delta_0) = \mathrm{Ar}(\delta(I)) = 2v - 2. \tag{13}$$

It now follows from $\mathrm{Ar}(\delta) \leq \mathrm{Ar}(\phi)$ and (9) that

$$v \leq 2p - 1. \tag{14}$$

Example 2, I, 15 shows that the equality sign can occur; in this case δ is the result of cutting the surface along the closed geodesic corresponding to the element b_1.

Theorem 10. *If the fixed element group of a non-identity automorphism contains more than a* *cyclic group, then it is a free group. The number* v *of free generators satisfies*

$$2 \leq v \leq 2p - 1.$$

Here $v = 2q + r - 1$, *where* q *is the genus and* r *is the number of boundary curves of the region* $\delta(I)$ *on* ϕ *covered by the kernel domain* $\Delta(I)$.

11. Type construction for a non-cyclic $H(I)$. Now one considers an interval of regularity i. Let h be the primary fixed element whose fundamental points bound i. In what follows, one may refer again to Figure 5 if one now assumes that $U(h)$ and $V(h)$ are no longer isolated singular points, but that all other singular points, and hence also the kernel region Δ, lie to the right of the axis of h in the figure. This axis is mapped onto itself by the group $\{h\}$ generated by h, and onto another boundary side of Δ by each other fixed element. Hence the half plane Φ_1 which lies to its left in Figure 5 is also mapped onto itself by $\{h\}$, and onto a half plane disjoint from Φ_1 by any other fixed element. A fundamental domain for $H(I)$ in Φ_1 is therefore the same as one for $\{h\}$ in Φ_1, and this lies between the lines α and $h\alpha$ in Figure 5. Now we repeat the consideration of §7: On the fundamental piece AhA there is only a finite, in fact even, number $2\mu_1$ of boundary fixed points, which are alternately attracting and repelling. If $\mu_1 > 0$, then the principal domain Ω extends beyond the kernel domain Δ along the boundary side of Δ considered, hence ω also extends beyond δ along the corresponding boundary curve of δ. In order to measure this excess, we have to measure the excess of Ω over Δ, as far as it lies in a fundamental domain of $\{h\}$. This is the $(\mu_1 + 4)$-gon in Figure 5, whose area was computed to be μ_1 in §7.

The same process in now repeated in all intervals of regularity not equivalent to i or each other. Their number is r, where r is the number of boundaries of δ (and a closed geodesic is counted twice as a boundary curve when it bounds δ from both sides). Thus one obtains r numbers $\mu_1, \ldots, \mu_r$. Each boundary curve of δ is associated with such a μ_i-value. However, two boundary curves of δ which coincide as a two-sided boundary curve have both their μ_i-values zero, as follows from Theorem 2.

One should not be deceived into thinking that δ has only two boundary curves in the case of Figure 10, while Δ_0 meets six boundary sides of Δ. If one imagines the lines carrying the crosscuts in Figure 10 as fully extended, then the result on the outer side of β is a fundamental domain like the one we constructed above for $\{h\}$ in Figure 5; because P_6 and P_6' are equivalent under the primary element whose axis is β. But such a fundamental domain results from taking together the pieces standing on the boundary sides α to α_4. One thinks of these now transferred to equivalent positions so that they all meet along α and are arranged next to each other. Then the outermost vertices on α are equivalent. Thus there are only two μ_i-values there, one for the boundary curve β_0 and one for the boundary curve α_0 of δ.

Now one sets

$$\mu = \mu_1 + \mu_2 + \cdots + \mu_r. \tag{15}$$

One obtains $\mathrm{Ar}(\omega)$ by adding to $\mathrm{Ar}(\delta)$ the excesses of ω over δ corresponding to the individual boundary curves of δ, whose areas are given by the individual μ_i. Hence (13) and (15) yield

$$\mathrm{Ar}(\omega(I)) = 2v - 2 + \mu \tag{16}$$

and, because of $\mathrm{Ar}(\omega) \leq \mathrm{Ar}(\phi)$ and (9),

$$2\nu + \mu \leq 4p - 2, \tag{17}$$

whence (14) is sharpened in case $\mu > 0$.

By Theorem 2, two regular boundary fixed points are never equivalent under a non-fixed element. By construction, each regular boundary fixed point is equivalent to exactly one of the 2μ points above, under a fixed element.

Theorem 11. *Under an automorphism I whose fixed element group H(I) contains more than a cyclic group, the number* $\nu \geq 2$ *of free generators of H and the number* $\mu \geq 0$ *of inequivalent regular boundary fixed points are subject to the restriction (17). Such automorphisms are divided into finitely many types, according to the values of* ν *and* μ.

This division into types may be refined, on the one hand by partitioning ν according to the genus q and the number of boundaries r of δ, by the formula $\nu = 2q + r - 1$, and on the other hand by partitioning μ into the r summands of (15), which correspond to the r inequivalent intervals of regularity.

As in §7, one can again construct the surface $K_1 = K \bmod H$, by first extending the fundamental domain $\mathfrak{H}$ of H to the exterior Ψ of E by reflection in E. $\mathfrak{H}$ is then the exterior of 2ν complete circles orthogonal to E, which carry the crosspieces of Figure 10. These correspond in pairs under the ν generating fixed elements. K_1 is therefore a closed surface of genus ν. On it there are r closed curves covered by the intervals of regularity. An I-inducing T-function induces a mapping of K_1 onto itself which belongs to the class of the identity[7], and whose restriction to each of the r curves depends only on I and has 2μ fixed points there, covered by the boundary fixed points in $\mathfrak{H}$.

12. Summary. An infinite cyclic group is the free group with one generator. We agree to speak of the group consisting of the identity alone as the *free group with zero generators*. Its fundamental domain is the whole of $\Phi + E$, or, when we include the exterior Ψ of E, the whole of K. Then the previous results may be summarized as follows:

Let I be a non-identity automorphism of the first kind and let t be an I-inducing T-function. Thus t preserves orientation. Then the fixed element group $H(I)$ is always a free group. Because of (14), the number ν of its free generators satisfies

$$0 \leq \nu \leq 2p - 1. \tag{18}$$

The surface

$$K_I = K \bmod H(I)$$

is a closed surface of genus ν, hence a sphere for $\nu = 0$ and a torus for $\nu = 1$, otherwise a surface of genus which is higher, but no greater than $2p - 1$. On K_I there are r closed curves, covered by the r classes of equivalent intervals of regularity. These may be called the "E-curves" of K_I. Their points are the only points of K_I which do not lie over points of ϕ. In the more general sense implied by this situation, one can call K_I a covering surface of ϕ. When $\nu = 0, r = 1$, since the E-curve on K_I is the whole of E. When $\nu = 1, r = 2$. When $\nu > 1$, $1 \leq r \leq 2p$, as follows from $\nu = 2q + r - 1$ and (14), and the equality sign can occur on either side. Thus in the case of the automorphism I in the Addition in Proof at the end of I, the element k is a fixed element

[7] See Theorem 12.

whose axis covers the single boundary curve of $\delta(I)$; so $r = 1$, while $q = 1$ and $v = 2$. On the other hand, suppose I_1 is the automorphism defined by

$$I_1 : \begin{cases} a_i \mapsto a_i b_i \\ b_i \mapsto b_i \end{cases} \qquad i = 1, 2, \ldots, p.$$

Comparison with Example 2, I, 15, immediately yields the fact that $\delta(I_1)$ is the planar region which results from ϕ by cutting along p disjoint closed geodesics covered by the axes of the elements b_i. Thus $r = 2p$. Here $q = 0$, so $v = 2p - 1$. In fact, $H(I_1)$ is generated by the $2p$ elements b_i and $a_i b_i a_i^{-1}$, $i = 1, \ldots, p$, exactly one of which may be eliminated by the defining relation of F, whence the others remain free.

The T-function tK induces a topological mapping of K_I onto itself (which we express directly as tK_I); because it satisfies the functional equation

$$tfx = f_I tx \tag{19}$$

for all x on K and all f in F, and hence in particular it satisfies the functional equation

$$thx = htx \tag{20}$$

for all x on K and all h in $H(I)$. Now we show in general that

Theorem 12. *The mapping tK_I belongs to the class of the identity.*

In order to prove this we have to show that there is a family of continuous (but not necessarily one-to-one) mapping functions $s_\lambda(x)$, depending continuously on a parameter λ, $0 \leq \lambda \leq 1$, such that $s_0(x) = t(x)$ and $s_1(x) = x$, and such that s_λ satisfies the functional equation

$$s_\lambda hx = hs_\lambda x, \tag{21}$$

corresponding to (20), for all x and all h in H. Because (21) is the necessary and sufficient condition for $s_\lambda(x)$ to determine a continuous mapping $s_\lambda K_I$ for each λ. However, we do not of course require s to satisfy a functional equation corresponding to (19). This is already excluded by the fact that $s_0(x) = t(x)$ satisfies the equation (19) and $s_1(x) = x$ satisfies the equation

$$s_1 fx = fs_1 x$$

for all f; whereas there are elements f for which $f_I \neq f$.

We shall construct s_λ in such a way that it, like s_0 and s_1, maps Φ into itself, E into itself, and Ψ into itself. To do this, we construct s_λ in $\Phi + E$ and then stipulate that the reflection in E of an arbitrary point be sent to the reflection in E of the image point.

When $v = 0$, then h in (21) can only be the identity. Thus there is really no functional equation to be satisfied. In this case the assertion is an immediate consequence of the Tietze deformation theorem, and s_λ can be chosen as a topological mapping.

Therefore, suppose $v > 0$. Let i be an interval of regularity, bounded by the fundamental points of the primary fixed element h. Let L be the axis of h and let Z be the non-Euclidean half plane bounded by L and i. The group $\{h\}$ maps Z onto itself. Z mod $\{h\}$ is a region homeomorphic to an annulus. One can again use Figure 5. It is now assumed that α is perpendicular to L. Then, e.g., the quadrilateral $AChChA$ of circular arcs is a fundamental domain for $\{h\}$ on Z. One maps it onto the unit square of a u-v-plane in such a way that the circles through $U(h)$ and $V(h)$ are sent to the lines $v = \text{const.}$, and their orthogonal trajectories to the lines $u = \text{const}$. Call this topological mapping τ. Then one sets

$$\tau(h(x)) = (u + 1, v)$$

when

$$\tau(x) = (u, v).$$

In this way, Z is mapped by τ onto the parallel strip $0 \leq v \leq 1$. We have a mapping ti of i onto itself which satisfies the equation $thx = htx$. Under the mapping $\tau t^{-1}\tau^{-1}$ of the line $v = 1$ into itself, the point $(u, 1)$ goes to the point $(u + \chi(u), 1)$, where $\chi(u)$ is a continuous periodic function of u with period 1. Now let Σ_ε be the mapping of the parallel strip onto itself defined by

$$\left.\begin{aligned} u' &= u + \varepsilon v \chi(u) \\ v' &= v \end{aligned}\right\} 0 \leq \varepsilon \leq 1$$

and let

$$\sigma_\varepsilon = \tau^{-1}\Sigma_\varepsilon \tau$$

be the topological mapping of Z onto itself which results from conjugation with τ^{-1}, and which depends continuously on the parameter ε. For all ε, σ_ε is the identity on L; σ_0 is the identity in all of Z, σ_1 is the mapping t^{-1} in i. Because of the periodicity of $\chi(u)$, σ_ε satisfies the equation

$$\sigma_\varepsilon hx = h\sigma_\varepsilon x \tag{22}$$

for all ε and all x in Z. Now if one no longer regards the h in (22) as the element determined earlier, but rather as an arbitrary fixed element, the definition of σ_ε becomes extended to all domains equivalent to Z under $H(I)$; these do not of course overlap. If there are more intervals of regularity (in other words, if $r > 1$), one chooses one of them and proceeds similarly with it and its equivalents under $H(I)$, r times. If $v > 1$, then it still remains to define σ_ε in the kernel domain Δ, and there one sets it equal to the identity for all values of ε. Now σ_ε is defined over all of $\Phi + E$, and it satisfies equation (22) there for all ε and all h in H.

Now one sets

$$t_\varepsilon(\Phi + E) = \sigma_\varepsilon t(\Phi + E);$$

t_ε satisfies (20), since t and σ_ε do. We have $t_0 = t$. The function t_1 satisfies (20) and is the identity on E. Thus the first part of the deformation problem is solved. On K_I it can be simply expressed by saying that the mapping can be deformed into the identity on all E-curves of K_I. Continuity requires only that one affect the immediate neighbourhood of these curves. We have confined this neighbourhood outside the kernel region; the deformation stops right at its boundary.

Now that deformation has produced a function t_1 which is the identity on E, so that the Euclidean distance from a point to its image tends to zero as E is approached, it is easy to obtain the identity by continued deformation, without violating (20). One need only move each image point in Φ with uniform velocity in a straight line (in the sense of the non-

Euclidean metric) to its preimage in unit time in order to conclude, as in I, 27, that (20) is satisfied at each instant, so that one is inducing a continuous deformation of the mapping on K_I. One now sees that the previous treatment of E and its neighbourhood was necessary in order to avoid discontinuity at E in the last process.

This completes the proof of Theorem 12.

Now one considers the regular boundary fixed points of t. By Theorem 5 these are isolated and bilaterally attracting or repelling on E. They are partitioned into finitely many equivalence classes with respect to H. The number, 2μ, of these classes is even. The restriction on μ, in conjunction with ν, is that

$$0 \leq 2\nu + \mu \leq 4p - 2. \tag{23}$$

This relation is the combination of (11), (12) and (17). We can define μ as half the number of boundary fixed points belonging to a fundamental domain $\mathfrak{H}$ of H. On K_I one can view μ as follows: The fixed points of the mapping tK_I underlie the class division defined in I, 31. The "principal class" of tK_I is defined to be that fixed point class of tK_I which is covered by the fixed point set of tK. Then 2μ is the number of fixed points of tK_I which belong to the principal class and lie on the E-curves. When an E-curve contains such a point, then it contains fixed points of the principal class only; because when an interval of regularity contains a fixed point then it can obviously contain no point which is moved by t to one of its own equivalents under H. Fixed points of the principal class on the E-curves are isolated fixed points of tK_I. This follows from their isolation on E and I, 34.

When $\mu + \nu > 1$, one speaks of a principal domain $\Omega(I)$. The area of the region ω on ϕ covered by Ω is

$$\operatorname{Ar}(\omega) = 2\nu + \mu - 2 \tag{24}$$

Cf. (10), (16) and §7. From (24) and (9) we get (23). $\operatorname{Ar}(\omega) = 0$ requires $\nu = 0$, $\mu = 2$, since $\nu + \mu > 1$. The covering of ω by Ω is simple or infinite sheeted according as $\nu = 0$ or $\nu > 0$.

When $\nu > 1$, one speaks of a kernel domain $\Delta(I)$. The area of $\delta = \Delta \bmod H$ is

$$\operatorname{Ar}(\delta) = 2(\nu - 1) > 0$$

by (13) and because $\nu > 1$.

The values of the characteristic numbers ν and μ give a division of all automorphisms of the first kind into types, the number of which is finite because of (23). This still holds when the types are further subdividied, for $\nu > 0$, as in §7 and §11.

The work I contains numerous examples of the different types, but no example in which $\nu > 0$ and $\mu > 0$ simultaneously. No example for this case is yet known to me.

CHAPTER II

Indices of the Various Types

13. Behaviour of the mapping in the neighbourhood of boundary fixed points. Let t be a T-function which induces the automorphism I, let P be a point of E which is regular with respect to $H(I)$, and let u be a neighbourhood of P containing no two points which are equivalent under a fixed element. Let O be the point $x = 0$, f an element of F, and suppose fO

is in u. We have $tfO = f_I tO$. The non-Euclidean distance between the point fO and its image under t therefore equals the non-Euclidean distance between the points O and $f^{-1}f_I tO$. Now let k be an element of F different from f, and suppose kO is in u. The element kf^{-1} sends fO to kO and hence it is not a fixed element, since both points are in u. It follows from

$$kf^{-1} \neq (kf^{-1})_I$$

that

$$k^{-1}k_I \neq f^{-1}f_I$$

and hence

$$k^{-1}k_I tO \neq f^{-1}f_I tO.$$

Now if C is an (arbitrarily large) positive constant, the circle of radius C round O contains only finitely many points equivalent to tO. Thus there are only finitely many points in u equivalent to O whose distances from their images under t are less than C. Now if r is the radius of a circle round O which includes the fundamental domain B of F, and if R is the radius of a circle round tO which includes tB, then each point x of u lies within distance r of the point fO, and tx lies within distance R of the point tfO. Thus as $x \to P$ the distance $e_t(x)$ between x and tx grows beyond all bounds. For P which are not fixed points of t this is obvious. The following theorem is therefore essentially a statement about regular *fixed points* of the boundary mapping:

Theorem 13a. *If P is a regular boundary point for t and if C is an arbitrary constant, then there is a neighbourhood u of P such that the non-Euclidean distance $e_t(x)$ between x and tx exceeds C for all x in u.*

The behaviour of the real function $e_t(x)$ of the complex variable x which gives the non-Euclidean distance between x and tx is quite different at a *singular* boundary point. Let $\varepsilon \geq 0$ be the minimum of $e_t(x)$ in Φ. Since

$$e_t(hx) = e_t(x)$$

for all fixed elements h, since $th = ht$, e_t takes its range of values in a fundamental domain $\mathfrak{H}$ of $H(I)$. All boundary points of $\mathfrak{H}$ belonging to E are regular, and by Theorem 13a, e_t grows beyond all bounds as they are approached. The function e_t is continuous and it therefore takes all values $\geq \varepsilon$ at points of $\mathfrak{H}$ not on E. Now the domains equivalent to $\mathfrak{H}$ under H accumulate at each singular point, hence we have:

Theorem 13b. *In each neighbourhood of a singular point, the distance function $e_t(x)$ takes every value which it assumes in Φ, infinitely often.*

Next to these theorems on the distance between x and tx we shall now place a theorem which says something about the "direction from x to tx" in the neighbourhood of a regular boundary fixed point.

Let P be a regular, bilaterally attracting fixed point on E. On the basis of Lemma 4, §4, we determine a sequence l_n of conjugate primary elements of F whose axes contract towards P, so that, as n increases, $V(l_n)$ converges monotonically to P from one side, and $U(l_n)$ converges likewise from the other. We assume that the sequence begins so late that all $U(l_n)$ lie in one fixed point free interval next to P, and all $V(l_n)$ lie in the other. Then, since P is bilaterally attracting, the axis of l_{nI} does not meet the axis of l_n, rather, it lies entirely on P's side of the latter axis. We now consider the pairs of axes of l_n and l_{nI} for all n. It can happen that the figures formed by two such pairs are equivalent, namely, when there is an element of F which simultaneously transforms l_m to l_n and l_{mI} to l_{nI} by conjugation. We partition the set of pairs of axis into such equivalence classes.

We first assume that each equivalence class contains only finitely many pairs of axes. Then one can choose a subsequence of l_n which yields inequivalent axis pairs. This new sequence is again denoted by l_n, $n = 1, 2, \ldots$; l_1 is denoted by l for short. By hypothesis, l_n is conjugate to l, and hence l_{nI} is conjugate to l_I. Let f_n be an element such that

$$l_n = f_n l f_n^{-1} = f_n l^{-\nu} l l^{\nu} f_n^{-1}.$$

The latter equation shows that f_n is only determined up to an arbitrary power of l. Now one transforms the axes of l_n and $l_{nI} = f_{nI} l_I f_{nI}^{-1}$ by conjugation with $l^{\nu} f_n^{-1}$, obtaining the axes of the elements

$$l \quad \text{and} \quad l^{\nu} f_n^{-1} f_{nI} \cdot l_I \cdot f_{nI}^{-1} f_n l^{-\nu} = l_n'.$$

Thus these form a figure equivalent to the pair of axes of l_n and l_{nI}. Now let A be a point on the axis of l, so that the segment from A to lA is a displacement segment on this axis. Then one can choose ν in such a way that the common perpendicular to the axes of l_n' and l meets the latter between A and lA. Then the shortest distance $d(l, l_n')$ between the axes of l and l_n' must grow beyond all bounds as n increases, since the axes of the l_n' are all different by hypothesis, and an infinite set of equivalent axes can only accumulate towards E. Thus the distance $d(l_n, l_{nI})$ between the axes of l_n and l_{nI} grows with n beyond all bounds.

Now the image of the axis of l under t is a Jordan curve in Φ which, since

$$tlx = l_I tx,$$

is mapped onto itself by l_I, and therefore it runs within a strip of constant distance from the axis of l_I. Let ρ be its maximum distance from the axis of l_I. The image of the axis of l_n under t then likewise accompanies the axis of l_{nI} at distance $\leq \rho$. Thus for all values of n for which $d(l_n, l_{nI}) > \rho$, the axis of l_n does not meet its image under t. Hence, for all axes beyond a certain point in the sequence, the image of the axis under t lies completely on P's side of the axis; the distance between axis and image therefore grows with n beyond all bounds.

If, on the other hand, the above assumption is not satisfied, then the original sequence of axis pairs contains at least one infinite equivalence class. We then choose this as a subsequence and again denote it by

$$l_1 = l, l_2, l_3, \ldots.$$

Let f_n be the (unique) element which simultaneously transforms l to l_n and l_I to l_{nI} by conjugation. It follows from

$$f_n l_I f_n^{-1} = l_{nI} = (f_n l f_n^{-1})_I = f_{nI} l_I f_{nI}^{-1}$$

(analogously to (7), §4) that

$$f_n^{-1} f_{nI} = l_I^{n'}. \tag{25}$$

Here n' is an exponent depending on n, and in fact, different values of n give different values

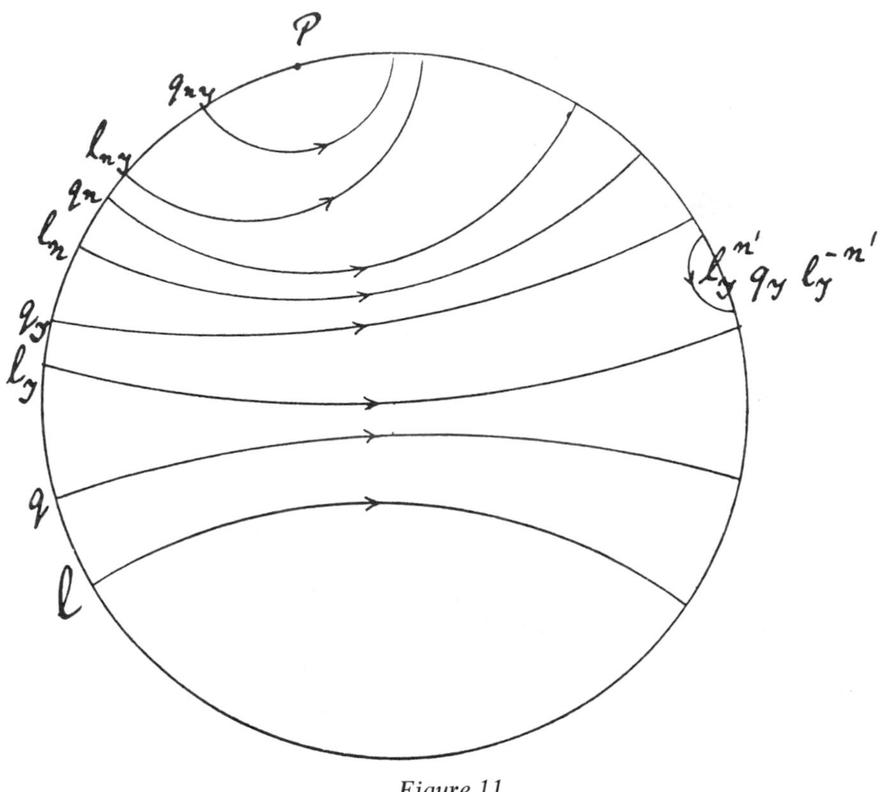

Figure 11

of n'. Assume on the contrary that $m \neq n$ and $m' = n'$. Then

$$f_m^{-1} f_{mI} = l_I^{m'} = l_I^{n'} = f_n^{-1} f_{nI};$$

hence

$$f_n f_m^{-1} = (f_n f_m^{-1})_I.$$

That is impossible, because $f_n f_m^{-1}$ transforms l_m into l_n, and it therefore has a fundamental point on P's side of these axes. But P is the only fixed point of the boundary mapping on this side, and it is not a fundamental point.

It follows that $|n'| \to \infty$ as $n \to \infty$.

Now let q (Figure 11) be an element of F whose axis separates the axes of l and l_I. Such an element exists by I, 5. Then the axis of q_I lies on P's side of the axis of l_I. Let

$$q_n = f_n q f_n^{-1}.$$

The axis of q_n separates the axes of l_n and l_{nI}. Thus the q_n are another sequence of elements whose axes contract towards P. Now

$$q_{nI} = (f_n q f_n^{-1})_I = f_{nI} q_I f_{nI}^{-1}$$

and hence by (25)

$$f_n^{-1} q_{nI} f_n = l_I^{n'} q_I l_I^{-n'}.$$

The axes of the latter elements accumulate towards the fundamental points of l_I as $|n'|$ increases, hence their distance from the axis of q grows beyond all bounds. Therefore, the distance $d(q_n, q_{nI})$ between the axes of q_n and q_{nI}, into which q and $l_I^{n'} q_I l_I^{-n'}$ are transformed by conjugation with f_n, also grows with n beyond all bounds. Then one concludes, exactly as above, that for all sufficiently large n the axis of q_n does not meet its image under t, and that the latter lies on P's side of q_n, at an arbitrarily great distance.

In exactly the same way, for a regular, bilaterally repelling fixed point Q one can find axes which contract towards Q and do not meet their images under t, with the images lying on the side of the axes opposite Q and at arbitrarily great distances.

Theorem 13c. *In any neighbourhood of a regular boundary fixed point P there are axes surrounding and contracting towards P which do not meet their images under t, and such that the images lie on P's side or the opposite according as P is bilaterally attracting or bilaterally repelling. Moreover, one can prescribe an arbitrarily large lower bound for the distance between an axis and its image.*

It should be stressed that one cannot conclude from this that any circle orthogonal to E which approaches a bilaterally attracting boundary fixed point sufficiently closely encloses its image under t. This is shown already by the more general example added to Example 11 in (I, 48). One takes I to be automorphism Λ' and $l = a_1 b_1 a^{-1}$. Then

$$l_I = b_1,$$

$$l_{I^2} = a_1^{-1} b_1 a_1,$$

$$\dots$$

$$l_{I^n} = a_1^{-1} b_1^{-1} \dots a_1^{-1} b_1^{-(n-2)} a_1^{-1} \cdot b_1 \cdot a_1 b_1^{(n-2)} \dots b_1 a_1.$$

Now one sets

$$l_n = l_{I^{n-1}}$$

with the above meaning, so

$$l_{nI} = l_{I^n}.$$

These two elements result from l and l_I by conjugation with

$$f_n = a_1^{-1} b_1^{-1} \dots a_1^{-1} b_1^{-(n-2)} a_1^{-1}.$$

Thus one has here the case of inequivalent axis pairs. By I, 24, Theorem 12 one can construct a T-function t which induces I and which sends a point on the axis of l to a point which lies on the side of the axis of l opposite to the axis of l_I. Then the axis of l meets its image under t, and hence so does the axis of each l_n, even though these axes contract towards the regular boundary fixed point P^+. In order to prove the property of the regular boundary fixed point P^+ given by Theorem 13c, one takes the element $\gamma = a_2 b_1$ of I, 48, Example 11 and uses all the $f_n \gamma f_n^{-1}$. Then γ corresponds to the q above; its axis separates the axes of l and l_I.

A *fixed fundamental point* cannot have the property expressed in Theorem 13c for P, because $e_t(x)$ is bounded on the whole fixed element axis ending at the point.

14. The indices of the different types. Now we have all the means necessary to determine the indices of the fixed point classes represented by the different types of T-functions. We can save space by referring to the work I, where the methods in question are presented in detail, partly in the general theory and partly in the worked examples.

Let t be a T-function inducing a non-identity automorphism I, let $\nu \geq 0$ be the number of free generators of the fixed element group $H(I)$, let $\mathfrak{H}$ be a fundamental domain of H, and let $\mu \geq 0$ be half the number of (regular) boundary fixed points in $\mathfrak{H}$. We then say that t (resp., I) is of "ν-μ-type." The part of the fixed point set of $t\Phi$ in $\mathfrak{H}$ is called the *fixed point class represented by* t (I, 31–33), and it is denoted by Π_t. No two points of Π_t are equivalent under F. Π_t lies inside a circle C which is concentric with E, but smaller, and hence in the intersection of C and $\mathfrak{H}$ (I, 34). Let Π be the fixed point set of $t(\Phi + E)$, and denote $\Phi + E - \Pi$ by W. If x is a point of W then its "direction point"

$$z(x, tx) = z_t(x)$$

is the point at which the non-Euclidean half line from x to tx meets E (I, 35). The "direction function" $\xi_t(x)$ is the real function, determined mod 1 by

$$\xi_t(x) = \frac{1}{2\pi i} \log z_t(x).$$

The functions z_t and ξ_t are defined only in W, and are continuous there.

The "index" $i(t)$ is the value of the integral $\int d\xi_t$ taken in the positive direction round a simple curve in W which encloses all of Π_t and no other points of Π. If the boundary of $\mathfrak{H}$ is fixed point free, the integral can be taken along the boundary of $\mathfrak{H}$. When $\nu = 0$, $\mathfrak{H} = \Phi$, hence E is the boundary of $\mathfrak{H}$. When $\nu > 0$ the boundary of $\mathfrak{H}$ consists of 2ν circular arcs in Φ orthogonal to E, the "Φ-boundary sides" of $\mathfrak{H}$, and 2ν arcs on E, the "E-boundary sides" of $\mathfrak{H}$. When $\nu = 1$ the Φ-boundary sides are chosen to be the α and $h\alpha$ in Figure 5. When $\nu > 1$ the Φ-boundary sides are the circular arcs orthogonal to E shown in Figure 10 carrying the crosspieces γ, γ', etc. Those pieces of the Φ-boundary sides which lie outside C are fixed point free. Nothing can be said about the pieces inside C. However, we shall suppose that t satisfies the hypothesis we called "Assumption α" in I, 37: Among the components of the complement of the fixed point set of the surface mapping $\tau\phi$ covered by $t\Phi$ there is one with the same genus p as ϕ. Then one can alter the pieces of the Φ-boundary sides of $\mathfrak{H}$ inside C so that they lie in W; the way of doing this was described in I, 47. When an E boundary side contains fixed points, then it can be replaced, e.g., by the arc of C which connects the same Φ-boundary sides. (C is chosen so large that it cuts all Φ-boundary sides.) In this way one obtains a simple closed curve which encloses just Π_t. The index $i(t)$ is therefore a well-defined integer, which we now determine.

One begins by considering a corresponding pair of Φ-boundary sides, say α and $h\alpha$ in Figure 5. Suppose x traverses α from A to B. We can choose α so that A and B, and hence also hA and hB, are not fixed points. Then $z_t(A)$ is the point tA and $z_t(B)$ is the point tB, hence

$$2\pi \int_A^B d\xi_t = \text{pos}(tA \rightarrow tB) + 2k\pi,$$

where k is an integer whose value is not known more precisely, and $\text{pos}(tA \rightarrow tB)$ denotes the

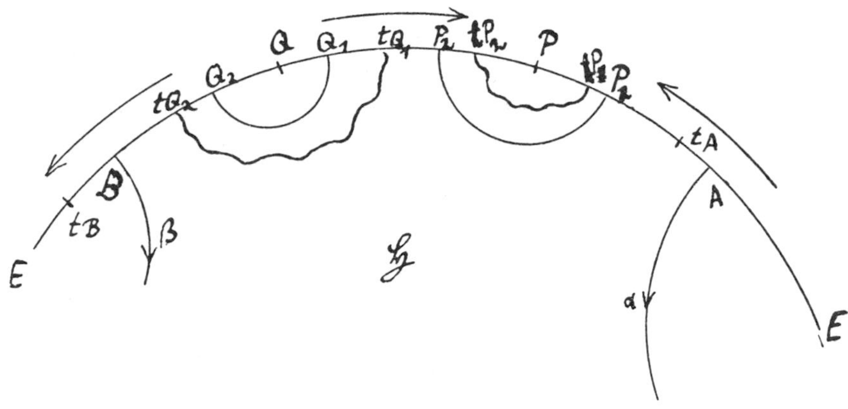

Figure 12

arc $<2\pi$ of E which runs in the positive direction from tA to tB (as in I, 47). Now

$$z_t(hx) = hz_t(x)$$

since h is a fixed element. Thus

$$2\pi \int_{hA}^{hB} d\xi_t = \text{pos}(thA \to thB) + 2k\pi$$

with the same k. Hence

$$2\pi\left(\int_A^B + \int_{hB}^{hA}\right) = \text{pos}(tA \to tB) - \text{pos}(thA \to thB)$$

$$= \text{pos}(tA \to tB) + \text{pos}(thB \to thA) - 2\pi.$$

This result holds for each of the v pairs of Φ-boundary sides.

Now one considers the E-boundary sides, say the one from A to B in Figure 12. Assume that it contains fixed points. By Theorem 5, there are finitely many of them and they are alternately attracting and repelling. Figure 12 supposes there is one attractor and one repeller; the direction of movement in the fixed point free intervals is shown by the arrows outside E. Then one uses Theorem 13c to enclose the attracting fixed point P by an axis P_1P_2 which encloses its own image, and to enclose the repelling fixed point Q by an axis Q_1Q_2 which is enclosed by its own image; in addition, one chooses those axes so close to P and Q that they are outside C. The parts of $\mathfrak{H}$ separated by these axes are then fixed point free in the interior of E.

Now (cf. I, 40)

$$2\pi \int_A^{P_1} = \text{pos}(tA \to tP_1),$$

$$2\pi \int_{P_2}^{Q_1} = \text{pos}(tP_2 \to tQ_1),$$

$$2\pi \int_{Q_2}^{B} = \text{pos}(tQ_2 \to tB).$$

Also,

$$2\pi \int_{P_1}^{P_2} = \text{pos}(tP_1 \to tP_2),$$

since $z_t(x)$ always remains on the arc P_1PP_2 while x traverses the axis P_1P_2. Finally

$$2\pi \int_{Q_1}^{Q_2} = \text{pos}(tQ_1 \to tQ_2) - 2\pi,$$

since $z_t(x)$ always remains outside the arc Q_1QQ_2 while x traverses the axis Q_1Q_2. It follows by addition that

$$2\pi \int_{A}^{B} = \text{pos}(tA \to tB) - 2\pi,$$

when one takes the integral from A along E to P_1, from P_1 along the axis to P_2, from P_2 along E to Q_1, from Q_1 along the axis to Q_2, and from Q_2 along E to B. One proceeds analogously at all boundary fixed points; each bilateral repeller contributes -2π.

Now one adds the contributions along the whole boundary of $\mathfrak{H}$. The sum of all the arcs pos( ) is exactly 2π. One divides by 2π and obtains

$$i(t) = 1 - \nu - \mu \tag{26}$$

in which each pair of corresponding Φ-boundary sides, and each bilaterally repelling boundary fixed point, gives rise to one negative unit.

Theorem 14. *Each T-function of ν-μ-type represents a fixed point class of index* $1 - \nu - \mu$.

15. Consequences. The result of the last paragraph produces a proof of the fact that was not treated in complete generality in the work I, namely that the index of a fixed point class represented by a T-function t depends only on the automorphism I induced by T—because ν and μ depend only on I. Now let $\tau\phi$ be a topological mapping of ϕ onto itself. By I, 32, the number of fixed point classes of $\tau\phi$ is finite. Let

$$t_1\Phi, t_2\Phi, \ldots, t_n\Phi$$

be n T-functions over $\tau\phi$ which represent the n fixed point classes of $\tau\phi$. Their respective indices

$$i_1, i_2, \ldots, i_n$$

depend only on the respective induced automorphisms

$$I_1, I_2, \ldots, I_n.$$

We are interested only in those T-functions whose class indices are $\neq 0$, since it is only these that one knows to represent non-empty fixed point classes for each mapping in the class of τ. These fixed point classes may be called *essential*, and the fixed point classes of index zero which are possibly present may be called *inessential*. Now the automorphism family of $\tau\phi$ depends only on the *class* of the mapping $\tau\phi$. In this family one chooses a complete set of representatives of the different isogredience classes into which the family is partitioned (I, 18 and I, 31). Only a finite number of these, say Z, have non-zero index. Thus Z is the number of essential fixed point classes, and the system of non-zero integers

$$i_1, i_2, \ldots, i_Z$$

is an invariant of the mapping *class* of $\tau\phi$. The same is then true of the index sum (I, 37)

$$\Xi = i_1 + i_2 + \cdots + i_Z. \tag{27}$$

Ξ can therefore be determined from those mappings in the class which satisfy the hypotheses used to derive the Alexander formula for Ξ in I, 43. In this way, the invariant meaning of the Alexander number is shown.

By I, 36–37, two surface mappings which are conjugate via a third surface mapping have the same system of invariants. The system of invariants is therefore attached to the complete system of conjugate mapping classes of ϕ.

The number Ξ is an invariant over a much broader system of mapping classes. Namely, Ξ is the same for two orientation-preserving mapping classes if the traces of their exponent sum matrices are the same; this is shown by the formula (50), I, 43, where Ξ_A is now to be read as Ξ.

The following additional consequences may now be drawn from (26).

(a) The greatest value of $i(t)$ is $+1$. This occurs only for $\nu = \mu = 0$, and hence only for T-functions which are fixed point free on E. Positive contributions to Ξ therefore consist only of positive units. Thus it follows from (27) that

$$Z \geq \Xi \tag{28}$$

which of course only says something when $\Xi > 0$.

(b) $i(t) = 0$ occurs only for $\nu = 1, \mu = 0$ and $\nu = 0, \mu = 1$. These are the cases in which the boundary mapping tE has exactly two fixed points; the latter determine either two periodic or two aperiodic fixed point free intervals. There are not necessarily fixed points of $t\Phi$ for either of these types. Each T-function not of the 1-0-type or 0-1-type necessarily represents a non-empty fixed point class.

(c) The remaining types have $\nu + \mu > 1$. By (23), §12 we have in this case

$$\nu + \mu \leq 4p - 2 - \nu,$$

hence by (26)

$$i(t) \geq 3 + \nu - 4p \geq 3 - 4p.$$

The extreme case $i(t) = 3 - 4p$ requires $\nu = 0, \mu = 4p - 2$. As mentioned at the end of §6, no example of this case is known to me.

(d) In the above list of types of T-functions the case of the identity automorphism is not taken into account. In this case the Birkhoff formula (I, 42) gives the index

$$i(t) = 2 - 2p > 3 - 4p$$

because $p > 1$.

Theorem 15. *The system of Z non-zero fixed point class indices $i_1, i_2, \ldots, i_Z$ is an invariant of the mapping class. Its sum Ξ is obtained by the Alexander formula. $Z \geq \Xi$. Each fixed point class index lies between the values $+1$ and $3 - 4p$, where $p > 1$ is the genus of the surface. The index $+1$ occurs only for T-functions which are fixed point free on E. Fixed point classes of index zero can only be represented by T-functions which have exactly two fixed point on E.*

16. Interpretation of the indices in terms of the Birkhoff formula. Let t be a T-function which induces a non-identity automorphism I of the first kind. One extends t by reflection in E to a topological mapping tK of the whole complex sphere K (I, 28). Suppose t is of ν-μ-type. Then (§12)

$$K_I = K \bmod H(I)$$

is a closed surface of genus ν. On K_I there are $r \geq 1$ E-curves (§12). By Theorem 12, t induces a mapping tK_I of K_I onto itself which belongs to the class of the identity. The principal class (§12) among the fixed point classes of tK_I—and by I, 42 this is the only *essential* fixed point class of tK_I—contains 2μ isolated fixed points on the E-curves of tK_I.

Reflection in E makes the axes P_1P_2 and Q_1Q_2 of Figure 12 into full circles γ_P and γ_Q orthogonal to E, the first of which completely encloses its image $t\gamma_P$, and the second of which is completely enclosed by its image $t\gamma_Q$ (when one reckons the interiors of these circles on K and K_I to be the sides containing P and Q, respectively). Now let σ be an arbitrary topological mapping of $t\gamma_P$ onto γ_Q, with orientation chosen so that, when x traverses $t\gamma_P$ with P on its left, σx traverses γ_Q with Q on its right. Now σt is a mapping of γ_P onto γ_Q and $t\sigma$ is a mapping of $t\gamma_P$ onto $t\gamma_Q$. One extends these in any way to a topological mapping of the annular region between γ_P and $t\gamma_P$ to the annular region between γ_Q and $t\gamma_Q$. Then one leaves the interiors of $t\gamma_P$ and γ_Q out of K_I and identifies each point x of $t\gamma_P$ with the point σx on γ_Q. Because of the orientation chosen for σ, the resulting closed surface is orientable and has genus higher than that of K_I by 1. One now chooses a new attracting fixed point and a new repelling fixed point on the E-curves, proceeds with them similarly, and continues the process μ times. The result is a closed orientable surface K_I^* of genus $\nu + \mu$, for which the induced mapping, which we also denote by tK_I^* from now on, belongs to the class of the identity. (In other words: We have identified each attractor among the 2μ isolated fixed points of the principal class of tK_I on the E-curves with a repeller, blown up the point of identification into a handle, and made the mapping tK_I fixed point free at the place of identification.)

Now by I, 42, tK_I^* has only one essential fixed point class, and this has index $2 - 2(\mu + \nu)$, since $\nu + \mu$ is the genus of K_I^*. None of the fixed points of this essential class lie on E. The fixed point sets on opposite sides of E are mirror images of each other. Consequently, the fixed point set of tK_I^*, and hence also of tK_I, which is covered by fixed points of $t\Phi$ has the index $1 - \nu - \mu$, in agreement with (26).

For $\mu = 0$ the surface K_I^* is of course the surface K_I itself.

Theorem 16. *The index of a fixed point class of an orientation-preserving mapping of a closed surface ϕ of genus $p > 1$ can always be obtained by application of the Birkhoff formula, either to the surface ϕ itself or to a closed auxiliary surface K_I or K_I^*. The auxiliary surface depends only on the class of the surface mapping in question, and by (23) its genus is bounded by the value* $4p - 2$.

CHAPTER III

Investigation of Complete Automorphism Families

17. Related *T*-functions and automorphisms. Let t be a T-function which satisfies the functional equation

$$tfx = f_I tx, \tag{29}$$

where I is a non-identity automorphism of the first kind. Suppose $t\Phi$ lies over $\tau\phi$. Assume that there are more than two fixed points of tE; then there is a principal domain $\Omega(t) = \Omega(I) = \Omega$. Now one chooses some element $l \neq 1$ from F and constructs the T-function lt related to t. It likewise lies over $\tau\phi$, and it induces the automorphism

$$J = I_{l^{-1}},$$

related to I, which sends f to $lf_I l^{-1}$, as (29) shows. Our next goal is to say something about the way in which the type of J depends on the choice of l, in particular, to determine the relative position of the principal domains $\Omega(I)$ and $\Omega(J)$, in case J also has more than two fixed boundary points.

Now when there is an element f of F such that (cf. I, 18)

$$l = ff_I^{-1}, \tag{30}$$

then by (29),

$$lt = ftf^{-1}$$

and

$$J = I_{0f} I I_{0f}^{-1},$$

where I_{0f} is the inner automorphism corresponding to f. In this case the automorphisms I and J, and likewise the T-functions t and lt, are what we called *isogredient* in I, 18. In this case we have

$$ltE = ftf^{-1}E$$

and hence the boundary fixed point set M satisfies

$$M(lt) = M(ftf^{-1}) = fM(t),$$

so that also

$$M^*(lt) = fM^*(t)$$

and

$$\Omega(J) = \Omega(lt) = f\Omega(t) = f\Omega(I).$$

In this case we already know from the first chapter that $\Omega(J)$ and $\Omega(I)$ do not overlap, since (30) shows that f is not a fixed element under I. In the next paragraph we shall show that these two principal domains also do not overlap in the general case.

If one has three elements l_1, l_2 and f such that

$$l_2 = fl_1 f_I^{-1},$$

then (29) gives

$$l_2 t = fl_1 f_I^{-1} t = f(l_1 t) f^{-1};$$

hence $l_1 t$ and $l_2 t$ are isogredient, and likewise so are the induced automorphisms J_1 and J_2. Then

$$\Omega(J_2) = f\Omega(J_1).$$

The corresponding fixed element groups are conjugate subgroups of F, since (cf. Theorem 8, I, 18)

$$H(J_2) = fH(J_1)f^{-1}.$$

Now we turn to the question of the type of lt for different choices of l, and first suppose that l is a fixed element under I. Then I, 17 contains the complete answer to our question, and it can be formulated as follows:

Theorem 17. *If $lt = tl$ and if the axis of l is not a boundary side of $\Omega(t)$, then lt is of type $\nu = 1$, $\mu = 0$. If the axis of l is a boundary side of $\Omega(t)$, then lt can belong to a type $\nu + \mu > 1$. In the latter case $\Omega(t)$ and $\Omega(lt)$ have the axis of l as a common boundary side and apart from this no point in common. The l in such a case can only be one particular power of the primary element for the axis. All other powers of this primary element lead again to the type $\nu = 1$, $\mu = 0$.*

From now on we shall assume that l is not a fixed element under I. The sequence of the elements l_{I^n} for all n was investigated in §2, with f in place of l. As we did there, we denote the endpoints of the fixed point free intervals which contain the fundamental points of l by

$$\left.\begin{aligned} U^+ &= \lim t^n U(l), & U^- &= \lim t^{-n} U(l) \\ V^+ &= \lim t^n V(l), & V^- &= \lim t^{-n} V(l) \end{aligned}\right\} n \to \infty$$

so the sequence l_{I^n} for $n \to \infty$, resp., $n \to -\infty$ is a fundamental sequence for V^+, resp., V^-, with U^+, resp., U^- as a possible exceptional point (§2). The fixed point free interval U^+U^- is again denoted by u, V^+V^- by v, and we set $u + U^+ + U^- = \bar{u}$, $v + V^+ + V^- = \bar{v}$.

In order to determine the type of lt, one has to investigate the fixed point set of the induced boundary mapping, and this is done by iteration of the mapping. For $n > 0$, repeated application of (29) gives (cf. I, 17 and I, 50):

$$\begin{aligned} (lt)^n &= ltlt \dots lt \\ &= ll_I t^2 \dots lt \\ &\dots \\ &= ll_I l_{I^2} \dots l_{I^{n-1}} t^n, \\ (lt)^{-n} &= t^{-1}l^{-1}t^{-1}l^{-1} \dots t^{-1}l^{-1} \\ &= l_{I^{-1}}^{-1} t^{-2} l^{-1} \dots t^{-1} l^{-1} \\ &\dots \\ &= l_{I^{-1}}^{-1} l_{I^{-2}}^{-1} \dots l_{I^{-n}}^{-1} t^{-n}. \end{aligned}$$

When one defines

$$\begin{aligned} &\left.\begin{aligned} l_n &= ll_I l_{I^2} \dots l_{I^{n-1}} \\ l_{-n} &= l_{I^{-1}}^{-1} l_{I^{-2}}^{-1} \dots l_{I^{-n}}^{-1} \end{aligned}\right\} n > 0, \\ &l_0 = 1 \end{aligned} \tag{31}$$

then one has, for all $n \gtreqless 0$,

$$(lt)^n = l_n t^n. \tag{32}$$

For all n one has the recursion formula

$$l_{n+1} = l \cdot l_{nI} = l_n \cdot l_{I^n}, \tag{33}$$

whence

$$l_{n+m} = l_m \cdot (l_n)_{I^m} = l_n \cdot (l_m)_{I^n} \tag{34}$$

for all n and m. If one writes (33) in the form

$$l_{I^n} = l_n^{-1} \cdot l \cdot l_{nI},$$

then one obtains the isogredience of l_{I^n} and l relative to I. This is the statement of Theorem 9, I, 18.

Corresponding to the equation (29)

$$f_I = tft^{-1},$$

one has

$$f_J - (lt)f(lt)^{-1} - lf_I l^{-1}$$

for each f, and for each n

$$f_{J^n} = (lt)^n f(lt)^{-n} = l_n t^n f t^{-n} l_n^{-1} = l_n f_{I^n} l_n^{-1}.$$

Thus

$$J^n = (I^n)_{l_n^{-1}}. \tag{35}$$

In particular, (34) and (35) give

$$(l_n)_{J^m} = l_m (l_n)_{I^m} l_m^{-1} = l_{n+m} l_m^{-1} \tag{36}$$

for all n and m.

18. Related principal domains. In the notation of the previous paragraph, what we now investigate is the boundary fixed point set of lt, and to do this we must distinguish three cases, depending on the position of the fundamental points of l.

(a) The two fixed point free intervals, u and v, which contain the fundamental points of l are disjoint; thus U^+, U^-, V^+ and V^- are four distinct points on E.

By (31), $l_1 = l$ and $l_2 = ll_I$. The positive fundamental points of l and l_I both lie in v, the negative ones both lie in u. Hence by Lemma 1, §1, $V(l_2)$ lies in v between $V(l)$ and $V(l_I)$, and $U(l_2)$ lies in u between $U(l)$ and $U(l_1)$. Moreover, $l_3 = l_2 l_{I^2}$. One therefore concludes similarly that $V(l_3)$ lies in v between $V(l_2)$ and $V(l_{I^2})$. $V(l_3)$ is closer to V^+ than $V(l_2)$ is, because $V(l_I)$ and *a fortiori* $V(l_{I^2})$ is closer to V^+ than $V(l_2)$ is. Likewise, $U(l_3)$ lies in u, closer to U^+ than $U(l_2)$ is. By means of (33) one can continue this line of reasoning indefinitely. Thus, with increasing n, $V(l_n)$ moves monotonically in v in the direction of V^+, and $U(l_n)$ moves monotonically in u in the direction of U^+. Let P be the limit point of $V(l_n)$ and let P' be the limit point of $U(l_n)$. P lies in $\bar{v}$ and P' lies in $\bar{u}$, so P and P' are different. The sequence l_n is a fundamental sequence for P with P' as the single possible exceptional point (§1). Since the axis of l_n converges to the line $P'P$, the displacement length of l_n grows with n beyond all bounds. Now one takes any point different from P', e.g., the origin O, and constructs the point sequence $l_n O$. By Lemma 3,

$$l_n O \to P$$

as $n \to \infty$, and hence

$$ltl_n O \to ltP$$

by the continuity of lt in $\Phi + E$. But now (29), (33) and Lemma 3 give

$$ltl_n O = ll_{nI} tO = l_{n+1} tO \to P,$$

since $tO \neq P'$. Thus $ltP = P$, and, as a fixed point of lt, P must be an interior point of v. This is because $t\bar{v} = \bar{v}$ and $lt\bar{v} = l\bar{v}$ is an interior subinterval of v. (The consequence that lt has at least one fixed point in v and, on the same grounds, at least one in u, was clear from the outset.) For the point P', (29), (33) and Lemma 3 give

$$l_n^{-1} O \to P',$$

$$tl_n^{-1} O \to tP',$$

$$tl_n^{-1} O = l_{nI}^{-1} tO = l_{n+1}^{-1} ltO \to P',$$

since ltO is a fixed point different from P. Thus $tP' = P'$, and therefore $P' = U^+$, since t has no fixed point in the interior of u.

Moreover, $l_{-1} = l_{I^{-1}}^{-1}$ by (31). Thus $V(l_{-1})$ is in u and $U(l_{-1})$ is in v. Also, $l_{-2} = l_{-1} l_{I^{-2}}^{-1}$. Thus $V(l_{-2})$ lies in u between $V(l_{-1})$ and $V(l_{I^{-2}}^{-1})$, etc. One concludes that, as $n \to \infty$, $V(l_{-n})$ is a sequence in u moving monotonically in the direction of U^-, with limit point Q say, and $U(l_{-n})$ is a sequence in v moving in the direction of V^-, with limit point Q'. One then has

$$l_{-n} O \to Q,$$

$$ltl_{-n} O \to ltQ,$$

$$ltl_{-n} O = l(l_{-n})_I tO = l_{-n+1} tO \to Q.$$

Hence $ltQ = Q$ and Q is an interior point of u. On the other hand,

$$l_{-n}^{-1} O \to Q',$$

$$tl_{-n}^{-1} O \to tQ',$$

$$tl_{-n}^{-1} O = (l_{-n}^{-1})_I tO = l_{-n+1}^{-1} ltO \to Q',$$

so $tQ' = Q' = V^-$.

According as the direction of movement under t in u and v is opposite or the same, one has the cases of Figures 13 and 14, in which u and v are drawn heavily.

Now let R be any fixed point of t. Then (32) gives

$$(lt)^n R = l_n t^n R = l_n R \to \begin{cases} P & \text{as} \quad n \to \infty \quad \text{if} \quad R \neq U^+ \\ Q & \text{as} \quad n \to -\infty \quad \text{if} \quad R \neq V^- \end{cases}.$$

Thus for each $R \neq U^+$, V^- it follows that the interval between P and Q containing R is fixed point free under lt. This applies to $R = V^+$ and $R = U^-$. Hence in Figure 13 P and Q are the only fixed points of lt. In Figure 14, the interval PQ on the right of the figure is fixed point free

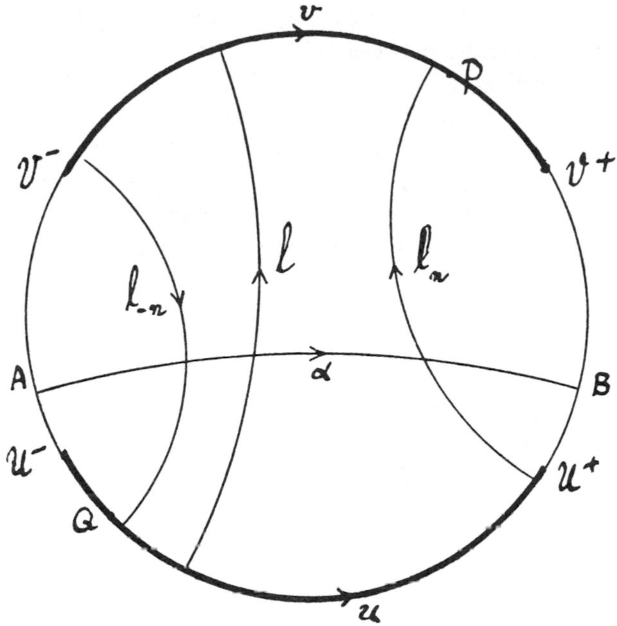

Figure 13

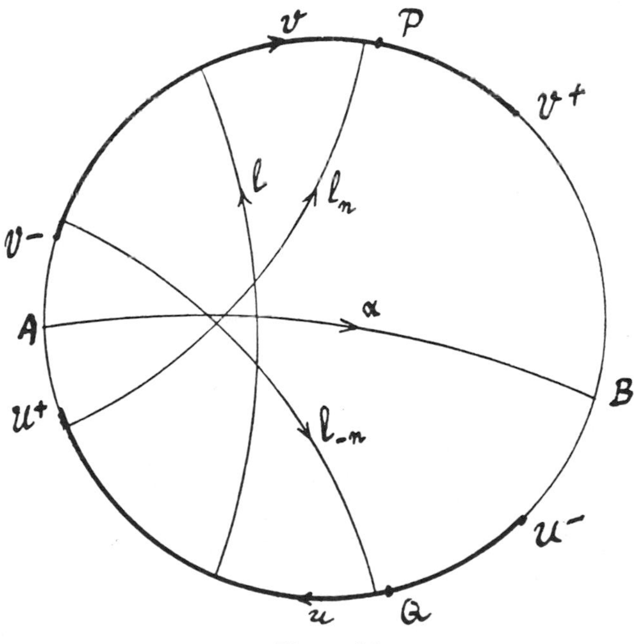

Figure 14

under lt. If the interval PQ on the left of Figure 14 contains a fixed point of t other that V^- and U^+ (which of course must lie between V^- and U^+), then this whole interval PQ is fixed point free under lt. In all cases

$$\left.\begin{aligned}(lt)^n V^- &= l_n V^- \to P\\ (lt)^{-n} U^+ &= l_{-n} U^+ \to Q\end{aligned}\right\} \quad \text{as} \quad n \to \infty \tag{37}$$

also holds here. Thus if lt is to have boundary fixed points in addition to P and Q, we must first of all have the case of Figure 14, and there must also be no other fixed point of t between V^- and U^+, so V^- and U^+ must be regular fixed points of t (since they are not the only fixed points of tE). Hence V^- must be bilaterally repelling and U^+ must be bilaterally attracting. Because of (37), all fixed points of ltE other than P and Q lie between V^- and U^+. The construction of $M^*(t)$ removes V^- from $M(t)$, and the construction of $M^*(lt)$ removes the bilaterally repelling fixed point Q from $M(lt)$. Thus $M^*(t)$ and $M^*(lt)$ are disjoint on E, and hence $\Omega(t)$ and $\Omega(lt)$ do not overlap.

P and Q are regular fixed points of lt, and hence not fundamental points. (In the case where they are not the only boundary fixed points of lt, this already follows from the fact that they are isolated.) In order to see this, one connects two points A and B on the two E-arcs which make up $E - \bar{u} - \bar{v}$ by an orthogonal arc α. See Figure 13 and 14. Since the point

$$l_n V^+ = l_n t^n V^+ = (lt)^n V^+$$

in both figures lies to the right of P for all $n > 0$, the same is the case with $l_n B$. The point $l_n A$ lies to the left of P. Therefore $l_n \alpha$ encloses P, and indeed more tightly as n increases. Now by (33)

$$ltl_n \alpha = l(l_n)_I t\alpha = l_n l_{I^n} t\alpha.$$

As $n \to \infty$, l_{I^n} forms a fundamental sequence for V^+ with U^+ as the single possible exceptional point. The Jordan arc $t\alpha$ connects the point tA between $U^\mp$ and V^- with the point tB between $U^\pm$ and V^+ (the upper sign for Figure 13, the lower for Figure 14). Thus, for sufficiently large n, $l_{I^n} t\alpha$ lies arbitrarily close to V^+, and the distance between $l_{I^n} t\alpha$ and α therefore grows with n beyond all bounds. The same is therefore true of the distance of the arcs $l_n \alpha$ and $ltl_n \alpha$ which arise from them by l_n. Thus P is enclosed arbitrarily tightly by orthogonal arcs whose images under lt lie on the side facing P, at arbitrarily great distances. Then, by the last remark in §13, P cannot be a fundamental point.

Theorem 18a. *If l is not a fixed element of the automorphism induced by t, and if the closures of the intervals u and v which are fixed point free under t and include the fundamental points of l are disjoint, then lt is in general of type $\nu = 0$, $\mu = 1$. Only when the interval endpoints V^- and U^+ are adjacent regular fixed points of t can lt have more than two boundary fixed points. The principal domain $\Omega(lt)$ which occurs in this case does not overlap $\Omega(t)$.*

In particular, lt is always of type $\nu = 0$, $\mu = 1$ when t has a type with $\nu > 1$ and the fundamental points of l lie in different intervals of regularity.

(b) $u = v$, so $U^+ = V^+$, $U^- = V^-$.

First assume that the points V^-, $U(l)$, $V(l)$, V^+ lie in this order on $\bar{v}$ (Figure 15). The interval from V^+ to $V(l)$ is mapped onto an interior subinterval of itself by lt, and the interval from

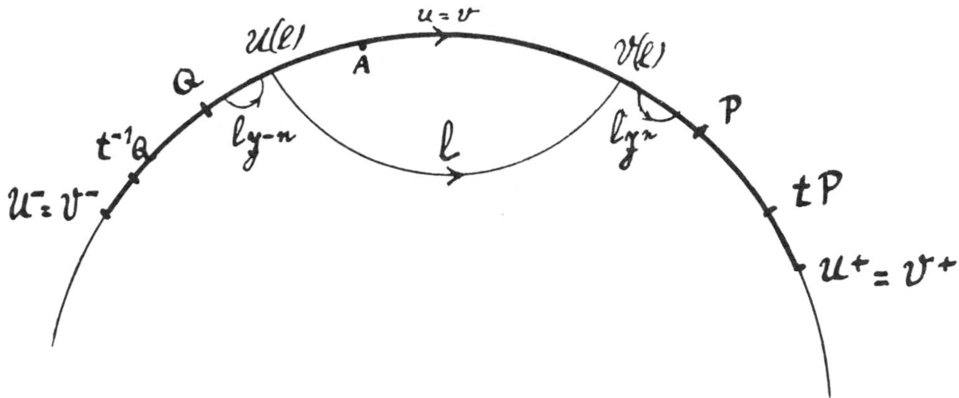

Figure 15

$U(l)$ to V^- is mapped onto an interval which properly includes the interval from $U(l)$ to V^-. Thus in each of these two intervals there is at least one interior point which is fixed by lt. Let A be any point of v between the fundamental points of l. In the figure, tA lies to the right of A. If tA still falls between the fundamental points of l, ltA lies to the right of tA. If tA lies to the right of $V(l)$, then ltA also lies to the right of $V(l)$. Thus in either case ltA lies to the right of A. The part of v between $U(l)$ and $V(l)$ is therefore fixed point free under lt, including its endpoints. Let PQ be the complete fixed point free interval for lt of which it is a part. Since lt induces the automorphism $J = I_{l^{-1}}$, the elements l_{J^n} form a fundamental sequence for P as $n \to \infty$, and a fundamental sequence for Q as $n \to -\infty$. The point tP lies between P and V^+, $t^{-1}Q$ between Q and V^-. Now, as $n \to \infty$, one has:

$$(lt)^n t^{-1} Q = (lt)^{n-1} l Q = l_{J^{n-1}} (lt)^{n-1} Q = l_{J^{n-1}} Q \to P,$$

$$(lt)^{-n} tP = (lt)^{-n} t (lt)^{-1} P = (lt)^{-n} l^{-1} P = l^{-1}_{J^{-n}} (lt)^{-n} P = l^{-1}_{J^{-n}} P \to Q.$$

Thus P and Q also bound a fixed point free interval on the other side, and they are therefore the only fixed points of lt. Hence lt is either of type $\nu = 1$, $\mu = 0$ or $\nu = 0$, $\mu = 1$. This is independent of whether or not tE has more than two fixed points.

Now suppose that the points V^+, $U(l)$, $V(l)$, V^- lie in this order on $\bar{v}$ (Figure 16). Here we make use of the assumption that tE has more than two fixed points. By §2 and §3, $lM(t)$ lies in v with at most the exception of lV^+. When lV^+ also lies in v, the interval $E - v$ is mapped by t onto itself and then by l onto a subinterval of v, hence lt maps it onto an entirely disjoint interval, which means that it is fixed point free under lt. However, when lV^+ falls outside v, §3 shows that the point V^+ must be bilaterally attracting, and V^- must be bilaterally repelling. Let A and B be the other endpoints of the intervals different from v which are next to V^+ and V^- and fixed point free under t. Then lV^+ falls between B and V^- and lA falls between V^- and $V(l)$. Thus lt maps the interval from V^+ to A onto the interval from lV^+ to lA, and the interval from A to V^- onto the interval from lA to lV^-. Both intervals, and hence also the whole interval $E - v$, are therefore fixed point free under lt.

In each case the whole fixed point set of ltE must therefore lie in v. If lt has more than two fixed points, this means that $\Omega(lt)$ and $\Omega(t)$ do not overlap.

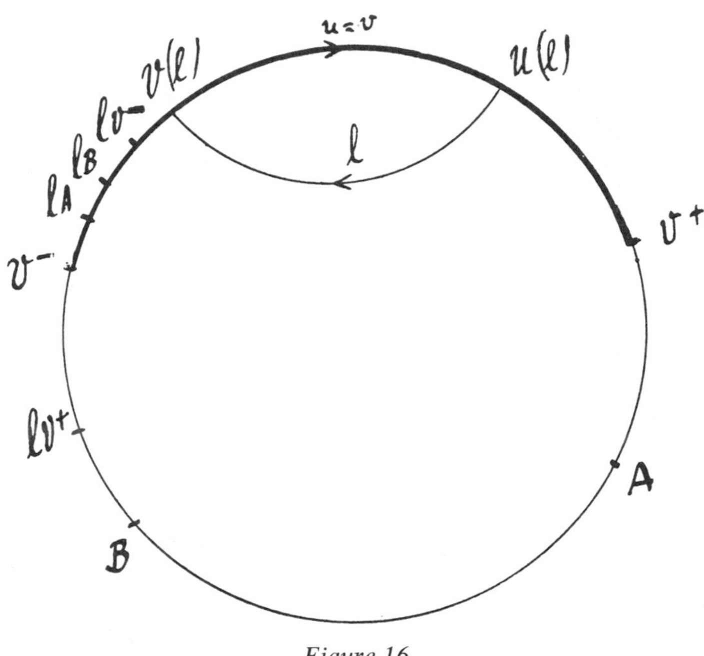

Figure 16

In the latter case one can also make more precise statements on the position of the boundary fixed point set $M(lt)$ of lt. One distinguishes between an l whose axis intersects that of l_I and one whose axis does not. In particular, ltE can be fixed point free, so that lt is of type $\nu = \mu = 0$. However, we shall not pursue the investigation of these situations in the present work.

Theorem 18b. *If the fundamental points of l lie in the same interval v which is fixed point free under t, and if the directions of movement under t and l are the same in the subinterval of v between the fundamental points, then lt is of type* $\nu = 1$, $\mu = 0$ *or* $\nu = 0$, $\mu = 1$. *If the directions of movement are opposite, and if tE has more than two fixed points, then ltE can also have more than two fixed points; in this case* $\Omega(t)$ *and* $\Omega(lt)$ *do not overlap.*

(c) The remaining case is where u and v are adjacent. Since t must have more than two boundary fixed points, Theorem 4 allows this only if $U^- = V^-$ or $U^+ = V^+$.

Suppose $U^- = V^-$ and let $j = E - u - v - U^-$ be the arc from U^+ to V^+ not containing U^-; j contains $M^*(t)$. By Theorem 2, lj lies entirely in v. Thus $ltj = lj$ is disjoint from j. The whole fixed point set of ltE, if it exists, therefore lies in $u + v$. Thus $\Omega(t)$ and $\Omega(lt)$ do not overlap.

Now suppose $U^+ = V^+$. One has Figure 17, which comes from Figure 13 by bringing the points U^+ and V^+ together. Exactly as in case (a) of this paragraph, one now concludes that, as $n \to \infty$, the sequence l_n is a fundamental sequence for P with U^+ as the possible exceptional point, and l_{-n} is a fundamental sequence for Q with V^- as the possible exceptional point. Here P is a point in v between $V(l)$ and V^+, and Q is a point in u between $U(l)$ and U^-. Now, as in

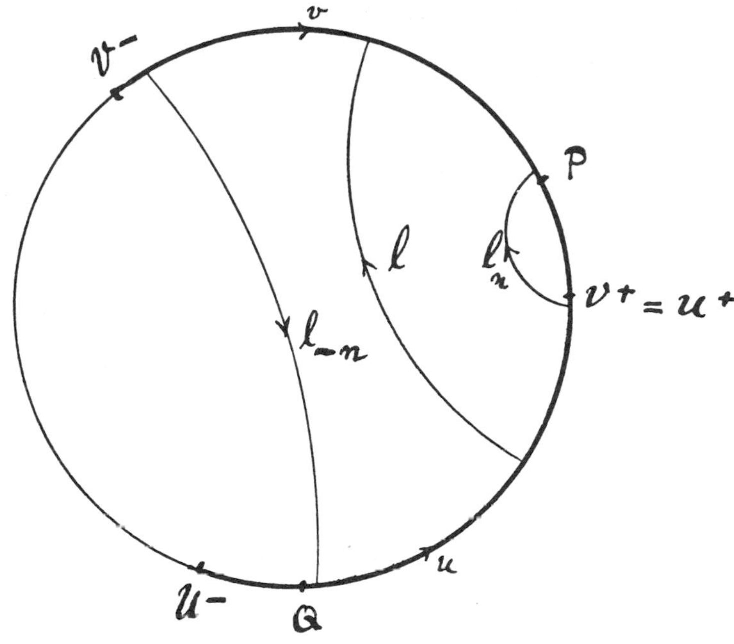

Figure 17

(a), we have

$$(lt)^n U^- = l_n t^n U^- = l_n U^- \to \begin{cases} P & \text{as} \quad n \to \infty \\ Q & \text{as} \quad n \to -\infty \end{cases}.$$

Thus the interval PQ containing U^- is fixed point free under lt. Also, as $n \to \infty$:

$$(lt)^{-n} V^+ = l_{-n} t^{-n} V^+ = l_{-n} V^+ \to Q.$$

Thus the subinterval of u from V^+ to Q is fixed point free under lt. Apart from the isolated bilaterally repelling fixed point Q, the whole fixed point set of ltE therefore lies on the arc from, but excluding, V^+ to, and including, P. $\Omega(lt)$ and $\Omega(t)$ therefore do not overlap.

Theorem 18c. *If the fundamental points of l lie in adjacent intervals which are fixed point free under t, and if t and lt each have more than two fixed points on E, then $\Omega(t)$ and $\Omega(lt)$ do not overlap.*

19. Consequences. When a T-function t has more than two boundary fixed points, then it belongs to a type for which $\nu + \mu > 1$, and hence by (26), §14, it has a negative index; and conversely. It now follows from Theorems 17 and 18 a-c that the principal domains of two related T-functions of negative index never overlap, and they can meet along a boundary side only in the case of Theorem 17. Isogredient T-functions have equivalent principal domains. If t_1 and t_2 are two related but not isogredient T-functions, then no interior point of $\Omega(t_1)$ is equivalent to a point of $\Omega(t_2)$; however, $\Omega(t_1)$ and $\Omega(t_2)$ can have equivalent boundary sides. The point set

$$\omega(t) = \omega(I) = \Omega(t) \bmod H(I)$$

covered by a principal domain $\Omega(t) = \Omega(I)$ may be called a "principal domain on ϕ." One can speak analogously of a "kernel domain on ϕ." One then has:

Theorem 19. *If $\tau\phi$ is an orientation-preserving mapping of the surface ϕ onto itself, and if one constructs the (finitely many, because of the finiteness of the class number) corresponding "principal domains on ϕ," i.e., the point sets on ϕ covered by the principal domains of the T-functions of negative index over $\tau\phi$, then these do not overlap. Two principal domains on ϕ can meet along a closed geodesic. In this case two T-functions whose principal domains meet along an axis in Φ covering the geodesic are conjugate to each other, via an element with this axis.*

One notices that the principal domains on ϕ depend only on the *class* of τ.

Now one considers a complete system of non-isogredient T-functions of negative index over $\tau\phi$. Let Z^- be the number of them. (Similarly, let Z^+ be the number with positive index, so that the "class number"

$$Z = Z^- + Z^+$$

is the number of essential fixed point classes in the mapping class of τ.) Let t_s be one such T-function. By (24), §12, one has that the area $A_s = \text{Ar}(\omega(t_s))$ of the corresponding principal domain on ϕ satisfies:

$$A_s = 2(v_s - 1) + \mu_s \qquad (v_s + \mu_s > 1) \tag{38}$$

when t_s is of v_s-μ_s-type. Now since the principal domains on ϕ do not overlap, (9), §5, gives:

$$\sum_{s=1}^{Z^-} A_s \leq \text{Ar}(\phi) = 4(p-1). \tag{38a}$$

One therefore has:

Theorem 20. *The numbers v and μ characterising the type of a fixed point class of negative index are subject to the restriction*

$$\sum_{s=1}^{Z^-} 2(v_s - 1) + \mu_s \leq 4(p-1), \tag{39}$$

where Z^- is the number of such classes and v_s and μ_s are the characteristic numbers of the sth class (so that $v_s + \mu_s > 1$).

This result has a purely group theoretic consequence: v_s is the number of free generators of the fixed element group $H(I_s)$ of the automorphism I_s induced by t_s. Isogredient T-functions have conjugate subgroups of F as fixed element groups. Now if one considers the set of those subgroups of F which appear as fixed element groups for the automorphisms in a complete automorphism family, then these divide into conjugacy classes of subgroups. If one omits the family which includes the identity automorphism, then one concludes from (39):

Theorem 21. *The number of mutually non-conjugate subgroups of F which appear as fixed element groups for an automorphism family and have more than one free generator is subject to the restriction*

$$\sum_s (v_s - 1) \leq 2p - 2, \tag{40}$$

where v_s denotes the number of free generators of the sth group.

In particular, there can be at most $2p - 2$ such groups, and then only when each has exactly two free generators.

20. Estimation of the class number from the index sum. Since

$$i_s = 1 - v_s - \mu_s,$$

one has that the number $Z = Z^+ + Z^-$ of essential fixed point classes ($v_s + \mu_s \neq 1$) satisfies

$$\Xi = \sum_{s=1}^{Z} (1 - v_s - \mu_s) = Z - \sum_{s=1}^{Z} (v_s + \mu_s),$$

where the latter sum is taken only over the classes for which $v_s + \mu_s > 1$. One therefore has

$$Z = \Xi + \sum_{s=1}^{Z^-} (v_s + \mu_s) \tag{40a}$$

if one numbers the classes in such a way that the subscript s first runs over the negative classes. Since the estimate $Z \geq \Xi$, which follows from (40a) and has already been expressed in (28), §15, says something only when $\Xi > 0$, and since $\Xi = 0$ implies only that $Z \geq 0$, we now seek another estimate which says something when $\Xi < 0$.

From (38) one sees that the area numbers A_s of negative classes satisfy

$$\begin{aligned}\Xi = \sum_{s=1}^{Z} i_s &= \sum_{s=1}^{Z^-} (1 - v_s - \mu_s) + \sum_{s=Z^-+1}^{Z^-+Z^+} 1 \\ &= \sum_{s=1}^{Z^-} (-A_s + v_s - 1) + Z^+ \\ &= \sum_{s=1}^{Z^-} (-A_s + v_s) - Z^- + Z^+;\end{aligned}$$

hence

$$Z = Z^- + Z^+ = -\Xi + 2Z^+ - \sum_{s=1}^{Z^-} A_s + \sum_{s=1}^{Z^-} v_s. \tag{41}$$

Then, because of (38a), this yields

$$Z \geq -\Xi - 4(p - 1)$$

and since there must be at least one class when $\Xi < 0$, one has:

Theorem 22. *The number Z of fixed point classes with non-zero index may be bounded from below in terms of the index sum Ξ obtained from the Alexander formula, as follows:*

$$\begin{aligned} &Z \geq \Xi \quad \text{for} \quad 0 \leq \Xi, \\ &Z \geq 1 \quad \text{for} \quad -4(p-1) \leq \Xi < 0, \\ &Z \geq |\Xi| - 4(p-1) \quad \text{for} \quad \Xi < -4(p-1). \end{aligned} \tag{42}$$

21. Frequency of the type $v = 0$, $\mu = 2$. Among the negative classes those are distinguished which correspond to a principal domain in the form of a line (§3). For them, $A_s = 0$ and hence $v_s = 0$, $\mu_s = 2$ and $i_s = -1$. Let their number be Z_2^-. Also, let Z_1^- be the number of negative classes with principal domains not in the form of lines. One numbers the Z essential classes in such a way that the subscript s first runs through these Z_1^- negative classes, then the remaining Z_2^- negative classes, and finally the Z^+ positive classes. Then one has

$$\begin{aligned}\Xi = \sum_{s=1}^{Z} i_s &= \sum_{s=1}^{Z_1^-} (-A_s + v_s - 1) + \sum_{s=Z_1^-+1}^{Z_1^-+Z_2^-} (-1) + \sum_{Z^-+1}^{Z} 1 \\ &= -\sum_{s=1}^{Z_1^-} A_s + \sum_{s=1}^{Z_1^-} v_s - Z_1^- - Z_2^- + Z^+.\end{aligned}$$

Each of the first Z_1^- classes uses up at least one unit of surface area of ϕ, hence by (38a)

$$Z_1^- \leq 4(p-1).$$

Then it follows from the last equation that

$$\begin{aligned}Z_2^- &\geq -\Xi - 8(p-1) + \sum_{s=1}^{Z_1^-} v_s + Z^+ \\ &\geq -\Xi - 8(p-1).\end{aligned} \tag{43}$$

Theorem 23. *When the index sum Ξ lies below $-8(p-1)$, then there are at least $|\Xi| - 8(p-1)$ classes of type $v = 0$, $\mu = 2$. When there are exactly $|\Xi| - 8(p-1)$ such classes, then there are also $4(p-1)$ classes of type $v = 0$, $\mu = 3$, and no other essential classes.*

Because $Z_2^- = -\Xi - 8(p-1)$ and (43) yield $Z^+ = 0$, $v_s = 0$ and $Z_1^- = 4(p-1)$, whence $A_s = 1$, $\mu_s = 3$ for $s = 1, \ldots, Z_1^-$.

One can easily give examples of mapping classes whose index sum is negative and has arbitrarily large absolute value. Thus, by forming powers of the exponent sum matrix of the automorphism denoted by I in I, 49, one obtains the following Ξ-values for $I, I^2, I^3, \ldots$:

$$4, 0, 10, -12, 24, -42, 88, -156, 298, \ldots,$$

etc., with alternating signs and increasing numerical values; here $p = 2$.

By Theorem 23, a negative value of Ξ the absolute value of which is large in comparison with $8(p-1)$ requires that "almost all" negative classes be of type $v = 0$, $\mu = 2$, and hence make a contribution to Ξ of one negative unit each.

The above-mentioned example was worked out in I, 49 and I, 51. The automorphism $J = I^4$ turned out to be of type $v = 0$, $\mu = 4$ hence $i = -3$ and $A = 2$. As mentioned in I, 51, the automorphism $J' = (I_c)^4$ related to J also turns out to be of this type, and one easily proves that J' and J are not isogredient. The two corresponding principal domains have a combined area of 4, and this is equal to $\mathrm{Ar}(\phi) = 4(p-1)$, since $p = 2$. Since $v = 0$ in both cases, the two principal domains have no boundary side in common. The complementary set on ϕ is a null set. These fixed point classes and their principal domains also belong to all powers of J. Now since the Ξ-values of the powers of J are all negative and grow in absolute value with n beyond all bounds, for sufficiently large n there must be arbitrarily many negative classes of J^n in addition to the two above (with index sum $-6 > -8(p-1)$). These additional classes must all be of type $\nu = 0$, $\mu = 2$, and their linear principal domains must lie in the null set on ϕ complementary to the two non-degenerate principal domains. Under odd powers of I, the endpoints on E of the linear principal domains are exchanged or permuted among themselves, so that they lead to positive classes.

22. Application to a special kind of mapping classes. To conclude, we now make a brief application to a special kind of mapping class, for which the solution of the general fixed point problem presented in I, 37 can be completely carried out.

Let I be an automorphism of type $\nu > 1$, $\mu = 0$. Then the kernel domain and the principal domain are identical. Let h be a primary element of F whose axis A is a side of the kernel domain $\Delta(I)$. Now suppose that there is an automorphism I_1 of type $\nu > 1$, related to I, whose principal domain meets $\Delta(I)$ along A. Then A is also a side of the kernel domain $\Delta(I_1)$, since its endpoints are fundamental points. Let t be a T-function which induces I. We seek a T-function for I_1. By Theorem 19 the latter results from t by a power of h:

$$t_1 = h^m t \tag{44}$$

(cf. Theorem 17.) Let j be the fixed point free interval of regularity for I which lies between the fundamental points of h. Now, since one is able to replace h by h^{-1}, one can assume that the direction of movement in j under t is from $V(h)$ to $U(h)$. All $h^m y$ with $m \leq 0$ are then fixed point free in j. Let P be a point in j and let α be the subinterval of j from P to hP. Then $t\alpha$ is also a subinterval of j. Moreover, for all exponents e,

$$th^e\alpha = h^e t\alpha$$

because $th = ht$. Now let $m_0 > 0$ be chosen so large that $h^{m_0}t\alpha$ falls between α and $V(h)$. Then $h^m t$ is fixed point free in j for all $m \geq m_0$. Hence only finitely many exponents

$$0 < m < m_0$$

have to be checked for (44), and hence a solution for (44) can be found, if there is one; e.g., by the method of boundary point expansion in I. There can be no more than one solution (cf. I, 17).

Now suppose that a solution has been found. Then there are two possibilities: First, t_1 can be isogredient to t, namely, when there is an f such that $h^m = ff_I^{-1}$. Then $t_1 = ftf^{-1}$ and $\Delta(I_1) = f\Delta(I)$. Thus these two kernel domains cover the same kernel domain δ on ϕ. A lies over a closed geodesic on ϕ which contributes to the boundary of δ from both sides. It may be established at the outset, from the boundary sides of $\Delta(I)$, whether δ has any such double boundary side. When δ is not the result of cutting the whole surface along one or more curves, one can choose A as a boundary side of $\Delta(I)$ which corresponds to a one-sided boundary curve of δ. Then one has the second possibility, namely, that t_1 is not isogredient to t. This gives two different, adjacent, kernel domains δ and δ_1 on ϕ. If now I_1 again belongs to a type with $\mu = 0$, so that $\Omega(I_1) = \Delta(I_1)$, and if $\delta + \delta_1$ still does not fill up ϕ, then one can again choose a free boundary side of $\delta + \delta_1$ and continue the process. The process can possibly be continued until the union $\delta + \delta_1 + \cdots + \delta_{N-1}$ of the kernel domains fills up ϕ. It ends at the moment when the kernel domains $\Delta, \Delta_1, \ldots, \Delta_{N-1}$ and their equivalents fill up Φ, which can be recognized from the correspondences between the boundary sides.

In other words, we are dealing here with *those mapping classes of ϕ whose kernel domains fill* *up* ϕ, in the sense that they are adjacent to each other across a certain number of disjoint, simple closed geodesics.

Now, how may these mapping classes be characterised? We shall construct a mapping $\tau\phi$ which corresponds to this case. It will be the identity on ϕ outside some arbitrarily thin annular regions which enclose the finitely many curves bounding the kernel region. These annuli will be chosen so thin that they do not meet each other. Inside an annulus the mapping takes the form of a "twist," i.e., it is similar to the following mapping of the annulus $R_1 \leq r \leq R_2$, described in polar coordinates (r, θ) by

$$r' = r + (R_2 - r)\frac{r - R_1}{R_2 - R_1},$$

$$\theta' = \theta + 2\pi m\frac{r - R_1}{R_2 - R_1}$$

which is fixed point free in the interior of the annulus and the identity on the boundaries. Here the integer m is the exponent m in equation (44) for each curve, and the sign of m, and hence the direction of twist, has to be correctly chosen at each boundary curve between kernel domains. The associated T-functions then give the identity mapping $tx = x$ in their kernel domains, outside an arbitrarily thin strip of constant width along the boundary sides. The fixed point sets of $\tau\phi$ are N multiply-connected regions, one in each of the N kernel domains; there are no other fixed points. The indices of the N classes are $1 - v_s$, where v_s is the v-number of the sth kernel domain. Incidentally, one sees easily that the mapping $\tau\phi$ can be deformed so that each of the N negative classes becomes represented by a single point, without introducing inessential fixed point classes in the process.

Example 12 of I, 48 belongs to this case. The two essential classes shown there correspond to kernel regions which together fill up ϕ and meet along the curves corresponding to the axes of the elements b, c and $b^{-1}c$. Both have $v = 2$, $\mu = 0$ and hence $i = -1$. In the mapping constructed there the 3 annuli can be chosen in such a way that their complement on ϕ consists only of the two curves in Figure 19 (I, 48). However, one can also choose them to be arbitrarily thin, so that the fixed point set fills almost the whole surface.

Theorem 24. *The general fixed point problem is completely soluble for those mapping classes in which the associated kernel domains fill up the whole surface, i.e., those which can be realized by twists along a system of disjoint, simple, non-homotopic curves.*

Investigations in the Topology of Closed Orientable Surfaces, III*

JAKOB NIELSEN

Contents

Introduction

The present work contains an extension of the investigations which the author has published in two works under the same title[1] in this journal. At the end of the introduction to II we raised the question whether a fixed point class of positive index can remain of this character under unlimited iteration of the mapping. This question is now *answered in the negative* by means of a theorem of H. Kneser (§1), and it is shown (§2), that the "characteristic exponent," i.e., the level of iteration at which the fixed point class acquires an index ≤ 0, *is bounded by a function of the genus.* The complete exploitation of this result is left for a later publication. The investigation of arbitrary mapping classes naturally draws attention first of all to *mapping classes of finite order.* A mapping class is said to be of finite order n when a mapping in the nth iterate of the class is deformable into the identity. One can understand the special role of mapping classes of finite order in the general case from the following consideration: A fixed point class of positive index is represented by a mapping of the universal covering surface which is fixed point free on the boundary circle E. Suppose the characteristic exponent, which

*Translation by John Stillwell of Untersuchungen zur Topologie der geschlossenen zweiseitigen Flächen, III. [Acta Math. 58 (1932), 87–167.]

[1] Untersuchungen zur Topologie der geschlossenen zweiseitigen Flächen, I, Acta Math. **50**, [N18] and II, Acta Math. **53**, [N20]. These two works are cited in the text as I and II, with paragraph numbers.

exists by §1, is n. Then the nth power has fixed points on E. As a result, the concepts of principal and kernel domain introduced in II also acquire significance for positive fixed point classes. Now suppose that the nth power has, e.g., a fixed element group with $v > 1$ free generators. Then we get a kernel domain which covers a bounded subsurface of lower genus than the given surface. The given mapping class may then be interpreted, *as far as its action on the subsurface is concerned*, as one of finite order.

Consequently, mapping classes of finite order occupy the major part of the present work. However, §§3–6 are concerned with the general case; they introduce a comprehensive mapping class invariant in the form of a group T of topological mappings of the boundary E of the universal covering surface. T contains the fundamental group F of the surface as a normal subgroup, and the mapping classes of finite order are characterised by finiteness of the quotient group T/F.

Now, one has examples of mapping classes of finite order in the *mappings of finite order*, i.e., "periodic" surface mappings, a power of which equals the identity mapping. These have been investigated by various authors, in particular L. E. J. Brouwer and W. Scherrer, and Brouwer has related them to the sheet permutations of regular branched coverings[2] by means of his concept of the "modular surface of an involution." This makes known numerical relations for branched coverings and the Hurwitz formula applicable. *However, does a mapping class of finite order always contain a mapping of finite order*? Independently of the answer to this question, it is shown how one can obtain the Hurwitz formula from the Alexander formula in the general case, as a *relation between certain group theoretic numbers* of the above class invariant T (§8), and how one can obtain further general properties of the mapping classes of finite order (§9), without speaking of branched coverings. But then it is shown in §11 how the group T itself leads, in a series of examples valid for each p, via its group diagram *to the associated branched covering*, and hence to periodic mappings in the corresponding mapping class.

All the properties looked for in these works are invariants of mapping classes; they are also the same for two classes which are *equivalent*, i.e., conjugate in the group of all mapping classes. A necessary condition for this equivalence is agreement in the characteristic polynomial of the exponent sum matrix for the induced automorphism of the fundamental group. However, this condition is not sufficient; even agreement in the matrix itself is not sufficient. What if the question is restricted to mapping classes of finite order? In this case there are only finitely many polynomials for a given genus p. At the end of this work it is shown, for three particular polynomials, that they in fact uniquely determine systems of equivalent mapping classes. Whether this is generally true for mapping classes of finite order depends on whether T always leads to a branched covering via the group diagram. When this is the case, it is understandable why, for mapping classes of finite order, the necessary algebraic conditions for the equivalence of two mapping classes are also sufficient.

I thank Dr. Werner Fenchel, Göttingen, for much advice and valuable help; during his stay in Copenhagen I was able to discuss with him most of the present work.

1. Impermanence of positive fixed point classes. Let $\tau\phi$ be a topological mapping of a closed

[2] Translator's note: I use the term "branched covering" where Nielsen uses "Riemann surface," since this seems to convey his meaning better to the modern reader.

orientable surface ϕ of genus $p > 1$ onto itself. The nth iterate of the mapping is denoted by $\tau^n\phi$, $n \gtreqless 0$. Take the interior Φ of the unit circle E in the complex x-plane as the image of the universal covering surface of ϕ. Let F be the group of hyperbolic linear fractional transformations which are the covering motions of Φ over ϕ. Let $t\Phi$ be a topological mapping of Φ onto itself which covers $\tau\phi$. Then $t\Phi$ may be continuously extended by a topological mapping tE of the boundary circle onto itself (I, 28). The resulting topological mapping $t(\Phi + E)$ of the closed disc satisfies the system of functional equations

$$t(f(x)) = f_I(t(x)),$$

where f runs through the group F and f_I is the element corresponding to f under the automorphism I of F induced by t (I, 22). We write this concisely as

$$tf = f_I t. \tag{1}$$

Then τ^n is covered by t^n, and the latter function satisfies the functional equations

$$t^n f = f_{I^n} t^n \tag{2}$$

in $\Phi + E$, where I^n is the nth iterate of the automorphism.

The fixed point class of τ represented by t (I, 31) has an index $j(t)$ (I, 37) which, by II, 14, 15, depends only on the structure of the fixed point set of the boundary mapping tE, and hence only on I, and it is subject to the restriction

$$3 - 4p \leq j(t) \leq 1.$$

Moreover, $j = 1$ only when tE is fixed point free. Now if tE has any fixed points at all, so that $j \leq 0$, then t^nE, for $n > 1$, obviously has exactly the same fixed point set as tE, hence

$$j(t^n) = j(t).$$

The remainder of this paragraph is devoted to the proof of the following fact, which is basic in what follows, and which shows that the situation is otherwise with the positive fixed point classes.

Theorem 1. *While each fixed point class of non-positive index for a surface mapping $\tau\phi$ maintains the same index as a fixed point class for $\tau^n\phi$, under arbitrarily high iterations, a fixed point class of positive index becomes of non-positive index under sufficiently high iterations of the mapping.*

The proof is indirect. We assume, contrary to the assertion, that the topological boundary mapping t^nE is fixed point free for each positive n, and we have to show that this assumption is not compatible with the functional equation (2).

It will be convenient to begin by recalling properties of a circle (Jordan curve) mapping which, together with all its powers, is fixed point free[3]. Let E be the circle and let t be the given mapping, so that t^nE is fixed point free for all $n \neq 0$. Let P be an arbitrary point of E and let M be the set of accumulation points of the point set t^nP. M does not depend on P, but only on t. M is closed and mapped into itself by $t : tM = M$. M is either identical with E (as in the case, e.g., when t is a rotation of E through an irrational multiple of π) or a perfect nowhere dense

[3] Cf. my work: "Om topologiske Afbildninger af en Jordankurve paa sig selv," Matematisk Tidsskrift B, 1928, [N19].

set on E. In the latter case the everywhere dense complementary set $E - M$ consists of a denumerable infinity of intervals, no two of which share an endpoint.

Now if E' is another circle and χ is a topological mapping of E onto E', then $t' = \chi t \chi^{-1}$ is a topological mapping of E' onto itself which, along with all its powers, is fixed point free. Also, its accumulation point set M' makes up all of E' if and only if M makes up all of E. However, if one changes the definition of χ, in the case where $M < E$, to allow χ to be merely a continuous mapping of E onto E', under which the everywhere dense set $E - M$ of intervals is mapped onto an everywhere dense point set with the same cyclic order on E', then it is still meaningful to speak of a mapping $t' = \chi t \chi^{-1}$ of E', in the sense that two points of that point set correspond under t' just in case the associated intervals on E correspond under t. Then t' is a mapping of E' which, together with all its powers, is fixed point free on E', and whose set of accumulation points makes up all of E'. Now H. Kneser has shown that the mapping χ of E onto E' may always be chosen in such a way that the mapping t' of E' is a rotation (through an irrational multiple of π)[4].

Now suppose χ has been so chosen. Also, let s denote are length on E, and let σ denote arc length on E'. Given an arbitrary subarc of E, we can speak of its s-measure and its σ-measure, where the latter is defined as the length of the image arc under χ. Then the σ-measure of a subarc is zero only when $M < E$ and the subarc lies entirely in one of the intervals making up $E - M$. On E, σ is a uniformly continuous, monotonic (and, when $M < E$, piecewise constant) function of s. Let $m(\lambda)$ be the infimum of the s-measures of subarcs of σ-measure λ, as the position of the subarc varies on E; then $m(\lambda) > 0$ for $\lambda > 0$ because of the uniform continuity of σ.

Because of the above choice of χ, the given mapping tE has the property that the σ-length of a subarc is not changed under the mapping t, and hence it also does not change under all its powers t^n. *Thus the s-lengths of the images of a subarc of positive σ-length λ under all powers t^n do not fall below the value $m(\lambda)$.*

It will now be shown that this property of a mapping tE which is fixed point free, along with all of its powers, is incompatible with the functional equation (2). For each element f of F there are two "congruence circles" $C(f)$ and $fC(f)$ with the same (Euclidean) radius $r(f)$. (Cf. I, end of 2, and II, 1; $C(f)$ is the locus of the points $|f'(z)| = 1$.) In a sequence of elements with equal non-Euclidean displacement lengths, for which the Euclidean distance between the fundamental points tends to zero, the r values also tend to zero (II, 1). From this, and from the fact that displacement lengths occurring in F have a positive lower bound (1, 5) it follows that there are only finitely many elements of F with congruence circle radius r above a given positive bound R, as the following argument shows: Let α be the axis of a transformation (not necessarily belonging to F) whose displacement length equals the smallest occurring in F and whose congruence radius equals R, and let Γ be a circle concentric with E which touches the axis of α, Then $r \leq R$ for all those elements of F whose axes do not pass inside Γ. Also, let l be the displacement length of a transformation (not necessarily belonging to F) whose congruence circle radius is R and whose axis is a diameter of E. Then $r \leq R$ for all elements of F whose displacement length is $\geq l$ (cf. II, 1). Among the conjugacy classes of F, there are only

[4] H. Kneser: "Reguläre Kurvenscharen auf den Ringflächen," Math. Ann. **91**, pp. 142–144. The theorem is connected with investigations of H. Poincaré, E. E. Levi and P. Bohl. A detailed presentation of the Kneser proof may be found in the above-mentioned work of mine.

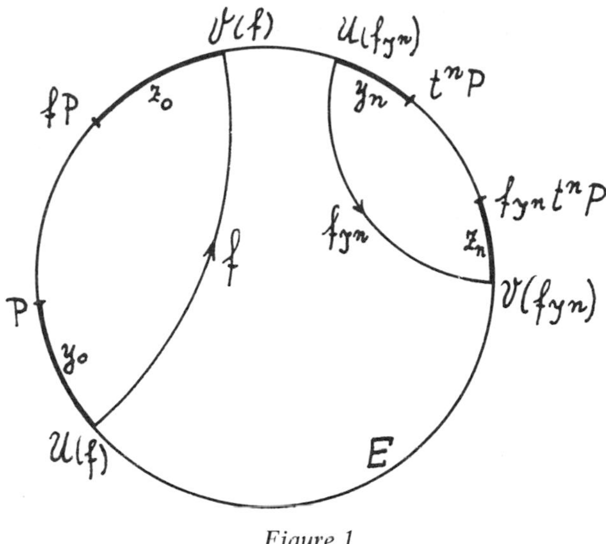

Figure 1

finitely many with displacement lengths $< l$ (I, 5). And only finitely many of the axes in a conjugacy class meet Γ. This proves the assertion. We can formulate it in the following:

Lemma. *In every infinite sequence of distinct elements of F the congruence circle radius tends to zero.*

Now let f be an arbitrary element of F, let P be a point of E different from the fundamental points of f, and let fP be its image under f (Figure 1). We let y_0 denote the subarc from $U(f)$ to P, and z_0 the subarc from fP to $V(f)$. The fundamental points $U(f)$ and $V(f)$ are sent to $U(f_{I^n})$ and $V(f_{I^n})$ by t^n. Also, P goes to t^nP and, by (2), fP goes to

$$t^n(fP) = f_{I^n}(t^nP),$$

so that y_0 is mapped onto y_n and z_0 onto z_n. Now, according as t^nP lies inside or outside the congruence circle for f_{I^n} with centre $U(f_{I^n})$, $f_{I^n}t^nP$ lies outside or inside the congruence circle with centre $V(f_{I^n})$. One of the two subarcs y_n, z_n therefore belongs entirely to a congruence circle of f_{I^n} (both do in the case where t^nP lies right on the congruence circle with centre $U(f_{I^n})$). Now the elements of the sequence f_{I^n}, $n = 0, 1, 2, \ldots$ are all different, since all t^nE ($n \neq 0$) are fixed point free by hypothesis. Thus the congruence circle radius of f_{I^n} tends to zero, by the lemma.

In at least one of the subarc sequences y_n and z_n the s-lengths must therefore have infimum zero. Hence we conclude from what was found above that at least one of the two subarcs y_0 and z_0 must have σ-length zero. It follows that the accumulation point set M does not make up all of E. We now also conclude that no fundamental point of F can belong to M. Suppose in fact that the fundamental point $V(f)$ of the element f belongs to M. Since no point of M is isolated, at least one of the subarcs into which E is divided by $U(f)$ and $V(f)$ contains M-points which accumulate towards $V(f)$. Among these one chooses a point P such that $U(f)P$ has positive σ-measure. The arc $fPV(f)$ has positive σ-measure, since $V(f)$ is an

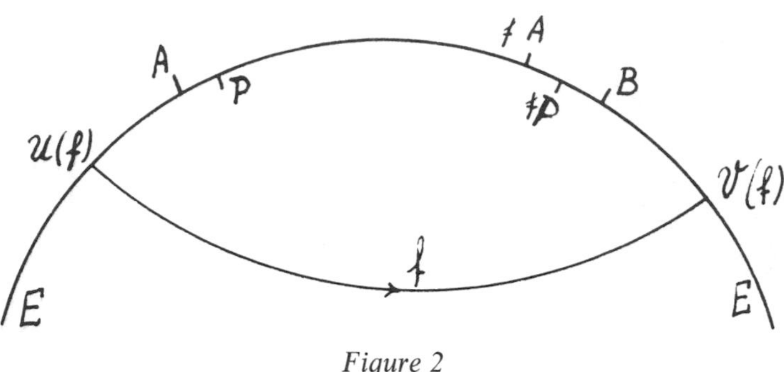

Figure 2

accumulation point of M-points on this side. One would then have subarcs y_0 and z_0 of the above kind with positive σ-measure, which we have just ruled out. Thus the fundamental points of each element f are interior points of intervals contained in $E - M$. We consider a subarc $U(f)V(f)$ of E which contains M-points. Let A and B be the endpoints on this subarc of the complementary intervals containing $U(f)$ and $V(f)$ respectively. If fA belongs to the arc AB (Figure 2), then one chooses a point fP between fA and B. The point P which results from application of f^{-1} then also belongs to AB, and the subarcs $U(f)P$ and $fPV(f)$ both have positive σ-measure. Thus we conclude that fA must be B or an interior point of the arc $BV(f)$. Since the same consideration applies to the other subarc $U(f)V(f)$, if it also contains M-points, in other words if the fundamental points of f do not belong to the same complementary interval, one sees that the whole set M is mapped by each element f of F into the closure of a complementary interval. Thus M and fM fall into two complementary subarcs of E. Now one can construct a non-Euclidean convex region on the closed set M by spanning each complementary interval by a circular arc orthogonal to E (in the same way as in the definition of the principal domain, II, 3). From what we have found we know that this region overlaps none of its equivalents under F, and hence it simply covers a certain region on the surface ϕ. However, it has an infinite non-Euclidean area, since M has infinitely many points. This contradicts the fact that ϕ has finite area, $4(p-1)$ (II, 5). The assumption that all t^nE $(n \neq 0)$ are fixed point free has therefore led to a contradiction.

This completes the proof of Theorem 1.

2. Restrictions on characteristic exponents due to genus. Once again, let t be a T-function for which tE is fixed point free. Then by Theorem 1 there is a smallest positive integer $n > 1$ such that t^nE has fixed points. This number n is called the *characteristic exponent* of t. Since tE depends only on the induced automorphism I, one can also call n the characteristic exponent of I. T-functions whose associated boundary mappings have fixed points are said to have characteristic exponent 1.

This definition of characteristic exponent deviates from that given in I, 11, when tE is fixed point free and t^nE has fixed points, but not fixed fundamental points, so that I^n has no fixed elements. The Theorem 2 there remains correct, and under the new definition its converse also holds: A fixed point free boundary mapping corresponds to an automorphism of the first kind with characteristic exponent > 1.

Now let P be an arbitrary fixed point of t^nE. Then the n points

$$P, tP, t^2P, \ldots, t^{n-1}P$$

are all different, since t^n is the lowest power of t which has fixed points on E. They divide E into n subarcs

$$B_1, B_2, \ldots, B_n$$

which are permuted by t, say

$$tB_i = B_{i+\rho},$$

where the subscripts are taken mod n and ρ is relatively prime to n. Now for each integer m we have identically:

$$t^n(t^m(x)) = t^m(t^n(x)).$$

Thus a suitable t^m converts the boundary mapping in the subarc B_1 to the boundary mapping t^n in B_i. In particular, t sends the fixed points of t^nE to fixed points of t^nE, and non-fixed points to non-fixed points; moreover, t sends isolated bilaterally repelling fixed points of t^n to others of the same kind. Hence t maps the point set M^*, defined in II, 3, as the boundary set of the *principal domain* $\Omega(I^n)$, one-to-one onto itself. Since any T-function also maps the set of fundamental points on E into itself, t maps the set of fixed fundamental points of t^n one-to-one onto itself, and hence also the set $S(I^n)$ of singular points of t^n (II, 5), which was defined to be the boundary set of the *kernel domain* $\Delta(I^n)$ in II, 8.

Now one subjects t^n to the type classification of II, 12. First suppose t^n is in a type with $\nu = 0$. Then t^nE has 2μ fixed points, μ of which are bilaterally attracting; we have (II, 6)

$$\mu \leq 4p - 2.$$

Since these must fall into n homeomorphic segments, from what we found above, n is a divisor of μ, and hence n is also subject to restriction by the genus,

$$n \leq 4p - 2.$$

Since $n > 1$, $\mu > 1$ also, and hence t^n has a negative index $1 - \mu$.

Next, suppose t^n is in a type with $\nu = 1$. Then there are exactly 2 singular points, and these must be exchanged by t. Hence it follows that

$$n = 2.$$

If $\mu = 0$, so that t^2 has two periodic fixed point free intervals, then the cyclic direction of movement under t^2 must be the *same* in both, since the mapping of one interval onto itself by t^2 is transformed by t into the mapping of the other by t^2. (Thus t^2E has no bilaterally repelling fixed point, and the idea of a (linear) principal and kernel domain is also upheld in this case.) If $\mu > 0$, then the regular boundary fixed points of t^2 are exchanged between the two intervals determined by the fixed element. Their number mod $H(I^2)$ is therefore the same in both intervals, and hence μ is even.

Finally, suppose $\nu > 1$. The fixed element group $H(I^n)$ is then a free group with $\nu \leq 2p - 1$ generators (II, 10), but not the whole of F. Since the fixed fundamental point set of t^n is mapped one-to-one onto itself by t, $H(I^n)$ is mapped isomorphically onto itself by the automorphism I. Now let

$$h_1, h_2, \ldots, h_\nu$$

be a system of independent generators of $H(I^n)$. I, which for the moment will be viewed only as an automorphism of $H(I^n)$, maps this onto a new system of independent generators

$$h'_1, h'_2, \ldots, h'_\nu,$$

where the latter elements can be represented as certain unique words in $h_1, \ldots, h_\nu$. One constructs the corresponding exponent sums: Let γ_{ik} be the exponent sum of h_k in the word for h'_i. Then one obtains a square $\nu \times \nu$ matrix:

$$\Gamma = (\gamma_{ik}), \qquad i, k = 1, \ldots, \nu.$$

Corresponding to the automorphism I^m of $H(I^n)$, we now have the matrix Γ^m, and corresponding to the automorphism I^n, which is the identity automorphism of $H(I^n)$, we have the $\nu \times \nu$ identity matrix E_ν:

$$\Gamma^n = E_\nu.$$

There is still the case in which I^n is the identity automorphism of F, $H(I^n) = F$, so that t^n is the identity mapping on E. Here $H(I^n)$ has $2p$ generators and it is not a free group, but since the exponent sum of each generator is zero in the defining relation of F, the exponent sum matrix Γ of I is again uniquely determined, and again

$$\Gamma^n = E_{2p}.$$

(Γ is the matrix already used, under a different name, at the end of I, 43, the trace of which appears in the Alexander formula.)

If one sets $\nu = 2p$ in the last-mentioned case, then the last two cases have led to the problem of determining all possible orders n of $\nu \times \nu$ integer matrices which one can solve most simply as follows: Let

$$\gamma(\lambda) = |\Gamma - \lambda E_\nu| = (-1)^\nu \lambda^\nu + c_{\nu-1}\lambda^{\nu-1} + \cdots + c_0 \tag{3}$$

be the characteristic polynomial of Γ, with roots

$$\lambda_1, \lambda_2, \ldots, \lambda_\nu$$

–the eigenvalues of Γ. Since $\Gamma^n = E_\nu$ has 1 as its only eigenvalue, the λ_i are n^{th} roots of unity. The coefficients of $\gamma(\lambda)$ are integers, the highest equal to ± 1. One splits $\gamma(\lambda)$ into factors

$$\gamma(\lambda) = \gamma_1(\lambda)\gamma_2(\lambda)\ldots\gamma_m(\lambda) \tag{4}$$

which are irreducible over the rationals. Then $\gamma_i(\lambda)$ is an integer polynomial with highest coefficient ± 1, whose roots form a complete system of primitive roots of unity of degree d_i, which is a divisor of n. The degree of $\gamma_i(\lambda)$ is therefore $\phi(d_i)$, where ϕ denotes the Euler function. Thus one has

$$\phi(d_1) + \phi(d_2) + \cdots + \phi(d_m) = \nu. \tag{5}$$

The determination of all possible orders n for a given degree ν now proceeds as follows: One determines the finitely many natural numbers d satisfying the condition

$$\phi(d) \leq \nu, \tag{6}$$

constructs the finitely many additive partitions of the number v into terms $\phi(d)$ as in (5), and for each such partition one determines the associated order n as the lowest common multiple of the numbers d used. For a given v there are finitely many possibilities for n.

Now since the given genus p of our surface allows only the possibilities $2, 3, \ldots, 2p$ for v, the last two of the four cases above end with only finitely many possibilities for n. Since the same was true in the first two cases, we have proved:

Theorem 2. *There is an upper bound, depending only on the genus of the surface, on the characteristic exponents of T-functions.*

This is at the same time a bound on the "order of permanence" of the fixed point classes of positive index on a surface of given genus.

3. Algebraic tools. In this paragraph we pursue the properties of exponent sum matrices (well known, for the most part) somewhat further than in the previous paragraph, but retaining the notation used there. Because of the relationship with the Alexander formula (I, 43), we focus on the traces

$$s_r = s(\Gamma^r)$$

of the matrices and their powers. Here

$$s_r = \lambda_1^r + \lambda_2^r + \cdots + \lambda_v^r$$

is the sum of the rth powers of the eigenvalues of Γ. Hence, by (3), one has the linear recurrence relation in the s_r:

$$(-1)^v s_{v+r} + c_{v-1} s_{v-1+r} + \cdots + c_0 s_r = 0. \tag{7}$$

The lowest degree polynomial

$$\chi(\lambda) = \lambda^\rho + k_{\rho-1}\lambda^{\rho-1} + \cdots + k_0,$$

which yields the zero matrix when Γ is substituted for λ, divides $\gamma(\lambda)$. Since $\rho < v$ is possible, this yields a shorter recurrence relation

$$s_{\rho+r} + k_{\rho-1} s_{\rho-1+r} + \cdots + k_0 s_r = 0. \tag{7a}$$

By means of the Alexander formula, this yields recurrence relations for the determination of index sums of powers of a given surface mapping.

Now let Γ be a matrix of finite order:

$$\Gamma^n = E_v.$$

Then the eigenvalues λ_i are n^{th} roots of unity, hence equal to ± 1 or pairwise complex conjugates, and therefore reciprocal. The equation $\gamma(\lambda) = 0$ is therefore a reciprocal equation, hence its coefficients read the same from both ends, though possibly with opposite signs.

Now one lets

$$\lambda^n - 1 = (\lambda - 1)(\lambda^{n-1} + \lambda^{n-2} + \cdots + \lambda + 1) = (\lambda - 1)Q_n(\lambda).$$

The roots of $Q_n(\lambda)$ have sum -1. When $\Gamma^n = E_v$ the λ_i are either 1 or roots of $Q_n(\lambda)$; the eigenvalues of Γ^r are the rth powers of the λ_i.

(1) Suppose first that n is a prime number, so that $Q_n(\lambda)$ is irreducible over the rationals, hence, since $\gamma(\lambda)$ has integer rational coefficients,

$$\gamma(\lambda) = (-1)^{\nu}(\lambda - 1)^{\alpha} Q_n(\lambda)^{\beta}$$

with $\alpha \geq 0$, $\beta > 0$. Also,

$$\alpha + (n-1)\beta = \nu$$

since ν is the degree of $\gamma(\lambda)$. The trace $s_1 = s(\Gamma)$ is then

$$s_1 = \sum \lambda_i = \alpha - \beta = \nu - n\beta.$$

When $1 < r < n$ the roots of Q_n are permuted when we raise to the power r, hence

$$\begin{aligned} s_1 &= s_2 = \cdots = s_{n-1} = \nu - n\beta, \\ s_n &= \nu. \end{aligned} \tag{8}$$

(This is of course only a very special consequence, following from well-known algebraic theorems, of the "extended Fermat theorem," which says that, for each integer matrix Γ, the characteristic polynomials of Γ and Γ^n are congruent modulo n when n is prime.)

From

$$\beta = \frac{\nu - \alpha}{n-1} \leq \frac{\nu}{n-1}$$

it also follows that

$$s_1 = \nu - n\beta \geq -\frac{\nu}{n-1}. \tag{9}$$

(2) In the case of a general n we go back to the decomposition (4) of $\gamma(\lambda)$ with condition (5) on the degrees of the individual factors. The d_i need not be different, hence $\gamma(\lambda)$ can contain powers of irreducible factors. We therefore write the relation (4) in a new notation:

$$\gamma(\lambda) = \prod_{d|n} \gamma_d(\lambda)^{\varepsilon_d} \tag{4a}$$

with $\varepsilon_d \geq 0$, where d runs through the divisors of n, n is the lowest common multiple of the divisors d which actually appear, and

$$\sum_d \varepsilon_d \phi(d) = \nu.$$

It is known that the sum $\mu(d)$ of the roots of γ_d satisfies

$\mu(d) = 0$ if d has a multiple prime factor,

$\mu(d) = 1$ if d has an even number of distinct prime factors,

$\mu(d) = -1$ if d has an odd number of distinct prime factors.

This sum of roots, multiplied by ε_d, gives the contribution to the trace of Γ due to the divisor d of n. Thus, for a given order n, there are only finitely many possibilities for $s(\Gamma)$, which are easily derived from the possible d-combinations. Here we have, in dependence only on the size

v of Γ,

$$-v \leq s(\Gamma) \leq v \tag{10}$$

since the eigenvalues of Γ have real parts between -1 and $+1$. In particular, $s(\Gamma) = v$ only for $\Gamma = E_v$ and $s(\Gamma) = -v$ only for $\Gamma = -E_v$.

In order to determine $s_r = s(\Gamma^r)$ from the d-combinations for Γ, one considers an irreducible factor $\gamma_d(\lambda)$ of $\gamma(\lambda)$. If $\delta = (r, d)$ is the highest common factor of r and d, and if $d = \delta d'$, then a primitive d^{th} root of unity is sent to a primitive d'^{th} root by raising to the power r, and thus to a root of $\gamma_{d'}$ of degree $\phi(d')$. The single factor γ_d in the characteristic polynomial of Γ therefore corresponds to $\phi(d)/\phi(d')$ factors $\gamma_{d'}$ in the characteristic polynomial of Γ^r. This allows the d-combination of Γ^r to be derived from that of Γ, and also enables the trace s_r to be determined. For each d relatively prime to r the contribution to the trace is unaltered. In particular,

$$s(\Gamma^r) = s(\Gamma) \tag{11}$$

for each r relatively prime to the order n of Γ.

Now we impose the condition $v = 2p$. The coefficient c_0 in $\gamma(\lambda)$ is the determinant of the matrix Γ, and hence ± 1, because of the fact that Γ^{-1} is integral. For matrices of automorphisms of F of the first kind it is therefore $+1$.[5] Thus the characteristic polynomial has the form

$$\gamma(\lambda) = \lambda^{2p} - s_1 \lambda^{2p-1} + \cdots - s_1 \lambda + 1, \tag{12}$$

with the same sequence of coefficients from both ends. We now consider an irreducible factor $\gamma_d(\lambda)$ of $\gamma(\lambda)$. When $d > 2$ the degree $\phi(d)$ of γ_d is even, and the roots of γ_d have product $+1$, since they are reciprocal in pairs. Since (12) now shows that the product of all the roots of $\gamma(\lambda)$ is likewise $+1$, the number ε_2 of linear factors $\gamma_2 = \lambda + 1$ with root -1 must be even. And since the total number of roots of $\gamma(\lambda)$ is $2p$, the number ε_1 of linear factors $\gamma_1 = \lambda - 1$ with eigenvalue 1 must also be even.

Finally we determine the trace in the case $\Gamma^n = E_{2p}$. An eigenvalue λ satisfies the equation

$$1 + \lambda + \lambda^2 + \cdots + \lambda^{n-1} = \begin{cases} n & \text{if } \lambda = 1 \\ 0 & \text{if } \lambda \neq 1 \end{cases}.$$

Thus by summing over all eigenvalues for the traces $s_r = s(\Gamma^r)$ we get the relation

$$s_0 + s_1 + s_2 + \cdots + s_{n-1} = \varepsilon_1 n$$

and hence, since $s_0 = 2p$,

$$s_1 + s_2 + \cdots + s_{n-1} = \varepsilon_1 n - 2p. \tag{13}$$

It remains only to prove the above assertion about the value of the determinant of Γ. If B denotes the bilinear form

$$B = x_1 y_2 - x_2 y_1 + x_3 y_4 - x_4 y_3 + \cdots + x_{2p-1} y_{2p} - x_{2p} y_{2p-1}$$

and if X and Y are two curves on the surface of genus p, then B gives the algebraic sum of their oriented intersections when x_i and y_i denote the exponent sums of $a_1, b_1, \ldots, a_p, b_p$ in X and

[5] A proof of this fact is included at the end of this paragraph.

Y respectively. If one subjects the surface to a topological mapping whose associated automorphism family has the exponent sum matrix $\Gamma = (\gamma_{ik})$, then B must either remain unaltered or change sign under application of the transformation Γ to both series of variables, according as the mapping preserves orientation or not, and hence according as the automorphism family is of first or second kind. Now if one lets (m, n) denote the value of B when the values of x_i, resp., y_i are replaced by the mth, resp., n^{th} column of Γ, then one finds by substitution that the necessary and sufficient condition for B to be reproduced by Γ with a factor η is that

$$(1, 2) = (3, 4) = \cdots = (2p - 1, 2p) = \eta$$

and all the remaining (m, n) vanish. (And one sees incidentally that Γ then has this property with respect to rows also.)

On the other hand, *any* $2p \times 2p$ determinant $|\Gamma|$ has an "expansion in terms of the bilinear form B," i.e., in terms of the (m, n). Namely, one lets $\{\alpha_1\alpha_2 \ldots \alpha_{2p}\}$ denote the number $+1$ or -1 according as $\alpha_1, \alpha_2, \ldots, \alpha_{2p}$ is an even or odd permutation of the numbers $1, 2, \ldots, 2p$, and sets

$$\begin{vmatrix} \gamma_{2i-1,\alpha_1} & \gamma_{2i-1,\alpha_2} \\ \gamma_{2i,\alpha_1} & \gamma_{2i,\alpha_2} \end{vmatrix} = (\alpha_1, \alpha_2)_i.$$

Thus

$$(\alpha_1, \alpha_2) = (\alpha_1, \alpha_2)_1 + (\alpha_1, \alpha_2)_2 + \cdots + (\alpha_1, \alpha_2)_p.$$

Then

$$|\Gamma| = |\lambda_{ik}| = \sum \{\alpha_1\alpha_2 \ldots \alpha_{2p}\} \gamma_{1\alpha_1}\gamma_{2\alpha_2}\gamma_{3\alpha_3} \cdots \gamma_{2p,\alpha_{2p}}$$

(where the sum is taken over all permutations $\alpha_1, \ldots, \alpha_{2p}$ of the numbers $1, \ldots, 2p$)

$$= \sum \{\alpha_1\alpha_2 \ldots \alpha_{2p}\} (\alpha_1, \alpha_2)_1(\alpha_3, \alpha_4)_2 \ldots (\alpha_{2p-1}, \alpha_{2p})_p$$

(sum over all permutations for which $\alpha_{2i-1} < \alpha_{2i}$)

$$= \sum_{(v)} \sum_{(\alpha)} \{\alpha_1\alpha_2 \ldots \alpha_{2p}\} (\alpha_1, \alpha_2)_{v_1}(\alpha_3, \alpha_4)_{v_2} \ldots (\alpha_{2p-1}, \alpha_{2p})_{v_p}.$$

Here $\sum_{(v)}$ is taken over all permutations $v_1, v_2, \ldots, v_p$ of the double row numbers $1, 2, \ldots, p$ and $\sum_{(\alpha)}$ over all column permutations with $\alpha_{2i-1} < \alpha_{2i}$ and

$$\alpha_1 < \alpha_3 < \alpha_5 < \cdots < \alpha_{2p-1}.$$

But then one can also allow the permutations (v) to be permutations with repetition. This is because a term in the double sum above, in which $v_r = v_s$ say, gives value zero when $\sum_{(\alpha)}$ is carried out for fixed (v), since one gets a determinant constructed from the double rows of Γ in which at least one double row appears twice. If one now expands $\sum_{(v)}$ in single terms, then the desired expansion results (as an identity valid for each $2p \times 2p$ determinant):

$$|\Gamma| = \sum \{\alpha_1\alpha_2 \ldots \alpha_{2p}\} (\alpha_1, \alpha_2)(\alpha_3, \alpha_4) \ldots (\alpha_{2p-1}, \alpha_{2p}),$$

where the sum is taken over all column permutations (α) for which $\alpha_{2i-1} < \alpha_{2i}$ and $\alpha_1 < \alpha_3 < \alpha_5 \ldots < \alpha_{2p-1}$.

Now if Γ is to transform B into ηB, then we see from the above that each term in the sum for $|\Gamma|$ contains a vanishing factor, except that for which $\alpha_1, \alpha_2, \ldots, \alpha_{2p}$ is in the natural order $1, 2, \ldots, 2p$, and in the latter each factor equals η. Thus

$$|\Gamma| = \eta^p.$$

Consequently we have the theorem:

The matrices for automorphisms of the first kind have determinant $+1$ for each p; those for automorphisms of the second kind have determinant $+1$ or -1 according as p is even or odd.

It may be remarked that this theorem is derivable from the general investigations of Cayley[6] on automorphisms of bilinear forms, and that one can likewise obtain it from the form of the generating automorphisms of the special bilinear form B, given by Poincaré[7]. However, the direct proof given here may perhaps also be of interest.

4. The group *T* corresponding to a mapping class. Let $\tau\phi$ be a mapping. When the associated T-functions include some with fixed point free boundary mappings, the first two paragraphs lead us to raise them to their characteristic exponents. We now make this step, by considering all powers τ^n, $-\infty < n < \infty$, and all their associated T-functions. Suppose then that t is a T-function over τ, and that t induces the automorphism I of F. Then t^n induces the automorphism I^n, and the functional equation (2) holds for all n. Now if

$$f_0 = 1, f_1, f_2, \ldots$$

is an enumeration of all elements of F, then one considers the system of T-functions

$$\begin{array}{cccc} & \cdots & & \\ t^{-2} & f_1 t^{-2} & f_2 t^{-2} & \cdots \\ t^{-1} & f_1 t^{-1} & f_2 t^{-1} & \cdots \\ 1 & f_1 & f_2 & \cdots \\ t & f_1 t & f_2 t & \cdots \\ t^2 & f_1 t^2 & f_2 t^2 & \cdots \\ & \cdots & & \end{array} \tag{14}$$

each row of which results from joining t^n to all elements of F. Because of (2) we have

$$f_i t^m f_k t^n = f_i f_{kI^m} t^{m+n},$$
$$(f_i t^m)^{-1} = t^{-m} f_i^{-1} = f_{iI^{-m}}^{-1} t^{-m},$$
$$f_i t^m \cdot f_k \cdot (f_i t^m)^{-1} = f_i \cdot f_{kI^m} \cdot f_i^{-1}.$$

These equations show that the above system (14) constitutes a group, which we denote[8] by T, and that T contains F as a normal subgroup whose cosets are the rows of the above system.

[6] See, e.g., the French edition of the Enzyklopädie I, 2, 4, p. 482.

[7] Palermo Rend. **18** (1904), p. 64.

[8] The letter T no longer denotes the group of all T-functions, as in I, 21, but the group of those T-functions which lie over a given surface mapping τ and its powers.

Now if one views T *only as a group of mappings of the boundary circle E* then T is associated with the mapping *class* of τ. We take this view in determining the quotient group T/F. In the general case, no power of tE falls in the group F; then all the elements of the system (14) are different mappings of E onto itself, and T/F is the free group on one generator (infinite cyclic group). Otherwise one determines the smallest positive exponent n for which t^nE is an f_iE. Then τ^n belongs to the class of the identity. When $n = 1$, τ itself is in the class of the identity and T is identical with F on E. When $n > 1$, T/F is the cyclic group of order n.

We let T_r denote the row in (14) beginning with t^r, which is the complete family of T-functions related to t^r. We call the elements of T_r the "functions of level r." Raised to the mth power, they give functions of level mr, but not in general all of T_{mr}. Thus in the case where T/F is of finite order n, the element $f_0 = 1$ of level n and 0 is not necessarily the nth power of an element of level 1; we shall discuss this more thoroughly later.

5. Correspondence of isogredience classes of different levels under powers. The element f_it^m of T induces the automorphism $I^m_{f_i^{-1}}$. The latter has fixed element group $H(I^m_{f_i^{-1}})$, which we also denote by $H(f_it^m)$; it is the subgroup of F consisting of all elements which commute with f_it^m. We also recall the definition of an isogredience class of T-functions (resp., automorphisms) given in I, 18: All T-functions which result from f_it^m by conjugation with arbitrary elements of F constitute an isogredience class; its elements are all of level m. In addition, one notes that the element t in (14) can be an arbitrary element of level one, and when one constructs the group T for τ^r one obtains the rows T_{mr} of the above schema, among which T_r is now of level one. Thus if one wants to investigate the behaviour of elements and isogredience classes of T under power iteration, it is no restriction to start from the element t.

Let $H = H(I) = H(t)$ be the fixed element group of I and let

$$F = H + g_1H + g_2H + \cdots$$

be a coset decomposition of F modulo H. Since the elements of H commute with t, the elements

$$t, g_1tg_1^{-1}, g_2tg_2^{-1}, \ldots \tag{15}$$

constitute the complete isogredience class of t. One raises the elements of (15) to the nth power and obtains

$$t^n, g_1t^ng_1^{-1}, g_2t^ng_2^{-1}, \ldots. \tag{16}$$

Since $H(t)$ is contained in $H(t^n)$, (16) contains all elements isogredient with t^n. Two equal elements appear in (16) if and only if $H(t^n)$ properly includes $H(t)$; ($g_it^ng_i^{-1} = g_kt^ng_k^{-1}$ implies $g_k^{-1}g_i \rightleftarrows t^n$, but by hypothesis, not $\rightleftarrows t$). Then tE is fixed point free, $H(t) = 1$, n is a multiple of the characteristic exponent of t, and I^n has fixed elements. In this case the elements g_i run through all of F and each element isogredient with t^n appears infinitely often in (16).

From what we have said above, it is evidently meaningful to speak of raising an isogredience class to a power, since this sends a complete class to a complete class. We now investigate the conditions under which two different isogredience classes of level one can coincide when raised to the nth power. We can choose representatives t and ft of the two classes in such a way that

$$(ft)^n = t^n. \tag{17}$$

Then f does not lie in $H(t)$, because then $f \rightleftarrows t$ would imply

$$(ft)^n = f^n t^n,$$

whence $f^n = 1$ by (17), and hence $f = 1$. Therefore f also does not lie in $H(ft)$. Because of (17), t^n commutes with ft, and since t^n clearly commutes with t, it also commutes with f. Thus f lies in $H(t^n)$, so $H(t^n)$ properly includes $H(t)$. Similarly, $H((ft)^n)$ properly includes $H(ft)$. We conclude that both t and ft are fixed point free on E. Also, they must both have the same characteristic exponent, because t^m and $(ft)^m$, for arbitrary m, have coincident n^{th} powers. They must therefore either both be fixed point free on E, or isogredient, and then necessarily equal. Thus they coincide when raised to their common characteristic exponent, and the above n is a multiple of the latter.

Now suppose that n is this characteristic exponent itself. We have

$$(ft)^n = ff_I f_{I^2} \ldots f_{I^{n-1}} t^n,$$

so f is a solution of the automorphism equation

$$xx_I x_{I^2} \ldots x_{I^{n-1}} = 1. \tag{18}$$

Thus there are as many isogredience classes of level one which coincide with the class of t^n when raised to their n^{th} powers as there are solutions of (18) non-isogredient relative to I. This number is finite, since the number of fixed point classes is finite. One should compare this with I, 44. The difference is that, here, one need not necessarily have $I^n = 1$. This is also expressed in the following fact: Let h be a solution of (18) and let k be an arbitrary element of F. Then

$$(khk_I^{-1})(khk_I^{-1})_I \ldots (khk_I^{-1})_{I^{n-1}} = kk_{I^n}^{-1};$$

hence it equals 1 only if k is a fixed element under I^n. Thus if one has two solutions h_1 and h_2 of (18) they are isogredient relative to I only when

$$h_2 = kh_1 k_I^{-1}$$

for a suitable *fixed element* k *of* I^n.

We formulate the main result of this paragraph as:

Theorem 3. *The elements of an isogredience class of arbitrary level yield, when raised to a power, all the elements of an isogredience class at the level of the power, and indeed without repetitions in general. However, there is infinite repetition in the case where the initial class is fixed point free on E and the power class has fixed elements.*

Two different isogredience classes can coincide, after being raised to a power, only when they have a common characteristic exponent $n > 1$, *of which the power is a multiple. Then they already coincide in the* n^{th} *power and have fixed elements there. The number of classes which coincide in this way is given by the number of non-isogredient solutions of the automorphism equation* (18).

6. Correspondence of isogredience classes under conjugation by elements of *T*. For the moment we let the symbol $[f_i t^r]$ denote the isogredience class of the element $f_i t^r$ of T. Now if g denotes an element of F we have

$$t \cdot gf_i t^r g^{-1} \cdot t^{-1} = g_I f_{iI} t^r g_I^{-1}.$$

If one lets g run through all of F, then g_I also runs through F, and one can write

$$t \cdot [f_i t^r] \cdot t^{-1} = [f_{iI} t^r].$$

It follows that conjugation of a complete isogredience class by an arbitrary element of T gives a complete isogredience class of the same level; the two classes may coincide or be different. In particular, the equation

$$t \cdot f_i t \cdot t^{-1} = t f_i = f_i^{-1} \cdot f_i t \cdot f_i$$

shows that a class of arbitrary level is mapped to itself when conjugated by an element of the same level. Classes of level one therefore commute with all of T, and the same is true of classes which are powers of classes of level one. However, there are still more classes which commute with T. Namely, let k be a fixed element of the automorphism induced by $f_i t$, so that $k \rightleftarrows f_i t$; then we also have

$$f_i t \rightleftarrows k(f_i t)^r,$$

so that $[k(f_i t)^r] \rightleftarrows T$. Now the converse also holds: If $[ft^r] \rightleftarrows T$, then ft^r may be written in the form $k(gt)^r$ with $k \rightleftarrows gt$. Because, by hypothesis, there is an element g of F such that

$$t \cdot ft^r \cdot t^{-1} = g^{-1} \cdot ft^r \cdot g;$$

hence $gt \rightleftarrows ft^r$. Now

$$(gt)^r = g_r t^r = g_r f^{-1} \cdot ft^r$$

when one sets, as in II, 17,

$$g_r = g g_I g_{I^2} \ldots g_{I^{r-1}}.$$

The element

$$k = f g_r^{-1} = ft^r \cdot (gt)^{-r}$$

commutes with gt, since ft^r and $(gt)^{-r}$ do. And

$$ft^r = k \cdot (gt)^r$$

as claimed. Thus k commutes also with ft^r.

Now let gt^r be an element of T, and suppose $[gt^r]$ does not commute with all of T. Thus $[gt^r]$ is not conjugate to itself via t. Let t^m, $m > 1$, be the lowest power of t which conjugates $[gt^r]$ to itself. We have $m \leq r$, since $[gt^r]$ is conjugate to itself via t^r. Then the elements

$$\begin{aligned} >^r, \\ t \cdot gt^r \cdot t^{-1} &= g_I t^r, \\ t^2 \cdot gt^r \cdot t^{-2} &= g_{I^2} t^r, \\ &\ldots \\ t^{m-1} \cdot gt^r \cdot t^{-(m-1)} &= g_{I^{m-1}} t^r \end{aligned}$$

belong to m different isogredience classes of level r which are permuted cyclically under conjugation by t. Consequently, m must divide r. Then these m classes are conjugate to themselves via exactly those elements of T which belong to τ^m, and hence via all elements of level mq. Hence by the above theorem there is a representation

$$gt^r = k \cdot (ht^m)^{r/m}, \qquad k \rightleftarrows ht^m, \tag{19}$$

where of course k can also be 1. Thus k also commutes with gt^r. For the sake of conciseness we can say that the class $[gt^r]$—and of course the other classes in the m term cycle to which it belongs—*takes root at level m*. Then the results of this paragraph may be summarized as follows:

Theorem 4. *The elements of a complete isogredience class of arbitrary level are sent, under conjugation by an arbitrary element of T, to the elements of a complete isogredience class of the same level. For any class of level r there is a number m, $1 \leq m \leq r$, which divides r, such that the class is conjugate to itself via t^m, but not via any lower power of t. A class of level r which "takes root at level m" in this way admits a representation of the form* (19). *It belongs to a cycle of m classes of level r which all take root at level m and which are permuted cyclically among themselves under conjugation by any element of T whose level is not a multiple of m.*

7. The case of finite quotient group T/F. Now we discuss these facts and their consequences in the case where the quotient group T/F is finite, and hence cyclic of a certain order n (as always, viewing T as a group of mappings of E alone). We assume $n > 1$. Then T_n is identical with $T_0 = F$. The implication of this for the surface mapping $\tau\phi$ is that $\tau^n\phi$ falls in the mapping class of the identity, and that this is not the case for any lower exponent. The implication for the automorphism I induced by t is that I^n is an inner automorphism of F (possibly the identity). Thus the matrix Γ of I satisfies $\Gamma^n = E_{2p}$.

The isogredience classes at the level of $T_n = F$ are none other than the conjugacy classes of F. One of these is distinguished in that it consists of the element $f_0 = 1$ alone. This class takes root at level one, since 1 commutes with t. The other classes can take root at various levels.

Now we take an element gt^r of level r, where $1 \leq r \leq n$. Let (n, r) be the greatest common divisor of n and r. Then $(gt^r)^{n/(n,r)}$ lies in T_n, and hence it is either 1 or an element $f \neq 1$ of F.

(a) The assumption

$$(gt^r)^{n/(n,r)} = f$$

implies

$$f \rightleftarrows gt^r$$

and hence gt^r has fixed elements. Each of its powers have these same fixed elements. But the $(n/(n, r))^{\text{th}}$ power is the operation f on E, hence it has only the two fundamental points of f as fixed points and its fixed element group is the infinite cyclic group $\{f^*\}$, where f^* is a certain primary element and f is a power of f^*. Then $\{f^*\}$ is also the fixed element group of gt^r, gt^rE has two periodic fixed point free intervals in which the directions of movement are opposite, and the index of the fixed point class of $\tau^r\phi$ corresponding to gt^r is zero (gt^r is of type $\nu = 1$, $\mu = 0$).

(b) The assumption

$$(gt^r)^{n/(n,r)} = 1$$

implies however that gt^r is of type $\nu = 0$, $\mu = 0$, hence fixed point free on E, so the result is a fixed point class of index $+1$ for $\tau^r\phi$.

These two possible behaviours are really assertions about the isogredience class of gt^r, since elements of the same isogredience class in this respect behave in the same way. In case (b) we can ask: At what level does the given isogredience class of T_r take root? Let it be level m. Then by Theorem 4, gt^r admits a representation (19), where k is a common fixed element of ht^m and gt^r. However, since gt^r is fixed point free on E, it has no fixed elements other than 1. Thus $k = 1$ and gt^r is the r/m^{th} power of an element ht^m of level m. The latter must itself be fixed point free on E, since one of its powers is. And it also takes root at level m, since it would otherwise be a pure power and gt^r would take root at a lower level. Our isogredience class of level r therefore commutes with all T_{mq}, and only these, and since it obviously commutes with T_n, m must divide n. By Theorem 4, our isogredience class is now a member of a cycle of m classes of level r, which are permuted cyclically under conjugation by t (the classes of gt^r, $g_I t^r$, $\ldots$, $g_{I^{m-1}} t^r$). Each of these therefore yields a fixed point class of index $+1$. In fact, such a cycle ht^m, $h_I t^m, \ldots, h_{I^{m-1}} t^m$ is already present on level m, and it asserts itself at all levels whose orders are multiples of m (in particular, the rth), until the nth, where everything is the identity boundary mapping.

Here we are dealing with m sets of points on the surface φ which are permuted cyclically by $\tau, \tau^2, \ldots, \tau^{m-1}$, and which first appear as m separate fixed point classes under τ^m. One could call them "conjugate fixed point classes of τ^m." Of course, τ^m may have other systems of m conjugate fixed point classes, and even other fixed point classes which already appear under lower powers of τ, corresponding to isogredience classes of T_m which are fixed point free on E and which take root at a lower level.

This reveals I, 45, and especially Theorem 23 there, in a clearer light, and extends it to all cases of a finite quotient group T/F: Again let n be the order of the mapping *class* of τ, i.e., τ^n belongs to the class of the identity. All powers $\tau, \tau^2, \ldots, \tau^{n-1}$ have only fixed point classes of index $+1$ or zero. A coincidence between initially different fixed point classes occurs only at the power τ^n, which has only one essential fixed point class, in fact a negative one. Distinct positive fixed point classes of τ assert themselves as distinct positive classes under $\tau^2, \ldots, \tau^{n-1}$. In addition to these, new positive classes can appear for each τ^m where m divides n, and indeed the number of classes first appearing for τ^m is a multiple of m. These also appear for $\tau^{2m}, \ldots,$ $\tau^{(n/m-1)m}$, and only for these. The number of classes for τ^r is therefore greater than or equal to the number of classes for τ, and consequently the traces of the matrices satisfy

$$s(\Gamma^r) \leq s(\Gamma), \tag{20}$$

or more generally,

$$s(\Gamma^{\mu m}) \leq s(\Gamma^m). \tag{20a}$$

8. The Hurwitz formula. We now use these results—still in the case of a cyclic quotient group of order n, hence a "mapping class of order n"—to calculate the sum $\mathfrak{S}$ of the class numbers of $\tau, \tau^2, \ldots, \tau^{n-1}$. An m-termed cycle of the above kind gives rise to m fixed point classes of index $+1$, which appear under $\tau^m, \tau^{2m}, \ldots, \tau^{(n/m-1)m}$ and nowhere else. Their total contribution to $\mathfrak{S}$ is therefore $(n/m - 1)m$. This contribution is now counted as often as the number A_m of such m-termed cycles. And then one has to let m run through all the divisors of n. Thus one obtains

$$\mathfrak{S} = \sum_{m|n} A_n(n - m).$$

On the other hand, τ^r has class number equal to the index sum, since the only fixed point classes are those with index 0 (which do not count) and index 1, hence the class number is $2 - s_r$ by the Alexander formula, where s_r is the trace of the rth power Γ^r of the matrix Γ of I. Consequently,

$$\mathfrak{S} = 2(n-1) - (s_1 + s_2 + \cdots + s_{n-1})$$
$$= 2n - 2 - \varepsilon_1 n + 2p$$

by (13). Recall that ε_1 was the multiplicity of the eigenvalue 1 in Γ, and ε_1 was even. It follows that

$$\sum_{m|n} A_m(n-m) = 2p - 2 - n(\varepsilon_1 - 2). \tag{21}$$

Now if one sets $n = m\zeta$ and agrees to omit the factors A_m in favour of *counting the values* ζ *as often as there are* n/ζ*-term cycles*, then (21) can be written

$$\sum_{\zeta|n} \frac{n}{\zeta}(\zeta - 1) = 2p - 2 - n(\varepsilon_1 - 2). \tag{21a}$$

This formula will be called the "Hurwitz formula" for our problem, because it coincides with the ordinary Hurwitz formula for *n-periodic* mappings of a surface (sheet permutations of a regular n-sheeted branched covering of a modular surface[9]); in that case the numbers $\zeta - 1$ are the orders of the various branch points, one sums over all the branch points, and ε_1 is twice the genus of the modular surface.

Of course it can also happen that all elements of T have fixed points on E, so that case (b) of the previous paragraph does not arise. Then $\tau, \tau^2, \ldots, \tau^{n-1}$ have no fixed point classes with non-zero index. All $A_m = 0$, and (21) shows that $\varepsilon_1/2 - 1$ divides $p - 1$, with the quotient being the order of T/F. One sees that in this case $n > 1$ implies $p > 2$. Indeed it is known that for $p = 2$ there are orientation preserving topological mappings of the surface with no fixed points. However, the class of such a mapping must, by the above, either be of infinite order, or, if it is of finite order n, there must already be essential fixed point classes appearing at powers lower than n.

The existence of mapping classes of finite order n (in fact n-periodic mappings) of all surfaces with $p > 2$, which are fixed point free for all powers up to the $(n-1)^{\text{th}}$, will be discussed in §11 with an obvious example (Example 2 there).

In §§4–8 we have spoken only of finite or infinite cyclic groups of mapping classes. It should be pointed out, however, that some of the considerations may be extended to other groups of mapping classes, in particular finite non-cyclic groups.

9. Properties of mapping classes of finite order. Let Γ be the matrix of a mapping class of finite order n. One factorizes its characteristic polynomial $\gamma(\lambda)$ as in (4), §2, resp., (4a), §3. The system of divisors d of n which actually appear, each listed ε_d times, where $\varepsilon_d > 0$ is its multiplicity, is called the "d-combination" of Γ. Let

$$d, d', d'', \ldots, \delta, \delta', \delta'', \ldots, 1, \ldots, 1$$

[9] See W. Scherrer: "Über periodische Transformationen von Flächen," Vierteljahrsschrift der Naturforschenden Gesellschaft in Zürich, Vol. **70** (1925).

be a d-combination, in which all $d > 1$, all $\delta > 1$, and each d is relatively prime to all δ. If D and Δ denote the lowest common multiples of the d and δ respectively, then D is relatively prime to Δ and $D\Delta = n$. Now one constructs the matrix Γ^D. The contribution which the eigenvalues 1 make to the trace s_D is the same as for s_1. The same holds for the contribution due to the divisors δ, since the exponent D is relatively prime to all δ. However, all eigenvalues of Γ corresponding to the divisors d are converted to $+1$ by raising to the power D. Thus $s_D > s_1$, contrary to (20). Hence

Theorem 5. *In the combination of divisors of n for the factorization of the characteristic polynomial of the matrix associated with a mapping class of finite order n, the divisors unequal to* 1 *cannot be partitioned into two relatively prime subsets.*

One always has

$$s(\Gamma) = s_1 \leq 2 \tag{22}$$

since the class number

$$Z = 2 - s_1 \geq 0.$$

Because of (10), §3, we have here, for $v = 2p$

$$Z \leq 2 + 2p$$

and equality is attained only for $n = 2$ and $\Gamma = -E_{2p}$. Moreover, when $n = 2$,

$$\begin{aligned} Z &\equiv 2 + 2p \pmod 4 \\ &\equiv 2 - 2p \pmod 4 \end{aligned}$$

because

$$s_1 = 2p - 2\varepsilon_2 \equiv 2p \pmod 4,$$

since ε_2 is even. If n is an odd prime, then (9), §3 and $v = 2p$ give

$$Z \equiv 2 - 2p \pmod n$$

and

$$Z \leq 2 + \frac{2p}{n-1}.$$

Hence, for some $k > 0$,

$$Z = 2 - 2p + kn \leq 2 + \frac{2p}{n-1},$$

which implies

$$n \leq \frac{2p}{k} + 1 \leq 2p + 1.$$

Then (13), §3 and (20), §7 give, for arbitrary n

$$\begin{aligned} (n-1)s_1 &\geq \varepsilon_1 n - 2p, \\ (\varepsilon_1 - s_1)n &\leq 2p - s_1. \end{aligned}$$

Thus for $\varepsilon_1 - s_1 > 0$ one has

$$n \leq \frac{2p - s_1}{\varepsilon_1 - s_1} = 1 + \frac{2p - \varepsilon_1}{\varepsilon_1 - s_1} \leq 2p + 1.$$

E.g., if $\varepsilon_1 > 0$, in which case $\varepsilon_1 \geq 2$, since it is even, and if the basic power of the mapping has essential fixed point classes, so that $s_1 \leq 1$, then $n \leq 2p - 1$.

If $\varepsilon_1 = 0$ and if $s_1 \geq 0$, then (13) shows that the inequality sign must hold in (20) for at least one $r < n$.

Theorem 6. *The order n of a mapping class of finite order is bounded by a function of the genus p (§2). The class number Z is determined by $Z = 2 - s_1$ and bounded by*

$$Z \leq 2 + 2p.$$

The equality sign occurs only for $n = 2$. If the order n is prime, then it is bounded by

$$n \leq 2p + 1.$$

In this case Z is bounded by

$$Z \leq 2 + \frac{2p}{n - 1}$$

and it satisfies the congruence

$$Z \equiv 2 - 2p \pmod{n},$$

which may be sharpened to

$$Z \equiv 2 - 2p \pmod{4}$$

when $n = 2$. If (for arbitrary n) the multiplicity ε_1 of the eigenvalue 1 in the matrix Γ is greater than the trace s_1, then in particular

$$n \leq 1 + \frac{2p - \varepsilon_1}{\varepsilon_1 - s_1} \leq 2p + 1.$$

10. Presentation of all matrix types of mapping classes of finite order for $p = 2$. When $p = 2$, so that $v = 2p = 4$, the divisors d of the orders which can occur according to (6), §2 are just those shown in the following table. The table also gives the sum $\mu(d)$ of the roots of $\gamma_d(\gamma)$ for each value of d, which is the sum of the primitive d^{th} roots of unity determined in §3.

d	$\phi(d)$	$\mu(d)$
1	1	1
2	1	−1
3	2	−1
4	2	0
5	4	−1
6	2	1
8	4	0
10	4	1
12	4	0

Now one has to construct d-combinations from these values which partition the number 4, according to (5), into the corresponding $\phi(d)$. We take into account what was shown at the end of §3, namely that $d = 2$ and $d = 1$ can only occur with even multiplicity. One then obtains first of all the possibilities shown in the table below.

The second column gives the order n as the lowest common multiple of the d used. The third column gives the trace s_1 as the sum of the corresponding $\mu(d)$. However, some of the possibilities found in this way conflict with properties already found for the matrices of mapping classes of finite order, and they must therefore be ruled out. Reasons are stated in the last column. Thus the combination (6,1,1) is excluded because the trace is > 2 in contradiction to (22). (The last combination (1,1,1,1) corresponds to the mapping class of the identity and has order $n = 1$. It therefore does not properly belong to the list, but it is convenient to count it in what follows.) The combinations (4,3) and (3,2,2) contradict Theorem 5. In ruling out other types it is necessary to square the matrix in question, which one can do directly from the d-combination by means of the process given in §3. The square of a matrix of type (6,2,2) is a matrix of type (3,1,1), and this has a greater trace, in contradiction to the condition $s_2 \leq s_1$ of (20), §7. Similarly, squaring (4,2,2) leads to (2,2,1,1) with a greater trace. Squaring (6,4) leads to (3,2,2) and is therefore excluded. Squaring (12) leads to (6,6) with a greater trace and is also excluded.

d-combination	Order n	Trace s_1	Contradicts
12	12	0	$s_2 \leq s_1$
10	10	1	
8	8	0	
5	5	-1	
6 6	6	2	
6 4	12	1	Theorem 5 (squared)
6 3	6	0	
6 2 2	6	-1	$s_2 \leq s_1$
6 1 1	6	3	$s_1 \leq 2$
4 4	4	0	
4 3	12	-1	Theorem 5
4 2 2	4	-2	$s_2 \leq s_1$
4 1 1	4	2	
3 3	3	-2	
3 2 2	6	-3	Theorem 5
3 1 1	3	1	
2 2 2 2	2	-4	
2 2 1 1	2	0	
1 1 1 1	1	4	

The remaining 12 types are compatible with the conditions previously established. This is made clear by the following table, which shows all the powers of these types and the way they reproduce. We take a fixed enumeration of these 12 types by Roman numerals, increasing with the order n. The columns numbered 1 to 10 give for the powers of each type: Below the type

Typus	*d* Kombination	n	1	2	3	4	5	6	7	8	9	10	Matrixgleichung
I	1 1 1 1	**1**	I 4										$\Gamma - E_4 = 0$
II	2 2 1 1	**2**	II 0 2	I 4									$\Gamma^2 - E_4 = 0$
III	2 2 2 2	**2**	III −4 6	I 4									$\Gamma + E_4 = 0$
IV	3 1 1	**3**	IV 1 1	IV 1 1	I 4								$\Gamma^3 - E_4 = 0$
V	3 3	**3**	V −2 4	V −2 4	I 4								$\Gamma^2 + \Gamma + E_4 = 0$
VI	4 1 1	**4**	VI 2 0	II 0 2	VI 2 0	I 4							$\Gamma^3 - \Gamma^2 + \Gamma - E_4 = 0$
VII	4 4	**4**	VII 0 2	III −4 6	VII 0 2	I 4							$\Gamma^2 + E_4 = 0$
VIII	5	**5**	VIII −1 3	VIII −1 3	VIII −1 3	VIII −1 3	I 4						$\Gamma^4 + \Gamma^3 + \Gamma^2 + \Gamma + E_4 = 0$
IX	6 3	**6**	IX 0 2	V −2 4	II 0 2	V −2 4	IX 0 2	I 4					$\Gamma^4 + \Gamma^2 + E_4 = 0$
X	6 6	**6**	X 2 0	V −2 4	III −4 6	V −2 4	X 2 0	I 4					$\Gamma^2 - \Gamma + E_4 = 0$
XI	8	**8**	XI 0 2	VII 0 2	XI 0 2	III −4 6	XI 0 2	VII 0 2	XI 0 2	I 4			$\Gamma^4 + E_4 = 0$
XII	10	**10**	XII 1 1	VIII −1 3	XII 1 1	VIII −1 3	III −4 6	VIII −1 3	XII 1 1	VIII −1 3	XII 1 1	I 4	$\Gamma^4 - \Gamma^3 + \Gamma^2 - \Gamma + E_4 = 0$

number of the power is the trace, and below that the number of positive fixed point classes. (The single negative class, of index -2, which belongs to type I, is not listed.) Finally, the last column gives the equation of lowest degree satisfied by the matrix of the type. (This is the equation denoted by $\kappa = 0$ in §3.) It is a divisor of the fourth degree characteristic equation of the matrix, and determines the recurrence relation for the traces of powers of the matrix, as shown in §3.

The examples in the next paragraph for arbitrary p, when specialised to $p = 2$, document all types in the table with the exception of IV and VI, whose existence remains uncertain.[10]

11. Examples of mapping classes of finite order. In this paragraph we shall give a series of typical examples of mapping classes of finite order for arbitrary p. When specialised to $p = 2$, they exemplify all types in the above table with the exception of IV and VI, which remain in doubt. The starting point each time will be a suitable abstract presentation of the fundamental group F which makes an automorphism family of finite order evident. Attention is then directed to the group T presented in §4, which contains F as a normal subgroup. It results by adding a generator t to F, satisfying a relation $t^n = f$, where n is the order of the mapping class and f denotes an element of F, possibly 1. An abstract presentation of T is obtained when one adds to the relation system for F the finitely many relations

$$tet^{-1} = e_I$$

in which e runs through a generating system for F and the correspondence $e \mapsto e_I$ defines the automorphism I of F induced by t. This abstract presentation of T is then simplified by transformations until it permits the "group diagram" in the sense of Dehn[11] to be constructed. The latter always turns out here to be a metrically regular graph in the non-Euclidean plane Φ, which admits T as a group of covering motions. A fundamental domain for the latter can be obtained as a single cell of the polygonal net dual to the group diagram, the vertices of which are the midpoints of group diagram cells. F is an invariant subgroup of this motion group T. The quotient of the plane Φ by F is the closed surface ϕ of genus p. The quotient of the plane by T is a closed surface M of a certain genus π. The surface $\phi = \Phi \bmod F$ is an n-sheeted regular branched covering of $\Phi \bmod T = M$, and the elements of the quotient group T/F correspond to the cyclic permutations of the n sheets. Thus it is shown for these examples that the given mapping class of finite order n contains members which are n-periodic, i.e., their n^{th} powers not only belong to the class of the identity, but are themselves the identity. And M is the associated "modular surface of the involution" in the sense of Brouwer[12]. Here positive fixed point classes are each represented by a single point (a branch point of ϕ over M) and fixed point classes of index 0 are completely avoided. Here the Hurwitz formula of §8 is the Hurwitz formula for the branched covering, and the genus π of M therefore has the value $\varepsilon_1/2$, where ε_1 is the multiplicity of the eigenvalue 1 in the matrix Γ of I.

One could object that as long as the aim is merely to present examples of mapping classes of finite order, the sheet permutations of regular branched coverings might as well be the starting point. To counter this, we emphasize that the following examples show above all how

[10] Editor's note: In a reprint of this memoir Nielsen claims in an added remark that they do not exist.

[11] M. Dehn, Über die Topologie des dreidimensionalen Raumes, Math. Ann. **69**, §1.

[12] L. E. J. Brouwer, Über topologische Involutionen, Amsterdam Proceedings, vol. XXI.

the previous group theoretic analysis leads naturally to the associated branched coverings in the present examples, This is important for the question (only partly solved in the paragraphs which follow) *whether a mapping class of finite order always contains a mapping of finite order.* The examples seek to open the way towards answering this question by striving to find a typical normal form for the abstract presentation of the group T and the presentation of the automorphism of F. This is also important for deciding *when two mapping classes are conjugate.*

Example 1. The presentation of the group F

$$F = \{a_1, b_1, \ldots, a_p, b_p\},$$
$$R = k_1 k_2 \ldots k_p = 1 \qquad (k_i = a_i b_i a_i^{-1} b_i^{-1})$$

reveals an evident automorphism of order p

$$I: \begin{cases} a_i \mapsto a_{i+1} \\ b_i \mapsto b_{i+1} \end{cases} \quad (i \bmod p).$$

Suppose it is induced by the mapping function $t(\Phi + E)$. For all powers t^r one has the trace $s_r = 0$ and hence the class number $Z_r = 2$ $(r = 1, 2, \ldots, p-1)$. The characteristic polynomial of t is

$$\gamma(\lambda) = (\lambda^p - 1)^2.$$

Thus $\varepsilon_1 = 2$ for the matrix Γ of I.

The group T (on E) is

$$T = \{a_1, b_1, \ldots, a_p, b_p, t\},$$
$$R = 1,\ t^p = 1,\ \left.\begin{aligned} ta_i t^{-1} &= a_{i+1} \\ tb_i t^{-1} &= b_{i+1} \end{aligned}\right\} \quad (i \bmod p).$$

One can use the latter equations for $i = 1, \ldots, p-1$ to eliminate $a_2, b_2, \ldots, a_p, b_p$. The last two,

$$ta_p t^{-1} = a_1, \qquad tb_p t^{-1} = b_1$$

are then satisfied identically because $t^p = 1$. Also

$$k_i = t^{i-1} k_1 t^{-(i-1)};$$

hence

$$R = (k_1 t)^p \cdot t^{-p},$$

so one obtains

$$T = \{a_1, b_1, t\},$$
$$t^p = (k_1 t)^p = 1.$$

If one introduces $t_1 = t^{-1} k_1^{-1}$ as a superfluous generator, then one has

$$T = \{a_1, b_1, t, t_1\},$$
$$t^p = t_1^p = t t_1 a_1 b_1 a_1^{-1} b_1^{-1} = 1,$$

and t and t_1 represent the two positive classes.

If one uses the central regular $4p$-gon in Φ to produce the group F, as in I, 3, then the mapping t in this example is simply a rotation of E through the angle $2\pi/p$, and hence an elliptic linear fractional transformation. Thus the mapping group T of E is given by a group of linear fractional transformations which contains F as an invariant subgroup. *Now one considers this group in* Φ. As a generating function for the present mapping class one can take a $1/p$ rotation of Φ about the midpoint O, because it satisfies the right functional equations. Now one considers the central fundamental domain of F:

Four of its successive sides correspond to a_1 and b_1 pairs. One connects the two free endpoints with the centre O of E. The resulting 6-gon is a fundamental domain of T. The two new sides are equivalent under T. This gives the polygonal group diagram of T. The surface ϕ is p-sheeted over M with two branch points of order $p-1$: That at O corresponds to the class of t, that at the vertices of the original $4p$-gon to the class of t_1. M has genus 1, corresponding to $\varepsilon_1 = 2$.

If r is a divisor of p, and if one considers the mapping class of t^r, then it is of order p/r and it also has two positive classes, given by t^r and $t_1^r = k_1 k_2 \ldots k_r t^r$. The eigenvalue 1 has multiplicity $2r$ and the modular surface M has genus r.

One can visualise this mapping class in space as follows: Let A be a diameter of a sphere. One attaches p "handles" to the sphere symmetrically with respect to rotation about A. Then the $1/p$ rotation about A gives the above mapping class, and the endpoints of A are the only fixed points under all powers.

Example 2. The presentation of F used in Example 1 also reveals the following automorphism

$$I: \quad \begin{aligned} &a_1 \mapsto a_1, \\ &b_1 \mapsto k_2^{-1} b_1, \\ &\left.\begin{aligned} a_i &\mapsto a_{i+1}, \\ b_i &\mapsto b_{i+1}, \end{aligned}\right\} \quad (i = 2, \ldots, p-1), \\ &a_p \mapsto a_1 a_2 a_1^{-1}, \\ &b_p \mapsto a_1 b_2 a_1^{-1} \end{aligned}$$

because the old generators may be expressed in terms of the new and R goes over to

$$a_1 k_2^{-1} b_1 a_1^{-1} b_1^{-1} k_2 k_3 k_4 \ldots k_p a_1 k_2 a_1^{-1} = a_1 k_2^{-1} a_1^{-1} \cdot R \cdot a_1 k_2 a_1^{-1}.$$

For $r = 1, 2, \ldots, p-1$ the powers I^r are

$$I^r: \quad \begin{aligned} &a_1 \mapsto a_1, \\ &b_1 \mapsto k_{r+1}^{-1} \ldots k_2^{-1} b_1, \\ &\left.\begin{aligned} a_i &\mapsto a_{i+r}, \\ b_i &\mapsto b_{i+r}, \end{aligned}\right\} \quad (i = 2, 3, \ldots, p-r), \\ &\left.\begin{aligned} a_{p-r+j} &\mapsto a_1 a_{j+1} a_1^{-1}, \\ b_{p-r+j} &\mapsto a_1 b_{j+1} a_1^{-1}, \end{aligned}\right\} \quad (j = 1, 2, \ldots, r). \end{aligned}$$

For $r = p - 1$ one has

$$a_1 \mapsto a_1 = a_1 a_1 a_1^{-1},$$
$$b_1 \mapsto k_p^{-1} \dots k_2^{-1} b_1 = a_1 b_1 a_1^{-1} \qquad \text{(because } R = 1\text{)},$$
$$\left.\begin{aligned} a_{1+j} &\mapsto a_1 a_{1+j} a_1^{-1}, \\ b_{1+j} &\mapsto a_1 b_{1+j} a_1^{-1}, \end{aligned}\right\} \qquad (j = 1, 2, \dots, p-1).$$

Thus I^{p-1} is the inner automorphism $(I_0)_{a_1^{-1}}$, so we are dealing with a mapping class of order $p - 1$. Let t be an I-inducing mapping function. Then

$$t^{p-1} = a_1$$

on E. The matrix of I is the identity matrix E_2 for the first two generators, and for the remainder it is like the matrix of Example 1 with $p - 1$ in place of p. Thus the characteristic polynomial is

$$\gamma(\lambda) = (\lambda - 1)^2(\lambda^{p-1} - 1)^2$$

and $\varepsilon_1 = 4$. The traces for the powers I^r, $r = 1, 2, \dots, p - 2$, are

$$s_r = 2$$

and hence the class numbers

$$Z_r = 0.$$

Thus the whole group T has no classes of positive index. One can also convince oneself of this directly. Because if ft^r is an arbitrary element of T and if σ is the exponent sum of a_1 in f, then σ does not change under application of I. Now

$$(ft^r)^{p-1} = ff_{I^r}f_{I^{2r}} \dots f_{I^{(p-2)r}}t^{r(p-1)} = ff_{I^r} \dots f_{I^{(p-2)r}}a_1^r = h$$

is an element of F, and its exponent sum in a_1 is

$$(p - 1)\sigma + r,$$

which is $\neq 0$ for $r = 1, 2, \dots, p - 2$. Thus h is not null homologous and therefore $h \neq 1$.

One has the following initial presentation for T:

$$T = \{a_1, b_1, \dots, a_p, b_p, t\},$$

$$R = 1, \qquad t^{p-1} = a_1, \qquad ta_1t^{-1} = a_1,$$
$$tb_1t^{-1} = k_2^{-1}b_1$$
$$\left.\begin{aligned} ta_it^{-1} &= a_{i+1} \\ tb_it^{-1} &= b_{i+1} \end{aligned}\right\} \qquad (i = 2, \dots, p-1),$$
$$ta_pt^{-1} = a_1 a_2 a_1^{-1},$$
$$tb_pt^{-1} = a_1 b_2 a_1^{-1}.$$

Now one can eliminate the generator a_1 by means of the second relation, and the generators $a_3, b_3, \dots, a_p, b_p$ by means of the relations containing the letter i. If one substitutes the resulting expressions everywhere, then the first, second last and last relation in the last group become satisfied identically. So only the second relation remains, and we write it

$$R' = tb_1t^{-1}b_1^{-1}a_2b_2a_2^{-1}b_2^{-1} = 1.$$

Substitution of $a_1, a_3, b_3, \ldots, b_p$ in R yields

$$\begin{aligned} R &= t^{p-1}b_1t^{-(p-1)}b_1^{-1} \cdot k_2 \cdot tk_2t^{-1} \cdot t^2k_2t^{-2} \cdot \ldots \cdot t^{p-2}k_2t^{-(p-2)} \\ &= t^{p-1}b_1t^{-(p-1)}b_1^{-1}(k_2t)^{p-1}t^{-(p-1)}. \end{aligned}$$

However, since $R' = 1$, we have

$$k_2t = b_1tb_1^{-1},$$

hence

$$R = t^{p-1} \cdot (b_1tb_1^{-1})^{-(p-1)} \cdot (b_1tb_1^{-1})^{p-1} \cdot t^{-(p-1)} = 1$$

is a consequence of $R' = 1$. Thus $T = \{t, b_1, a_2, b_2\}$ has the single relation $R' = 1$. T is therefore isomorphic to the fundamental group of a surface M of genus 2, and the group diagram is known. (T is again representable by a group of linear fractional transformations.)

F is the invariant subgroup of T (with index $p - 1$) whose elements have an exponent sum in t divisible by $p - 1$. Conversely, one gets from T to F by means of $t^{p-1} = a_1$ and the conjugations

$$\left.\begin{aligned} t^ra_2t^{-r} &= a_{2+r}, \\ t^rb_2t^{-r} &= b_{2+r}, \end{aligned}\right\} \qquad (r = 1, 2, \ldots, p-2),$$

noticing that $t^rb_1t^{-r}$ can be expressed in terms of the others since $R' = 1$. Then t induces the automorphism I in F.

T is the covering transformation group of the universal covering surface of the closed surface M of genus 2—this agrees with $\varepsilon_1 = 4$—and, because of the index $p - 1$ of F in T, the surface ϕ of genus p is a $(p - 1)$-sheeted covering surface of M. Its sheets are permuted cyclically by traversal of the path on M corresponding to the element t; one can view them as cyclically connected along the closed curve of M which corresponds to the element b_1 coupled with t. The covering of M by ϕ is unbranched[13]. Thus the mapping class in question contains a $(p - 1)$-periodic mapping which is fixed point free under all powers.

For $p = 2$, t itself is in the class of the identity and ϕ is identical with M. Thus we have a mapping class of order > 1 only for $p > 2$. On this point, cf. the end of §8. One can visualize this mapping class in space as follows: On a torus which is rotationally symmetric about an axis A one places $p - 1$ handles, also symmetrically with respect to A. Then the $1/(p - 1)$ rotation about A is the mapping class in question for the surface of genus p.

When r divides $p - 1$, the mapping class of t^r has modular surface of genus $r + 1$ and, as one easily confirms, the eigenvalue 1 has multiplicity $2r + 2$.

Example 3. First we shall give the group F a new presentation, suitable for the present

[13] For unbranched covering surfaces and their connection with the subgroups of finite index in the fundamental group, see also Ingebrigt Johansson: Topologische Untersuchungen über unverzweigte Überlagerungsflächen. Norsk Vid. Ak. Skr. Mat. Naturv. Kl. 1931, in particular p. 57.

example. Proceeding from the presentation used in the first two examples, one sets

$$\begin{aligned} c_1 &= a_1 b_1 a_1^{-1} \\ c_2 &= b_1^{-1} a_2 b_2 a_2^{-1} \\ c_3 &= b_2^{-1} a_3 b_3 a_3^{-1} \\ &\dots \\ c_p &= b_{p-1}^{-1} a_p b_p a_p^{-1}. \end{aligned}$$

Then $F = \{a_1, \dots, a_p, c_1, \dots, c_p\}$, because the b_i may be expressed in terms of these generators, and indeed

$$b_i = a_i^{-1} a_{i-1}^{-1} \dots a_1^{-1} \cdot c_1 a_1 c_2 a_2 \dots c_i a_i.$$

Now one introduces two superfluous generators c_{p+1} and a_{p+1} by the defining equations

$$\begin{aligned} c_1 c_2 \dots c_p c_{p+1} &= 1, \\ a_1 a_2 \dots a_p a_{p+1} &= 1 \end{aligned}$$

and makes the abbreviations

$$\left.\begin{aligned} (c)_i &= c_1 c_2 \dots c_i, \\ (a)_i &= a_1 a_2 \dots a_i, \\ (ca)_i &= c_1 a_1 c_2 a_2 \dots c_i a_i, \end{aligned}\right\} \qquad i = 1, 2, \dots, p+1.$$

Then $(c)_{p+1} = (a)_{p+1} = 1$ by the defining equations above. Now

$$b_i = (a)_i^{-1} (ca)_i;$$

hence

$$k_i = a_i b_i a_i^{-1} b_i^{-1} = (a)_{i-1}^{-1} (ca)_{i-1} \cdot c_i \cdot (ca)_i^{-1} (a)_i$$

and consequently

$$R = k_1 k_2 \dots k_p = (c)_p (ca)_p^{-1} (a)_p.$$

Then since

$$(c)_p = c_{p+1}^{-1}, \qquad (a)_p = a_{p+1}^{-1},$$

we get

$$R = a_{p+1} a_{p+1}^{-1} c_{p+1}^{-1} (ca)_p^{-1} a_{p+1}^{-1} = a_{p+1} \cdot (ca)_{p+1}^{-1} \cdot a_{p+1}^{-1}.$$

Thus

$$\begin{aligned} F &= \{c_1, \dots, c_{p+1}, a_1, \dots, a_{p+1}\}, \\ (c)_{p+1} &= (a)_{p+1} = (ca)_{p+1} = 1 \end{aligned} \tag{23}$$

is the presentation of F in the new set of $2p + 2$ generators. We still agree to interpret the subscripts on the c_i, a_i and the symbols $(\ \ )_i$ modulo $p+1$. Then the original definition of the bracket quantities $(\ \ )_i$ remains correct for $i > p+1$, by (23).

The presentation (23) for F now reveals the following automorphism of F, denoted by H.

$$H: \begin{matrix} c_i \mapsto a_i \\ a_i \mapsto c_{i+1} \end{matrix}.$$

Because

$$(c)_{iH} = (a)_i,$$
$$(a)_{iH} = c_i^{-1}(c)_{i+1},$$
$$(ca)_{iH} = c_i^{-1}(ca)_i c_{i+1};$$

hence

$$(c)_{p+1} \mapsto (a)_{p+1},$$
$$(a)_{p+1} \mapsto c_1^{-1}(c)_{p+1} c_1,$$
$$(ca)_{p+1} \mapsto c_1^{-1}(ca)_{p+1} c_1.$$

It follows that we have

$$H^{2r-1}: \begin{matrix} c_i \mapsto a_{i+r-1} \\ a_i \mapsto c_{i+r} \end{matrix}, \qquad H^{2r}: \begin{matrix} c_i \mapsto c_{i+r} \\ a_i \mapsto a_{i+r} \end{matrix},$$

whence $H^{2(p+1)} = 1$. Let h be an H-inducing mapping function (previously denoted by t). Then $h^{2p+2} = 1$ *on E*.

To construct the traces and characteristic polynomial one has to go back to the independent generators $c_1, \ldots, c_p, a_1, \ldots, a_p$ in the homology group, and hence to eliminate c_{p+1} and a_{p+1} by means of the relations $(c)_{p+1} = 1$ and $(a)_{p+1} = 1$. One then gets the presentation for F

$$F = \{c_1, \ldots, c_p, a_1, \ldots, a_p\}, \qquad (24)$$
$$(ca)_p (c)_p^{-1} (a)_p^{-1} = 1$$

in which each generator has exponent sum 0 in the relation. The only c_i or a_i which contribute to the traces are those which go into c_{p+1} or a_{p+1}. Now c_{p+1} contains all c_i with exponent -1 and no a_i, while a_{p+1} contains all a_i with exponent -1 and no c_i. Therefore

$$s_{2r-1} = 0$$

and

$$s_{2r} = -2 \qquad (\text{for } r \not\equiv 0 \bmod p+1).$$

The odd powers of h therefore have the class number $Z_{2r-1} = 2$ and the even powers have the class number $Z_{2r} = 4$. One finds the characteristic polynomial of H to be:

$$\gamma(\lambda) = \lambda^{2p} + \lambda^{2p-2} + \cdots + \lambda^2 + 1;$$

hence $\varepsilon_1 = 0$.

Before we investigate the associated group T, we read off a second automorphism S from the presentation (23) of F:

$$S: \begin{array}{l} c_i \mapsto (ca)_{i-1} c_i^{-1} (ca)_{i-1}^{-1} \\ a_i \mapsto (ca)_i a_i^{-1} (ca)_i^{-1} \end{array},$$

which one checks by substituting to derive the equations

$$\begin{aligned} (c)_{iS} &= (a)_i (ca)_i^{-1}, \\ (a)_{iS} &= (c)_i (ca)_i^{-1}, \\ (ca)_{iS} &= (ca)_i^{-1}. \end{aligned}$$

When one now replaces i everywhere by $p+1$, one sees that the relation system goes over to an equivalent system. It also follows immediately from the last equation that $S^2 = 1$. Thus if S is induced by a mapping function s, then $s^2 = 1$ on E.

If one goes back to (24), then the last two transformation formulae for S can be omitted. One sees that S has the matrix $-E_{2p}$, hence $s_1 = -2p$, $Z_1 = 2 + 2p$ and

$$\gamma(\lambda) = (\lambda + 1)^{2p}$$

is the characteristic polynomial, whence $\varepsilon_1 = 0$.

We mention briefly, in this connection, that one can establish two other automorphisms of order 2 in a similar way:

$$G_1: \begin{array}{l} c_i \mapsto (c)_i c_i^{-1} (c)_i^{-1} \\ a_i \mapsto (c)_i a_i (c)_i^{-1} \end{array}, \qquad G_2: \begin{array}{l} c_i \mapsto (a)_{i-1} c_i (a)_{i-1}^{-1} \\ a_i \mapsto (a)_{i-1} a_i^{-1} (a)_{i-1}^{-1} \end{array}.$$

One checks that these satisfy the relations

$$\begin{aligned} G_2 &= H^{-1} G_1 H, \\ G_1^2 &= G_2^2 = 1, \\ S &= G_1 G_2 = G_2 G_1 = G_1^{-1} H^{-1} G_1 H. \end{aligned}$$

It follows that S, as a product of two automorphisms of the same kind, is an automorphism of the first kind. However, G_1 and G_2 are of the second kind. The determinant of their matrix is obviously $(-1)^p$. Cf. the theorem at the end of §3.

The investigation of the groups T in the present work concerns only automorphisms of the first kind. S and H are of this kind (H has a characteristic exponent > 2, cf. I, 11). We consider the behaviour on E of the group of mappings U which results from adding the generators s and h to F:

$$U = \{c_i, a_i, h, s\} \qquad (i = 1, \ldots, p+1).$$

Since F is conjugate to itself via h and s, F is a normal subgroup of U. Substituting S in H one finds

$$HS: \begin{array}{l} c_i \mapsto (ca)_i \cdot a_i^{-1} \cdot (ca)_i^{-1} \\ a_i \mapsto (ca)_i \cdot c_{i+1}^{-1} \cdot (ca)_i^{-1} \end{array},$$

and substituting H in S,

$$SH: \begin{array}{l} c_i \mapsto c_1^{-1}(ca)_{i-1}c_i \cdot a_i^{-1} \cdot c_i^{-1}(ca)_{i-1}^{-1}c_i = c_1^{-1}(ca)_i \cdot a_i^{-1} \cdot (ca)_i^{-1}c_1 \\ a_i \mapsto c_1^{-1}(ca)_i c_{i+1} \cdot c_{i+1}^{-1} \cdot c_{i+1}^{-1}(ca)_i^{-1}c_1 = c_1^{-1}(ca)_i \cdot c_{i+1}^{-1} \cdot (ca)_i^{-1}c_1 \end{array}.$$

Let C_1 denote the inner automorphism of F which sends each element f to $c_1 f c_1^{-1}$. It is induced by the element c_1. Then, as one sees

$$SH = HSC_1^{-1}.$$

Thus the corresponding mapping functions satisfy

$$hs = c_1^{-1}sh$$

because, e.g.,

$$hsfs^{-1}h^{-1} = hs_S h^{-1} = (f_S)_H = f_{SH}.$$

The relation just found,

$$shs^{-1}h^{-1} = c_1,$$

shows that the commutator of the two generators of U which are outside F is inside F, hence the quotient group U/F is abelian. It has order $4p + 4$, with the cosets of s and h as basis elements. This gives the following presentation of U:

$$\begin{aligned} &U = \{c_i, a_i, h, s\}, \\ &(c)_{p+1} = (a)_{p+1} = (ca)_{p+1} = h^{2p+2} = s^2 = 1, \qquad hc_ih^{-1} = a_i, \\ &shs^{-1}h^{-1} = c_1, \qquad ha_ih^{-1} = c_{i+1}, \\ &sc_is^{-1} = (ca)_{i-1}c_i^{-1}(ca)_{i-1}^{-1}, \\ &sa_is^{-1} = (ca)_i a_i^{-1}(ca)_i^{-1}. \end{aligned} \tag{25}$$

The relation system (25) is complete, because one can obviously use it to commute s and h among themselves and with all elements of F, this bringing each product

$$\textstyle\prod_1(c_i, a_i, h, s)$$

into the form

$$\textstyle\prod_2(c_i, a_i) \cdot h^x \cdot s^y$$

with $0 \leq x \leq 2p + 1$ and $0 \leq y \leq 1$. And the first three relations of (25) are complete for products of the form $\prod_2$.

By conjugating

$$c_1 = shs^{-1}h^{-1} = shsh^{-1}$$

with powers of h one obtains the whole series of c_i and a_i:

$$\begin{aligned} c_i &= h^{2i-2}shsh^{-(2i-1)}, \\ a_i &= h^{2i-1}shsh^{-2i}. \end{aligned} \tag{26}$$

Thus U is already generated by h and s. One obtains a complete relation system by substituting the expressions (26) in all the relations of (25). The first result of this is that the relations expressing the automorphism H become satisfied identically. Also, since $s^2 = 1$,

$$(ca)_i = sh^{2i}sh^{-2i}.$$

If one substitutes this in the relations expressing the automorphism S, they are also satisfied identically. Then if one substitutes $i = p + 1$ in the last equation, one finds the relation $(c)_{p+1} = 1$ satisfied identically, because $h^{2p+2} = 1$. Thus it remains only to substitute in the first two relations of (25). One finds

$$(c)_i = (sh)^{2i}h^{-2i},$$

$$(a)_i = h(c)_i h^{-1},$$

and hence both yield the equation $(sh)^{2p+2} = 1$ when $i = p + 1$. Consequently,

$$U = \{h, s\},$$
$$h^{2p+2} = s^2 = (sh)^{2p+2} = 1. \tag{27}$$

If one introduces a new generator $k = h^{-1}s$ then one gets

$$U = \{s, h, k\},$$
$$s^2 = h^{2p+2} = k^{2p+2} = shk = 1. \tag{28}$$

We stay with the presentation (27). F is the normal subgroup of U whose elements have an even exponent in s and an exponent sum in h divisible by $2p + 2$. The associated quotient group is represented by the cosets of the elements

$$h^x \cdot s^y, \qquad 0 \leq x < 2p + 2, \quad 0 \leq y < 2.$$

It corresponds to an abelian, but non-cyclic group of mapping *classes* of the surface of genus p. The order of this group is $4p + 4$. The traces and class numbers for powers of h have already been determined above. Since s has the matrix $-E_{2p}$, its trace changes sign with each additional factor s. Thus the mapping classes $h^{2r-1}s$ have trace 0 and class number 2, and the mapping classes $h^{2r}s$ have trace 2 and class number 0.

The group diagram of U as presented by (27) may be given an immediate metrically regular realisation in the hyperbolic plane Φ, as shown in Figure 3. It consists of regular $(2p + 2)$-gons and $(4p + 4)$-gons with equal sides, and magnitude such that the corner angle of the $(2p + 2)$-gon plus twice the corner angle of the $(4p + 4)$-gon equals 2π. The sides of the $(2p + 2)$-gon and every second side of the $(4p + 4)$-gon are h-sides and carry arrows, the remaining sides are s-sides and are not oriented, because $s^2 = 1$. The cells of the dual polygonal group diagram—shown dotted in the figure—are congruent triangles (or better, quadrilaterals with an angle π). The generating motions are the rotation h about the point P in the figure through an angle $\pi/(p + 1)$, and the half turn s about Q. Φ mod U results from the fundamental triangle and has the connectivity of the sphere. *This exhibits the group U as a group of linear fractional transformations.* F is a normal subgroup of U of index $4p + 4$. Φ mod F is therefore a $(4p + 4)$-sheeted branched covering of the "sphere" Φ mod U. A fundamental domain of F is formed by the $4p + 4$ triangles which result from the two in the figure under rotation by all the

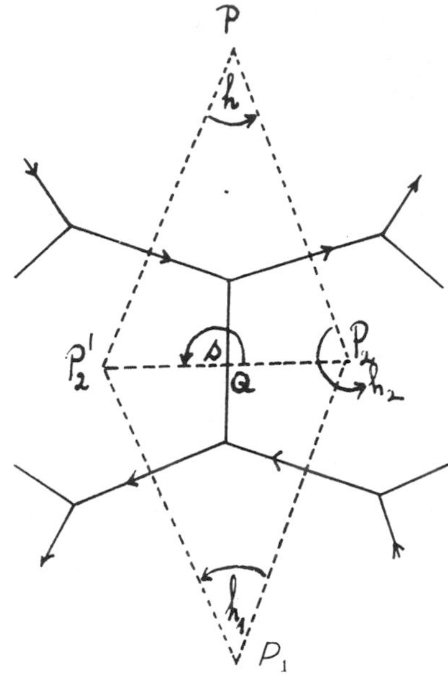

Figure 3

powers of h. The $4p + 4$ free boundary sides of this starlike figure around P correspond in pairs under the $2p + 2$ motions c_i and a_i ($i = 1, \ldots, p + 1$), as one can easily see from the presentation (26).

U/F is not cyclic. In order to obtain examples of cyclic mapping class groups, one extracts cyclic subgroups. We take the following three. As subgroups of U, they are all representable by linear fractional transformations, and hence each corresponds to a branched covering.

(1) The automorphism H yields the group

$$T(H) = \{c_i, a_i, h\}.$$

It is the normal subgroup of U whose elements have an even exponent sum in s, and hence it is generated by h and

$$h_1 = shs.$$

If one substitutes this in the last relation of (27), one obtains

$$(sh)^{2p+2} = (shsh)^{p+1} = (h_1 h)^{p+1} = 1.$$

The relation $s^2 = 1$ has now to be omitted, but one has to add the relations which result from the others by conjugation with s. The last relation then gives nothing new; however, the first relation of (27) yields $h_1^{2p+2} = 1$. Consequently,

$$\begin{gathered} T(H) = \{h, h_1\}, \\ h^{2p+2} = h_1^{2p+2} = (h_1 h)^{p+1} = 1. \end{gathered} \tag{29}$$

Next to this presentation we write the favourite normal form, which comes from introducing

$h_2 = h_1^{-1}h^{-1}$:

$$T(H) = \{h, h_1, h_2\},$$
$$h^{2p+2} = h_1^{2p+2} = h_2^{p+1} = hh_1h_2 = 1. \tag{30}$$

A fundamental domain for $T(H)$ consists of the two triangles shown in Figure 3; h_1 is the rotation about P_1, h_2 is the rotation about P_2 through an angle $2\pi/(p+1)$. The modular surface $M = \Phi \bmod T(H)$ has genus 0, corresponding to $\varepsilon_1 = 0$ (see above). The surface $\phi = \Phi \bmod F$ is $(2p+2)$-sheeted over M with 2 branch points of order $2p+1$ at P and P_1, and a branch point where the sheets are joined in two separate sets of $p+1$ sheets, at P_2 (and P_2'). The elements h and h_1 represent the two positive classes for the first power of the mapping class, h_2 (and the conjugate hh_2h^{-1}) represents the two classes appearing for the even powers. The last relation in (30) expresses the contractibility of a path on M enclosing the three branch points.

(2) In $T(H)$ we now take the normal subgroup $T(H^4)$. For even p one has $(h^4)^{p/2} = h^{2p} = h^{-2}$, hence $T(H^4) = T(H^2)$. We therefore assume p is *odd*. Then 4 divides $2p+2$. $T(H^4)$ is the normal subgroup of $T(H)$ whose elements have an exponent sum in h divisible by 4. We had

$$h_1 = shs = c_1h.$$

Now one sets

$$h^4 = u,$$
$$h_1^4 = (c_1h)^4 = c_1a_1c_2a_2h^4 = u_1,$$
$$(h_1h)^2 = (c_1h^2)^2 = c_1c_2h^4 = u_2,$$
$$(hh_1)^2 = (hc_1h)^2 = (a_1h^2)^2 = a_1a_2h^4 = u_3.$$

It follows from the first equation and the last two that

$$c_2 = c_1^{-1}u_2u^{-1},$$
$$a_2 = a_1^{-1}u_3u^{-1}. \tag{31}$$

Substituting these in the second equation gives the relation

$$c_1a_1c_1^{-1}u_2u^{-1}a_1^{-1}u_3u_1^{-1} = 1. \tag{32}$$

Now

$$T(H^4) = \{c_i, a_i, h^4\} = \{c_1, a_1, c_2, a_2, h^4\}$$

since

$$c_{i+2} = h^4c_ih^{-4},$$
$$a_{i+2} = h^4a_ih^{-4}.$$

Thus, by (31) again,

$$T(H^4) = \{c_1, a_1, c_2, a_2, u\} = \{c_1, a_1, u, u_2, u_3\} = \{c_1, a_1, u, u_1, u_2, u_3\},$$

where (32) can be taken as the defining equation for u_1. Substitution in (29) gives the following relations for these generators:

$$u^{(p+1)/2} = u_1^{(p+1)/2} = u_2^{(p+1)/2} = 1. \tag{33}$$

In order to obtain a complete relation system for $T(H^4)$, one has to complement (32) and (33) with the relations which result from them by conjugation with the elements h, h^2 and h^3 not lying in $T(H^4)$. We need not explicitly use $h_1 = c_1 h$, since c_1 is in F and hence in $T(H_4)$; h and h_1 are therefore interchangeable. Now one finds

$$\begin{aligned} huh^{-1} &= hh^4h^{-1} = h^4 = u, \\ hu_1h^{-1} &= h(c_1a_1c_2a_2h^4)h^{-1} = a_1c_2a_2c_3h^4 = c_1^{-1}u_1h^{-4}c_3h^4 = c_1^{-1}u_1c_1, \\ hu_2h^{-1} &= h(h_1h)^2h^{-1} = (hh_1)^2 = u_3, \\ hu_3h^{-1} &= ha_1a_2h^4h^{-1} = c_2c_3h^4 = c_1^{-1}\cdot c_1c_2h^4\cdot h^{-4}c_3h^4 = c_1^{-1}u_2c_1. \end{aligned}$$

Thus only the last relation of (33) yields anything new, namely $u_3^{(p+1)/2} = 1$. And one finds, because of (31), that the left-hand side of (32) goes to itself under conjugation by h_1. Consequently,

$$\begin{aligned} T(H_4) &= \{c_1, a_1, u, u_1, u_2, u_3\}, \\ u^{(p+1)/2} &= u_1^{(p+1)/2} = u_2^{(p+1)/2} = u_3^{(p+1)/2} = 1, \\ c_1a_1c_1^{-1}u_2u^{-1}a_1^{-1}u_3u_1^{-1} &= 1. \end{aligned} \tag{34}$$

In order to find a symmetrical form of the last relation, one sets

$$\begin{aligned} u_1 &= v_1, \\ u_3^{-1} &= v_2, \\ a_1ua_1^{-1} &= v_3, \\ a_1u_2^{-1}a_1^{-1} &= v_4 \end{aligned}$$

and obtains

$$\begin{aligned} T(H^4) &= \{c_1, a_1, v_1, v_2, v_3, v_4\}, \\ v_1^{(p+1)/2} &= v_2^{(p+1)/2} = v_3^{(p+1)/2} = v_4^{(p+1)/2} = 1, \\ v_1v_2v_3v_4 &= c_1a_1c_1^{-1}a_1^{-1}. \end{aligned} \tag{35}$$

H^4 has $\varepsilon_1 = 2$ for odd p, as one sees easily from the above polynomial $\gamma(\lambda)$ for H. A fundamental domain for $T(H^4)$ results from the double triangle of Figure 3 by application of h^0, h^1, h^2 and h^3, and gives a modular surface M of genus 1. The surface ϕ is a $(p+1)/2$-sheeted covering of this with 4 branch points of order $(p+1)/2 - 1$, corresponding to the first four relations of (35). The last relation says that a circuit around all the branch points is equivalent to a complete traversal of a canonical cut system for the modular surface.

(3) Finally we consider the subgroup $T(SH^2)$ of U. One has

$$SH^2: \begin{array}{l} c_i \mapsto a_1^{-1}c_1^{-1}(ca)_i \cdot c_{i+1}^{-1} \cdot (ca)_i^{-1}c_1a_1 \\ a_i \mapsto a_1^{-1}c_1^{-1}(ca)_{i+1} \cdot a_{i+1}^{-1} \cdot (ca)_{i+1}^{-1}c_1a_1 \end{array}$$

and this automorphism of F is induced by $t = h^2 s$. To construct the associated matrix one has to eliminate c_{p+1} and a_{p+1}, and when one writes $c_1, \ldots, c_p$ first and then $a_1, \ldots, a_p$, one obtains twice the $p \times p$ matrix

$$\begin{matrix} 0 & -1 & 0 & \ldots & 0 \\ 0 & 0 & -1 & \ldots & 0 \\ & & \ldots & & \\ 0 & 0 & 0 & \ldots & -1 \\ 1 & 1 & 1 & \ldots & 1 \end{matrix}$$

along the main diagonal and zeros else. This gives the characteristic polynomial

$$\gamma(\lambda) = (1 - \lambda + \lambda^2 - \cdots + (-1)^p \lambda^p)^2,$$

which yields $\varepsilon_1 = 0$ for even p and $\varepsilon_1 = 2$ for odd p.

Now,

$$t^2 = (h^2 s)^2 = h^2 s h^2 s = h^2 h_1^2 = h^2 (c_1 h)^2 = c_2 a_2 h^4.$$

Thus $T(H^4)$ is contained in $T(SH^2)$. For even p, h^2, and hence also $h^{-2} t = s$, lies in $T(SH^2)$ along with h^4, and the latter group is therefore the normal subgroup of U whose elements have even exponent sums in h. In what follows, $T(SH^2)$ will only be presented for *odd* p. In this case we can use the presentation (34) found above for the normal subgroup $T(H^4)$, and we retain the notation used there. One has $c_2 a_2 h^4 = a_1^{-1} c_1^{-1} u_1$, hence

$$t^2 = a_1^{-1} c_1^{-1} u_1. \tag{36}$$

Thus

$$t^{p+1} = (a_1^{-1} c_1^{-1} u_1)^{(p+1)/2}$$

belongs to F, and the mapping class of $t = h^2 s$ is of order $p + 1$. (For even p it is of order $2p + 2$.) Now one needs only to say how the generators of the normal subgroup $T(H^4)$ transform under conjugation by t. One finds, by conjugating first with s and then twice with h, that

$$\begin{aligned} t c_1 t^{-1} &= u u_2^{-1} c_1, \\ t a_1 t^{-1} &= c_1^{-1} u_2 u_1^{-1} c_1 a_1, \\ t u t^{-1} &= a_1^{-1} c_1^{-1} u_1 c_1 a_1, \\ t u_1 t^{-1} &= u, \\ t u_2 t^{-1} &= a_1^{-1} u_3 a_1, \\ t u_3 t^{-1} &= c_1^{-1} u_2 c_1. \end{aligned} \tag{37}$$

Thus $T(SH^2)$ results from (34) by adding the generator t and the relations (36) and (37). Finally, one can use (36) to eliminate u_1, and then use (37) to eliminate u, u_2 and u_3, and one then obtains the following simple presentation by substitution:

$$T(SH^2) = \{c_1, a_1, t\},$$
$$(c_1 a_1 t^2)^{(p+1)/2} = (c_1 t a_1 t)^{(p+1)/2} = 1. \tag{38}$$

The surface ϕ is $(p + 1)$-sheeted over a modular surface of genus 1, with two branch points. At each branch point, the sheets are joined in two sets of $(p + 1)/2$ each. The odd powers of the mapping are fixed point free, and the even powers have four fixed points[14].

Example 4. We begin by rewriting F in a new presentation. Proceeding from the presentation in Example 1, one sets

$$V_0 = 1,$$
$$V_r = b_r V_{r-1} = b_r b_{r-1} \dots b_2 b_1 \qquad (r = 1, 2, \dots, p)$$

and defines $2p$ new generators $d_1, d_2, \dots, d_{2p}$ by

$$d_{2r-1} = V_{r-1}^{-1} a_r,$$
$$d_{2r} = b_r a_r^{-1} V_{r-1} = b_r a_r^{-1} b_r^{-1} V_r.$$

Then

$$d_{2r} d_{2r-1} = b_r;$$

hence

$$d_{2r} d_{2r-1} \dots d_2 d_1 = V_r$$

and therefore the a_i are also expressible in terms of the d_i. On the other hand,

$$d_{2r-1} d_{2r} = V_{r-1}^{-1} a_r b_r a_r^{-1} b_r^{-1} V_r;$$

hence

$$k_r = a_r b_r a_r^{-1} b_r^{-1} = V_{r-1} d_{2r-1} d_{2r} V_r^{-1}$$

and

$$R = k_1 k_2 \dots k_p = d_1 d_2 d_3 d_4 \dots d_{2p-1} d_{2p} \cdot V_p^{-1}$$
$$= d_1 d_2 \dots d_{2p} d_1^{-1} d_2^{-1} \dots d_{2p}^{-1}.$$

Consequently one has the presentation for F:

$$F = \{d_1, d_2, \dots, d_{2p}\},$$
$$d_1 d_2 \dots d_{2p} d_1^{-1} d_2^{-1} \dots d_{2p}^{-1} = 1. \tag{39}$$

This presentation reveals the automorphism

$$I: \begin{array}{ll} d_i \mapsto d_{i+1}^{-1} & (i = 1, 2, \dots, 2p - 1) \\ d_{2p} \mapsto d_1 & \end{array} \tag{40}$$

[14] This last example corresponds to the orientation-preserving mapping in §2 of my work: "Über fixpunktfreie topologische Abbildungen geschlossener Flächen", Math. Ann. **81** (1920), p. 94, [N6].

since this sends the relation (39) back to itself after a cyclic sequence of $2p + 1$ steps. One finds

$$I^{2p}: d_i \mapsto d_i^{-1} \qquad (i = 1, 2, \ldots, 2p); \tag{40a}$$

hence $I^{4p} = 1$. The traces of the matrix of I are obviously all zero with the exception of $s_{2p} = -2p$, hence the class numbers of the powers of this mapping class are all 2 with the exception of $Z_{2p} = 2 + 2p$. The characteristic polynomial of I is

$$\gamma(\lambda) = \lambda^{2p} + 1;$$

hence $\varepsilon_1 = 0$.

Now let t be an I-inducing mapping function. Then one has the following presentation for the group $T(I)$ on E:

$$\begin{aligned} T &= \{d_1, d_2, \ldots, d_{2p}, t\}, \\ d_1 d_2 \ldots d_{2p} d_1^{-1} d_2^{-1} &\ldots d_{2p}^{-1} = 1, \\ t^{4p} &= 1, \\ t d_i t^{-1} &= d_{i+1}^{-1} \qquad (i = 1, 2, \ldots, 2p - 1), \\ t d_{2p} t^{-1} &= d_1. \end{aligned} \tag{41}$$

Now one observes that, since $Z_1 = 2$, there is exactly one positive class on level one apart from t. Because of (40) and (40a),

$$I^{2p+1}: \begin{matrix} d_i \mapsto d_{i+1} & (i = 1, 2, \ldots, 2p - 1) \\ d_{2p} \mapsto d_1^{-1} & \end{matrix}; \tag{40b}$$

hence

$$\begin{aligned} (d_1 t^{2p+1})^{4p} &= d_1 d_{1_{I^{2p+1}}} d_{1_{I^{2(2p+1)}}} \ldots d_{1_{I^{(4p-1)(2p+1)}}} \cdot t^{(2p+1)4p} \\ &= d_1 d_2 \ldots d_{2p} d_1^{-1} d_2^{-1} \ldots d_{2p}^{-1} \cdot t^{4p(2p+1)} \\ &= 1 \end{aligned}$$

because of the first two relations of (41). One introduces

$$t_1 = d_1 t^{2p+1}$$

as a new generator of T and then, by (40b), one can express all d_i in terms of t and t_1:

$$\begin{aligned} d_i &= t^{(2p+1)(i-1)} d_1 t^{-(2p+1)(i-1)} \\ &= t^{(2p+1)(i-1)} t_1 t^{-(2p+1)i} \qquad (i = 1, 2, \ldots, 2p). \end{aligned} \tag{42}$$

When (42) is substituted in the last relation of (41), together with $t^{4p} = 1$, this relation becomes satisfied identically. Substitution in the second last relation gives just

$$(t_1 t^{-1})^2 = 1$$

for all values of i, and substitution in the first relation of (41), together with those just found, gives $t_1^{4p} = 1$. Consequently

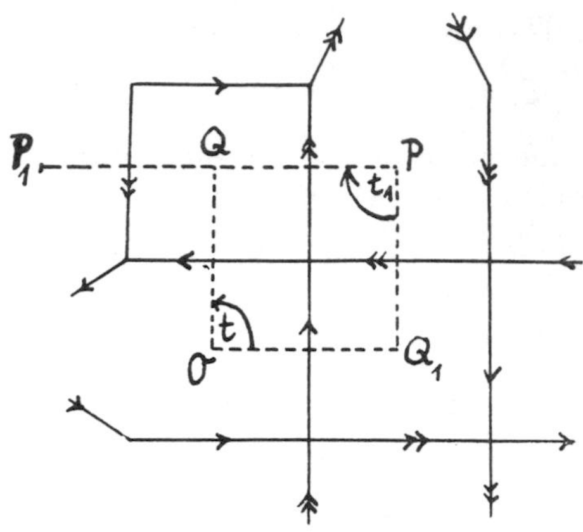

Figure 4

$$T = \{t, t_1\},$$
$$t^{4p} = t_1^{4p} = (t_1 t^{-1})^2 = 1. \tag{43}$$

If one writes

$$t_1 = z_1, \qquad t^{-1} = z_2, \qquad tt_1^{-1} = z_3,$$

then one has the normal form

$$T = \{z_1, z_2, z_3\},$$
$$z_1^{4p} = z_2^{4p} = z_3^2 = z_1 z_2 z_3 = 1, \tag{44}$$

in which z_1 and z_2 represent the two positive classes at the ground level and z_3 represents the cycle of $2p$ classes appearing with the $2p^{\text{th}}$ power.

The group diagram of (43) admits a metrically regular presentation in the hyperbolic plane Φ and is shown in Figure 4. At each vertex there are two regular quadrilaterals and two regular $4p$-gons, all with the same side length. The sides with simple and double arrows correspond to t and t_1 respectively. The dual polygonal group diagram is indicated in the figure by dotted lines. Its cells are congruent quadrilaterals in which one pair of opposite angles (at Q and Q_1) have magnitude $\pi/2$, the others (at O and P) have magnitude $\pi/2p$. The elements t and t_1 are $1/4p$ turns about O and P respectively. In this way, T is represented as a group of linear fractional transformations. $M = \Phi \bmod T$ has the connectivity of the sphere, corresponding to $\varepsilon_1 = 0$. One obtains a fundamental domain for F by combining the quadrilateral cell shown (the fundamental domain for T) with its images under $t, t^2, \ldots, t^{4p-1}$. The resulting star of quadrilaterals about O forms a regular $4p$-gon with corner angle $\pi/2p$. PP_1 is one of its sides (the angles at Q are indeed right). The motion

$$t_1 t^{-(2p+1)} = d_1$$

obviously carries the side opposite to PP_1 of this $4p$-gon to PP_1. Because of (42), the remaining d_i behave likewise for the remaining pairs of opposite sides. (The displacement direction of these hyperbolic transformations alternates with the rotation about O, with the exception of the sequence $d_{2p}d_1$.) Thus we get a closed surface of genus p, by identification of opposite sides of a $4p$-gon. The surface $\phi = \Phi \bmod F$ is $4p$-sheeted over $M = \Phi \bmod T$, with branch points of order $4p - 1$ at O and P, and a branch point at Q where the $4p$ sheets are joined in $2p$ separate pairs.

Example 5. We alter the presentation (39) of F by introduction of the superfluous generator

$$d_{2p+1} = d_1^{-1} d_2^{-1} \dots d_{2p}^{-1}$$

and arrive at the presentation

$$\begin{aligned} F &= \{d_1, d_2, \dots, d_{2p}, d_{2p+1}\}, \\ d_1 d_2 \dots d_{2p} d_{2p+1} &= 1, \\ d_{2p+1} d_{2p} \dots d_2 d_1 &= 1. \end{aligned} \tag{45}$$

This reveals the automorphism

$$I : d_i \mapsto d_{i+1}^{-1}, \tag{46}$$

where the subscript i is taken mod $2p + 1$, because this sends each of the defining relations of (45) to the other. One has

$$I^{2p+1} : d_i \mapsto d_i^{-1} \tag{47}$$

and hence $I^{4p+2} = 1$. To construct the traces and characteristic polynomial one has to go back to the independent generators $d_1, \dots, d_{2p}$ in the homology group and to note then that d_{2p+1} has exponent -1 in each of them. Because of (47), $s_{2p+1} = -2p$, hence $Z_{2p+1} = 2 + 2p$. For all other powers the only contribution to the trace comes from the element which goes to $d_{2p+1}^{\pm 1}$, and this is $+1$ for odd powers and -1 for even powers. The former therefore have trace and class number equal to 1, the latter have trace -1 and class number 3. One finds the characteristic polynomial to be

$$\gamma(\lambda) = \lambda^{2p} - \lambda^{2p-1} + \cdots + \lambda^2 - \lambda + 1.$$

Again $\varepsilon_1 = 0$.

Now let t be an I-inducing mapping. Then one has the following presentation for the group $T(I)$ on E.

$$\begin{aligned} T &= \{d_1, d_2, \dots, d_{2p+1}, t\}, \\ d_1 d_2 \dots d_{2p+1} &= 1, \\ d_{2p+1} d_{2p} \dots d_1 &= 1, \\ t^{4p+2} &= 1, \\ t d_i t^{-1} &= d_{i+1}^{-1} \qquad (i \bmod 2p+1). \end{aligned} \tag{48}$$

In order to present T more simply, one observes that since $Z_1 = 1$ there is no positive class on the ground level except that of t, so another should be sought on an even level, e.g., the $(2p + 2)^{\text{th}}$. Because of (47) and (46),

$$I^{2p+2} : d_i \mapsto d_{i+1}; \tag{49}$$

hence

$$(d_1 t^{2p+2})^{2p+1} = d_1 d_2 \dots d_{2p+1} t^{(4p+2)(p+1)} = 1.$$

One sets

$$d_1 t^{2p+2} = t_1,$$

then

$$d_1 = t_1 t^{-(2p+2)}$$

and, using (49)

$$\begin{aligned} d_i &= t^{(2p+2)(i-1)} d_1 t^{-(2p+2)(i-1)} \\ &= t^{(2p+2)(i-1)} t_1 t^{-(2p+2)i} \end{aligned}$$

for $i = 1, 2, \dots, 2p + 1$. Substitution in the last relation of (48) gives

$$(t_1 t^{-1})^2 = 1. \tag{50}$$

Substitution in the first relation gives

$$t_1^{2p+1} \cdot t^{-(2p+2)(2p+1)} = 1,$$

and the power of t appearing here equals 1. Substitution in the second relation gives

$$(t^{-1} t_1 t^{-1})^{2p+1} = 1.$$

But, again using (50),

$$t^{-1} t_1 t^{-1} = t_1^{-1}$$

which comes back to $t_1^{2p+1} = 1$. One therefore has

$$\begin{aligned} T &= \{t, t_1\}, \\ t^{4p+2} &= t_1^{2p+1} = (t_1 t^{-1})^2 = 1 \end{aligned} \tag{51}$$

or in normal form:

$$\begin{aligned} T &= \{z_1, z_2, z_3\}, \\ z_1^{4p+2} &= z_2^{2p+1} = z_3^2 = z_1 z_2 z_3 = 1. \end{aligned} \tag{52}$$

The element z_1 represents the positive class on the ground level, z_2 the cycle of two other classes in the even powers, and z_3 the cycle of $2p$ more classes in the $(2p + 1)^{\text{th}}$ power.

The group diagram of (51) is obtained in exactly the same way as that for (43) in Example 4. It can be illustrated by the same Figure 4. Now one has only to think of the polygon with midpoint O as a regular $(4p + 2)$-gon, and that with midpoint P as a regular $(2p + 1)$-gon. Correspondingly, the motion t is the $1/(4p + 2)$ turn about O, and t_1 is the $1/(2p + 1)$ turn about P. $M = \Phi \bmod T$ has genus 0, corresponding to $\varepsilon_1 = 0$. The surface $\phi = \Phi \bmod F$ again results from the star of quadrilaterals around O. Here we have the construction of a closed surface of genus p by identification of opposite sides of a $(4p + 2)$-gon. In contrast to the previous example, the polygon vertices now represent *two* different points on the surface, while the vertices of the $4p$-gon represent only *one*. The covering of M by ϕ is $(4p + 2)$-sheeted, with a branch point of order $4p + 1$ at O, a branch point at P where the sheets are joined in two separate sets of $2p + 1$, and a branch point at Q where the sheets are joined in $2p + 1$ separate pairs.

Specialisation of the examples to $p = 2$. We now specialise the characteristic polynomials obtained in the preceding examples to $p = 2$.

(1) For $p = 2$, Example 1 has the characteristic polynomial

$$\gamma(\lambda) = (\lambda^2 - 1)^2 = (\lambda - 1)^2(\lambda + 1)^2$$

which represents the d-combination 2 2 1 1 of type II in the table on page 423.

(2) As we have already observed, Example 2 is proper only for $p > 2$. For $p = 2$ the result is the class of the identity, and hence the type I of the table. For $p = 2$,

$$\gamma(\lambda) = (\lambda - 1)^4.$$

(3) The characteristic polynomial of H in Example 3 is

$$\gamma(\lambda) = \lambda^4 + \lambda^2 + 1 = (\lambda^2 - \lambda + 1)(\lambda^2 + \lambda + 1)$$

for $p = 2$, hence it gives the d-combination 6 3 and the type IX.

(4) S in Example 3 gives

$$\gamma(\lambda) = (\lambda + 1)^4;$$

hence the d-combination 2 2 2 2 of type III.

(5) SH^2 in Example 3 gives

$$\gamma(\lambda) = (\lambda^2 - \lambda + 1)^2;$$

hence the d-combination 6 6 of type X.

(6) Example 4 gives

$$\gamma(\lambda) = \lambda^4 + 1;$$

hence $d = 8$ and the type is XI.

(7) Example 5 gives

$$\gamma(\lambda) = \lambda^4 - \lambda^3 + \lambda^2 - \lambda + 1;$$

hence $d = 10$ and the type is XII.

At the same time, this shows the existence of the types V, VII and VIII, because they are the squares of types IX (or X), XI and XII.

Examples are lacking only for the types IV and VI, and none are known. There are various indications that these types cannot appear among mapping classes of finite order when $p = 2$.

12. Questions arising from the examples. The examples in the previous paragraph give rise to a series of fundamental questions.

(I) Suppose N classes of topological mappings of a surface of genus p onto itself are given, constituting a group. *Is it possible to choose a representative mapping of each class so that these N mappings constitute a group?*

(II) Or, specialised to cyclic groups: *Does each mapping class of finite order contain a mapping of finite order?*

If question II is answered affirmatively, then the invariants of mapping classes of finite order become fully illustrated by finite-sheeted branched coverings. The class number Z, which in all cases is a lower bound for the minimum number of fixed points under mappings of the given class, is then equal to this minimum number. Instead of having to deal with mapping classes of finite order n, one can deal with n-periodic mappings of the surface from the outset, as Brouwer[15] and Scherrer[16] have done. Brouwer's concept of the modular surface then leads to the representation of the given surface as a branched covering; and one can apply the known relations between the constants of branched coverings (such as the formula of Hurwitz and that of Scherrer, *loc. cit.*) to the n-periodic mappings. The present work goes in the opposite direction, by showing how one, *independently of the answer to question* II, can derive a relation like the Hurwitz formula as a *relation between certain group theoretic quantities* for mapping classes of finite order, and in the examples of the previous paragraph it is shown how the *boundary circle group* T, which is an invariant of the mapping class, can lead via its *group diagram* to the associated branched covering. Question II then appears as equivalent to the following:

(III) *Is the boundary circle group T corresponding to a mapping class of finite order always representable as a group of linear fractional transformations?*

A more precise formulation is obtained as follows. The boundary circle E contains the fundamental group F as an everywhere dense set of fundamental points. An automorphism I of F of the first kind induces a topological mapping t of E onto itself. If I^n is an inner automorphism, then t^n is an element of F. Question: Can E be mapped topologically onto a circle E' in such a way that the conjugate mapping t' of E' is a linear fractional transformation, the conjugate group F' remains a group of such transformations, and the functional equations are preserved; in other words, can the conjugate group T' be a group of linear fractional transformations on E'? Also, is this mapping of E onto E' uniquely determined up to the obvious conjugation with linear fractional transformations of E' onto itself?

One could attempt to construct such a topological mapping of E onto E' directly by means of the functional equation. The examples take the other route of finding an abstract presentation of the group T which makes its group diagram easy to survey. This leads to the questions:

(IV) *Is the normal form found for the abstract presentation of the group T in the examples generally valid for mapping classes of finite order?*

(V) *If an automorphism I of F has some power which is an inner automorphism, is there a canonical presentation for I, obtainable by change of generators and regulated by the characteristic polynomial?*

(In analogy with the rational canonical forms of the matrices of linear fractional transformations, obtainable by change of variables.) And finally:

[15] L. E. J. Brouwer: Amsterdam Proceedings, vol. XXI, nos. 5, 9 and 10 (1918–1919) and Comptes rendus (Paris) vol. **168**, p. 677 and p. 845 (1919), and vol. 171, p. 89 (1920).

[16] W. Scherrer: "Über topologische Involutionen" and "Über periodische Transformationen von Flächen". Vierteljahrschrift der Naturforschenden Gesellschaft in Zürich, vol. 70 (1925).

(VI) *Are two mapping classes of finite order with the same characteristic polynomial equivalent?*

I.e., when two mapping classes A_1 and A_2 of finite order have the same characteristic polynomial, is there an exponent ρ (relatively prime to their order) and a mapping class A_3 such that

$$A_2 = A_3 A_1^{\rho} A_3^{-1}?$$

13. Theorems on automorphisms of order 2. Let I be an automorphism of the first kind and order 2. So $I^2 = 1$, and if t is an I-inducing mapping function, $t^2 = 1$ on E. Let $f \neq 1$ be an element of F which leads to a mapping function of order 2:

$$1 = (ft)^2 = ff_I t^2 = ff_I;$$

hence

$$f_I = f^{-1}. \tag{53}$$

Thus the axis of f is reversed under I, i.e., its fundamental points are exchanged by the boundary mapping t. All elements whose axes are reversed by I satisfy (53). In particular, this holds for the axis of the primary element f^* of which f is a power. Now

$$f^{*2r} = f^{*r}(f^{*-r})_I \approx 1,$$
$$f^{*2r+1} = f^{*r}f^*(f^{*-r})_I \approx f^*,$$

where the sign $\approx$ denotes the isogredience of two elements of F relative to I. (Thus we write $g \approx f$ for elements g, f of F when there is an element h of F such that

$$gt = hfth^{-1} = hfh_I^{-1}t;$$

hence

$$g = hfh_I^{-1}.$$

(Non-isogredience is denoted by $\not\approx$.) This says nothing about whether $f^* \approx 1$ or not, but at any rate an axis which is reversed by I gives rise to at most one isogredience class of the mapping function of index $+1$, other than the class of t.

Suppose now that f is primary, $\not\approx 1$, and that it satisfies (53). Suppose also that the conjugate gfg^{-1} satisfies (53). Then

$$(gfg^{-1})_I = gf^{-1}g^{-1};$$

on the other hand,

$$(gfg^{-1})_I = g_I f_I g_I^{-1} = g_I f^{-1} g_I^{-1}.$$

Equating these two expressions gives

$$g^{-1}g_I \rightleftarrows f.$$

Thus the element $g^{-1}g_I$ is on the one hand ≈ 1, on the other hand it is a power of f and hence necessarily an even power, because

$$f^{2r+1} \approx f \not\approx 1.$$

It follows from

$$g^{-1}g_I = f^{2r} = f^r(f^{-r})_I$$

that gf^r is a fixed element under I. But since $I^2 = 1$, I has only the fixed element 1, hence $g = f^{-r}$ and $gfg^{-1} = f$. Under the hypothesis on f, this implies that no axis equivalent to the axis of f is reversed by I, or in other words: f is the only element in its conjugacy class which satisfies (53).

The number of positive classes of the mapping class corresponding to I is given by the number of solutions of (53) which are non-isogredient relative to I. One solution is $f = 1$, hence t itself represents a positive class. All other solutions can be chosen as primary, by what we found above, and all solutions belong to different conjugacy classes of F.

Let P be an arbitrary point of E and let tP be its image. The two arcs determined by P and tP on E are exchanged by t. Thus if Q is any other point, P and Q are not separated by tP and tQ. If g is an element of F, the axes of g and g_I therefore do not intersect. Also, if f is a solution of (53) then the axes of $f^r gf^{-r}$ and $f^s g_I f^{-s}$ do not intersect for any r and s, because

$$f^s g_I f^{-s} = f^{s+r}t \cdot f^r gf^{-r} \cdot (f^{s+r}t)^{-1}$$

and $(f^{s+r}t)^2 = 1$.

Now suppose that a primary solution f of (53), with $f \not\approx 1$, has an axis which covers a self-intersecting geodesic on the surface ϕ. Let k be an element of F for which the axis of kfk^{-1} intersects the axis of f. I sends it to $k_I f^{-1} k_I^{-1}$; the axis of this element also intersects the axis of f, and indeed in the opposite direction. The axes of kfk^{-1} and $k_I fk_I^{-1}$ intersect the axis of f in the same direction. The element $k_I fk_I^{-1}$ does not result from kfk^{-1} when conjugated by a power f^r of f, otherwise it would be reversed under conjugation by $f^r t$, and $f^r t$ already reverses its equivalent, f. If one considers the axes of the elements

$$f^r kfk^{-1}f^{-r}$$

for arbitrary r and

$$f^s k_I fk_I^{-1}f^{-s}$$

for arbitrary s, then the axes within a single system do not intersect because of their position relative to the axis of f, and an axis of the first system does not intersect one of the second by the above, since one is sent to the other by conjugation with $f^i t$. Thus this set of axes divides Φ into strips across the axis of f, and a displacement segment on the axis of f crosses exactly two strips. The two systems are exchanged by t and there is exactly one pair which is exchanged and bounds a strip, so that it is separated by no other axes. Let this pair consist of

$$f^r kfk^{-1}f^{-r} = mfm^{-1}$$

and

$$f^{-r} k_I fk_I^{-1}f^r = m_I fm_I^{-1}.$$

One notices that the conjugating factor m is not completely determined by this requirement, since it can be replaced by mf^i with arbitrary i. We now make use of this fact by using the axis pair $m^{-1}fm$ and $m_I^{-1}fm_I$, which cuts the axis of f in the sense opposite to that of the previous pair, to again produce all $f^i m^{-1}fmf^{-i}$ and $f^j m_I^{-1}fm_I f^{-j}$ and proceed exactly as above: i may

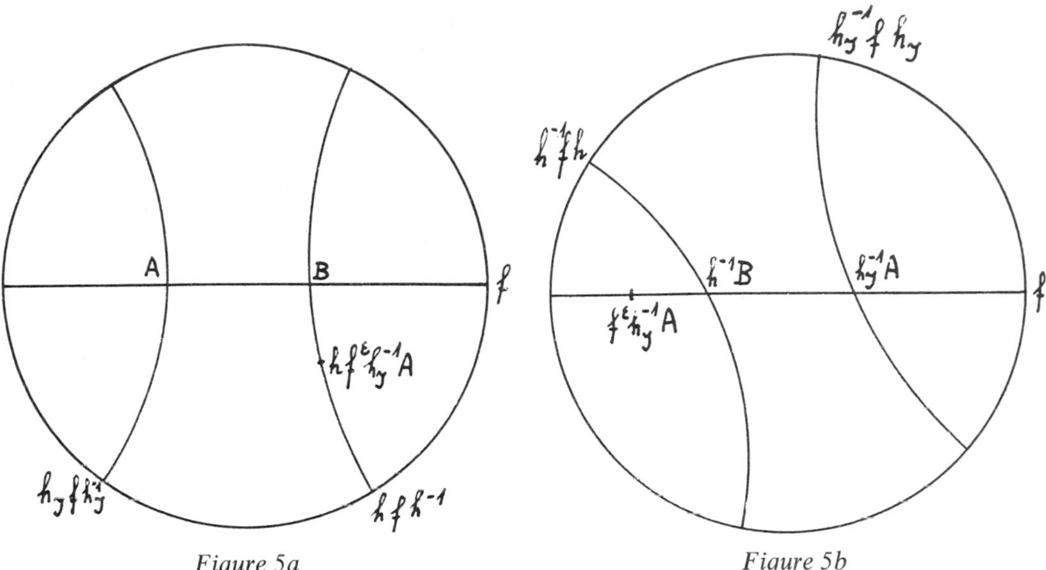

Figure 5a *Figure 5b*

be chosen so that the axis of $f^i m^{-1} f m f^{-i}$ bounds a strip in conjunction with its image under I, so that the two are not separated by any other axes. In summary, we have converted the original conjugating factor k to a new one

$$h = mf^{-i} = f^r k f^{-i}$$

so that the following situation obtains, as illustrated by Figures 5(a) and 5(b): The axes of hfh^{-1} and $h_I f h_I^{-1}$ intersect the axis of f at two points B and A whose distance from each other is smaller than the displacement length l of f. The axes of $h^{-1}fh$ and $h_I^{-1} f h_I$ intersect the axis of f at the points $h^{-1}B$ and $h_I^{-1}A$, whose distance from each other is also $< l$. Apart from this, the position of the former axes relative to the latter can be arbitrary; e.g., they can intersect each other. In order to avoid assumptions being made about them, they are shown in separate figures. The elements h and h^{-1} play equivalent roles and one can therefore assume that

$$AB \leq (h_I^{-1}A)(h^{-1}B).$$

Now one extends the larger of the two to l by adding the complementary segment from $h^{-1}B$ to $f^{\varepsilon}h_I^{-1}A$, where $\varepsilon = +1$ or -1, depending on the positive direction of the axis f. The image of this segment of the axis under h lies on the axis of hfh^{-1} from B to $hf^{\varepsilon}h_I^{-1}A$. Its length is

$$l - (h^{-1}B)(h_I^{-1}A) \leq l - AB.$$

The rectilinear distance therefore satisfies

$$A(hf^{\varepsilon}h_I^{-1}A) < AB + B(hf^{\varepsilon}h_I^{-1}A) \leq l.$$

The endpoints of the segment on the left-hand side correspond under the element $hf^{\varepsilon}h_I^{-1}$. Hence this element also has a displacement length $< l$. And

$$f' = hf^{\varepsilon}h_I^{-1} \approx f^{\varepsilon} \approx f.$$

Possibly the new element f' is not primary. Then since

$$f' \approx f \not\approx 1,$$

we necessarily have

$$f \approx f' \approx f'^*,$$

where f'^* is the primary element with the same axis as f', and hence has a smaller displacement length. Now if f'^* also corresponds to curve type with double points, one can reapply the same reasoning. But the group F realizes only a finite number of displacement lengths below a given bound (1, 5), hence the process must terminate. We therefore have:

Theorem 7. *A class of positive index, different from the class of t, may be represented by an ft such that f is primary and corresponds to a simple curve type.*

Now we consider two different classes of this kind. Suppose then that $f_I = f^{-1}$, $g_I = g^{-1}$, $f \not\approx g$, $f \not\approx 1$, $g \not\approx 1$, f and g are primary, and f as well as g corresponds to a simple curve type on ϕ.

The axes of f and g intersect, because both are reversed by t. The corresponding geodesics on ϕ therefore have at least one point of intersection. (These two geodesics are different, because f and g do not belong to the same conjugacy class of F, as shown above.) Now suppose that they have *more than one* point of intersection. Then there are axes lying as shown in Figure 6. A displacement segment of f reaches from A to fA. By hypothesis, this segment is cut by the axis of an element hgh^{-1}, at a point C say. The latter axis runs in the strip between the axes of g and fgf^{-1}, since g corresponds to a simple curve type. The latter two axes go to the axes of g and $f^{-1}gf$ under I, and hence the axis of $h_I g h_I^{-1}$ runs in the resulting strip, and cuts the axis of f at a point B. We first investigate whether B and C can be equivalent, so that they lie over the same surface point. They lie on the geodesic corresponding to f, which is simple by hypothesis. Thus the only possibility is $C = fB$. Since g is also simple, this implies

$$hgh^{-1} = fh_I g h_I^{-1} f^{-1};$$

hence

$$h^{-1} f h_I \rightleftarrows g$$

and therefore

$$h^{-1} f h_I = g^r.$$

But then we would have $f \approx g$ or $f \approx 1$, and both are excluded by hypothesis. Thus B and C are not equivalent. Apart from A then, the intersections of the geodesics corresponding to f and g appear in pairs, and hence the number of intersections is odd. *It follows that f, as well as g, is not null homologous.*

The point fB is different from C but between A and fA. Suppose that fB falls between A and C. Then $BC > l(f)$, where $l(f)$ denotes the displacement length of f. The point $f^{-1}C$ then falls between A and B and

$$(f^{-1}C)(fB) < l(f).$$

The elements $fh_I g h_I^{-1} f^{-1}$ and $f^{-1} hgh^{-1} f$ can therefore take the place of the pair used previously. In other words: One is entitled to choose the conjugating factor h so that $BC < l(f)$.

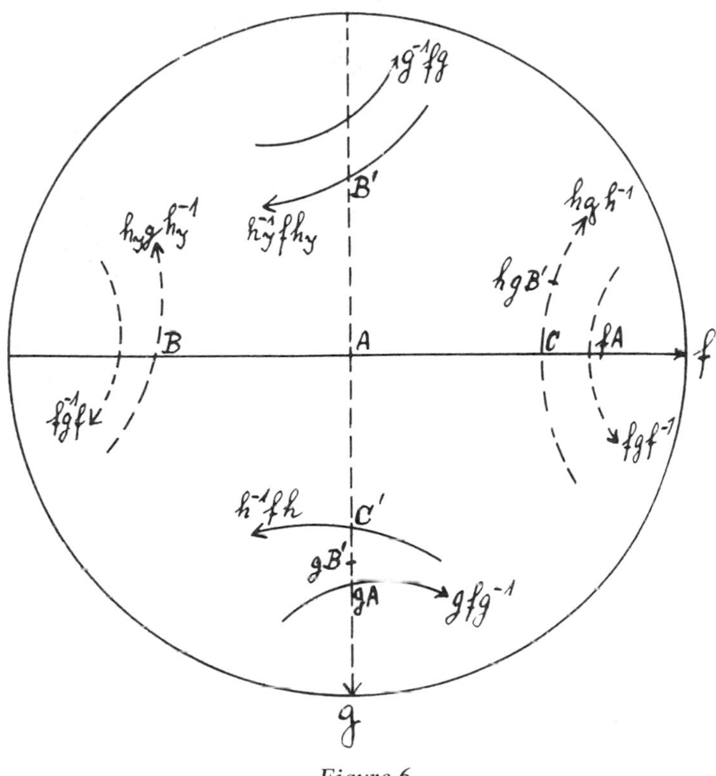

Figure 6

The motion h^{-1} sends C to a point $C' = h^{-1}C$ on the axis of g. Now one notices that the above determines h only up to an arbitrary power of g:

$$hgh^{-1} = hg^r \cdot g \cdot g^{-r}h^{-1}.$$

One can therefore choose h so that $h^{-1}C = C'$ falls between A and gA. Then the axis of $h^{-1}fh$ falls in the strip between the axes of f and gfg^{-1}; so the axis of $h_I^{-1}fh_I$ falls in the strip between the axes of f and $g^{-1}fg$ and cuts the axis of g at the point $B' = h_I^{-1}B$. One could also choose h so that C' falls between A and $g^{-1}A$; then B' falls between A and gA.

$B'C' < l(g)$ for one of these cases, and that is the case we take. (The figure assumes the first case, but the following argument applies to both.)

The preceding argument treats f and g equally. We can therefore assume

$$BC \leq B'C'.$$

Then choose $\varepsilon = \pm 1$ so that

$$C'(g^\varepsilon B') = l(g) - B'C'.$$

(The figure shows the case where one should choose $\varepsilon = +1$.) This piece of axis is then displaced by the motion h, so that it lies on the axis of hgh^{-1} and reaches from $hC' = C$ to

$hg^\varepsilon B' = hg^\varepsilon h_I^{-1}B$. The latter point is the image of B under the motion

$$g' = hg^\varepsilon h_I^{-1} \approx g$$

and we have

$$l(g') \leq B(hg^\varepsilon h_I^{-1}B) < BC + C(hg^\varepsilon h_I^{-1}B) = BC + l(g) - B'C' \leq l(g).$$

Thus one can replace g by an element g' with smaller displacement length without leaving its isogredience class. If g' is not primary, then one replaces it by the corresponding primary element (we have $g' \approx g \not\approx 1$) which reduces the displacement length even further. If it is not simple, by Theorem 7 one can replace it by a simple member of its class, which again reduces the displacement length. If one again has a primary simple element g_1 of the class, and if the latter still has more than one intersection with f, then one applies the above process to f and g_1, and continues. The process must terminate, since the sum of the displacement lengths of f and g decreases continually. If one has more than two isogredience classes, then one can apply the process to any two of them. Thus one obtains:

Theorem 8. *It is possible to find a complete system*

$$f_1 = 1, f_2, f_3, \ldots, f_Z$$

of non-isogredient solutions of the equation (53), *such that $f_2, \ldots, f_Z$ correspond to geodesics on the surface ϕ which are non-bounding and simple, and any two of which have exactly one point of intersection.*

14. Mapping classes with the characteristic polynomial $(\lambda + 1)^{2p}$. Now suppose we have a mapping class of order 2 with the maximum number $Z = 2p + 2$ of positive fixed point classes. The existence of such classes is shown by the automorphism S of Example 3, §11, as well as the automorphism I^{2p} of Example 4 ($= I^{2p+1}$ of Example 5). The corresponding trace is then $s = -2p$, the matrix itself is $-E_{2p}$, and the characteristic polynomial is $(\lambda + 1)^{2p}$.

Let t be an element of order 2 in the associated group T, so $t^2 = 1$ on E. The mapping t induces the automorphism I and $I^2 = 1$. The equation

$$xx_I = 1 \tag{54}$$

then has $2 + 2p$ non-isogredient solutions relative to I. If one chooses $x = 1$ as one of them, then the result of the last paragraph allows us to choose the remainder

$$f_1, f_2, \ldots, f_{2p+1} \tag{55}$$

so that

1. these elements are primary,
2. the corresponding closed geodesics on the surface are simple and not null homologous,
3. any two of these geodesics meet at exactly one point.

Now one considers a point O of a surface ϕ of genus p and n simple closed curves u_i which cross each other at O and are otherwise disjoint, as shown by the dashed lines in Figure 7 for even n and in Figure 8 for odd n. Bands are drawn on the surface which form strip neighbourhoods of these curves. The n bands are viewed as beginning and ending at a disc neighbourhood of O. This disc, together with the n bands, forms a subsurface ϕ^* of ϕ. We now distinguish

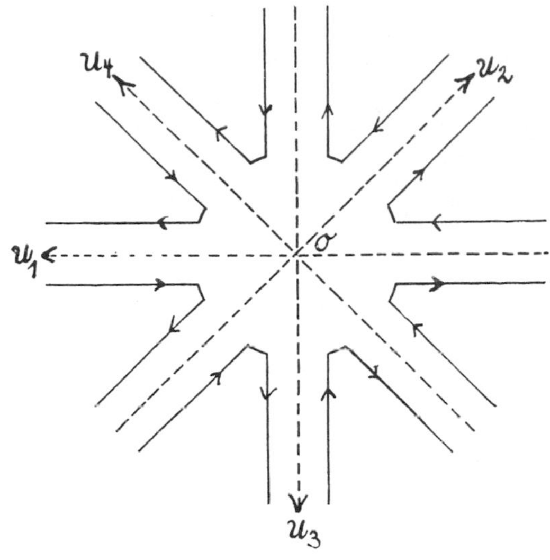

Figure 7

between even and odd n, and refer to the figures:

(a) If $n = 2r$, then ϕ^* has only one boundary curve and with suitable ordering of the u_i, and suitable traversal, the boundary curve of ϕ^* is given by the element

$$R = u_1 u_2 \ldots u_{2r} u_1^{-1} u_2^{-1} \ldots u_{2r}^{-1} \tag{56}$$

of the fundamental group. R is null homologous, and the complementary surface $\phi - \phi^*$ has the same boundary curve R. If this complementary surface is a disc, then $R = 1$, $r = p$, and the group

$$U = \{u_1, u_2, \ldots\}$$

is the group F as presented in Example 4. If the complementary surface is not a disc, then $R \neq 1$, $r < p$ and U is a proper subgroup of F, in fact it is the fundamental group of the subsurface ϕ^* of genus r and one boundary curve, and hence free. The complementary surface $\phi - \phi^*$ has genus $p - r$ and one boundary curve. In this case one can consider the closed simple geodesic corresponding to the element R of U. It is null homologous, and therefore separates ϕ, into components homeomorphic to ϕ^* and $\phi - \phi^*$. We may as well assume that the boundary curve is already this geodesic. Its lift in the covering surface Φ of ϕ, together with all its images under the subgroup U, are circles orthogonal to E which bound the universal covering surface Φ^* of ϕ^* in Φ. Φ^* is the convex hull of the set of axes for the subgroup U.

(b) If $n = 2r + 1$ then ϕ^* has two boundary curves, and these correspond to the elements

$$\begin{aligned} R_1 &= u_1 u_2 \ldots u_{2r+1}, \\ R_2 &= u_{2r+1} u_{2r} \ldots u_1 \end{aligned} \tag{57}$$

of U. If $\phi - \phi^*$ consists of two discs with these boundary curves, then $R_1 = R_2 = 1$, $r = p$, and U is the group F as presented in Example 5. If $R_1 \neq 1$ and $R_2 \neq 1$ in F then U is the (free)

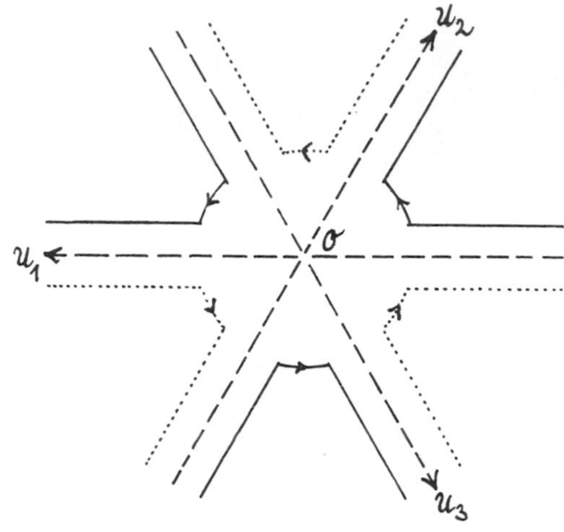

Figure 8

fundamental group of a surface of genus r with two boundary curves. The axes of R_1 and R_2, together with all their images under U then bound a convex region in Φ, the convex hull of the set of axes of U, which is the universal covering surface of ϕ^*.

We now construct such a curve system u_i on the surface ϕ step by step, using the solution system (55). The element f_1 generates the group $U_1 = \{f_1\}$; f_2 is not an element of U_1, since it is now a power of f_1. Let O be the intersection of the axes of f_1 and f_2 in Φ. We take the corresponding geodesics as the first curves u_1 and u_2 on ϕ. The result is a perforated torus with boundary curve R, according to (56), because $p > 1$. The boundary curve is the axis of the element

$$R = f_1 f_2 f_1^{-1} f_2^{-1}$$

and this, together with its images under U, bounds a convex region Ω_2 which is the convex hull of the set of axes of

$$U_2 = \{f_1, f_2\}$$

in Φ. Ω_2 is the universal covering surface of the perforated torus.

Now we add f_3, and first convince ourselves that f_3 is not an element of U_2. Suppose on the contrary that

$$f_3 = x\Pi(f_1, f_2)y,$$

where x and y are either $f_1^{\pm 1}$ or $f_2^{\pm 1}$. This representation is unique, since U_2 is free. Now on the one hand,

$$f_{3I} = x^{-1}\Pi(f_1^{-1}, f_2^{-1})y^{-1},$$

on the other hand

$$f_{3I} = f_3^{-1} = y^{-1}\Pi^{-1}x^{-1}.$$

Then, by the uniqueness of representation in U, $x = y$ and hence

$$f_3 = x\Pi x = x\Pi x_I^{-1} \approx \Pi.$$

Thus, without leaving its isogredience class, f_3 can be replaced by a shorter word $\Pi(f_1, f_2)$. The latter can be shortened similarly, and finally we get $f_3 \approx 1$ or $f_1^{\pm 1}$ or $f_2^{\pm 1}$, which are all contrary to the hypothesis on (55).

Thus f_3 does not belong to U_2, and $f_3\Omega_2$ is disjoint from Ω_2. The axis of f_3 intersects Ω_2, since it intersects the axes of f_1 and f_2, but of course the intersection will not in general occur at O. The axis of f_3 also intersects $f_3\Omega_2$, but not any region equivalent to Ω_2 between Ω_2 and $f_3\Omega_2$, because a displacement segment of the axis of f_3 which begins in Ω_2 ends in $f_3\Omega_2$, and by hypothesis it cannot meet axes equivalent to those of f_1 and f_2 between these regions. Thus if one connects the point O, which lies in Ω_2, to the point f_3O, which lies in $f_3\Omega_2$, by a non-Euclidean line segment, then the latter meets none of its equivalents. The result is a geodesic segment on ϕ which issues from O and returns to O (in general forming a corner) without meeting itself or the other two curves already present until it crosses them at O.

Now we use these three curves as u_1, u_2 and u_3 and we have two boundary curves, corresponding to (57). *With possible renumbering of the first three f,*

$$R_1 = f_1 f_2 f_3,$$

$$R_2 = f_3 f_2 f_1,$$

and these elements of F are not both 1, since ϕ has genus $p > 1$. But then they are necessarily both $\neq 1$, since

$$R_{1I} = R_2^{-1}.$$

Thus

$$U_3 = \{f_1, f_2, f_3\}$$

is again free, and the axes of R_1 and R_2 bound a doubly-perforated torus on ϕ. Let its covering surface, which is the convex hull of the set of axes of U_3, be Ω_3. Then one concludes, exactly as above, that f_4 does not belong to U_3, that its axis meets the axes of f_1, f_2 and f_3 in Ω_3, that $f_4\Omega_3$ is disjoint from Ω_3 and that the two are not separated by any of their equivalents. The segment connecting O to f_4O is then introduced as the u_4 curve. If $p > 2$, one now has a subsurface of ϕ with genus 2, one boundary curve, and a free fundamental group

$$U_4 = \{f_1, f_2, f_3, f_4\}$$

etc. The process terminates with the introduction of f_{2p}, since ϕ^* then attains the genus p of the surface ϕ. Then one has, with possible renumbering of $f_1, \ldots, f_{2p}$:

$$U_{2p} = F = \{f_1, f_2, \ldots, f_{2p}\},$$

$$f_1 f_2 \cdots f_{2p} f_1^{-1} \cdots f_{2p}^{-1} = 1.$$

Comparison with Example 4 shows that these curves from a $4p$-gon on ϕ with opposite sides incident in pairs. The element f_{2p+1} of (55) has not yet been mentioned. This leaves us free to choose it to be the element

$$f_{2p+1} = f_1^{-1} f_2^{-1} \cdots f_{2p}^{-1},$$

which corresponds to the diagonal of the $4p$-gon, and hence to a primary, simple curve type. It is not isogredient to 1 or any of the f_i already used. One sees this as follows: If

$$g = xhx_I^{-1},$$

then g and h have exponent sums of equal parity in all generators, because x and x_I^{-1} have equal exponent sums in all generators as a result of $f_{iI} = f_i^{-1}$. But f_{2p+1} does not satisfy this condition either with 1 or any f_i. Finally, f_{2p+1} also satisfies (54), as follows from the relation of F.

The automorphism denoted here by I is transformed by the automorphism

$$f_i \mapsto d_i \qquad (i = 1, 2, \ldots, 2p)$$

of F into the automorphism denoted by I^{2p} in Example 4.

Theorem 9. *All mapping classes of the first kind and order* 2 *with* $(\lambda + 1)^{2p}$ *as characteristic polynomial, and hence* $2 + 2p$ *positive fixed point classes, are conjugate.*

To present the group T in *normal form* one has to set

$$t_1 = f_1^{-1}t$$
$$t_2 = f_2 t$$
$$\cdots$$
$$t_{2p-1} = f_{2p-1}^{-1}t$$
$$t_{2p} = f_{2p}t$$
$$t_{2p+1} = f_1 f_2 \ldots f_{2p}t.$$

Then one can express all f_i in terms of t and the t_i and one obtains

$$T = \{t, t_1, t_2, \ldots, t_{2p+1}\},$$
$$t^2 = t_1^2 = \cdots = t_{2p+1}^2 = tt_1t_2\ldots t_{2p+1} = 1.$$

This gives an affirmative answer to questions (II)–(VI) *of* §12 *in the case of the mapping classes considered here.*

15. Mapping classes with characteristic polynomial $\lambda^{2p} + 1$ and with characteristic polynomial $\lambda^{2p} - \lambda^{2p-1} + \cdots - \lambda + 1$. Suppose we have a mapping class of finite order with characteristic polynomial $\lambda^{2p} + 1$. Example 4, §11, shows that there are such mapping classes. Eigenvalues, order, traces and class numbers are the same as in Example 4. Thus the order is $4p$, and there are two positive classes in all powers except the $2p^{\text{th}}$, where there are $2p + 2$. Let t be a mapping function on the ground level with $t^{4p} = 1$ on E, and let I be the induced automorphism, so $I^{4p} = 1$.

I^{2p} comes under the general case of the previous paragraph, since it has order 2 and matrix $-E_{2p}$. The equation

$$xx_{I^{2p}} = 1 \tag{58}$$

therefore has $2 + 2p$ non-isogredient solutions relative to I^{2p}. However, for our present purpose these must be specially chosen. To do this we go back to the general theory of §6. Two of the $2 + 2p$ positive classes at level $2p$ take root at level one; the remaining $2p$ constitute a cycle which appears at level $2p$. From a solution system of (58) with the properties described in Theorem 8, one selects an element f such that ft^{2p} does not take root at level one, but represents a class appearing at level $2p$. This means that f is primary and corresponds to a simple curve which is not null homologous. Then by §6 the elements

$$f, f_I, f_{I^2}, \ldots, f_{I^{2p-1}} \tag{59}$$

represent the $2p$-termed cycle. They are all *primary*, and correspond to *simple, non-null homologous* curve types, since they result from a curve of this type by powers of a topological surface mapping. However, we need a new proof that they also can be chosen so as to have the last property mentioned in Theorem 8, *that any two of them have exactly one point of intersection.* The proof is a generalisation of the argument based on Figure 6. The displacement length on the axis of f is again denoted by $l(f)$; one also sets

$$L(f) = l(f) + l(f_I) + \cdots + l(f_{I^{2p-1}}).$$

Any two axes of the system (59) intersect, since both are reversed under I^{2p}. Thus any two of the corresponding geodesics on ϕ have at least one point of intersection. Now suppose that two of them have more than one intersection. The same then holds for the axis pairs, viewed as geodesics on ϕ, which result from the given pair under powers of I. We can therefore assume that we are dealing with f and f_{I^r}, $1 \leq r \leq 2p - 1$. Now one considers Figure 6 and reads f_{I^r} in place of g and I^{2p} in place of I. Otherwise, all the arguments about this figure made in §13 still apply, except that we drop the assumption $BC \leq B'C'$ made on grounds of symmetry there. In order to indicate that B' and C' lie on the axis of f_{I^r} we write them as B_r and C_r, and we write B and C as B_0 and C_0. Nothing is assumed about the relative sizes of B_0C_0 and B_rC_r, but the hypothesis $B_0C_0 < l(f)$, $B_rC_r < l(f_{I^r})$ remains in force. Thus we come to Figure 9 and consider the segment from C_0 to $f^{\varepsilon}B_0$ $(\varepsilon = \pm 1)$ which extends B_0C_0 to length $l(f)$. If we displace this by h^{-1}, then it falls on the axis of $h^{-1}fh$ and reaches from $h^{-1}C_0 = C_r$ to $h^{-1}f^{\varepsilon}B_0 = h^{-1}f^{\varepsilon}h_{I^{2p}}B_r$. One therefore has the inequality

$$l(h^{-1}f^{\varepsilon}h_{I^{2p}}) < B_rC_r + l(f) - B_0C_0.$$

Now if one replaces all elements by their images under I^s, one obtains a picutre of axes in completely analogous positions, hence, when one denotes the points on the axis of f_{I^s} corresponding to the points B_0 and C_0 by B_s and C_s, then one obtains the inequality

$$l(h_{I^s}^{-1}f_{I^s}^{\varepsilon}h_{I^{2p+s}}) < B_{r+s}C_{r+s} + l(f_{I^s}) - B_sC_s$$

for all s. When $s = 2p$, $f_{I^s} = f^{-1}$ and the points B_0 and C_0 are simply exchanged, thus, as far as their (absolute) lengths are concerned, segments need only be considered with subscripts mod $2p$. Then the last inequality yields

$$\sum_{s=0}^{2p-1} l(h_{I^s}^{-1}f_{I^s}^{\varepsilon}h_{I^{2p+s}}) < \sum_{s=0}^{2p-1} l(f_{I^s})$$

since the BC segments cancel in pairs. Now

$$f' = h^{-1}f^{\varepsilon}h_{I^{2p}} \approx f$$

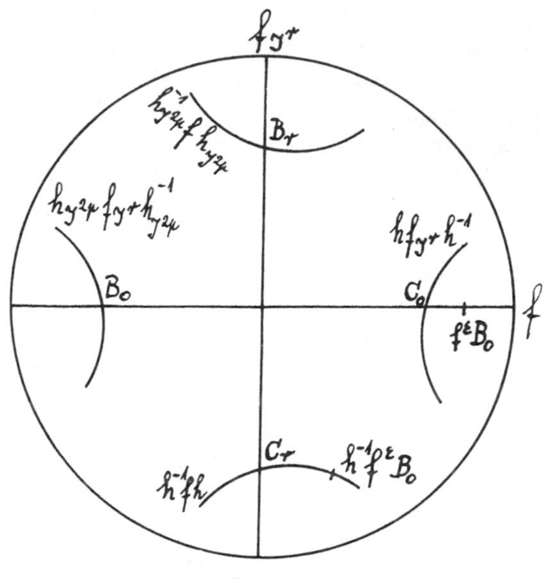

Figure 9

relative to I^{2p}, and hence it represents the latter class, and the f'_{I^s} therefore represent its conjugate classes. The last inequality now shows *that the quantity L becomes smaller* when f is replaced by $f' : L(f') < L(f)$.

If the new element f' is not primary, one replaces it by the corresponding primary, which means a further reduction in the quantity L. But what if f' does not correspond to a simple curve type? Then we generalise the argument based on Figures 5(a) and 5(b) of §13, again writing f in place of f'. We obtain Figures 10(a) and 10(b) with the relations $A^* = h_{I^{2p}}^{-1} A$, $B^* = h^{-1}B$, and we make no assumption about the relative lengths of AB and A^*B^*. One has

$$hf^\eta A^* = hf^\eta h_{I^{2p}}^{-1} A = f'A,$$
$$h^{-1}f^\varepsilon A = h^{-1}f^\varepsilon h_{I^{2p}} A^* = f''A^*$$

when one introduces

$$f' = hf^\eta h_{I^{2p}}^{-1} \qquad (\eta = \pm 1),$$
$$f'' = h^{-1}f^\varepsilon h_{I^{2p}}. \qquad (\varepsilon = \pm 1).$$

The figures then show that

$$l(f') < AB + l(f) - A^*B^*,$$
$$l(f'') < A^*B^* + l(f) - AB;$$

hence

$$l(f') + l(f'') < 2l(f).$$

Application of I^s gives axes in analogous positions, hence

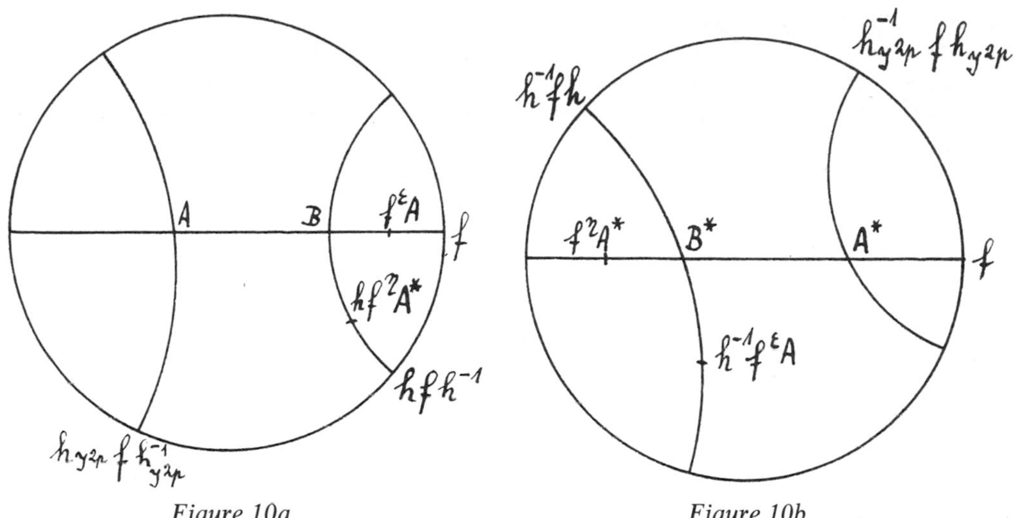

Figure 10a *Figure 10b*

$$l(f'_{I^s}) + l(f''_{I^s}) < 2l(f_{I^s})$$

and summation from $s = 0$ to $s = 2p - 1$ gives

$$L(f') + L(f'') < 2L(f).$$

Thus either

$$L(f') < L(f)$$

or

$$L(f'') < L(f).$$

And both $f' \approx f$ and $f'' \approx f$ relative to I^{2p}. Hence one can again reduce the quantity L by exchanging with another member in the class of f.

One continues until one has, after a finite number of steps, a primary simple f. If the new system (59) still contains two members with more than one intersection, then one repeats the first process and exchanges accordingly. Since there are only finitely many possible values of L below a given bound, the process comes to an end.

Thus we can now assume that the system (59) of $2p$ elements has the property of representing $2p$ closed geodesics on ϕ, any two of which intersect exactly once. But then, as shown in the previous paragraph, it is possible to draw $2p$ simple closed curves through a point O on the surface which meet each other only where they cross at O. And these curves correspond to the system of elements (59), with O as initial and final point. We denote the curves by the same letters as the elements (59). Let $f_{I^r}^{\pm 1}$ be the first in counterclockwise order round O after f^{+1}. Then r is relatively prime to $4p$. We now consider the power

$$J = I^r \quad \text{or} \quad J = I^{2p+r}$$

according as $f_{I^r}^{+1}$ or $f_{I^r}^{-1}$ follows f^{+1}. In both cases J is a primitive power of I. The positive directions of the elements

$$f, f_J, f_{J^2}, \ldots, f_{J^{2p-1}} \tag{60}$$

then follow each other in counterclockwise order around O. Comparison with Figure 7 then yields the relation

$$ff_J^{-1}f_{J^2}f_{J^3}^{-1} \ldots f_{J^{2p-1}}^{-1}f^{-1}f_J \ldots f_{J^{2p-1}} = 1 \tag{61}$$

for the elements of (60). The substitution

$$f_{J^{2r}} \mapsto d_{2r+1},$$
$$f_{J^{2r+1}} \mapsto d_{2r+2}^{-1},$$

where the d_i are the generators in Example 4, then defines an automorphism Λ of F, because it carries the relation (61) into the relation (39). And now

$$\Lambda J \Lambda^{-1} = \bar{I},$$

where $\bar{I}$ (in contrast to the present I) denotes the automorphism I of Example 4. At the same time we have shown that

$$\Lambda I \Lambda^{-1} = \Lambda J^{\rho} \Lambda^{-1} = \bar{I}^{\rho}$$

when ρ is a solution of the congruence

$$\text{resp.,} \qquad \left.\begin{aligned} \rho r &\equiv 1 \\ \rho(2p + r) &\equiv 1 \end{aligned}\right\} \qquad \text{mod } 4p.$$

$\bar{I}^{\rho}$ is a primitive power of $\bar{I}$. Thus the group $T(I)$ may be brought into the same normal form as the isomorphic group $T(\bar{I})$ in Example 4. Everything in Example 4 is therefore valid for the mapping classes considered here, and hence questions (II)–(VI), §12, have an affirmative answer for them.

Now we briefly mention how one can deal similarly with the mapping classes which appeared in Example 5, §11, with the characteristic polynomial

$$\lambda^{2p} - \lambda^{2p-1} + \cdots + \lambda^2 - \lambda + 1.$$

Here the order is $4p + 2$, the ground level has one positive class, in addition a cycle of two classes appears in the second power, and a cycle of $2p + 1$ classes appears in the $(2p + 1)^{\text{th}}$ power. One again begins by representing the latter by suitable elements

$$f, f_I, f_{I^2}, \ldots, f_{I^{2p}},$$

and drawing a corresponding system of $2p + 1$ simple curves which cross at a point on the surface. Two relations result for these elements, corresponding to Figure 8. And then one sees, as above, that a primitive power of the given automorphism of order $4p + 2$ is conjugate to the automorphism treated in Example 5. Thus questions (II)–(VI) of §12 also have an affirmative answer for these mapping classes.

Permissions

Birkhäuser Boston would like to thank the original publishers of the papers of Jakob Nielsen for granting permission to reprint specific papers in this collection:

[N-1] Reprinted from Inaugural-Dissertation, Kiel, ©1913 by the University of Kiel, Germany.
[N-2] Reprinted from Math. Ann., 78, ©1917 by Springer-Verlag, Heidelberg.
[N-3] Reprinted from Math. Ann., 79, ©1918 by Springer-Verlag, Heidelberg.
[N-6] Reprinted from Math. Ann., 81, ©1920 by Springer-Verlag, Heidelberg.
[N-7] Reprinted from Math. Ann., 82, ©1921 by Springer-Verlag, Heidelberg.
[N-8] Reprinted from The Mathematical Scientist, 6, ©1981 by the Australian Mathematic Society.
[N-9] Reprinted from Mat.Tidsskr. B, ©1923 by The Danish Mathematical Society.
[N-10] Reprinted from Mat. Tidsskr. B, ©1924 by The Danish Mathematical Society.
[N-11] Reprinted from Math. -fys. Medd. Danske-Vid. Selsk. 5, nr.12, ©1924 by The Royal Danish Academy of Sciences and Letters.
[N-12] Reprinted from Math. Ann., 91, ©1924 by Springer-Verlag, Heidelberg.
[N-14] Reprinted from Mat. Tidsskr. B, ©1925 by The Danish Mathematical Society.
[N-15] Reprinted from Den sjette skandinav, Matematikerkongres, Kobenhavn, ©1925 by Gjellerup & Gad.
[N-18] Reprinted from Acta. Math., 50, ©1927 by The National Mathematical Societies in Denmark, Finland, Norway and Sweden.
[N-19] Reprinted from Mat. Tidsskr. B, ©1928 by The Danish Mathematical Society.
[N-20] Reprinted from Acta Math., 53, ©1929 by The National Mathematical Societies in Denmark, Finland, Norway and Sweden.
[N-21] Reprinted from Acta Math., 58, ©1923 by The National Mathematical Societies in Denmark, Finland, Norway and Sweden.

Contemporary Mathematicians

Gabor Szego: Collected Papers
Edited by *Richard Askey*
0-8176-3063-5 (complete set)
0-8176-3056-2 (Volume 1: 1915-1927)
0-8176-3060-0 (Volume 2: 1927-1943)
0-8176-3061-9 (Volume 3: 1943-1972)

Lars Valerian Ahlfors: Collected Papers
0-8176-3077-5 (complete set)
0-8176-3075-9 (Volume 1: 1929-1955)
0-8176-3076-7 (Volume 2: 1955-1979)

The Selected Papers of Theodore S. Motzkin
Edited by *D. Cantor, et al.*
0-8176-3087-2

Fritz John: Collected Papers
Edited by *Jurgen Moser*
0-8176-3265-4 (complete set)
0-8176-3266-2 (Volume 1)
0-8176-3267-0 (Volume 2)

Kurt Otto Friedrichs: Collected Mathematical Papers
Edited by *Cathleen Morawetz*
0-8176-3270-0 (complete set)
0-8176-3268-9 (Volume 1)
0-8176-3269-7 (Volume 2)

Jakob Nielsen: Collected Papers
Edited by *Vagn Lundsgaard Hansen*
0-8176-3152-6 (complete sets)
0-8176-3140-2 (Volume 1)
0-8176-3151-8 (Volume 2)

Universal Algebra and Lattice Theory: Selected Papers of Garrett Birkhoff
Edited by *Gian-Carol Rota* and *Joseph Oliveira*
0-8176-3114-3

Shizuo Kakutani: Selected Papers
Edited by *Robert Kallman*
0-8176-3279-4 (complete set)
0-8176-3277-8 (Volume 1)
0-8176-3278-6 (Volume 2)

Alfred Tarski: Collected Papers
Edited by *Steven R. Givant* and *Ralph McKenzie*
0-8176-3284-0 (complete set)
0-8176-3280-8 (Volume 1)
0-8176-3281-6 (Volume 2)
0-8176-3282-4 (Volume 3)
0-8176-3283-2 (Volume 4)